Nuclei Off the Line
of Stability

ACS SYMPOSIUM SERIES 324

Nuclei Off the Line of Stability

Richard A. Meyer, EDITOR
Lawrence Livermore National Laboratory

Daeg S. Brenner, EDITOR
Clark University

Developed from a symposium sponsored by
the Division of Nuclear Chemistry and Technology
at the 190th Meeting
of the American Chemical Society,
Chicago, Illinois,
September 8–13, 1985

American Chemical Society, Washington, DC 1986

Library of Congress Cataloging-in-Publication Data

Nuclei off the line of stability.
(ACS symposium series, ISSN 0097-6156; 324)

Includes bibliographies and indexes.

1. Nuclear structure. 2. Chemistry, Physical and theoretical.

I. Meyer, Richard A., 1933-   . II. Brenner, Daeg S., 1939-   . III. American Chemical Society. Division of Nuclear Chemistry and Technology. IV. Series.

QC793.3.S8N84   1986   539.7′4   86-25905
ISBN 0-8412-1005-5

ACS Symposium Series

M. Joan Comstock, *Series Editor*

FOREWORD

The ACS SYMPOSIUM SERIES was founded in 1974 to provide a medium for publishing symposia quickly in book form. The format of the Series parallels that of the continuing ADVANCES IN CHEMISTRY SERIES except that, in order to save time, the papers are not typeset but are reproduced as they are submitted by the authors in camera-ready form. (The papers in this book were prepared with a different text area, reference style, and format from those usually used in the ACS SYMPOSIUM SERIES.) Papers are reviewed under the supervision of the Editors with the assistance of the Series Advisory Board and are selected to maintain the integrity of the symposia; however, verbatim reproductions of previously published papers are not accepted. Both reviews and reports of research are acceptable, because symposia may embrace both types of presentation.

CONTENTS

Section I: Motivation for Further Exploration

PART 1: PROGRESS IN THEORY

Unit A: Models Using Boson Techniques

Unit B: Shell Model

Section II: Experimental Exploration of Current Issues
PART 1: INTRUDER STATES IN NUCLEI

PART 2: OCTUPOLE MODES IN NUCLEI

Section III: New Facilities and Techniques

PART 1: MAJOR FACILITIES

PART 2: TECHNIQUES

Unit A: In-Beam

Unit B: Radioactive Targets and Radiochemistry

PART 3: TECHNIQUE COMPARISON
(INVESTIGATIONS OF ^{138}SM)

Section IV: Survey of Current Research

INDEXES

PREFACE

THE FUNDAMENTAL CONSTITUENTS OF NUCLEI, protons and neutrons, and their interactions with each other and with particles of their own kind within the atomic nucleus have been studied intensively for more than half a century. In attempts to understand and explain the structure of the nucleus, scientists have borrowed proven concepts from atomic chemistry and physics and used these as points of departure. Thus, nuclei, extraordinarily small entities of enormous density, have been characterized in terms of shells, as analogous to the periodicity of atoms, and as having physical properties such as shape and surface tension. Elementary modes of collective excitation, rotations and vibrations, have been suggested as analogous to molecular systems. These concepts have served the nuclear scientist well; like borrowed clothes, they tend to be familiar, economical, and well-worn. Unfortunately, these concepts do not fit nuclei quite as well as they fit their original owners. A major focus of this book is the recent exciting progress toward finding a new set of "clothes" for the nucleus.

The symposium upon which this book is based was originally planned as a timely snapshot of an interdisciplinary field, one that has traditionally involved both chemists and physicists. Many see our discipline at a critical juncture; one path could lead to an exciting future for a somewhat neglected discipline; the other could lead to lower expectations. The nuclear structure research community as a whole looks toward an exciting future and was anxious to gather together for a comprehensive presentation of recent research. Usually, a two-day symposium would have sufficed, but as word spread and interest grew, the symposium expanded to four and one-half days and numbered nearly one hundred presentations. Participants agreed that a very high level of excitement in nuclear structure research has been brought about by recent developments in both theory and experiment.

In organizing this book, we decided to arrange presentations in four broad categories: Section I, Motivation for Further Exploration; Section II, Experimental Exploration of Current Issues; Section III, New Facilities and Techniques; and Section IV: Survey of Current Research. Section I provides a rich but focused summary of motivations for future work in the field. Revolutionary advances in theory and sparkling achievements in experimentation describe the state of nuclear structure research. The introduction of symmetries and supersymmetries has encouraged experimenters to confirm some of the new predictions that flow from boson models. Reexamination of the existing nuclear structure data base has brought to light simplifying correlations that in turn promise to improve our ability to accurately extrapolate nuclear parameters into regions not yet accessible to experimental exploration. In this regard, the region of neutron-rich isotopes is especially significant because of its importance to astrophysics and reactor technology.

Section II provides an examination of the current status of nuclear structure research and presents but a sample of the many exciting and clever experiments underway or completed recently. Four topics are highlighted: intruder states in nuclei, octupole modes in nuclei, high-spin-state investigations using heavy ions, and nuclear moments. This selection reflects the opinion of independent reviewers and our belief that these four areas provide a comprehensive set of new data for testing new theories. In many instances, critical testing of new theories will be possible only if new facilities and techniques are forthcoming.

Section III contains discussions of major new facilities—some under construction, some in intensive design stages, and some still in the idea stage. These and other major facilities are critical to maintaining the rapid pace of advancement in nuclear structure research and represent, for the most part, a history of multinational participation. The coordination and phased development of new facilities avoid duplication of effort and enhance overall productivity of the worldwide research community. Section III also looks at a selection of new techniques and smaller facilities that provide a sample of imaginative approaches to current and future experiments. Three very different approaches to the study of a particular nucleus, ^{138}Sm, are included to illustrate the complementarity of different facilities and techniques in approaching the study of specific cases.

Finally, the call for papers for the symposium provided us with a broad selection of very fine manuscripts. Unfortunately, because of space requirements we could publish in full only a portion of those presentations. Abstracts of the other papers are included in Section IV. These papers can be obtained by writing to Richard A. Meyer or Daeg S. Brenner.

Many individuals contributed to the success of the symposium upon which this book is based. We especially thank Richard W. Hoff, Chair, and Joseph R. Peterson, Treasurer, of the Division of Nuclear Chemistry and Technology of the American Chemical Society for their valuable assistance and trouble-shooting. We also gratefully acknowledge financial support from the National Science Foundation, the Division of Nuclear Chemistry and Technology, and Aptec Nuclear, Inc., in presenting the symposium. Their support enhanced significantly the quality and scope of the symposium.

RICHARD A. MEYER
Division of Nuclear Chemistry
Lawrence Livermore National Laboratory
Livermore, CA 94550

DAEG S. BRENNER
Office of Academic Affairs
Clark University
950 Main Street
Worcester, MA 01610

August 25, 1986

INTRODUCTION
By Richard A. Meyer and Daeg S. Brenner

MUCH RESEARCH ON NUCLEAR STRUCTURE and dynamics in recent years has investigated the behavior of nuclear matter under extreme conditions. For example, the study of rapidly rotating nuclei has led to a simple understanding of how a nucleus adjusts its structure in order to carry high angular momentum. By examining the deexcitation process, scientists have gained detailed knowledge of shape changes that accompany a loss of excitation energy. Another way of stressing nuclei is to produce them in such a way as to create an unstable imbalance involving constituent protons and neutrons. Studies of decay processes of these far-from-stability nuclides have revealed important subtleties in nuclear structure. "Intruder" states, nuclear configurations that arise from an exchange of particles (holes) with an adjacent shell or subshell, are one such discovery. The coexistence of normal and intruder states implies a coexistence of different shapes within a given nucleus.

For many years researchers have known that nuclei can be excited into vibrational modes of motion that are not reflection symmetric. The simplest of these asymmetric modes, the octupole vibration, has been charted extensively. Only recently, however, has new evidence suggested that some nuclei have reflection asymmetric, or pear-shaped, ground-state configurations. Although there is disagreement as to whether these nuclei are pear-shaped or pimpled, it is becoming clear that a more detailed mapping of the nuclear surface is necessary to explain both the spectroscopic properties and the masses of heavy elements.

The field of nuclear structure and dynamics has developed for practical as well as esoteric reasons. Two of our most important customers for nuclear knowledge, nuclear engineers and astrophysicists, have longstanding interests in nuclear data and suggestions for nuclear spectroscopists. On the practical side, we often study properties of nuclides of particular interest to the nuclear power industry. Unfortunately, our way of studying these nuclei does not often lead to results useful to scientists concerned with improving our knowledge of the reactor source term, which predicts the buildup of heat in a reactor following a loss-of-coolant accident.

Astrophysicists are also avid customers for nuclear information and theories. Calculations of elemental abundances resulting from r-process nucleosynthesis and related mechanisms depend on the systematics of nuclear properties. Nuclear properties such as masses, half-lives, and delayed neutron emission probabilities affect these predictions in complicated ways. In the absence of data very far from stability, models are used to extrapolate relevant parameters. This uncomfortable situation would be eased if more data were available for neutron-rich nuclides. Thus, we have two very important reasons—one cosmic, one practical—to pursue studies of nuclear properties to the frontier of spontaneous nuclear disintegration.

In our opinion, however, the overriding reasons for renewed excitement in the nuclear structure community are the successes in theory arising from formulations of nuclear interactions in terms of pairs of interacting bosons, or, in the case of odd-A nuclei, boson-fermion interactions. These new approaches to understanding and calculating nuclear structures and behavior have borrowed the concepts of symmetry from high-energy physics and from crystallography. Although these algebraic models provide less of a sense of the tangible than older geometric models, they have two very important advantages: they are readily employed by experimenters as well as theorists, and they can be applied to regions of shape transition in a natural way. The second advantage is extremely important because the interacting boson model (IBM) is the only model of nuclear structure that provides a global formalism for calculating nuclear structure. The computational successes of the model have stimulated a reexamination of the systematic patterns of nuclear structure. Examination of the trends of a variety of nuclear data as a function of the product of the number of valence neutrons and protons appears to provide an improved framework for reliable extrapolation of nuclear properties into regions far from stability.

Finally, because models such as the IBM consider only valence particles—those beyond the nearest closed shell or, in some instances, subshell—researchers are interested in mapping, experimentally, the locations of shells and subshells into regions far from stability. An improved knowledge of shell and subshell gaps at the extremes of nuclear stability will provide important benchmarks for testing nuclear models.

Section I: Motivation for Further Exploration

Part 1: Progress in Theory

Unit A: Models Using Boson Techniques
Chapters 1-10
Unit B: Shell Model
Chapters 11-13

In this section are papers that deal with the extraction of the fundamental concepts that govern the behavior of nuclei from a vast assortment of nuclear information. The ultimate goal is to develop nuclear models that accurately reproduce the existing data base and that can be shown to reliably predict unknowns. The rapid development of and great excitement associated with models using boson techniques is illustrated by the preponderance of papers dealing with various aspects of the subject. Papers dealing directly or indirectly with the model appear in subsequent sections as well.

The uses of nuclear data and models justify much of the effort invested in research. Many applications in areas such as nuclear medicine and environmental science are not dealt with in this book. Instead, the "needs and uses" justification for nuclear research has focused on subjects related to astrophysics, particularly processes of elemental synthesis, and on the needs of the nuclear power industry. In both instances clear and significant needs exist for systematic data on very unstable nuclides far from stability and on nuclear models that predict level densities, reaction cross sections, decay lifetimes, and Q values. Together, the topics covered in this section provide a strong justification for further exploration of the nuclear landscape.

1
Supersymmetry in Nuclei
Recent Developments

F. Iachello

A. W. Wright Nuclear Structure Laboratory, Yale University, New Haven, CT 06511

Recent developments in applications of supersymmetry to the study of complex spectra are discussed. In particular, extensions of supersymmetry to nuclei with an odd number of protons and neutrons are presented.

1.Introduction

Since its introduction in nuclear physics in 1980 [IAC80], supersymmetry has been extensively used to describe properties of complex nuclei [BAL81a,BAL81b,BAL83,SUN83a,SUN83b,CAS84,IAC85]. In its original formulation, supersymmetry was used to link properties of nuclei with an even number of protons and neutrons (even-even nuclei) with those of nuclei with an odd number of protons and an even number of neutrons (odd-even nuclei) or viceversa (even-odd nuclei). This formulation was based on a description of properties of nuclei in terms of a set of collec-tive variables (bosonic in nature) and a set of single particle variables (fermionic in nature). In treating the collective vari-ables, no distinction was made between protons and neutrons (interacting boson model-1). It has been recognized since, that the proton-neutron degree of freedom plays an important role in the description of some properties of nuclei and thus it must be explicitly introduced (interacting boson model-2). In addition to providing a better understanding of the properties previously described, the introduction of the proton-neutron degree of freedom leads to the possibility of describing properties of nuclei with an odd number of protons and neutrons (odd-odd nuclei). In this lecture, I will discuss some properties of odd-odd nuclei as obtained from supersymmetry.

0097–6156/86/0324–0002$06.00/0

2.The model

In the interacting boson model-2, low lying collective states of nuclei are described in terms of 12 dynamical bosons [ARI77,OTS78], six proton and six neutron bosons. The six proton and neutron bosons have angular momentum $J=0$ (s-boson) and $J=2$ (d-boson). It is convenient to introduce creation $(d_\mu^\dagger, s^\dagger)$ $(\mu=\pm 2, \pm 1, 0)$ and annihilation $(d\mu, s)$ operators. When proton (π) and neutron (ν) degrees of freedom are added, the creation and annihilation operators assume an extra label (π, ν), $d_{\pi,\mu}^\dagger$, $s_\pi^\dagger$, $d_{\nu,\mu}^\dagger$, $s_\nu^\dagger$. For sake of simplicity, I shall denote the 12 operators by $b_{\alpha\rho}^\dagger$, where $\alpha=1,\ldots,6$; $\rho=\pi,\nu$. The corresponding annihilation operators will be denoted by $b_{\alpha\rho}$. The two-dimensional variable ρ is called F-spin. It is worthwhile noting that the proton and neutron bosons correspond to correlated pairs of protons and neutrons. Thus, for each nucleus, their number is fixed to the number $N_\pi(N_\nu)$ of proton (neutron) pairs outside the major closed shells. When only collective degrees of freedom are considered, the Hamiltonian is built out of the 36+36 generators of the group $U_\pi^{(B)}(6) \otimes U_\nu^{(B)}(6)$,

$$G_{\alpha\rho,\alpha'\rho}^{(B)} = b^\dagger_{\alpha\rho} \, b_{\alpha'\rho} \tag{1}$$

and is given by

$$H^{(B)} = E_0^{(B)} + \sum_{\alpha,\alpha',\rho} \varepsilon_{\alpha,\alpha'}^{(\rho)} G_{\alpha\rho,\alpha'\rho}^{(B)} +$$

$$+ \frac{1}{2} \sum_{\alpha,\alpha',\beta,\beta',\rho,\tau} u_{\alpha,\alpha',\beta,\beta'}^{(\rho,\tau)} G_{\alpha\rho,\beta\rho}^{(B)} G_{\alpha'\tau,\beta'\tau}^{(B)} \tag{2}$$

In this equation, $E^{(B)}_0$ is assumed to be invariant under $U_\pi^{(B)}(6) \otimes U_\nu^{(B)}(6)$.

In addition to collective degrees of freedom, one has also single particle degrees of freedom . These are the unpaired particles. The unpaired particles are fermions occupying single particle levels. The degeneracy of a single particle level with angular momentum j is $(2j+1)$. If the unpaired particle can occupy several single particle levels, the total degeneracy is $\Omega = \sum_i (2j_i+1)$. It is convenient also here to introduce creation and annihilation operators for fermions, $a^\dagger_i$ $(i=1,\ldots,\Omega)$, a_i. If the proton-neutron degree of freedom is considered, the creation and annihilation operators acquire an extra label, $a^\dagger_{ir}$ $(i=1,\ldots,\Omega;r=1,2)$. The extra label is the isospin label which distinguishes protons from neutrons. The Hamiltonian describing the fermions can be written in terms of the $\Omega^2 + \Omega^2$ generators of $U_\pi^{(F)}(\Omega) \otimes U_\nu^{(F)}(\Omega)$,

$$G^{(F)}_{ir,i'r} = a^\dagger_{ir} a_{i'r} \tag{3}$$

and is given by

$$H^{(F)} = E^{(F)}_0 + \sum_{i,i',r} \epsilon^{(r)}_{ii'} G^{(r)}_{ir,i'r} +$$

$$+ \frac{1}{2} \sum_{i,i',k,k',r,t} v^{(r,t)}_{ii'kk'} G^{(F)}_{ir,kr} G^{(F)}_{i't,k't} \tag{4}$$

The quantity $E^{(F)}_0$ is assumed to be invariant under $U^{(F)}(\Omega) \otimes U^{(F)}(\Omega)$. States in each nucleus can thus be characterizaed by the number of unpaired fermions they contain. This number will be denoted by M_π for protons and M_ν for neutrons.

Finally, one has an interaction between protons and neutrons. This interaction is written as

$$V^{(BF)} = \sum_{\alpha,\alpha',i,i',\rho,r} W^{(\rho,r)}_{\alpha\alpha'ii'} G^{(B)}_{\alpha\rho,\alpha'\rho} G^{(F)}_{ir,i'r} \qquad (5)$$

The algebraic structure of the total Hamiltonian

$$H = H^{(B)} + H^{(F)} + V^{(BF)} \qquad (6)$$

is then that of $U_\pi^{(B)}(6) \otimes U_\nu^{(B)}(6) \otimes U_\pi^{(F)}(\Omega) \otimes U_\nu^{(F)}(\Omega)$

3.Symmetries

In general, the eigenvalues of H, Eq.(6), must be found numerically. This is a very difficult problem, since the introduction of the proton-neutron degree of freedom produces many states. For a medium mass and heavy nucleus, the size of the matrices to diagonalize exceeds the capabilities of present-day computers. The use of symmetries is thus crucial in understanding the structure of these nuclei. Since the group structure of Eq.(6) is rather rich, symmetries can be used here in a variety of ways. I will concentrate my attention in what follows to odd-odd nuclei. Applications to even-even nuclei have been presented elsewhere [BAL81a,BAL81b,BAL83,SUN83a,SUN83b].

The idea that symmetries may be of practical use in odd-odd nuclei was introduced by Hübsch, Paar and Vretnar [HUB85] and by van Isacker, Jolie, Heyde and Frank [VAN85]. The first class of symmetries I want to discuss is what has been called Bose-Fermi (or spinor) symmetries [IAC81,BIJ84,BIJ85]. Consider the case in which all degrees of freedom have a common dynamic symmetry, for example $SU(4) \approx O(6)$. Then, the various groups appearing in the group chain originating from $U_\pi^{(B)}(6) \otimes U_\nu^{(B)}(6) \otimes U_\pi^{(F)}(\Omega) \otimes U_\nu^{(F)}(\Omega)$ can be combined together. Writing the Hamiltonian in terms of Casimir invariants of the combined groups, leads to mass or energy formulas.

Example 1

An example is provided by the group chain

$$U_\pi^{(B)}(6) \otimes U_\nu^{(B)}(6) \otimes U_\pi^{(F)}(4) \otimes U_\nu^{(F)}(4) \supset$$

$$O_\pi^{(B)}(6) \otimes O_\nu^{(B)}(6) \otimes SU_\pi^{(F)}(4) \otimes S\underline{U}_\nu^{(F)}(4) \supset$$

$$O_{\pi+\nu}^{(B)}(6) \otimes SU_{\pi+\nu}^{(F)}(4) \supset SU^{(B+F)}(4) \supset$$

$$Sp^{(B+F)}(4) \supset SU^{(B+F)}(2) \supset O(2) \qquad (7)$$

In this example as in the discussion of the previous section, it has been assumed that the dimensions Ω of the fermionic spaces for protons and neutrons are identical, $\Omega_\pi = \Omega_\nu = 4$. A somewhat more elaborate case is that in which this condition is not met and one must consider different dimensions, Ω_π, Ω_ν, with corresponding group structure $U_\pi^{(B)}(6) \otimes U_\nu^{(B)}(6) \otimes U_\pi^{(F)}(\Omega_\pi) \otimes U_\nu^{(F)}(\Omega_\nu)$.

Example 2

As an example of this situation, consider

$$U_\pi^{(B)}(6) \otimes U_\nu^{(B)}(6) \otimes U_\pi^{(F)}(4) \otimes U_\nu^{(F)}(12) \supset$$

$$U_\pi^{(B)}(6) \otimes U_\nu^{(B)}(6) \otimes U_\pi^{(F)}(4) \otimes U_\nu^{(F)}(6) \otimes SU_{S\nu}^{(F)}(2) \supset$$

$$O_\pi^{(B)}(6) \otimes O_\nu^{(B)}(6) \otimes SU_\pi^{(F)}(4) \otimes SU_\nu^{(F)}(4) \otimes SU_{S\nu}^{(F)}(2) \supset$$

$$O_{\pi+\nu}^{(B)}(6) \otimes SU_{\pi+\nu}^{(F)}(4) \otimes SU_{S\nu}^{(F)}(2) \supset$$

$$SU^{(B+F)}(4) \otimes SU_{S\nu}^{(F)}(2) \supset Sp^{(B+F)}(4) \otimes SU_{S\nu}^{(F)}(2) \supset$$

$$SU^{(B+F)}(2) \otimes SU_{S\nu}^{(F)}(2) \supset SU(2) \supset O(2) \qquad (8)$$

This particular chain could be applied to the odd-odd nuclei in the Os-Ir-Pt-Au region. In this region, the even-even nuclei have been described by an $O^{(B)}(6)$ chain, the odd-proton nuclei by an $SU_\pi^{(F)}(4)$ chain [BAL81a,BAL81b] and the odd-neutron nuclei by an $SU_\nu^{(F)}(6) \otimes SU_{S\nu}^{(F)}(2)$ chain [BAL83,SUN83a,SUN83b]. Combining these chains together, one can make predictions about the structure of the odd-odd nuclei in this region [VAN85].

4.Supersymmetries

In the same way in which one proceeds to enlarge the symmetry from a Bose-Fermi symmetry to supersymmetry in the case in which no distinction is made between proton and neutron degrees of freedom, one may consider here as well more elaborate situations. The simplest of these situations is that in which proton bosons and proton fermions (and neutron bosons and neutron fermions) are combined together into a representation of a super-group. If we denote by $U_\pi(6/\Omega)$ and $U_\nu(6/\Omega)$ the appropriate super-groups, we can then discuss states in terms of representations of $U_\pi(6/\Omega) \otimes U_\nu(6/\Omega)$. The corresponding algebras are generated by the operators

$$G^{(B)}_{\alpha\rho,\alpha'\rho} = b^\dagger_{\alpha\rho} b_{\alpha'\rho} \quad , \quad \rho = \pi,\nu$$

$$G^{(F)}_{ir,i'r} = a^\dagger_{ir} a_{i'r} \quad , \quad r = \pi,\nu$$

$$F^\dagger_{\alpha\rho,ir} = b^\dagger_{\alpha\rho} a_{i'r} \quad , \quad (\rho=r)=\pi,\nu \tag{9}$$

$$F_{ir,\alpha\rho} = a^\dagger_{ir} b_{\alpha\rho} \quad , \quad (\rho=r)=\pi,\nu$$

The representations of $U_\pi(6/\Omega)$ and $U_\nu(6/\Omega)$ that appear are the totally supersymmetric representations $[N_\pi\}$ and $[N_\nu\}$, characterized by the Young supertableaux

$$[N] = \overbrace{\boxdot \boxdot \ldots \boxdot}^{N\text{-boxes}} \tag{10}$$

One can the consider the decomposition

$$U_\pi(6/\Omega) \otimes U_\nu(6/\Omega) \supset U_{\pi+\nu}(6/\Omega) \tag{11}$$

Contrary to the case in which no distinction is made between protons and neutrons, when one considers the Kronecker product of two supersymmetric representations, one obtains representations of U(6/Ω) which are no longer totally supersymmetric. Rules for taking Kronecker products of representations of supergroups are discussed by Bars [BAL81c,BAL81d,BAL82,BAR85]. They are similar to those which apply to Kronecker products of representations of normal Lie groups. For example,

$$\boxdot \otimes \boxdot = \boxdot\boxdot \oplus \begin{array}{c}\boxdot \\ \boxdot\end{array} \tag{12}$$

More details are given in [BAR85]. In a specific supermultiplet the numbers $\dot{N}_\pi$ and N_ν are given by

$$N_\pi = N_\pi + M_\pi$$

$$N_\nu = N_\nu + M_\nu \tag{13}$$

where $N_\pi(N_\nu)$, $M_\pi(M_\nu)$ are the numbers of proton (neutron) pairs and unpaired nucleons respectively.

Example 1

As a specific example, we can consider the case

$$U_\pi(6/4) \otimes U_\nu(6/4) \supset U(6/4) \supset U^{(B)}(6) \otimes U^{(F)}(4) \supset$$

$$O^{(B)}(6) \otimes SU^{(F)}(4) \supset Spin(6) \supset$$

$$Spin(5) \supset Spin(3) \supset Spin(2) \tag{14}$$

As in Eq.(6), this example relies on the fact that the dimensions of the fermionic spaces (Ω) for protons and neutrons are identical, $\Omega_\pi = \Omega_\nu = 4$. A more elaborate case is that in which the dimension of the fermionic space for protons is different from that for neutrons. The decomposition Eq.(11) is no longer possible and one must consider each system (protons and neutrons) separately.

Example 2

An example of this situation is

$$U_\pi(6/4) \otimes U_\nu(6/12) \supset$$

$$U_\pi^{(B)}(6) \otimes U_\pi^{(F)}(4) \otimes U_\nu^{(B)}(6) \otimes U_\nu^{(F)}(12) \supset$$

$$U_\pi^{(B)}(6) \otimes U_\pi^{(F)}(4) \otimes U_\nu^{(B)}(6) \otimes U_\nu^{(F)}(6) \otimes SU_{S\nu}^{(F)}(2) \supset$$

$$O_\pi^{(B)}(6) \otimes SU_\pi^{(F)}(4) \otimes O_\nu^{(B)}(6) \otimes SU_\nu^{(F)}(4) \otimes SU_{S\nu}^{(F)}(2) \supset$$

$$O(6) \otimes SU(4) \otimes SU_{S\nu}^{(F)}(2) \supset$$

$$Spin(6) \otimes SU_{S\nu}^{(F)}(2) \supset Spin(5) \otimes SU_{S\nu}^{(F)}(2) \supset$$

$$Spin(3) \otimes SU_{S\nu}^{(F)}(2) \supset SU(2) \supset O(2) \tag{15}$$

This case covers that discussed in the Example 2 of the previous section. While the example 1 can hardly be found in practice, since it rarely occurs in heavy nuclei that protons and neutrons occupy the same single particle orbitals, the example discussed here may be obtaibed in realistic situations. The occurrence of a supersymmetry will manifest itself as a particular relationship between the properties of even-even, even-odd, odd-even and odd-odd nuclei. For example, the nuclei

$$^{192}_{78}Pt_{114} \qquad N_\pi=2,\ N_\nu=6,\ M_\pi=0,\ M_\nu=0$$

$$^{193}_{79}Au_{114} \qquad N_\pi=1,\ N_\nu=6,\ M_\pi=1,\ M_\nu=0$$

$$^{193}_{78}Pt_{115} \qquad N_\pi=2,\ N_\nu=5,\ M_\pi=0,\ M_\nu=1$$

$$^{194}_{79}Au_{115} \qquad N_\pi=1,\ N_\nu=5,\ M_\pi=1,\ M_\nu=1 \qquad (16)$$

will all belong to the same supermultiplet [2] $\otimes$ [6] with $N_\pi=2$, $N_\nu=6$. The experimental spectrum of ^{194}Au is not sufficiently well known to be able to conclude whether or not the supersymmetry applies. Van Isacker et al [VAN85] have applied a variation of the chain Eq.(15) to the nuclei

$$^{196}_{78}Pt_{118} \qquad N_\pi=2,\ N_\nu=4,\ M_\pi=0,\ M_\nu=0$$

$$^{197}_{79}Au_{118} \qquad N_\pi=1,\ N_\nu=4,\ M_\pi=1,\ M_\nu=0$$

$$^{197}_{78}\text{Pt}_{119} \qquad N_\pi = 2, \ N_\nu = 3, \ M_\pi = 0, \ M_\nu = 1$$

$$^{198}_{79}\text{Au}_{119} \qquad N_\pi = 1, \ N_\nu = 3, \ M_\pi = 1, \ M_\nu = 1 \qquad\qquad (17)$$

considered as members of the supermultiplet [2] $\otimes$ [4]. On the basis of their analysis, one expects the situation shown in Fig.1

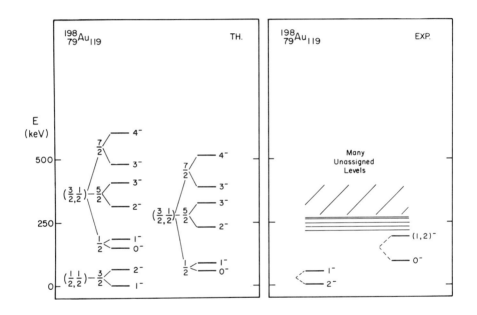

Figure 1. The spectrum of ^{198}Au, as predicted by the $U_\pi(6/4)$ $\otimes$ $U_\nu(6/12)$ supersymmetry [VAN85].

for the spectrum of the odd-odd nucleus ^{198}Au. It would be interesting to see to what extent this situation is met in practice.

5.Conclusions

    The concept of supersymmetry in nuclei has been extended in
the last year to include the proton-neutron degree of freedom
[HUB85,VAN85]. With this extension, it becomes possible now to
predict prperties of odd-odd nuclei. Thus, in addition to providing the first experimental example of supersymmetry in physics
[CAS84,IAC85], this concept appears now to be able to make
predictions for yet unknown quantities. The experimental determination of the predicted spectra will indicate to what extent
supersymmetry is useful in nuclear physics.

Acknowledgments

    I wish to thank V. Paar, P. van Isacker, and A. Frank for
pointing out the importance of supersymmetry in odd-odd nuclei
and A.B. Balantekin and I. Bars for many discussions on the
subject. This work was performed in part under DOE Contract No.
DE-AC-02-76-ER-03074.

References
[ARI77] A. Arima, T. Otsuka, F. Iachello and I. Talmi, Phys.
Lett. 66B,205(1977)
[BAL81a] A.B. Balantekin, I. Bars and F. Iachello, Nucl. Phys.
A370,284(1981)
[BAL81b] A.B. Balantekin, I. Bars and F. Iachello, Phys. Rev.
Lett. 47,19(1981)
[BAL81c] A.B. Balantekin and I. Bars, J. Math. Phys.
22,1149(1981)
[BAL81d] A.B. Balantekin and I. Bars, J. Math. Phys.
22,1810(1981)
[BAL82] A.B. Balantekin and I. Bars, J. Math. Phys. 23,1239(1982)
[BAL83] A.B. Balantekin, I. Bars, R. Bijker and F. Iachello,
Phys. Rev. C27,176(1983)

[BAR85] I. Bars, in "Supersymmetry in Physics", V.A. Kostelecki and D.K. Campbell eds., Physica 15D,42(1985)

[BIJ84] R. Bijker and V.K.B. Kota, Ann. Phys. (N.Y.) 156,110(1984)

[BIJ85] R. Bijker and F. Iachello, Ann. Phys. (N.Y.) 161,360(1985)

[CAS84] R.F. Casten and D.H. Feng, Physics Today 37,No.11,26(1984)

[HUB85] T. Hübsch, V. Paar and D. Vretnar, Phys. Lett. 151B,320(1985)

[IAC80] F. Iachello, Phys. Rev. Lett. 44,772(1980)

[IAC81] F. Iachello and S. Kuyucak, Ann. Phys. (N.Y.) 136,19(1981)

[IAC85] F. Iachello in "Supersymmetry in Physics", V.A. Kostelecki and D.K. Campbell eds., Physica 15D,85(1985)

[OTS78] T. Otsuka, A. Arima, F. Iachello and I. Talmi, Phys. Lett. 76B,139(1978)

[SUN83a] H.Z. Sun, A. Frank and P. van Isacker, Phys. Rev. C27,2430(1983)

[SUN83b] H.Z. Sun, A. Frank and P. van Isacker, Phys. Lett. 124B,275(1983)

[VAN85] P. van Isacker, J. Jolie, K. Heyde and A. Frank, Phys. Rev. Lett. 54,653(1985)

RECEIVED August 26, 1986

2

Boson–Fermion Symmetries and Dynamical Supersymmetries for Odd–Odd Nuclei

A. B. Balantekin[1], T. Hübsch[2], and V. Paar[3]

[1]Oak Ridge National Laboratory, Oak Ridge, TN 37831
[2]University of Maryland, College Park, MD 20742
[3]Prirodoslovno-matematicki fakultet, University of Zagreb, 41000 Zagreb, Yugoslavia

The concept of boson-fermion symmetries and supersymmetries is applied to odd-odd nuclei.

The approach to even-even and odd-even nuclei based on nuclear symmetries in IBM/IBFM has received much attention in recent years [ARI76, ARI78, ARI79, BAL81, BAL83, IAC79, IAC80].

In this report we discuss the extension of this concept to odd-odd nuclei. Odd-odd nuclei provide richer and more complex structure, and the residual proton-neutron interaction appears explicitly in the boson-fermion interaction.

Odd-odd nuclei are described as mixed system of bosons and fermions (proton and neutron) by the Hamiltonian

$$H = H_B + H_F + V_{BF} \tag{1}$$

Here, H_B is identical to the IBM Hamiltonian, H_F includes one-fermion and fermion-fermion interaction terms and V_{BF} is the boson-fermion interaction. The computer code for diagonalizing Hamiltonian (1) for odd-odd nuclei has been recently written [VRE84]. As a residual proton-neutron force the surface delta-, spin- and tensor-interaction were included. Computations have been performed for some particular cases [BRA84, VRE84, PAA84, PAA85, MEY85].

Particularly, the Hamiltonian (1) was diagonalized in the case of a proton particle j_p and a neutron particle j_n coupled to the SU(3) core. The computed energy pattern exhibits two regular low-lying bands based on the states of angular momenta $J = j_p + j_n$ and $J = |j_p - j_n|$ [PAA84, PAA85, PAA85a]. In comparison to the rotational model, these two bands are the truncated analogs of the Gallagher-Moszkowski bands based on the Nilsson

0097–6156/86/0324–0014$06.00/0

states $\Omega_p = j_p$, $\Omega_n = j_n$ [PAA84] . It was shown [PAA85a] that the wave functions of the states of $J = j_p + j_n$ band can be brought to the form of the approximate SUSY wave function of the analogs of Nilsson states in odd-A nuclei [SUN84].The energies of this band follow the $J(J+1)$ energy rule, with the same moment of inertia as for the boson core.

The band structure in odd-odd nuclei has been also explored in the case of O(6) boson core [BRA84].

Here we discuss the boson-fermion system for odd-odd nuclei using the concept of dynamical symmetry and supersymmetry. The dimension of the fermionic subspace is $n = n_\pi + n_\nu$, where $n_\pi (n_\nu)$ is the total number of components of the angular momenta of the unpaired protons (neutrons). One can construct the $(n_\pi + n_\nu)^2$ generators of the group $U_F(n_\pi + n_\nu)$, where the subscript F reminds that a fermionic realization is employed. In general, the group structure of the Hamiltonian is

$$U_B(6) \times U_F(n_\pi + n_\nu) \tag{2}$$

There is a second, alternative approach. One could assume that the unpaired neutron and the unpaired proton form a quasibound state. The total number of components of the angular momenta of this quasi-bound state is given by $n_\pi n_\nu$. Then we introduce a pair of new bosonic creation and annihilation operators associated with each level of this subsystem, $c_I^+, c_J, I, J = 1,2,\ldots,n_\pi n_\nu$. The $(n_\pi n_\nu)^2$ operators $\bar{G}_{IJ} = c_I^+ c_J$ constitute a bosonic realization of the $U(n_\pi n_\nu)$ algebra. The group structure of the corresponding Hamiltonian is

$$U_B(6) \times U_F(n_\pi n_\nu) \tag{3}$$

where the appropriate representation of $U(n_\pi n_\nu)$ is the fundamental representation $(1,0,0,\ldots,0)$.

To find analytical solutions to the eigenvalue problem of either Hamiltonian, associated with (2) or (3), the key idea is the use of the isomorphisms between groups in the two chains, one starting with $U_B(6)$ and the other one starting with either $U_F(n_\pi + n_\nu)$ or $U_F(n_\pi n_\nu)$. (This idea is along the line of approach to boson-fermion symmetries for odd-even nuclei, which was introduced in ref. [IAC80].) Groups obtained by joining two chains transform simultaneously bosons into bosons and fermions into fermions.

Some special solutions associated with the group structure (3) have been studied by two of us [HÜB84, PAA85, HÜB85a].

In the recent work [BAL85] we give a detailed study of various level schemes for odd-odd nuclei obtained by analytical solutions of Hamiltonians associated with either (2) or (3). The isomorphisms between the two group chains are elaborated for the case where the unpaired nucleons occupy some or all of levels with $j = 1/2, 3/2, 5/2$ and the analytical expressions for the corresponding energy eigenvalues are given.

In a further step, such solutions are embedded into a supergroup and new chains arising from such embedding are given.

Now we look for correlations in the spectra of the four neighboring nuclei: the even-even nucleus with (Z,A), the odd-even nucleus with $(Z,A+1)$, the even-odd nucleus with $(Z+1,A+1)$, and the odd-odd nucleus with $(Z+1,A+2)$. Such correlations arise if the Hamiltonian associated with (2) has a supergroup structure $U(6/n_\pi + n_\nu)$. Consequently, the properties of these four nuclei could be related by the symmetry operations of this supergroup. In particular, we have obtained analytical expressions for the eigenvalues of this Hamiltonian in terms of the eigenvalues of the Casimir operators of chains of supergroups starting with $U(6/n_\pi + n_\nu)$ and terminating with $Spin^{B+F}(3)$ [BAL85] . Such a situation was termed a dynamical supersymmetry in the previous investigations of odd-even nuclei [BAL81].

The concept of dynamical supersymmetries was successfully used to connect the properties of even-even nuclei with the neighboring odd-even nuclei. If we want to study, say, the correlation between an even-even nucleus and the next, odd-proton nucleus, the first decomposition in the supergroup chain is

$$U(6/n_\pi + n_\nu) \supset U(6/n_\pi) \times U(n_\nu) . \tag{4}$$

One can then continue the chain with the decompositions of the supergroup $U(6/n_\pi)$ as was done in ref. [BAL81].

The appropriate representation of the group $U(n_\nu)$ in the case of an odd-proton nucleus is a singlet, hence the existence of the group $U(n_\nu)$ in the chain does not affect any of the quantum numbers. A similar situation arises in the case of correlations between an even-even nucleus and the neighboring odd-neutron nucleus.

In this case we start with, what we call, the canonical decomposition

$$U(6/n_\pi + n_\nu) \supset U_B(6) \times U_F(n_\pi + n_\nu) \supset U_B(6) \times U_F(n_\pi) \times U_F(n_\nu), \tag{5}$$

and continue the chains in various possible forms [BAL85]. The difference to the case of boson-fermion dynamical symmetries is that several nuclei are now placed in the same supermultiplet. Consequently, parameters appearing in the energy formulae take the same values for all nuclei in the same supermultiplet.

As an illustration of dynamical supersymmetry with canonical decomposition we construct a supermultiplet starting from the even-even nucleus ^{194}Pt. In this region the proton shell is dominated by $j = 3/2$ and the neutron shell by $j = 1/2, 3/2, 5/2$. Hence $n_\pi = 4$ and $n_\nu = 12$. The relevant representation of the appropriate supergroup $U(6/16)$ is the one with $\mathcal{N} = 7$. Various nuclei are placed in the tensor product representations as follows:

$$U(6/16) \supset U_B(6) \times U_F(16) \supset U_B(6) \times U_F^{(\pi)}(4) \times U_F^{(\nu)}(12)$$

$|7\} = ([7],\{0\})\qquad = ([7],\{0\},\{0\})$
$$\underline{\ \ ^{194}Pt\ \ }$$

$+\ ([6],\{1\})\qquad = ([6],\{1\},\{0\}) + ([6],\{0\},\{1\})$
$$\qquad\qquad\qquad\ \ ^{195}Au\qquad\qquad\quad ^{195}Pt$$

$+\ ([5],\{1^2\})\qquad = ([5],\{1^2\},\{0\}) + ([5],\{1\},\{1\}) + ([5],\{0\},\{1^2\})$
$$\qquad\qquad\qquad\ \ ^{196}Hg^* \qquad\qquad\quad ^{196}Au \qquad\qquad ^{196}Pt^*$$

$+\ \ldots \tag{6}$

In the above expression asterisk over a given symbol denotes the two-quasiparticle states in that nucleus. Eq.(6) is illustrated below

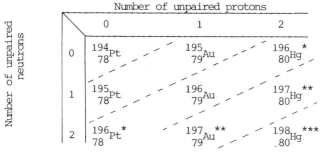

Asterisk * denotes two-quasiparticle states in even-even and ** denotes
three-quasiparticle states in odd-even nuclei. Dashed line separates nuclei
belonging to the same representation of the direct product group
$U_B(6) \times U_F(n_\pi + n_\nu)$.

A new possibility of supersymmetry arises when $n_\pi = n_\nu = n$. In this
case, using fermionic creation and annihilation operators it is possible to
construct the generators of the symplectic group Usp (2n). Consequently a
supergroup chain starting with decomposition into the orthosymplectic group

$$U(6/2n) \supset Osp(6/2n) \qquad (7)$$

might be relevant in such cases. The properties of orthosymplectic supergroups
are studied in refs. [JAR79,BAL82]. The appealing aspect of this case is that
it emphasize^{s}the residual force between the unpaired neutron and the unpaired
proton, while retaining the supersymmetry scheme. The main problem, however,
is that this group decomposition does not conserve the boson number N.

However, there is an intermediate situation with the decomposition

$$U(6/2n) \supset SU_B(6) \times SU_F(2n) \supset SU_B(6) \times Sp_F(2n) \qquad (8)$$

The representation of Sp(2n) contains the n-dimensional fundamental representation $(1,0,0,...,0,0)$ of SU(n), which we denote □ and its conjugate representation $(1,1,1,...,1,0)$ which we denote ▣ .

In general, the proper bases to describe nuclei with one unpaired
nucleon, would be given by linear combinations of the bases of SU(n):

$$|isotope> = \cos\theta |□> + \sin\theta |▣> \qquad (9)$$
$$|isotone> = -\sin\theta |□> + \cos\theta |▣> \qquad (10)$$

where $|□>$ denotes the basis of the representation □ and $|▣>$ denotes
that of ▣ . Here we consider the case $\theta = 0$, but the results can easily
be generalized for the finite θ case. (The choice $\theta = 0$ would be physically transparent if, say, the unpaired neutrons are particles and the unpaired protons are holes, or vice versa, since it is reasonable to consider
conjugate representations for holes.)

A good place to look for such a supersymmetry is again the Pt-Au region. In this region the odd neutron and proton occupy mostly levels with
$j = 1/2,3/2,5/2$. In this case we have $n_\pi = n_\nu = n = 12$. The resulting
supersymmetry has then a U(6/24) structure. Again if we start from the even-
even nucleus ^{194}Pt with $\mathcal{N} = 7$ and O(6) core, we employ the tensor product
representation

$$U(6/24) \supset U_B(6) \times U_F(24) \supset U_B(6) \times Sp_F(24) \supset U_B(6) \times SU_F(12) \qquad (11)$$

which gives the corresponding energy formula [BAL85]. A typical spectrum of
the low-lying states in Sp(24) scheme is shown in fig.1 and compared to the
experimental states of ^{196,198}Au.

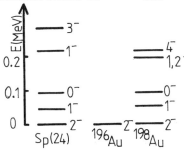

Fig.1 Sp(24) scheme for low-lying
states in comparison to the
experimental data in
^{196,198}Au.

A comparison of the Sp(24) spectrum with the experimental spectra of ^{196}Au and ^{198}Au seems encouraging. This is particularly interesting since the previous attempts to describe such nuclei using the canonical chain could not account for the observed ground state spin [BAL,BAR84]. Experimental studies of the odd-odd nuclei is very desirable to decide whether or not the Sp (24) chain is applicable in this region.

As another illustrative application of supersymmetry extension to odd-odd nuclei we consider the spectrum of odd-odd $^{62}_{29}Cu_{33}$ if one assumes that the even-even nucleus $^{64}_{30}Zn_{34}$, odd-even nucleus $^{63}_{29}Cu_{34}$ and odd-odd nucleus $^{62}_{29}Cu_{33}$ correspond to the members of the same supermultiplet. In a simplified presentation with odd proton and odd neutron restricted only to j = 3/2 configurations and with U_B(5) boson core, the resulting energy formula [BAL85] gives the spectrum presented in fig.2.

We note that this spectrum is obtained using the supersymmetry relation to the neighboring nuclei ^{63}Cu and ^{64}Zn, without adjusting any parameter to ^{62}Cu.

We note that the idea of symmetry and supersymmetry was further extended to hypernuclei by two of us [HÜB85,HÜB85a, PAA85].

Concluding, we point out that the application of the idea of dynamical symmetry and supersymmetry to odd-odd nuclei leads to a series of energy formulas, and gives a new insight into the nuclear structure. Therefore, more detailed studies of odd-odd nuclei are highly desirable.

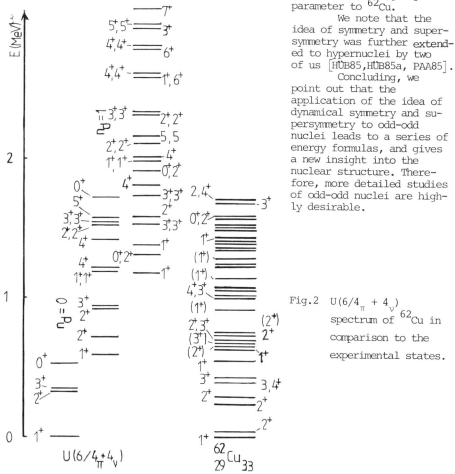

Fig.2 $U(6/4_\pi + 4_\nu)$ spectrum of ^{62}Cu in comparison to the experimental states.

Acknowledgments

V. Paar was assisted by the National Science Foundation under Grant no. YOR 80/001. Research by A. B. Balantekin was sponsored by the Division of Nuclear Physics, U.S. Department of Energy, under contract no. DE-AC05-840R21400 with Martin Marietta Energy Systems, Inc.

References

[ARI76] A.Arima and F.Iachello, Ann.Phys. 99 253 (1976)

[ARI78] A.Arima and F.Iachello, Ann.Phys. 111 201 (1978)

[ARI79] A.Arima and F.Iachello, Ann.Phys. 123 468 (1979)

[BAL] A.B.Balantekin, I.Bars and F.Iachello, unpublished

[BAL81] A.B.Balantekin, I.Bars and F.Iachello, Nucl. Phys. A370 284 (1981)

[BAL81a] A.B.Balantekin and I.Bars, J.Math.Phys. 23 1239 (1982)

[BAL83] A.B.Balantekin, I.Bars, R.Bijker and F.Iachello, Phys.Rev. C27 1761 (1983)

[BAL85] A.B.Balantekin,T.Hübsch and V.Paar, to be published

[BAR84] I.Bars, in Bosons in nuclei, Ed.D.H.Feng,S.Pittel and M.Vallieres (World Scientific, 1984)

[BRA84] S.Brant,V.Paar and D.Vretenar, Z.Phys. A319 355 (1984)

[HÜB84] T.Hübsch and V.Paar, Z.Phys. A319 11 (1984)

[HÜB85] T.Hübsch and V.Paar, Z.Phys. A320 351 (1985)

[HÜB85a] T.Hübsch and V.Paar, Phys.Lett. 151B 1 (1985)

[HÜB85b] T.Hübsch, V.Paar and D.Vretenar, Phys.Lett. 151B 320 (1985)

[IAC79] F.Iachello and O.Scholten, Phys.Rev.Lett. 43 679 (1979)

[IAC80] F.Iachello, Phys.Rev.Lett. 44 772 (1980)

[JAR79] P.D.Jarvis and H.S.Green,J.Math.Phys. 20 2115 (1979)

[MEY85] R.A.Meyer,S.Brant, V.Paar and D.Vretenar, to be published

[PAA84] V.Paar, in In-Beam Nuclear Spectroscopy, Ed. Zs. Dombradi and T. Fenyes (Akademiao Kiado, Budapest, 1984) Vol.2, p.675

[PAA85] V.Paar, in Capture Gamma-Ray Spectroscopy and Related Topics-1984, Ed. S.Raman (American Institute of Physics, New York, 1985) p.70

[PAA85a] V.Paar,D.K.Sunko and D.Vretenar, to be published

[SUN84] D.K.Sunko and V.Paar, Phys.Lett. 146B 279 (1984)

[VRE84] D.Vretenar, S.Brant and V.Paar, Computer code OIBM/OTQM, unpublished (1984)

RECEIVED May 2, 1986

3

Experimental Tests of Boson–Fermion Symmetries and Supersymmetries Using Coulomb Excitation with Heavy Ions

M. Loiselet, O. Naviliat, R. Holzmann, and J. Vervier

Institut de Physique Nucléaire, B-1348 Louvain-la-Neuve, Belgium

Accurate lifetime measurements by the recoil-distance method following Coulomb excitation with ^{40}Ar ions have been performed in ^{106,108}Pd, ^{103}Rh, ^{107,109}Ag. The results are compared with the predictions of the Core Particle Weak Coupling model on the one hand, and of various boson-fermion symmetries and supersymmetries, on the other hand. Good agreement with the latter but not with the former is obtained. This suggests the influence of a finite number of bosons in the relevant nuclei.

The presently available experimental data on the negative-parity levels in the nuclei ^{103}Rh, ^{107,109}Ag up to 1.5 MeV [NDS85] include : a $1/2^-$ ground state ; two $(3/2^-, 5/2^-)$ doublets ; one $(7/2^-, 9/2^-)$ doublet ; as well as a few other levels. This is suggestive of a Core Particle Weak Coupling (CPWC) description of these nuclei [DES61], in which the ground state, first and second 2^+, and first 4^+ levels, respectively, of the even-even cores, ^{102}Ru, ^{106,108}Pd, respectively, are weakly coupled to the odd proton in a $2p1/2^-$ shell model orbit. Such a model predicts equal B(E2)'s for the ground state decays of the first $3/2^-$ and $5/2^-$ levels in the odd-A nuclei ; in addition, the ratios $R_{oe} \equiv [B(E2 ; 5/2^- - 1/2^-)]/[B(E2 ; 2^+ - 0^+)]$ and $R'_{oe} \equiv [B(E2 ; 9/2^- - 5/2^-)]/[B(E2 ; 4^+ - 2^+)]$ should be both equal to 1, regardless of the nature of the core. Furthermore, if the core is described as an harmonic vibrator, the ratios : $R_{ee} \equiv [B(E2 ; 4^+ - 2^+)]/[B(E2 ; 2^+ - 0^+)]$ and $R_{oo} \equiv [B(E2 ; 9/2^- - 5/2^-)]/[B(E2 ; 5/2^- - 1/2^-)]$ should both be equal to 2 ; if the core is described as a γ-unstable vibrator, $R_{ee} = R_{oo} = 10/7 = 1.429$. The ratios R_{ee}, R_{oo}, R_{oe} and R'_{oe} deal with the first levels of the indicated spins. Different predictions result from a description of these nuclei in the general framework of the boson models, wherein : the even-even nuclei correspond to the dynamical symmetries SU(5) (corresponding to an harmonic

0097–6156/86/0324–0020$06.00/0

vibrator) [ARI76] or O(6) (a γ-unstable vibrator) [ARI79] of the Interacting
Boson Model (IBM) ; the odd-A nuclei, to the boson-fermion (BF) symmetries
SU(5) x 1/2 [BIJ84], O(6) x 1/2 [IAC81], SU(5) x 1/2, 3/2, 5/2 [VAN84a, VER85]
or O(6) x 1/2, 3/2, 5/2 [VAN84a, SUN84, BIJ85] of the Interacting Boson Fermion Model (IBFM) ; and the even-even and odd-A nuclei considered together,
to the associated SUperSYmmetries (SUSY) [BAL81], in the present case, the
U(6/2) and U(6/12) supersymmetries [BIJ84, VAN84a, BIJ85]. These boson models
predict $R_{oe} < 1$, $R'_{oe} < 1$, $R_{ee} < 2$, $R_{oo} < 2$; the precise theoretical values
depend on the numbers N of bosons for the relevant nuclei, on the IBM symmetry
(SU(5) or O(6)) considered, and on the shell model orbits included (2p1/2$^-$
alone for 1/2, or 2p1/2$^-$, 2p3/2$^-$, 1f5/2$^-$ for 1/2, 3/2, 5/2). The predicted
deviations from 1 and 2 are of the order of 15 to 35 % for the nuclei investigated.

The experiments whose results are reported in the present paper have
been carried out in order to test the above-mentioned predictions of the BF
symmetries and supersymmetries with respect to the B(E2)'s, as opposed to
those of the CPWC model. As the difference between the two kinds of models
mainly lie in the number of bosons, which is finite and well defined in the
former, and infinite in the latter [FAE83], these experiments may be considered as testing the influence of a finite number of bosons in the relevant nuclei. The B(E2)'s of interest have been determined through accurate measurements of the lifetimes of the first 9/2$^-$ and 5/2$^-$ levels in ^{103}Rh, ^{107,109}Ag,
and first 2$^+$ in ^{106,108}Pd, using Coulomb excitation with ^{40}Ar ions to populate these states, and the recoil-distance method (RDM) [ALE78]. Furthermore,
special cares have been taken in the accurate determination of the corrections to the zero-order results from the RDM, especially those which originate from the perturbation of the angular distributions of the γ-rays by the
hyperfine fields acting on the ions recoiling into the vacuum [ALE78].
Preliminary results of these experiments have already been reported [LOI84],
and a full account of the experimental methods and data analysis will be given
elsewhere [LOI85]. In addition to RDM measurements, angular distribution
experiments have also been carried out, in order to determine the above-
mentioned corrections. The results obtained so far are presented in Table 1,
together with the B(E2)'s which can be deduced from them. As the nucleus
^{102}Ru has not been investigated in the present work, previous results on its
B(E2 ; 2$^+$ - 0$^+$) [BOC79, LAN80, HIR82] have been used. For the same reason,
we have adopted previously obtained data [NDS85] on B(E2 ; 4$^+$ - 2$^+$) in the

Table 1

Nuclei	^{102}Ru - ^{103}Rh	^{106}Pd - ^{107}Ag	^{108}Pd - ^{109}Ag
$\tau(2_1^+)$	-	18.3(5)	34.2(10)
$\tau(9/2_1^-)$	8.1(6)	2.3(3)	2.6(3)
$\tau(5/2_1^-)$	112.1(34)	51.4(24)	47.6(20)
B(E2 ; $2_1^+ - 0_1^+$)	0.125(3)	0.126(4)	0.154(5)
B(E2 ; $9/2_1^- - 5/2_1^-$)	0.178(13)	0.179(23)	0.223(26)
B(E2 ; $5/2_1^- - 1/2_1^-$)	0.111(3)	0.108(5)	0.125(6)
B(E2 ; $4_1^+ - 2_1^+$)	0.186(26)	0.217(28)	0.278(39)

The lifetimes τ are given in ps., and the B(E2)'s, in $e^2.b^2$.

even-even nuclei. The values of R_{ee}, R_{oo}, R_{oe} and R'_{oe}, which can be deduced from the results of Table 1 and from [NDS85], are given in Table 2. We stress that, since all nuclei (except ^{102}Ru) have been studied by the same RDM method, with the same apparatus, and using the same procedure for calculating the corrections to the results, the <u>ratios</u> R_{oo} and R_{oe} between the B(E2)'s, which matter for the comparison with the various nuclear models as will now be shown, should be quite reliable.

The experimental ratios R_{ee}, R_{oo}, R_{oe} and R'_{oe} given in Table 2 are systematically smaller than 2, 2, 1 and 1, respectively ; in some cases, the differences amount to several standard deviations, especially for R_{oe}. This means that the specific versions of the collective models considered here, i.e. an harmonic vibrator core and the CPWC model, do not agree with these experimental data. The deviations from 2 of R_{ee} and R_{oo} could be due to 2 reasons, at least within the framework of the models considered in the present paper : either the cores cannot be described as pure harmonic vibrators, but rather correspond to intermediate situations between harmonic and γ-unstable vibrators ; or the number of bosons in the even-even and odd-A nuclei is indeed finite [VER83]. A combination of the 2 explanations is of course also possible. Concerning R_{oe} and R'_{oe}, their deviations from 1 represent disagreements with the CPWC model, regardless of the description of the cores, harmonic, γ-unstable or intermediate vibrators [VER83]. In the framework of the collective model, these results show that the core-particle coupling is not weak, although the level energies, and in particular the splittings of the $(3/2^-, 5/2^-)$ and $(7/2^-, 9/2^-)$ doublets, suggest that it is so.

The predictions of the various boson models referred to above concerning R_{ee}, R_{oo}, R_{oe} and R'_{oe} are compared in Table 2 with the experimental data. These models thus offer possible explanations to the deviations from 2, 2, 1 and 1, respectively, of these 4 quantities, in particular for R_{oe} and R'_{oe}. These results may be considered as experimental evidences for the influence of a finite number of bosons in the relevant nuclei. A detailed comparison between the experimental values and the predictions of the various versions of the boson models considered here, especially for the accurately known R_{oe} ratios, yields the following conclusions. There is a good agreement, within the experimental uncertainties, between the experimental data and the parameter-free predictions of the symmetries SU(5), SU(5) x 1/2 and U(6/2), which have been applied to these nuclei previously [VER82], except, maybe, for R_{oe} in

Table 2

Nuclei N	^{102}Ru – ^{103}Rh 7 6	^{106}Pd – ^{107}Ag 7 6	^{108}Pd – ^{109}Ag 8 7
R_{ee} exp.	1.49(21)	1.72(23)	1.81(26)
SU(5)	1.714	1.714	1.750
0(6)	1.336	1.336	1.354
R_{oo} exp.	1.60(12)	1.66(23)	1.78(22)
SU(5) x 1/2	1.667	1.667	1.714
0(6) x 1/2	1.310	1.310	1.336
SU(5) x 1/2, 3/2, 5/2	1.714	1.714	1.750
0(6) x 1/2, 3/2, 5/2	1.336	1.336	1.354
R_{oe} exp.	0.888(32)	0.857(48)	0.812(47)
U(6/2) ⌠SU(5) x 1/2	0.857	0.857	0.875
⌊ 0(6) x 1/2	0.779	0.779	0.802
U(6/12) ⌠ SU(5),0(6) ⌠e_b=e_f	1	1	1
⌊x 1/2,3/2,5/2 ⌊e_f=0	0.735	0.735	0.766
R'_{oe} exp.	0.96(15)	0.83(15)	0.80(15)
U(6/2) ⌠SU(5) x 1/2	0.833	0.833	0.857
⌊ 0(6) x 1/2	0.764	0.764	0.791
U(6/12) ⌠ SU(5),0(6) ⌠e_b=e_f	1	1	1
⌊x 1/2,3/2,5/2 ⌊e_f=0	0.735	0.735	0.766

^{108}Pd - ^{109}Ag ; this is not the case for O(6) and O(6) x 1/2, whose predictions are systematically lower than the experimental results. It has been suggested that the even-even Ru and Pd isotopes can be described by an intermediate situation between SU(5) and O(6) [STA82], and that the odd-A Rh isotopes similarly correspond to an intermediate situation between SU(5) x 1/2, 3/2, 5/2 and O(6) x 1/2, 3/2, 5/2 [VAN84b]. The comparison performed in Table 2 shows that the predictions of such descriptions could be made to agree with the experimental data by a suitable choice of the parameter ξ which governs the SU(5) to O(6) transition [STA82], and of the boson and fermion effective charges, e_b and e_f, respectively [VAN84b]. More specific statements on these points will probably be made when other B(E2)'s in the odd-A nuclei will be determined, resulting partly from a more detailed analysis of the experiments reported in the present paper [LOI85], in particular of the angular distribution measurements.

References

[ALE78] T.K. Alexander, and J.K. Forster, Adv. Nucl. Phys. X Ch. 3 (1978).

[ARI76] A. Arima, and F. Iachello, Ann. Phys. (N.Y.) 99 253 (1976).

[ARI79] A. Arima, and F. Iachello, Ann. Phys. (N.Y.) 123 468 (1979).

[BAL81] A.B. Balantekin, I. Bars, and F. Iachello, Phys. Rev. Lett. 47 19 (1981) ; Nucl. Phys. A370 284 (1981).

[BIJ84] R. Bijker, and V.K.B. Kota, Ann. Phys. (N.Y.) 156 110 (1984).

[BIJ85] R. Bijker, and F. Iachello, Ann. Phys. (N.Y.) 161 360 (1985).

[BOC79] A. Bockisch et al., Zeits. f. Phys. A292 265 (1979).

[DES61] A. de-Shalit, Phys. Rev. 122 1530 (1961).

[FAE83] A. Faessler, Nucl. Phys. A396 291c (1983).

[HIR82] J.H. Hirata, and O. Dietzsch, Proc. Int. Conf. Nucl. Struct., Amsterdam, 1982, vol. I 131 (1982).

[IAC81] F. Iachello, and S. Kuyucak, Ann. Phys. (N.Y.) 136 19 (1981).

[LAN80] S. Landsberger et al., Phys. Rev. C21 588 (1980).

[LOI80] M. Loiselet et al., Phys. Lett. 146B 187 (1984).

[LOI85] M. Loiselet et al., to be published.

[NDS85] Nuclear Data Sheets, last issues on the relevant mass chains as to September 1 (1985).

[STA82] J. Stachel, P. Van Isacker, and K. Heyde, Phys. Rev. C25 650 (1982).

[SUN84] H.Z. Sun et al., Phys. Rev. C29 352 (1984).

[VAN84a] P. Van Isacker, A. Frank, and H.Z. Sun, Ann. Phys. (N.Y.) 157 183 (1984).

[VAN84b] P. Van Isacker et al., Phys. Lett. 149B 26 (1984).

[VER82] J. Vervier, and R.V.F. Janssens, Phys. Lett. 108B 1 (1982).

[VER83] J. Vervier, Phys. Lett. 133B 135 (1983).

[VER85] J. Vervier et al., Phys. Rev. (to be published).

RECEIVED April 11, 1986

Manifestations of Fermion Dynamical Symmetries in Collective Nuclear Structures

Da Hsuan Feng

Department of Physics and Atmospheric Science, Drexel University, Philadelphia, PA 19104

A Fermion dynamical symmetry model which can account for both the *low* as well as *high* spin nuclear collective phenomena is presented.

The phenomenological Interacting Boson Model (IBM), introduced about a decade ago by Arima and Iachello[1], has linked the collective phenomena in the low energy region for nuclei (E < 2 MeV and J < 10, say) with the concept of dynamical symmetries. The profound program of the IBM is remarkably simple and can be understood as follows. By simulating the L=0 and 2 valence coherent nucleon pairs as s and d bosons, the IBM has the highest symmetry $U^B(6)$ which, together with the lowest symmetry $O^B(3)$ physically demanded by rotational invariance, resulted in three limiting *multi-chain* dynamical symmetries: $U^B(5)$, $SU^B(3)$ and $O^B(6)$. Using the generalized coherent states of Perelomov[2] and Gilmore[3] (See also the book recently edited by Klauder and Skagerstam[4]), one can show that each of these chains depicts a particular type of geometrical motion (vibrational: $U^B(5)$, rotational: $SU^B(3)$ and γ-soft: $O^B(6)$)[5]. It should be mentioned that everyone of these dynamical symmetries of the IBM is realized in nuclear structure. This point is well discussed by the many talks in this symposium, especially the overview talk of Iachello[1]. Hence, the IBM treats all the collective motions on equal footing: each has its own characteristics and its own set of eigenstates and more importantly, its own geometrical interpretation. For example, a special characteristic of the $O^B(6)$ limit is to predict the staggering of the states in the γ-band while the $SU^B(3)$ limit does not. Clearly, the multi-chain concept of the IBM, which is perhaps the *most important lesson* one learns from the model, is a clear

0097–6156/86/0324–0027$06.00/0

departure from the conventional usage of dynamical symmetry of Elliott and its many subsequent developments[6] where there is only *one* dynamical symmetry chain ($SU^F(3)$ or pseudo-$SU^F(3)$).

Although the multi-chain dynamical symmetries concept of the IBM is successful in anchoring the various types of collective structures, the full microscopic justification for each chain is still not entirely transparent. Thus, on a purely theoretical level, we deem it important to know whether the multi-chain dynamical symmetries of the IBM are:

(a)   merely a fortuituous consequence of the boson assumption for the coherent valence fermion pairs; or

(b)   inherent in the "raw" fermionic shell structure, i.e. the most general shell model hamiltonian with one and two body interactions, simplified only by the same "physics input" of the IBM.

Also, there are experimental reasons as to why these questions require serious consideration. For example, it was recently noted by Casten and von Brentano[7] that nuclei with A≈130 (a large number of the Xe and Ba isotopes), just as the previously studied A≈ 196 system (Pt isotopes)[8], are very well described by the IBM's $O^B(6)$ limit. Of course, the boson assumption of the IBM prevents an obvious way to explore the reason (or reasons) why these two mass regions should possess the same dynamical symmetry even though they manifestly have different underlying shell structures. Thus, such experimental observations clearly hasten the necessity to seek answers to the above raised questions.

There is another equally important reason as to why it is necessary to seek the dynamical symmetries from the fermionic point of view. One knows that the physics of nuclei in states of large angular momentum constitutes an important branch of nuclear structure physics as well. Experiments are now routinely done yielding detailed spectroscopy of states in the range of J = 35 ~ 45 (typically E is about 10 MeV or so)[9]. No comprehensive theory of nuclear structure can ignore these facts - such a theory must adequately address both the low as well as the high spin phenomena. Therefore, to propose a fermionic dynamical symmetry model of collective nuclear structure which is only applicable in the low energy, low spin region must be inadequate by definition.

Motivated by these considerations, we have recently proposed a multi-chain fermionic dynamical symmetry model (FDSM) which was developed to specifically address the above raised questions. Our starting point is the Ginocchio SO(8) model[10](since from now on only fermion groups will be mentioned, we shall drop the use of the F superscript to denote them). In our opinion, Ginocchio was the first person to seriously pursue the concept of multi-chain dynamical symmetries from a fermionic viewpoint. The main ingredients of the Ginocchio model can be summarized as follows. If one were to take the fermion pair (i.e. a^+a^+ type of operators) with λ=0(S) and 2(D) and certain multipole operators (i.e. a^+a type of operators), both types are constructed from recoupling the single particle j's into pseudo-spin (i) and pseudo-orbit (k), then we will obtain either the Sp(6) or SO(8) algebras. The SO(8) and Sp(6) algebras have the following group chains :

$$SO(8) \supset SO(5)xSU(2) \supset SO(5) \supset SO(3) \tag{1a}$$

$$\supset SO(6) \qquad \supset SO(5) \supset SO(3) \tag{1b}$$

$$\supset SO(7) \qquad \supset SO(5) \supset SO(3) \tag{1c}$$

and

$$Sp(6) \supset SU(3) \qquad \supset SO(3) \tag{2a}$$

$$\supset SU(2)xSO(3) \quad \supset SO(3) \tag{2b}$$

The eigenstates of (1a) and (1b) are identical with the U(5) vibrational and O(6) γ-soft limits of the IBM respectively, while (1c) has no IBM counterpart. The eigensates of (2a) and (2b) are identical with the SU(3) rotational and U(5) vibrational limits of the IBM respectively.

There are many interesting features of the Ginocchio model which we do not have the space to cover here. The two most notable ones are: (a) the SO(8) ⊃ SO(6) has similar characteristics of the O(6) limit of the IBM. This is the first fermionic (dynamical symmetry) description of the γ-soft collective mode in nuclei. This mode has so far not been understood microscopically. (b) Although there exists a "rotational" chain Sp(6) ⊃ SU(3), it was ruled out on the grounds that the most important representation (2N,0) (where N is the pair number) is disallowed due to the Pauli principle (more about this later when we get into the description of our model). What is perhaps most lacking in the Ginocchio model is how one might link such algebraic structures to the shell structures, without which the model cannot be used for the study of real nuclei(i.e. toy model)

The model which we have developed is called the Fermion Dynamical Symmetry Model (FDSM)[11] which is the subject matter of two recent preprints. The FDSM begins with a shell model Hamiltonian in one major valence shell.

$$H = \sum_j \varepsilon_j n_j + V_p + V_q \tag{3}$$

where $V_{p(q)}$ is the pairing (multiple) part of the residual interaction. Our model makes three assumptions. (a) The dominant parts of the pairing interaction V_p are monopole and quadrupole. (b) The two body matrix elements of $V_{p(q)}$ are proportional to the degeneracy of each major valence shell. (c) Terms involving the single particle energies and the multipole interactions are approximated so that the Hamiltonian is a function of the generators of a tractable Lie algebra. Clearly, the first assumption is a result of the successes of the IBM which we have used as input physics. The second assumption is the simple generalization of the usual pairing assumption in nuclear physics and the third is derived from the belief that a dynamical symmetry in the Hamiltonian corresponds to a certain collective mode of the system, which is perhaps the most important lesson one learns from the IBM.

As a result of our assumption (a), we can reclassify all the single particle levels in the major valence shell (see **Fig.** 1) in terms of the pseudo orbit (k) and pseudo spin (i), thus allowing us to make contact with the Ginocchio model for symmetry classifications. The result is that for the normal parity levels, we will have either the Sp(6) or SO(8) symmetry and for the "intruder" level, it is an SU(2) algebra, meaning that for the intruder, only monopole pair is allowed in the lowest energy levels. Thus we have either Sp(6)xSU(2) or SO(8)xSU(2), depending on which major shell we are referring to.   For example, for the major shell with normal single particle levels $s_{1/2}$, $d_{3/2}$, $d_{5/2}$, $g_{7/2}$ (k=2, i=3/2) and intruder single particle $h_{11/2}$ (k=0, i=11/2) level, the symmetry is (uniquely) SO(8)xSU(2).   Let me mention in passing that this is fully consistent with the experimental observation of a large number of stable O(6) nuclei for this mass region[7]. As we have mentioned earlier, although one could construct an Sp(6) ⊃ SU(3) chain in the Ginocchio model, it was nevertheless discarded because the (2N,0) representation is ruled out for most deformed nuclei due to the Pauli principle (N ≤ Ω/3 where Ω, the full degeneracy of a major physical shell, was nebulously specified). This so-called "major flaw" of the model was also reiterated by Hecht[6]. However, in our model, the Pauli restriction applies only for the normal parity levels, i.e. $N_1 \le \Omega_1/3$ where N_1 and Ω_1 are the pair number and the total degeneracy of the normal levels, respectively. There is no requirement in our model that the total number of pairs in one major physical shell, defined as $N = N_0 + N_1$ must be ≤ Ω/3.   Hence N can be approximately Ω/2 (a condition for most deformed nuclei) while $N_1 \le \Omega_1/3$. Physically, this means that the Pauli principle squeezes out the remaining N_0 pairs to the unique parity levels. In a forthcoming paper, we will discuss in detail an analysis based on the Nilsson scheme. The result indicates that for the strongly deformed nuclei, N_1 is generally ≤ $\Omega_1/3$ thus eliminating the reason for rejecting the Sp(6) chain. Thus, we now have, within the context of the FDSM, a handle on how the multi-chain dynamical symmetries, all treated on equal footing just as ther IBM, can manifest themselves.

As we have emphasized earlier, any comprehensive theory of nuclear structure must adequately address the high spin phenomena. By simulating the coherent fermion pairs as bosons, the IBM essentially ruled out the possibility of discussing high spin phenomena without additional ingredients.   In our model, however, this is accomplished in a rather simple fashion. Note that the SO(8) [Sp(6)] and the SU(2) possess the quantum numbes u and υ_0 where they denote the generalized seniority (for S and D pairs in the normal levels) and seniority (for the S pairs in the intruder level) respectively. For u = υ_0= 0, one can find a one to one association with the IBM dynamical symmetries for the FDSM. For example, in the SU(3) limit, for the ground band, one can obtain the usual rigid rotation situation. On the other hand, when either u ≠ 0 or υ_0 ≠ 0 or both, we have the possibility of broken (either generalized or S) pair(s) which corresponds to the well known phenomena in high spin physics, i.e. rotational alignment, Coriolis antipairing, multiple band crossing and the associated backbendings. The ability to incorporate such features is a consequence of this being a fermion (not boson) model!

We present, as an example of this work, the results for the nucleus ^{232}Th as predicted by the FDSM. It is of course known that for this nucleus, the neutrons occupy the 8th major shell while the protons occupy the 7th major shell. Since our results are based on dynamical symmetries, no interband mixing is introduced. In **Fig**. 2, the results of the yrast states plotted as a function of J(J+1) (although only the J value is indicated) are given. It is seen very clearly that roughly between J=10 to 12, the band switches from the ground band (with u = υ_0= 0) to the i = 7/2 pair band ("broken" proton normal pairs with u=2 and υ_0 = 0). Similarly, another band crossing occurs between J=18 to 20. The new band is the i=9/2 pair band ("broken" neutron normal pairs with u=2 and υ_0 = 0). Finally, for a much higher J (24 or so), the new band is associated with the "breaking" of the proton S pair in the j=i=13/2 orbit. These band crossing phenomena is even more transparent in the B(E2)s', plotted as a function of J^{12}, as can be seen from **Fig**. 3. (The data is joined by the dash line to guide the eye). For the j=13/2 band crossing, there is a sharp drop of the B(E2) values. The physical reason for this to occur is quite clear, whenever there is a band crossing (i.e., the breaking of a pair of coherent fermions), the "core" of the rotor will slow down because of angular momentum conservation. This slowing down of the core (the core is the primary contributor of the B(E2)s) is reflected in the data. For the FDSM, the same physics applies also for the low J band crossing phenomena. Our B(E2) calculations, which ignore band mixing, show very clearly this predicted behavior, i.e., whenever band crossing occurs, there is a reduction of the B(E2) (The dotted line in Fig. 3). The reduction is in fact zero when there is no band mixing and with just a small amount of band mixing, the B(E2) can be fitted rather well. Also, whether these primordial structures can remain depends very much on the strength of band mixing. For the actinides (^{232}Th, U isotopes, for example), such structures prevail. Hence, since the FDSM incorporates fermion degrees of freedom, it gives a better description of high-spin B(E2) values than the algebraic boson model.

So far, we have concentrated only on the even-even nucleus. Of course, our model encompass the even-odd as well as odd-odd nuclei as well (by the study of other seniority states).

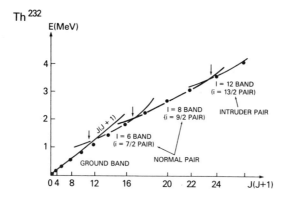

Figure 1. Dynamical symmetries of the shell model. Details may be found in Refs. 10-11.

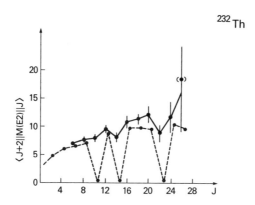

Figure 2. Yrast states of ^{232}Th. The data are taken from Ref. 12.

No	1	2	3			4	5		6		7			8		
n	0	1	2			3	3	4	4	5	5	5	6	6	6	7
k	0	1	1			0	1	0	2	0	1	1	0	1	1	0
i	1/2	1/2	3/2			7/2	3/2	9/2	3/2	11/2	1/2	7/2	13/2	3/2	9/2	15/2
CONFIGURATION	$s_{1/2}$	$p_{1/2}$ $p_{3/2}$	$s_{1/2}$ $d_{3/2}$ $d_{5/2}$			$f_{7/2}$	$p_{1/2}$ $p_{3/2}$ $f_{5/2}$	$g_{9/2}$	$s_{1/2}$ $d_{3/2}$ $d_{5/2}$ $g_{7/2}$	$h_{11/2}$	$p_{1/2}$ $p_{3/2}$	$f_{5/2}$ $f_{7/2}$ $h_{9/2}$	$i_{13/2}$	$s_{1/2}$ $d_{3/2}$ $d_{5/2}$	$g_{7/2}$ $g_{9/2}$ $i_{11/2}$	$j_{15/2}$
SYM			G_6 G_8 G_3				G_6 G_8 G_3		G_8		G_6			G_6		
Ω_0	0	0	0			0	5		6		7			8		
Ω_1	1	3	6			4	6		10		15			21		
Ω	1	3	6			4	11		16		22			29		
$\mathcal{n}$	2	8	20			28	50		82		126			184		

$G_6 = (Sp_6^k \times SO_3^i) \times (S\mathcal{U}_2 \times S\mathcal{O}_3)$

$G_3 = (SU_3^k \times SO_6^i) \times (S\mathcal{U}_2 \times S\mathcal{O}_3)$

$G_8 = (SO_8^i \times SO_3^k) \times (S\mathcal{U}_2 \times S\mathcal{O}_3)$

Figure 3. The theoretical and experimental B(E2) values. The data are taken from Ref. 12.

In conclusion, just as the IBM, the FDSM contains, for each low energy collective mode, a dynamical symmetry. For no broken pairs, some of the FDSM symmetries correspond to those experimentally known and studied previouly by the IBM. *Thus all the IBM dynamical symmetries are recovered.* In addition, as a natural consequence of the Hamiltonian, the model describes also the coupling of unpaired particles to such modes. Furthermore, since the model is fully microscopic, its parameters are calculable from effective nucleon-nucleon interactions. The uncanny resemblance of these preliminary results to well-established phenomenology leads us to speculate that fermion dynamical symmetries in nuclear structure may be far more pervasive than has commonly been supposed.

Acknowledgments

This work is supported by the National Science Foundation. It is indeed a pleasure for me to thank my colleagues and collaborators in China and the US in this work. They are **Cheng-Li Wu** of Jilin University, **Xuan-Gen Chen** of the Nanjing Military College, **Jin-Quan Chen** of Nanjing University and **Michael W. Guidry** of the University of Tennessee, all of whom I have known and learned from for a number of years. It must also be mentioned that during the development of this work, **Joe Ginocchio** of the Los Alamos Scientific Laboratory has played the role of a "knowledge provider" and critical evaluator of our ideas. Finally, I must thank the two organizers of this symposium, Dr. R. A. Meyer and Prof. Daeg Brenner, for inviting me to participate in my first (and hopefully not last) ACS meeting.

References

1. A. Arima and F. Iachello, see Professor Iachello's contribution in this volume.
2. A. M. Perelomov, Commun. Math. Phys. **26**, 222 (1972).
3. F. T. Arecchi, E. Courtens, R. Gilmore and H. Thomas, Phys. Rev. **A6**, 2211(1972);
   R. Gilmore, Rev. Mex. de Fisica **23**, 143 (1974).
4. J. N. Ginocchio and M. W. Kirson, Phys. Rev. Lett. **44**, 1744(1980);
   A. E. L. Dieperink, O. Scholten and F. Iachello, Phys. Rev. Lett **44**, 1747(1980);
   D. H. Feng, R. Gilmore and S. R. Deans, Phys. Rev. **C23**, 1254(1981).
5. COHERENT STATES, ed. by J. R. Klauder and B.-S. Skagerstam, **World Scientific, Singapore**, 1985.
6. J. P. Elliott, Proc. Roy. Soc. **A245**, 128(1958); **A245**, 562(1958);
   K. T. Hecht, Lectures delivered at the VIIIth symposium on Nuclear Physics, Oaxtepec, Mexico, Jan. 1985 and Phys. Rev. C (to be published).
   J. P. Draayer, Lectures delivered at the VIIIth symposium on Nuclear Physics, Oaxtepec, Mexico, Jan. 1985.
7. R. F. Casten and P. von Brentano, Phys. Lett. **B152** , 22(1985).
8. J. A. Cizewski et al., Phys. Rev. Lett. **40**, 167(1980).
9. For a comprehensive discussion, see R. Bengtsson and J. D. Garrett in COLLECTIVE PHENOMENA IN ATOMIC NUCLEI, edited by T. Engeland, J. Rekstad and J. S. Vaagen (**World Scientific, Singapore**, 1985).
10. J. N. Ginocchio, Ann. of Phys. **126**, 234(1980).
11. C.-L. Wu, D. H Feng, X.-G. Chen, J.-Q. Chen and M. W. Guidry, Preprint, 1985. Submitted to Phys. Lett. and Ann. of Phys.
12. This is the so-called "GSI catastrophe" which shows that as a function of J, the IBM predictions of the B(E2) values will eventually drop off. See H. Ower, Ph.D. thesis, University of Frankfurt, 1980.

RECEIVED May 2, 1986

On the Microscopic Theory of the Interacting Boson Model

M. Sambataro

Istituto Nazionale di Fisica Nucleare-Sez. di Catania, Corso Italia 57, Catania, Italy

We present a new mapping procedure from fermion onto boson spaces. The procedure is based on the Generator Coordinate Method. We show an application of this technique to systems of nucleons moving in a single j-orbit. The extension of this microscopic investigation to many j-orbit systems is also examined.

One of the most interesting aspects of the Interacting Boson Model concerns its connections with the underlying fermion space. The understanding of the mechanism through which boson-like features arise from effective nuclear hamiltonians provides, in fact, a way to relate collective spectra to the fermion motion.

Since the original study of Otsuka, Arima and Iachello (OTS78), much work has been carried out on this subject. In spite of that, until now, the description of well deformed nuclei can not be considered satisfactory. We present here a new mapping technique which is directed toward the treatment of these systems.

The microscopic derivation of a boson hamiltonian from a fermion one is basically a two step process. In the first step, one has to select the collective subspace of the shell model space. For the IBM this means truncating the shell model space to the space of collective S-D pairs. In the second step, this space has to be mapped onto the s-d boson space.

In the realistic case of nucleons moving in many j-orbits, the first step implies the non-trivial task of fixing the internal structure of the collective pairs. In order to focus on the mechanism of the mapping, then, we will first discuss the case of nucleons in a single j-orbit, where this problem does not exist. The discussion concerning more realistic systems will be postponed to the second part of this contribution.

The mapping procedure

The procedure we are going to expose is based on the Generator Coordinate Method (GCM). The GCM looks for solutions of the Schrödinger equation of the form

$$|\psi\rangle = \int d\alpha \, f(\alpha) \, |\phi(\alpha)\rangle \tag{1}$$

0097–6156/86/0324–0035$06.00/0

where $|\phi(a)>$, the generating functions, depend on some parameters $\{a\}$, the generator coordinates. The weight function $f(a)$ is determined by imposing

$$\int \frac{<\Psi|H|\Psi>}{<\Psi|\Psi>} = 0 \qquad (2)$$

This condition leads to the Hill-Wheeler equation

$$\int da' \left(<\phi(a)|H|\phi(a')> - E <\phi(a)|\phi(a')> \right) f(a') = 0 \qquad (3)$$

The basic idea of the mapping procedure is the following one. Let $|\Psi(a)>$ be fermion states depending on some variables α and $|\phi(\beta)>$ be boson states depending on some other variables β. Let us suppose that one can establish a relation between the variables α and β, i.e. find a function $\beta = F(\alpha)$, such that

$$<\Psi(\alpha)|\Psi(\alpha')> = <\phi(\beta)|\phi(\beta')> \qquad (4)$$

Let us suppose, furthemore, that in correspondence to a given fermion hamiltonian H_F we find a boson hamiltonian H_B so that also

$$<\Psi(\alpha)|H_F|\Psi(\alpha')> = <\phi(\beta)|H_B|\phi(\beta')> \qquad (5)$$

In this case, the Hill-Wheeler equations in the fermion and boson spaces are equivalent and lead to identical eigenvalues.

As an application of this method let us choose as fermion generating functions the intrinsic states

$$|\Psi(a)> = \frac{1}{\sqrt{N_F(a)}} \left(S^+ + a\, D_0^+ \right)^N |0> $$

$$S^+ = \frac{1}{\sqrt{2}} \left[a_j^+ a_j^+ \right]^{(0)} \qquad (6)$$

$$D_0^+ = \frac{1}{\sqrt{2}} \left[a_j^+ a_j^+ \right]_0^{(2)}$$

The example which we are going to discuss refers to the case $N=3$ and $j=23/2$.

As boson generating function we choose

$$|\phi(\beta)> = \frac{1}{\sqrt{N_B(\beta)}} \left(s^+ + \beta\, d_0^+ \right)^N |0> \qquad (7)$$

The overlaps $<\Psi(\alpha)|\Psi(\alpha')>$ and $<\phi(\beta)|\phi(\beta')>$, with $\beta = \alpha = 1.3$ and $\beta' = \alpha'$, are shown in fig. 1; (a) and (b), respectively. They behave similarly but in the limits $\alpha' \gg \alpha$ and $\beta' \gg \beta$ they have different values. We get a relation between α and β by imposing

$$\lim_{\alpha' \to \infty} <\Psi(\alpha)|\Psi(\alpha')> = \lim_{\beta' \to \infty} <\phi(\beta)|\phi(\beta')> \qquad (8)$$

Using this relation, the overlap $<\phi(\beta)|\phi(\beta')>$ becomes very close to the fermion one, so we have plotted only one curve in this case; (a), in fig. 1.

As an example of fermion hamiltonian we take the quadrupole-quadrupole hamiltonian

$$H_F = [a_j^+ \tilde{a}_j]^{(2)} \cdot [a_j^+ \tilde{a}_j]^{(2)} \tag{9}$$

The behaviour of $\langle \Psi(\alpha) | H_F | \Psi(\alpha) \rangle$ is shown in fig.2, (a). We wish to investigate whether such a behaviour can be reproduced by a boson s-d hamiltonian. We also wish this hamiltonian to be hermitian and at most two body. The most general hamiltonian of this kind can be written as (IAC82)

$$H_B' = \varepsilon \, n_d + \delta_0 \, P^+ \cdot P + \delta_1 \, L \cdot L + \delta_2 \, Q \cdot Q + \delta_3 \, T_3 \cdot T_3 + \delta_4 \, T_4 \cdot T_4 + C(N) \tag{10}$$

where C(N) is a function depending on the boson number N.

We find that a special case of this hamiltonian

$$H_B = 0.12 \, P^+ \cdot P - 0.075 \, Q \cdot Q + 0.04 \, T_3 \cdot T_3 - 1.38 \tag{11}$$

gives a reasonable fit of this behaviour (fig. 2, (b)). The same is verified also for non-diagonal matrix elements.

The spectra of H_F and H_B in the fermion S-D space and in the boson s-d space, respectively, are given in fig. 3. As expected on the basis of the discussion given above, these spectra look quite similar. A good agreement is found for binding energies as well (B.E.=-3.38 MeV for fermions, B.E.=-3.34 MeV for bosons). More details on this mapping procedure are given in Sambataro et al.(SAM85a)

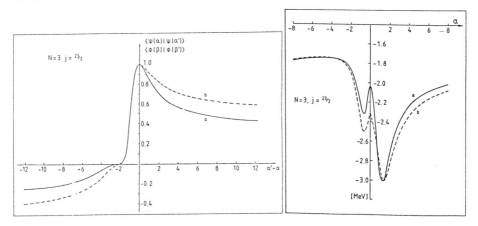

Figure 1 (left). Comparison between fermion (a) and boson (b) overlaps, as described in the text. Reproduced with permission from (SAM85a). Copyright 1986 North-Holland Physics Publishing Company.

Figure 2 (right). Diagonal fermion (a) and boson (b) matrix elements between the states described in the text. Reproduced with permission from (SAM85a). Copyright 1986 North-Holland Physics Publishing Company.

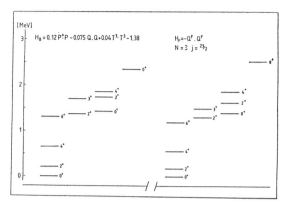

Figure 3. Comparison between the spectra generated by a quadrupole-quadrupole hamiltonian within the S-D subspace (N = 3, j = 23/2) and by the IBM hamiltonian obtained with this mapping procedure. Reproduced with permission from (SAM85a). Copyright 1986 North-Holland Physics Publishing Company.

Microscopic calculations for many j-orbit systems

     Working in the fermion space, even in a considerably reduced shell model space like the S-D subspace, is quite complicated if nucleons are supposed to move in many j-orbits. In recent calculations performed in truncated shell model spaces, for nucleons in a single j-orbit (CAT85), a computational procedure has been set up to evaluate overlaps and matrix elements. The procedure has used   recursion formulas which have been based on the commutation properties of the fermion operators. In principle, it is possible to extend this procedure to the case of nucleons occupying many j-orbits. However, in this case, the increased complexity of the commutation relations involved does not leave much hope that calculations can be performed without running into many difficulties. In this contribution we outline an approximation which considerably simplifies this procedure.
     In the single j-orbit case, let us consider states of the kind

$$| N, \{\lambda \mu\} > = A^{+}_{\lambda_1 \mu_1} A^{+}_{\lambda_2 \mu_2} \cdots A^{+}_{\lambda_N \mu_N} |0> \qquad (12)$$

where

$$A^{+}_{\lambda \mu} = \frac{1}{\sqrt{2}} \left[a^{+}_{j} a^{+}_{j} \right]^{(\lambda)}_{\mu} \qquad (13)$$

As far as the evaluation of overlaps is concerned, we observe that one can generate recursion formulas according to which overlaps between states (12) with N pairs are expressed in terms of overlaps of specially generated states with N-1 pairs

$$< N, \{\lambda \mu\} | N, \{\lambda' \mu'\} > = \sum_{\ell=1}^{N} I(\lambda_N \mu_N, \lambda_\ell \mu_\ell) < N-1, \{\lambda \mu\}_N | N-1, \{\lambda' \mu'\}_\ell > + \qquad (14)$$

$$\sum_{J} \sum_{i>\ell=1}^{N-1} D(\lambda_N \mu_N, \lambda_i \mu_i, \lambda_\ell \mu_\ell, J) \; < N-1, \{\lambda \mu\}_N \mid N-1, \{\lambda' \mu'\}_{i,\ell}^{J} >$$

In this formula, $\mathcal{I}(\lambda_i \mu_i, \lambda_j \mu_j) = \delta_{\lambda_i \lambda_j} \delta_{\mu_i \mu_j}$, D coefficients are defined by

$$\left[[A_{L m}, A_{Kq}^+], A_{R_2}^+ \right] = \sum_{J} D(Lm, Kq, R_2, J) \; A_{J \; q+2-m}^+ \tag{15}$$

and the symbol $\{\lambda \mu\}_{\cdots}^{\cdots}$ stands for a new set $\{\lambda' \mu'\}$ obtained from $\{\lambda \mu\}$ by removing the pairs whose indices are indicated in the subscript and substituting them with pairs of multipolarity indicated in the superscript (and appropriate projection) if any. Similar recursion formulas can be derived for one-body and two-body matrix elements.

In extending the previous formalism to many j-orbits, one can define a set of basis states like (12) but where the operators are now replaced by

$$M_{\lambda \mu}^+ = \sum_{(jj')} \gamma_\lambda (jj') A_{\lambda \mu}^+ (jj') \tag{16}$$

However, the application of recursion formulas like (14) becomes quite complicated due to the fact that commutators like, for instance,

$$\left[[M_{\lambda_1 \mu_1}, M_{\lambda_2 \mu_2}^+], M_{\lambda_3 \mu_3}^+ \right] \tag{17}$$

do not generate pairs belonging to the original collective space.

To investigate possible approximate schemes to the procedure just discussed, we first observe that, taking into account the properties of the structure coefficients $\gamma_\lambda (jj')$, one can write

$$A_{J m}^+ (jj') = \sum_\nu \gamma_J^{(\nu)} (jj') \; M_{J m}^{+ (\nu)} \tag{18}$$

where $M_{J m}^{+ (\nu)}$ are linearly independent correlated pairs of the type (16). By replacing (18) in commutators like (17), it is possible to express these commutators in terms of the said pairs.

The approximation we explore is the one in which among all correlated pairs appearing in the commutators like (17), we retain only one pair for each multipolarity, i.e. the collective pair which defines the collective subspace. This approximation appears consistent with the assumption, already implicit in the choice of the basis states, that only these pairs are essential for a description of the low-lying nuclear states. The important advantage of this approximation is that recursion formulas of the type (14), in the case of nucleons moving in many j-orbits, become as easy to use as those in single j-orbit.

To test this approximation we have evaluated the expectation value of a pairing interaction in a state with N pairs, in the shell 50-82, and generalized seniority v=0. Exact and approximated energies obtained with a minimization procedure are

shown in fig.4 for different values of N. Also approximated
occupation numbers compare well with the exact ones. All the
details of these calculations are contained in Sambataro et al.
(SAM85b).

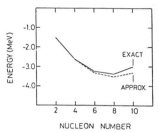

Figure 4. Comparison between exact and approximated expectation
values of a pairing hamiltonian in states with generalized seniority
v = 0. Reproduced with permission from (SAM85b). Copyright 1986
North-Holland Physics Publishing Company.

Conclusions

In this contribution, we have presented a new mapping
procedure from fermion onto boson spaces. This procedure has
been tested for a fermion quadrupole-quadrupole interaction in
the case of nucleons moving in a single j-orbit. The energy spectrum generated by this interaction within the S-D subspace has
been found to compare rather well with the spectrum of an SU(3)-
O(6) IBM hamiltonian whose parameters have been obtained with
this procedure.
An important point which will be the object of future investigation concerns the evaluation of the renormalization effects
on the boson hamiltonian introduced by the truncation of the
fermion space.
Finally, we have outlined an approximated method to perform calculations in the fermion space, for nucleons moving in
many j-orbits. The method has been tested for a pairing hamiltonian and has provided encouraging results for both energies and
occupation numbers.

REFERENCES

(CAT85) F.Catara,A.Insolia,M.Sambataro,E.Maglione and A.Vitturi,
        Phys. Rev. C32 634 (1985)
(IAC82) F.Iachello,in: Lecture Notes in Physics, Ed. J.S.Dehesa,
        Vol.161 (Springer,Berlin,1982).
(OTS78) T.Otsuka,A.Arima and F.Iachello, Nucl. Phys. A309 1 (1978)
(SAM85a)M.Sambataro,H.Schaaser,D.M.Brink,Phys.Lett.167B 145 (1986)
(SAM85b)M.Sambataro and A.Insolia, Phys.Lett. 166B 259 (1986)

RECEIVED June 25, 1986

Realistic and Model Applications of the Boson Expansion Theory

Taro Tamura

Department of Physics, University of Texas at Austin, Austin, TX 78712

The structure of the boson expansion theory is explained
based on simple models. It gives a good idea on how this
theory would work when it is applied to realistic cases.

Our group at the University of Texas has been engaged in the boson
expansion theory (BET), during the past decade or so. Our work started
with two papers [1,2], which we shall henceforth refer to as KT1 and KT2.
The BET of KT1 and KT2 are very similar in their forms, but differ
critically in what they describe. The BET of KT2 was used extensively
for realistic calculations, with notable success in fitting a number
of data of collective motions in a variety of nuclei [3].

Subsequently, we renewed our formal study of BET as a whole, resulting in a paper [4], which we shall call KT3 henceforth. In both KT1 and
KT2, a method called a commutator method was used, while the method used
in KT3 was (a generalized version of) that of MYT [5]. In parallel with
or following KT3, a few additional papers were published, clarifying
further the general structure of BET [6,7].

One aspect which characterises our realistic calculations [2,3] is
the use of the BCS theory, making our theory be sometimes called BCS+BET.
As is well known, the BCS theory causes certain types of errors. However,
it is also well known that one will not commit too serious errors, if one
uses BCS, as we did, within the bound of its applicability.

It is, nervertheless, highly desirable, to construct a theory that
maintains the simplicity of the BCS theory, yet removes the (major part of
the) BCS errors. A theory, recently developed by Li [8], which uses the
number-conserving quasi-particle (NCQP) method, is such a theory. We are
now working on replacing BCS+BET by NCQP+BET, and on resuming realistic
calculations. This time we shall be able to go much beyond the bounds
imposed by the BCS approximation.

It is also valuable to investigate on how our new form of BET would
perform, when it is applied to simple models. During the past year or so
we indeed did this, the models taken up being the single-j shell model
(1j-SM) and the Ginocchio model [9].

0097-6156/86/0324-0041$06.00/0
© 1986 American Chemical Society

In the following, we discuss mostly the use of our BET for these simple models. We do this because it will help the reader to understand the essence of our BET, and also because these models were studied by Otsuka et. al. [10,11], and by Arima et. al. [12]. (They did this with a stated purpose of justifying the interacting-boson approximation (IBA) [13].) It will be seen below that, in spite of the use of the same models, the conclusions we arrive at differ significantly from those given by these authors.

SINGLE-j SHELL MODEL

The 1j-SM is characterised by the fact that it is very easy to construct in it a basis states in the seniority scheme. Let a (basis) state $|n,v;\alpha\rangle$ contain n particles, v of which contributing to make this state have seniority v. (α stands for additional quantum numbers.) This state then has S pairs (of Cooper type) of which number equals $k=(n-v)/2$. The states with $n=v$ (and hence $k=0$) are called highest senioriy (HS) states.

Once the states are thus constructed in the seniority scheme, a step called seniority reduction (SR) can follow. An example of it is given as

$$\langle n+2,v';\alpha'|C_{2\mu}^{\dagger}|n,v;\alpha\rangle$$

$$= \sqrt{2}\ U_{-1,1}\ V_{0,0} \quad \langle v+2,v+2;\alpha'|B_{2\mu}^{\dagger}|v,v;\alpha\rangle \quad \delta_{v';v+2}$$

$$+ \quad (U_{0,0}^{2} - V_{0,0}^{2}) \quad \langle v,v;\alpha'|C_{2\mu}^{\dagger}|v,v;\alpha\rangle \quad \delta_{v';v} \qquad (1a)$$

$$+\sqrt{2}\ U_{-1,-1}\ V_{0,-2}\ \langle v-2,v-2;\alpha'|B_{2\widetilde{\mu}}^{\dagger}|v,v;\alpha\rangle \quad \delta_{v';v-2}$$

In (1a), $B_{2\mu}^{\dagger}$ is a pair creatrion opeartor, $C_{2\mu}^{\dagger}$ is a scattering operator, and the U and V coefficients are defined as ($\Omega = j + 1/2$)

$$U(i,j)=[(2\Omega-n-v-i-j)/(2\Omega-2v-2j)]^{1/2}$$

$$\qquad\qquad\qquad\qquad\qquad\qquad\qquad\qquad (1b)$$

$$V(i,j)=[(n-v+i-j)/(2\Omega-2v-2j)]^{1/2}$$

As seen, the SR is to express a matrix element of an operator, between states which are not (usually) of HS nature, by a sum of matrix elements of related operators taken between HS states. The SR formula of (1) is well known; see, e.g., Lawson [14]. We gave it, however, in a form which is different from what is usually employed. Namely we expressed the coefficients on the rhs of (1a) in terms of the U and V factors. The reason we chose this form is that we want to emphasize the fact that (1) is very similar to what we get when we perform the usual Bogoliubov transformation, i.e., use BCS. In fact, we obtain

$$C_{2\hat{\mu}}^{\dagger}= \sqrt{2}\ UV\ B_{2\mu}^{\dagger} + (U^2-V^2)\ C_{2\mu}^{\dagger} + \sqrt{2}\ UV\ B_{2\widetilde{\mu}} \qquad (2a)$$

$$U = [(2\Omega-n)/(2\Omega)]^{1/2}\ ; \quad V = [n/(2\Omega)]^{1/2} \qquad (2b)$$

The operators on the rhs of (2a) are quasiparticle pair operators.

As seen, (1) and (2) are in fact in the same algebraic forms. The
U and V factors in (2b), however, lack the delicate i, j and v depndence,
which the U and V factors in (1b) have. This results in the number non-
conservation problem of BCS.

Note that the bra and ket states that appear on the rhs of (1a) contain
S pairs, while the HS states on the rhs do not. This means that SR has
eliminated completely the S pairs from our description, their initial presence being, nevertheless, accurately remembered by the emergence of the U
and V factors. The complete elimination of the S-pairs is also the case with
BCS, the only difference being that U and V factors now have somewhat poorer
memory. The sigfificance of the similarity of (1) and (2) lies in this
complete elimination of the Cooper pairs in both SR and BCS.

Let us now consider bosonizing [4] the pair operators, based on the
reduced matrix elements that appear on the rhs of (1). We then have an expansion which contains no s bosons in it. This BET, which we may call SR+BET, is
exact (assuming that the boson expansion is carried out to a desired order).

An exact BET can thus be constructed without s bosons, even when the
seniority scheme is used, a fact which may surprise those who know the OAI
work [10]. It appears that it has been normaly belived that OAI had s
bosons, because it used the seniority scheme, while, e.g., our BET did not,
because BCS was used. What we showed above is that such a belief is
unfounded. We may also note that, in the 1j-SM, the NCQP+BET, which replaces
BCS+BET, is exact and agrees exactly with SR+BET.

OAI request that the fermion (S,D) space must always be mapped onto the
boson (s,d) space; and obtain s-bosons. We showed, however, that an exact
BET exist without s-bosons. The s-bosons in OAI may then be considered
essentially as mathematical artifacts. OAI claimed that they derived IBA, by
creating s-bosons in this way. Then the s-bosons in IBA may also be mathematical artifacts.

OAI further claimed that the lowest order expansion was sufficiently
good. However, one sees that the OAI tables does not actually show that
such is the case. Higher order terms are definitely needed [15], indicating
that a microscopic IBA is not as simple as is the phenemenological IBA.

We also noted [15] that the OAI and OAIT [11] theories are crucially
different. In OAI, the coefficients multiplying s and d boson operators are
constant; thus OAI can be an IBA. In OAIT, however, the coefficients depend
on v, and thus OAIT is not an IBA. (With OAIT, the SU(6) symmetry, which is
the key ingredient of the phenomenological IBA, may largely be lost.) In any
case, the (somewhat better looking) OAIT numerical results, rather than the
(poorer) OAI results, were presented in OAI tables, without mentioning at
all that this was done.

Sometime ago, Arima [12] argued that the use of our BET in 1j-SM caused
errors of about a factor of 3 in some B(E2) values. We did the same calculation ourselves recently [16], and found that the errors were in the range
of 10% or so; never as large as Arima stated. We then noticed that Arima

started with the BET of KT1, and then truncated to the d-boson component.
However, this should not have been done. When trucation is done, the
commutator equations of KT1 must be reconstructed, and then resolved.
And this is exactly what was done in KT2. It appears that Arima failed to
notice this fact and presented very misleading results.

GINOCCHIO MODEL

Ginocchio model is a degenerate 4-j SM, the j values ranging from
$\ell-3/2$ to $\ell+3/2$, for a given ℓ. In spite of this 4-j shell nature, the
specific combination of the 4 shells makes this model very similar to
1j-SM. Thus, again the seniority scheme can be constructed easily. In
addition to this, the (S,D) space becomes a closed space. This makes
the algerbra of the Ginocchio model even simpler than that of 1j-SM.

In Fig.1, we show the results of our recent calculations [17]. This
is the case of the so-called SO(6) limit [12], and the energy spectra
predicted by various boson theories are compared with the exact fermion
spectrum. It is seen first that the NCQP+BET result agrees almost perfectly with the exact spectrum. (The remaining discrepancy was caused
by a finite boson expansion.) The OAIT also performs fairly well though
not as good as does NCQP+BET. The performance of BCS+BET is much poorer
compared with these two, the energies getting too high with increased
spins. The spectrum we denote in Fig.1 as OAI also behaves rather poorly,
this time, however, underpredicting the high-spin state energies.

As we emphasized towards the end of Sec.II, OAI is an IBA but OAIT is
not. Nevertheless, Arima et. al. presented [12] the relatively better OAIT
result as an evidence to justify IBA. However, this presentation is again
very misleading. The OAI results should have been presented as IBA, as we
have done so in our Fig.1.

In the above, we have repeatedly mentioned the bound of applicability of
the BCS theory. It means more specifically that the BCS theory is to be
used only when the (effective) space size Ω, and the particle number n, are
both sufficiently larger than is the quasi-particle number (which is the
senioprity v in the Ginocchio model).

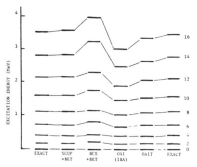

Fig.1. Level spectra
predicted by various
boson theories in the
SO(6) limit of the
Ginocchio model.

Thus, we do not expect from the beginning that the BCS+BET will fit the high-spin states in Fig.1. (Note that in Fig.1, Ω = 22 and n=16. For I=16 state we thus have v = n.) However, we expect that the lower states are fit satisfactorily, and Fig.1 shows that this is indeed the case. Note that this fact assures to a good extent that our realistic calculations were done quite meaningfully. We have considered [2,3] only lowlying states, and thus have stayed within the bound of applicability of the BCS theory.

DISCUSSIONS

In the present paper we have discussed mostly cases with simple models. In spite of this restriction, we seem to have a rather clear view, on where we stand now with our BET, including its applications to realistc cases.

We have already shown [2,3] that our BCS+BET worked rather nicely in fitting a number of data. And Sec.III of the present paper showed that BCS+BET is in fact a dependable theory. We also showed in both Secs.II and III that NCQP+BET can remove the BCS errors. In any case, our future realsitic calculations will be done in terms of NCQP+BET.

An important remark we want to make here is that it is rather easy to construct NCQP+BET even for realistic cases with non-degenerate many-j shells. The same is not the case, however, with OAI and OAIT (and SR+BET). In order to construct the many-j shell versions of these theories, one must be able to handle the generalized seniority scheme. However, this is a very difficult task to achieve.

In constructing NCQP+BET, we first construct BCS+BET. Since we use BCS first, we eliminate Cooper pairs completely, as we emphasized in Sec.II. Thus NCQP+BET is void of s-bosons. Note that the switch from BCS+BET to NCQP+BET means to modify (improve) the U and V factors. No change is made of the basic structure of BET.

As we remarked, the many-j shell version of OAI (or OAIT), and hence the microscopic version of IBA (of many-j shell nature), which have been coveted for, seem hard to come by. Note, however, that we showed in Sec.II that SR+BET (=NCQP+BET), OAI and OAIT were essentially equivalent, in spite of the subtle differences which were produced by different tastes. Then, it may make sense to consider that the <u>many-j shell NCQP+BET is indeed the many-j shell version of all the above theories.</u>

In other words, to construct the many-j shell version NCQP+BET may be considered as an excellent way of <u>bypassing</u> the difficult task of handling the generalized seniority. However, an important remark to be repeated here is that the NCQP+BET is void of s-bosons (and hence is not of the IBA form). This is not a problem for us. We have already shown [2,3], e.g., that we can produce vibration, gamma-unstable, rotation and other transisional situations, without having s-bosons. We do not need the SU(6) symmetry of IBA for such purposes.

In conclusion, we believe we are in a rather good shape with our BET, particularly since NCQP [8] has become available. We expect to start shortly to produce numerical results that could be compared with a variety of experimental data. As we stressed above, we can now go beyond the bounds of the BCS theory. Also, the rather general framework of BET, constructed in KT3, now allows us to take into account the non-collective, as well as collective components in a rather systematic way. Further accumulation of data, pertaining to both kinds of levels, is thus highly hoped for.

ACKNOWLEDGMENTS

The author is very much indebted to V. G. Pedrocchi and C.-T. Li for their invaluable cooperations. This work was supporterd in part by the U. S. Department of Energy.

REFERENCES

1. T. Kishimoto and T. Tamura, Nucl. Phys. $\underline{A192}$ 246 (1972).
2. T. Kishimoto and T. Tamura, Nucl. Phys. $\underline{A270}$ 317 (1976).
3. T. Tamura, K. J. Weeks and T. Kishimoto, Nucl. Phys. $\underline{A347}$ 359 (1980).
   K. J. Weeks, T. Tamura, T. Udagawa and F.J.W. Hahne, Phys. Rev. C $\underline{24}$ 703 (1981).
4. T. Kishimoto and T. Tamura, Phys. Rev. C $\underline{27}$ 341 (1983).
5. T. Marumori, M. Yamamura and A. Tokunaga, Prog. Theo. Phys. $\underline{31}$ 1009 (1964).
6. V. G. Pedrocchi and T. Tamura, Prog. Theo. Phys. $\underline{68}$ 820 (1982).
   V. G. Prdrocchi and T. Tamura, Phys. Rev. C $\underline{28}$ 4510 (1983).
7. T. Tamura, Phys. Rev. C $\underline{28}$ 2154 (1983); C $\underline{28}$ 2480 (1983).
   V. G. Pedrocchi and T. Tamura, Phys. Rev. C $\underline{29}$ 1461 (1984).
8. C.-T. Li, Nucl. Phys. $\underline{A417}$ 37 (1984).
9. J. N. Ginocchio, Ann. Phys. (N.Y.) $\underline{126}$ 234 (1980).
10. T. Otsuka, A. Arima and F. Iachello, Nucl. Phys. $\underline{A309,}$ 1 (1978).
11. T. Otsuka, A. Arima, F. Iachello and I. Talmi, Phys. Lett. $\underline{76B}$ 139 (1978).
12. A. Arima, Nucl. Phys. $\underline{A347}$ 339 (1980).
    A. Arima, N. Yoshida and J. N. Ginocchio, Phys. Lett. $\underline{101B}$ 209 (1981).
13. A. Arima and F. Iachello, Ann. Phys. (N.Y.) $\underline{99}$ 253 $(\overline{1976})$; $\underline{111}$ 201 (1978).
14. R. D. Lawson, "Theory of Nuclear Shell Model", Clarendon Press, Oxford (1980).
15. T. Tamura, C.-T. Li and V. G. Pedrocchi, Phys. Rev. C (in press).
16. C.-T. Li, V. G. Pedrocchi and T. Tamura, Phys. Rev. C (in press).
17. C.-T. Li, V. G. Pedrocchi and T. Tamura, (preprint).

RECEIVED May 2, 1986

Mixed-Symmetry Interpretation of Some Low-Lying Bands in Deformed Nuclei and the Distribution of Collective Magnetic Multipole Strength

O. Scholten[1], K. Heyde[2], P. Van Isacker[2], and T. Otsuka[3]

[1]Michigan State University, East Lansing, MI 48824
[2]Institute for Nuclear Physics, Proeftuinstraat 86, B-9000 Gent, Belgium
[3]Japan Atomic Energy Research Institute, Tokai, Ibaraki-ken, Japan

The effect of neutron-proton symmetry breaking on the distribution of M1 strength in the SU(3) limit of the Interacting Boson Model (IBA-2) is studied. A possible alternative choice for the Majorana force is investigated, with a structure that resembles more closely that which is calculated in microscopic theories. It is found that the specific choice for the Majorana interaction has important consequences for the magnetic strength distribution function. In addition it allows for an alternative interpretation of the second excited $K^{\pi}=0^{+}$ band in rare earth nuclei, as a mixed-symmetry state.

Experimental evidence is accumulating [Boh84] on the existence and properties of states in deformed nuclei that are not fully symmetric in the neutron and proton degrees of freedom. The occurrence of these mixed-symmetry modes have been predicted in various geometrical models [LoI79, Suz77] and in the version of the Interacting Boson Model where neutron and proton degrees of freedom are explicitly taken into account (IBA-2) [Die83, Ari83].

In the IBA-2 model [Ari83], the structure of the collective states in even-even nuclei is calculated by considering a system of interacting neutron and proton s and d bosons. We will focus attention on the Majorana force,

$$M_{\nu\pi} = \xi_2 (s_\nu^\dagger d_\pi^\dagger - d_\nu^\dagger s_\pi^\dagger)^{(2)} \cdot (s_\nu \tilde{d}_\pi - \tilde{d}_\nu s_\pi)^{(2)} - 2 \sum_{k=1,3} \xi_k (d_\nu^\dagger d_\pi^\dagger)^{(k)} \cdot (\tilde{d}_\nu \tilde{d}_\pi)^{(k)};$$

which serves to raise the excitation energy of the states that are not fully symmetric in the neutron and proton degrees of freedom. This Majorana force contains three parameters, ξ_1, ξ_2, and ξ_3. In most phenomenological applications, as for example given in refs. [Van84, Sch85], for simplicity, these parameters have been chosen equal, in which case the Majorana force

0097-6156/86/0324-0047$06.00/0

pushes up the mixed-symmetry states by an equal amount, i.e. this force is a true Majorana force.

While for the ε, κ and $\chi_{\nu,\pi}$ parameters extensive microscopic calculations exist [Ots78,Sch83], the microscopic origin of the Majorana force is at best only partially understood [Sch83,Dru85]. In particular there is very little, or no, evidence that the three parameters ξ_1, ξ_2, and ξ_3 should be chosen equal [Dru85]. Microscopic calculations for the parameters of the IBA-2 hamiltonian, as reported in ref. [Dru85], indicate that the parameters ξ in the Majorana force should not be taken equal, but that the choice $\xi_1=\xi_3=b$ and $\xi_2=0$. is more realistic. For this reason we made a phenomenological investigation of the observables that depend strongly on the structure of the Majorana force. In this work we compare the results of two calculations, one using the standard choice for the Majorana force, $a=\xi_1=\xi_3=\xi_2$ (calculation I), and one using $b=\xi_1=\xi_3$ and $\xi_2=0.0$ (calculation II) as suggested by microscopic calculations [Sch85,Dru85]. The difference between the two choices for the Majorana force is most intriguing in the SU(3) limit of the IBA model on which we will therefore focus our attention.

As an example of a typical deformed rare earth nucleus we will discuss in some more detail the nucleus ^{156}Gd. The parameters of the IBA-2 hamiltonian have been taken from ref. [Sch80]. The strength of the Majorana force, as determined from the position of the 1^+ level near $E_x \cong 3$ MeV [Boh84], can be taken as $a=0.15$ MeV (calculation I) or as $b=0.3$ MeV (calculation II). In Fig. 1 the calculated excitation energies of the bandheads using these two different choices for the Majorana force are compared with some known bandheads in the experimental spectrum of ^{156}Gd. Even though in both calculations the strength of the Majorana force is chosen such that the $K^\pi=1^+$ band lies near $E_x \cong 3$ MeV, the spectrum of mixed-symmetry states in the two calculations is totally different. In calculation (I) the $K^\pi=1^+$ band is the lowest mixed-symmetry band, while in calculation (II) the lowest mixed-symmetry band lies near the position of the β and γ band and all other mixed-symmetry states occur at the position of the $K^\pi=1^+$ bandhead or higher. The lowest mixed-symmetry $K^\pi=0^+$ band in calculation (II) results in energy very near an experimentally well known $K^\pi=0_3^+$ band.

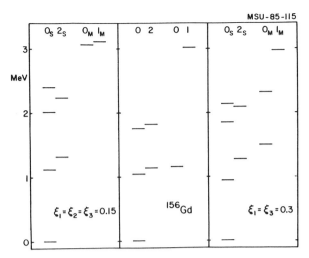

Fig. 1 A comparison between calculated bandhead energies and those observed in ^{156}Gd. In the calculation presented on the left-hand side (calculation I) the Majorana force is parameterized (see text) by a=0.15 MeV while in the calculation on the right-hand side (II) b=0.3 MeV is used. Above an excitation energy of 2.5 MeV, the symmetry assignment may not be accurate due to the F-spin nonscalar nature of the hamiltonian.

In this section some of the E2 decay properties of the low-lying mixed symmetry $K^{\pi}=0_3^+$ band are discussed. In the IBA-2 model, E2 transitions are calculated using the operator

$$T^{E2} = e_{\pi}Q_{\pi}^{(2)} + e_{\nu}Q_{\nu}^{(2)}$$

where $\overline{Q_{\pi}^{(2)}}$ and $Q_{\nu}^{(2)}$ are the proton and neutron quadrupole operators. In phenomenologic calculations one usually assumes that the neutron and proton boson effective charges are equal, $e_{\pi}=e_{\nu}$. Shell-model, and recent phenomenological [Ver85] calculations tend to favor a value of $1.5e_{\nu}\leq e_{\pi} \leq 2.0e_{\nu}$. For this reason we present in table 1 some B(E2) values for the mixed-symmetry states, calculated for two different choices for the boson effective charges, $e_{\pi}=2e_{\nu}$ and $e_{\pi}=e_{\nu}$. Even though the neutron and proton boson effective charges differ by a factor 2, the transitions from the $K^{\pi}=0_3^+$ band to the other low-lying bands is only a fraction of the $2_1^+ \rightarrow 0_1^+$ transition and of the order of a few s.p.u. The transitions within the band are strong but somewhat weaker than transitions within the g.s. band. These predictions are very similar to what one expects in the traditional 2 qp interpretation

Table 1. Calculated B(E2) values for various transitions involving members of the mixed-symmetry $K^\pi = 0^+$ band for two different choices for the boson effective charges.

Transition	$e_\nu = e_\pi /2$	$e_\pi = e_\nu$
$2_1^+ \rightarrow 0_1^+$	100.	100.
$0_{MS}^+ \rightarrow 2_1^+$	0.36	0.86
$0_{MS}^+ \rightarrow 2_\beta^+$	0.85	1.11
$0_{MS}^+ \rightarrow 2_\gamma^+$	2.13	3.38
$2_{MS}^+ \rightarrow 0_1^+$	0.005	0.12
$2_{MS}^+ \rightarrow 0_\beta^+$	0.10	0.05
$2_{MS}^+ \rightarrow 0_{MS}^+$	78.15	70.86

Source: Reproduced with permission from [Sch85]. Copyright 1985 American Physical Society.

of these bands. Even for the case of equal boson effective charges the transitions leading to the mixed-symmetry states do not vanish, indicating a considerable amount of F-spin breaking. Since in ^{156}Gd the number of neutron and proton bosons ($N_\pi = 7$, $N_\nu = 5$) is not equal, the quadrupole-quadrupole interaction acting only between neutrons and protons introduces a strong F-spin breaking term in the hamiltonian. In the present case the interaction matrix element between the pure 0_β^+ and 0_{MS}^+ is of the order of 200 keV.

Another important difference between the two choices for the Majorana interaction shows up in the calculated M1 strength distribution. In the IBA-2 model, the M1 transition operator is written as

$$T^{M1} = \sqrt{30/4\pi} \; (\; g_\pi \; (d_\pi^\dagger \tilde{d}_\pi)^{(1)} + g_\nu \; (d_\nu^\dagger \tilde{d}_\nu)^{(1)} \;)$$

where g_π and g_ν are the boson g-factors. On the basis of the collectivity of the bosons one can argue that the spin contributions to g_π and g_ν essentially cancel and that therefore the g-factors are equal to the orbital nucleon g-factors, $g_\pi \simeq 1.0\ \mu_N$ and $g_\nu \simeq 0.0\ \mu_N$ [Sam84].

In table 2 the M1 transition strength leading to the lowest 1^+ states is given. In calculation I all strength is concentrated in the first 1^+ level, as is expected in the pure SU(3) limit [Sch85]. In calculation II the strength is distributed over several levels, resulting in a considerably lower strength for the 1_1^+ level. This could be part of the explanation of the observed splitting of the M1 strength in experiment [Boh84]. It is also interesting to note that the experimentally observed strength of $1.3 \mu_N^2$ is close to the prediction of calculation II.

Table 2. Some calculated B(M1↑) values, in units of μ_N^2, using $g_\pi = 1\mu_N$ and $g_\nu = 0$. The difference between the two calculations is explained in the text.

I		II	
E_x	B(M1)	E_x	B(M1)
3.1	2.43	3.0	1.69
4.2	0.00	3.8	0.55
4.3	0.00	4.1	0.05
4.4	0.00	4.3	0.00

In this contribution we have investigated the effect, on calculated observables in the IBA-2 model, of an alternative choice for the Majorana force, as suggested by microscopic calculations [Dru85]. This choice has the peculiar feature of producing in the SU(3) limit a spectrum in which there appears a mixed-symmetry $K^\pi = 0^+$ band at approximately the same energy as that of the β and γ bandheads. In addition this gives rise to an appreciable spreading of the collective M1 strength, a prediction which can be tested, for example in inelastic electron scattering.

Acknowledgments

We would like to thank Drs. A.E.L. Dieperink, S. Pittel, and D. Warner for helpful discussions. This work was supported by the U.S. National Science Foundation under grant no. PHY-84-07858 and by NATO grant no. RG85/0036.

References

[Ari83] A. Arima and F. Iachello, Adv. in Nucl. Phys. 13, 139 (1983).

[Boh84] D. Bohle et al., Phys. Lett. 137B, 27 (1984).

D. Bohle et al., Phys. Lett. 148B, 260 (1984).

C.W. de Jager, in Interacting Boson-Boson and Boson-Fermion Systems, p.225, ed. O. Scholten, (World Scientific, Singapore 1984)

[Die83] A.E.L. Dieperink, Prog. Part. Nucl. Phys. 9, 121 (1983).

[Dru85] C.H. Druce, S. Pittel, B.R. Barrett and P.D. Duval, Phys. Lett. 157B, 115 (1985).

[LoI79] N. Lo Iudice and F. Palumbo, Nucl. Phys. A326, 193 (1979).

[Ots78] T. Otsuka, A. Arima and F. Iachello, Nucl. Phys. A309, 1 (1978).

T. Otsuka in Interacting Boson-Boson and Boson-Fermion Systems, p. 3, ed. O. Scholten, (World Scientific, Singapore 1984)

[Sam84]   M. Sambataro, O. Scholten, A.E.L. Dieperink and G. Piccitto, Nucl. Phys. $\underline{A423}$, 333 (1984).

[Sch80]   O. Scholten, Ph.D. Thesis, Univ. of Groningen (1980)

[Sch83]   O. Scholten, Phys. Rev. $\underline{C28}$, 1783 (1983).

[Sch85]   O. Scholten, K. Heyde, P. Van Isacker, J. Jolie, J. Moreau, M. Waroquier and J. Sau, Nucl. Phys. $\underline{A438}$, 41 (1985).

[Suz77]   T. Suzuki and D.J. Rowe, Nucl. Phys. $\underline{A289}$, 461 (1977).

[Van84]   P. Van Isacker, K. Heyde, J. Jolie, J. Moreau, M. Waroquier and O. Scholten, Phys. Lett. $\underline{144B}$, 1 (1984).

          P.D. Duval, S. Pittel, B.R. Barrett and C.H. Druce, Phys. Lett. $\underline{129B}$, 289 (1983).

[Ver85]   J. Vervier, Contr. "XXIII International Winter meeting on Nuclear Physics", Bormio, Italy, 1985

[Sch85]   O. Scholten, K. Heyde, P. Van Isacker, and T. Otsuka, Phys. Rev. $\underline{C30}$, 1730, 1731 (1985).

RECEIVED May 19, 1986

Interacting Boson Model-2 for High-Spin States

Raymond A. Sorensen and Kevin Fowler

Carnegie-Mellon University, Pittsburgh, PA 15213

The N-P Interacting Boson Model is extended to include bosons of spins
4, 6, 8,.. in addition to the usual S and D bosons, in order to treat
nuclear states of high spin within the IBM formalism.

Up until about ten years ago, most calculations of nuclei assumed them to
be composed of neutrons and protons with their spacial and spin degrees of
freedom, and with two body interactions between them. Recent developments
indicate a quark substructure to nucleons so there are now attempts to find
and calculate nuclear phenomena requiring these extra degrees of freedom. On
the other hand, there are a large number of nuclear calculations being
performed today using many _fewer_ degrees of freedom than those represented by
the neutrons and protons, namely the interacting boson model (IBM)
calculations [ARI78].

The IBM calculations use only two nuclear constituents, identical S and D
bosons, or in the IBM-2, proton (π) and neutron (ν) S and D bosons. While
this model cannot contain all the properties of models with more degrees of
freedom, it has the advantage of simplicity and also seems to be able to
account for many details of the quadrupole collective motion in a single
formalism. Thus one might hope to be able to use the IBM to extrapolate from
known properties of nuclei near the stability line to nuclei far from
stability. The parameters are the total numbers of π and ν bosons, N_π and N_ν,
and their one and two boson interactions. The usefulness of this theory for
extrapolation depends on these parameters being independent of N and Z, or
known functions of N,Z.

Two deficiencies of the IBM are that the force parameters are in general
not independent of N,Z so that separate fits to individual nuclei are often
required. And second, the usual theory is limited to low spin states. In
this paper we describe an extension of the usual IBM-2 designed to be
applicable to deformed nuclei including the high spin states [SOR85]. The
model, which contains bosons with L = 4,6,8...as well as the usual L=0, S
bosons and L = 2, D bosons, has many more degrees of freedom than the usual
IBM, but remains much simpler than the treatment in terms of fermions. The
higher spin bosons are supposed to simulate the effects of high spin nucleon
pairs such as the aligned pairs that are important in backbending nuclei. We
discuss general features of the model for strongly backbending nuclei, and
then present a fit to ^{168}Yb. We then study to what extent the model with the
same force parameters can fit the spectra of the neighboring nuclei.

The Hamiltonian
The boson creation operators are $\gamma_{\rho i}^\dagger$ where $\rho = \pi, \nu$ indicates proton
or neutron and i = L,M the angular momentum of the basis bosons. We
consider even L, even parity for the bosons, which are supposed to represent
pairs of like particles (or holes). The numbers $N_\rho = \Sigma_i n_{\rho i}$ are assumed to
equal 1/2 the number of proton or neutron particles or holes from the
nearest closed shell, and are thus well defined functions of N and Z, to the

0097-6156/86/0324-0053$06.00/0

extent that subshells are not considered. The number operators are defined
as $n_{\rho i} = \gamma^\dagger_{\rho i} \gamma_{\rho i}$.

For the Hamiltonian, we use

$$H = \Sigma_i \, \epsilon_{\pi i} \, n_{\pi i} + \Sigma_i \, \epsilon_{\nu i} \, n_{\nu i} + \kappa Q_\pi \cdot Q_\nu, \tag{1}$$

where $\epsilon_{\rho i}$ is the single boson energy and Q the quadrupole moment operator.
Only the proton-neutron quadrupole force is used. This Hamiltonian
includes the important features needed to describe collective states in
deformed nuclei. For the one body energies $\epsilon_{\rho i}$ we set the i=L=0 state lower
than the rest to simulate the fact that a like particle pair favors spin zero
due to the effects of the pairing force. The Q·Q force between unlike
particles simulates the quadrupole deforming schematic force of the pairing
plus quadrupole model. We include only the P-N Q·Q force for simplicity and
because that is the most important deforming component.

The quadrupole operator Q is defined as

$$Q_{\rho\mu} = \Sigma_{jj'} \, X^\rho_{jj'} \left[\gamma^\dagger_{\rho j} \, \tilde{\gamma}_{\rho j'} \right]^2_\mu. \tag{2}$$

For the $X_{jj'}$ parameters we use $X_{jj'} = X_{jj'}[SU(3)] \, \alpha_j \, \alpha_{j'}$. If the α's are
unity, this is the SU(3) Quadrupole operator of Elliott [ELL58]. The α
parameters allow us to shift the force away from the pure SU(3) limit.

The Hartree-Bose Approximation.
For this extended IBM, exact solutions are not possible, and we are
forced to use a mean field approximation. The system is cranked to generate
the high spin states. This HB approximation is simpler than the
corresponding Hartree Fock (HF) method, since in the HB case in the lowest
state all the bosons (of each type) are in a single boson condensate rather
than in a set of occupied states as in the HF case. Finding the self
consistent solution is a simple numerical problem and requires little
computer time. The ground state is of the form:

$$|\Phi_0\rangle = (N_\pi! \ N_\nu!)^{-1/2} \, (B^\dagger_{\pi o})^{N_\pi} \, (B^\dagger_{\nu o})^{N_\nu} \, |o\rangle, \tag{3}$$

where $B^\dagger_0$ is the condensate boson creator, which will depend on the cranking
rotational frequency or angular momentum.

For the rare earth nuclei the self consistent field is deformed and the
cranking procedure (of adding a term ωj_x to the Hamiltonian) produces a
collective rotation. This method gives states $|\Phi_0\rangle$ = GSB (ground state
band), which are not eigenstates of the angular momentum, but have a
predetermined value of the x component of angular momentum. The total
angular momentum I is identified as $I = \langle j_x \rangle$. An interesting effect occurs
in the HB calculations if the parameters, ϵ's and α's, are chosen to produce
a backbending spectrum. The effect is that as the angular momentum is
increased, ΔI, the angular momentum spread of the state $|\Phi_0\rangle$ increases
rapidly at and above the backbend. This suggests that the single boson
condensate state in this region is not a good approximation to the exact wave
function, which would be an angular momentum eigenstate. One solution to
this problem is to consider Bands with one (or a few) Excited Bosons
replacing one or more of the condensate bosons. Such states are labeled

1-BEB, 2-BEB etc. We find that such states have much improved angular momentum properties and are not much more difficult to calculate in the HB approximation. In the region above the backbend, where ΔI for $|\Phi_0\rangle$ is increasing rapidly, ΔI for the state with an excited boson decreases to a minimum. At a still higher spin, the 2-BEB state has the lowest ΔI value, as seen in Figure 1.

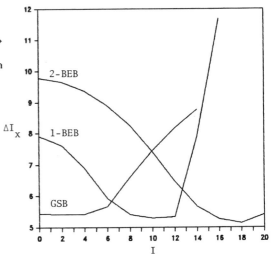

Figure 1. The rms deviation ΔI_x in the x component of the angular momentum, vs. I for a strongly backbending nucleus. The three curves are for the GSB, 1-BEB, and 2-BEB.

Angular Momentum Projection

The HB method can give good results particularly for cases in which the properties of the states do not change too rapidly with spin. For back bending nuclei, for which the energy does not vary smoothly with angular momentum, a better method is needed if quantitative results are required. To treat these cases we have projected the HB states to states of good angular momentum. The projection of these HB states is somewhat simpler than that for fermions in the HF approximation, and we are able to project even our non-axial high spin states to angular momentum eigenstates [SOR77].

The result is that a good rotor spectrum in the HB approximation is not much changed by projection, but for a back bending nucleus the projected and unprojected energy spectra are significantly different. Figure 2 on the next page shows the energy vs. angular momentum plot for the backbending model of Figure 1; and Figure 3 shows the usual plot of moment of inertia $\mathcal{J}$ vs. ω^2 for the same nucleus. The parameters used are shown in the tables below together with those of ^{168}Yb.

<table>
<tr><td colspan="5">**Table 1.**   Backbender</td></tr>
<tr><td>L</td><td>ϵ_π</td><td>ϵ_υ</td><td>α_π</td><td>α_υ</td></tr>
<tr><td>0</td><td>0</td><td>0</td><td>1.0</td><td>1.0</td></tr>
<tr><td>2</td><td>0.5</td><td>0.5</td><td>0.9</td><td>0.9</td></tr>
<tr><td>4</td><td>3.0</td><td>3.0</td><td>0.8</td><td>0.6</td></tr>
<tr><td>6</td><td></td><td>3.0</td><td></td><td>0.6</td></tr>
<tr><td>8</td><td></td><td>0.5</td><td></td><td>0.9</td></tr>
</table>

$\kappa = -0.028$, $N_\pi = 6$, $N_\upsilon = 8$

Table 2.	^{168}Yb			
L	ϵ_π	ϵ_υ	α_π	α_υ
0	0	0	1.0	1.0
2	1.0	1.0	0.9	0.9
4	1.6	1.6	0.8	0.8
6		1.9		0.8

$\kappa = -0.015$, $N_\pi = 6$, $N_\upsilon = 8$

Results of the Calculations

In Figure 2, the upper three bands are the GSB, 1-BEB, and 2-BEB resp. and the lowest one is the projected band. Note that as I approaches the backbend region, the 1-BEB is almost equal in energy to the GSB. At higher energy still, all three self consistent bands are nearly degenerate. We interpret this as indicating that the "excited" band at low spin is simulating an aligned nucleon pair, which at higher spin crosses the ground band. This interpretation [STE72] is reinforced by the calculated wave function, which shows that the excited boson has most of the angular momentum at the backbend. The second crossing resembles two aligned pairs.

The shape of the yrast line qualitatively resembles the projected energy curve, but as seen in Figure 3, the projected and unprojected bands differ significantly. The angular momentum projection is done from the GSB.

From Table 1 it is seen that the backbend or band crossing is produced by having a single high spin boson at low energy. The α parameters are also chosen to enhance the Q·Q interaction of that boson. In contrast, a boson spectrum which is more regular, with the boson energy increasing monotonically with L, such as that of Table 2, leads to a smooth upbending spectrum to which the unprojected HB mean field approximation is rather good. Also, in that case the ΔI value is lowest for the GSB and the excited 1-BEB does not come so close to the GSB as the energy is increased.

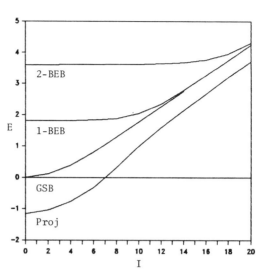

Figure 2. Plots of Energy vs. Angular Momentum for a model backbender with parameters of Table 1.

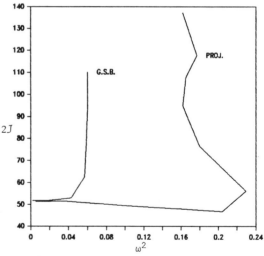

Figure 3. Moment of Inertia vs. the square of the angular velocity for the backbender. The GSB and the projected bands are shown.

For the remainder of the paper we will discuss the model fit to the spectrum of ^{168}Yb, for which the HB approximation is reasonably good. We will show that the same parameter set can give a good fit to the collective character of the neighboring nuclei as well.

The energy spectrum is fit quite well and only the $\mathcal{I}$ vs. ω^2 curve is shown in Figure 4. Note that the small deviations shown would be barely seen in a plot of E vs. I. With the many ϵ and α parameters available, it is not surprising that a good fit can be made, and no effort was made for further improvement.

However, the test of the model becomes much more severe if it is required that this model, with its parameters unchanged, fit the spectra of neighboring nuclei as well. Thus, in the remaining two figures the dependence on N and Z of the calculated energy spectra are compared with the experimental data.

As a measure of the shape or the collectivity of the nuclei we use the ratio E_4/E_2, where E_2 and E_4 are the excitation energies of the spin 2 and spin 4 states. The value of this quantity is 3 1/3 for a nucleus with a pure rotational spectrum, and 2 for a harmonic vibrational spectrum.

In Figures 5 and 6 this ratio of energies is plotted for the HB approximation for the model of Table 2, which was chosen to fit the spectrum of ^{168}Yb. With the same values of the parameters, but with N_π and N_ν chosen to correspond to the neighboring nuclei, the

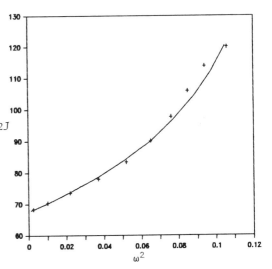

Figure 4. Moment of inertia vs. the Angular velocity for ^{168}Yb. The line is the model of Table 2 and the (+)'s are the experimental data.

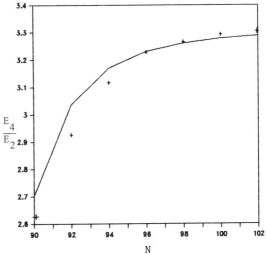

Figure 5. The E_4/E_2 ratio as function of N. The model is compared to the experimental values, which are marked with a (+). The model parameters are fit at N = 98, Z = 70.

calculated ratios are compared
with the experimental values.

From the figures, it is
clear that the same set of
parameters makes a reasonable
fit to the N, Z dependence of
the nuclear collectivity.
Another widely used measure of
nuclear collectivity is the
energy of the first 2+ state.
No figure is presented, but the
comparison of the model and the
experimental 2+ energies shows a
fit to the N, Z dependence of
comparable quality to that of
the energy ratios.

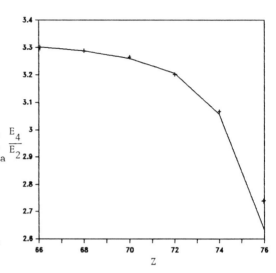

Figure 6. The E_4/E_2 ratio as a
function of Z. (See fig. 5)

Discussion

An extended IBM-2 model has been presented which is able to treat the
spectra of deformed nuclei up to high spin. There are more parameters than
in the usual IBM-2, which enable a good fit to individual yrast spectra to be
obtained. It is also found that the model, with parameters fit to a
particular nucleus, is able to fit properties of neighboring nuclei as well,
suggesting that the model may be useful for extrapolating the spectra away
from the line of nuclear stability. Of course such extrapolation will be
limited by the fact that the magic numbers defining N_π and N_ν may be N
and Z dependent, and also the presence low in the spectrum of a high spin
pair (modeled by a low energy, high spin ϵ value) will be dependent on
N and Z. Nevertheless, this method should be useful in extending the range
of nuclear phenomena which can be treated by the Boson models.

Acknowledgments

We wish to acknowledge discussions with S. Pittel and with Philip Duval,
who wrote the basic code for the HB GSB calculation.

References

[ARI78] A. Arima and F. Iachello, Ann. Phys. (NY) **111** 201 (1978)
[ELL58] J. P. Elliott, Proc Roy Soc. **A245** 128 (1958)
[SOR77] R. A. Sorensen, Nucl. Phys. **A259** 452 (1977)
[SOR85] R. A. Sorensen, K. P. Fowler and S. Pittel, preprint (1985)
[STE72] F. S. Stephens and R. S. Simon, Nucl. Phys. **A138** 257 (1972)

RECEIVED May 2, 1986

L = 2 Fermion Pairs in Nuclear Dynamic Symmetry Models

Steven A. Moszkowski

Department of Physics, University of California at Los Angeles, Los Angeles, CA 90024

Ginocchio has treated S (L=0) and D (L=2) Fermion pairs in a many-Fermion system by introducing pseudo-spin i and pseudo-orbital angular momentum k (whose sum is the single particle angular momentum j). For nucleons in degenerate orbits, the wavefunction of a pair with L > 0 is model dependent. We discuss here properties of D pairs according to three symmetry models: 1) Surface Delta Interaction, 2) Dynamic Symmetry with i-i Coupling, and 3) Dynamic Symmetry with k-k Coupling. As an example, for nucleons in a degenerate shell with j = 1/2, 3/2, and 5/2 of the same parity, the total pair degeneracy Ω = 6, and i = 3/2 , k = 1. In each case, the quadrupole operator (which turns an S pair into a D pair) is among the generators of an appropriate group. In the semi-classical limit where i, k, and the single particle j's are large compared to the pair angular momentum L, the wavefunction amplitudes for all three coupling schemes are given in terms of Wigner d functions with the same indices but different angles.

I. INTRODUCTION

A few years ago, Ginocchio [GIN80] studied a schematic Hamiltonian with a pairing plus quadrupole-quadrupole interaction for the purpose of establishing a correspondence between the states of nucleon pairs and of bosons. He used the method of pseudo-spins which had been previously developed by Arima et al. [ARI69] and also by Hecht and Adler [HEC69]. In this method one separates the single nucleon angular momentum J into a pseudo-orbital angular momentum k and a pseudo-spin i:

$$\vec{j} = \vec{k} + \vec{i}$$

As was emphasized by all these authors, k is _not_ the real angular momentum and i is _not_ the intrinsic spin, and, in fact, i will in general be larger than 1/2. Ginocchio was able to obtain solutions to the Hamiltonian which

0097-6156/86/0324-0059$06.00/0

decouple the S and D Fermion pairs from the rest of the shell model space. These then correspond to s and d bosons. An extension of the Ginocchio model to pairing and multipole interactions of arbitrary order has been recently made by Wu et al. [WU85]. The idea is to provide a unique shell model classification of single particle states with respect to k and i, and to produce tractable dynamical symmetries. There are two possible pairing coupling schemes, called k-k coupling and i-i coupling, each of which specifies the wavefunction of a Fermion pair of given L.

In this paper, we point out that a somewhat different model, where a Surface Delta Interaction [PLA66] acts between nucleons in degenerate orbits, also makes predictions for these fermion pair amplitudes. For the cases discussed in this paper, the SDI results are mostly intermediate between those for i-i and k-k coupling. For an S-pair, all three models give identical results. Also, for i = 1/2, the results are the same as if i is a conventional intrinsic spin. We then discuss the more realistic case where k = 1 and i = 3/2. This allows for j = 1/2, 3/2, and 5/2, e.g. degenerate s and d shells, or degenerate p1/2, p3/2, and f5/2 shells.

We also discuss the results in the semi-classical limit where i, k, and the single particle j's are large compared to the pair angular momentum L. In this limit, the wavefunction amplitudes are just special cases of Wigner d functions, namely associated Legendre polynomials. For the same sets of single particle angular momenta, the three coupling schemes i-i, k-k, and SDI, give the same expressions, except that the angle appearing in the associated Legendre polynomial is different.

II. PAIR WAVEFUNCTION FOR DEGENERATE ORBITS ACCORDING TO DYNAMICAL SYMMETRY OR SURFACE DELTA INTERACTION

A. General Expressions

We assume that each nucleon has a pseudo-spin i and pseudo-orbital angular momentum k. These couple to form the single particle angular momenta j, j' (in [j]) of the two interacting nucleons. The wavefunction of a pair of nucleons coupled to a total angular momentum L (and z component μ) is then given by:

$$L_\mu = M \sum_{jj'} \begin{pmatrix} k & i & j \\ k & i & j' \\ K & I & L \end{pmatrix} \{a_j{}^+ a_{j'}{}^+\}_\mu^L$$

where the quantity in brackets is a unitarized 9-j coefficient, and M is a normalization constant.

There are two possible kinds of coupling schemes: a.) i-i coupling. For this case, where $K = 0$, we obtain:

$$L_\mu = \sum_{jj'} (-)^{j-1/2} d_{jj'L}^{ii} \{a_j^+ a_{j'}^+\}_\mu^L$$

where $\hat{j} = \sqrt{(2j+1)}$ and the coefficients are given by:

$$d_{jj'L}^{ii} = \hat{k}^{-1} \hat{j} \hat{j}' \left\{ \begin{matrix} i & j' & k \\ j & i & L \end{matrix} \right\}$$

Alternatively, we may have b.) k-k coupling ($I = 0$) in which case:

$$L_\mu = \sum_{jj'} (-)^{j-1/2} d_{jj'L}^{kk} \{a_j^+ a_{j'}^+\}_\mu^L$$

where

$$d_{jj'L}^{kk} = \hat{i}^{-1} \hat{j} \hat{j}' \left\{ \begin{matrix} k & j' & i \\ j & k & L \end{matrix} \right\}$$

For two particles in degenerate orbits interacting via an SDI, coupled to any given L, only a single state is shifted in energy. The wavefunction of this state is:

$$L_\mu = \sum_{jj'} (-)^{j-1/2} c_{jj'L} \{a_j^+ a_{j'}^+\}_\mu^L$$

where

$$c_{jj'L} = N \hat{j} \hat{j}' \left(\begin{matrix} j & j' & L \\ 1/2 & -1/2 & 0 \end{matrix} \right)$$

and N is a normalizalion constant.

In the models discussed here, the quadrupole operator (which changes an S-pair to a D-pair) is among the generators of one of the following symmetry groups:

$$i\text{-}i \text{ coupling} \leftrightarrow SO_{2(2i+1)}{}^{(i)}$$
$$k\text{-}k \text{ coupling} \leftrightarrow Sp_{2(2k+1)}{}^{(k)}$$

The SDI corresponds to the symmetry SDI $\leftrightarrow Sp_{(2i+1)(2k+1)}{}^{(j)}$. (Note that if $i = 1/2$, then SDI and k-k coupling correspond to the same symmetry group $Sp_{2k+1}{}^{(j)}$, so that they give the same results.)

B. S-Pair ($L = 0$)

For an S- pair, it is evident that we must have $j = j'$ and $I = K = 0$. We find that all three coefficients c_{jj0}, d_{jj0}^{ii}, and d_{jj0}^{kk} are identical and equal to $\hat{j} [\sum_j (2j+1)]^{-1/2}$.

C. Pseudo-Spin i = 1/2

For this case, dynamic symmetry again yields the same wavefunctions as the SDI, since for i = 1/2, the pseudo-spin is the usual intrinsic spin. We obtain:

$$d_{jj'L}^{kk} = 2^{-1/2} \; \hat{j} \; \hat{j}' \left\{ \begin{array}{ccc} k & j' & 1/2 \\ j & k & L \end{array} \right\}$$

which has the same dependence on j and j' as does $c_{jj'L}$. This follows from the following well-known identity:

$$(2k+1) \left(\begin{array}{ccc} k & k & L \\ 0 & 0 & 0 \end{array} \right) \left\{ \begin{array}{ccc} k & j' & 1/2 \\ j & k & L \end{array} \right\} = - \left(\begin{array}{ccc} j & j' & L \\ 1/2 & -1/2 & 0 \end{array} \right)$$

Note that for i = 1/2, $d_{jj'L}^{ii}$ vanishes (if $L \geq 2$), so we cannot construct a D or G pair with ii coupling.

D. D-Pair (L = 2) for k = 1, i = 3/2

For this case, we can have j = 1/2, 3/2 and 5/2, corresponding, for example, to a degenerate oscillator s,d shell or p1/2, p3/2, f5/2 shell. We list here the values of the coefficients $d_{jj'2}^{ii}$ and $d_{jj'2}^{kk}$, and also $c_{jj'2}$, which correspond to the following symmetry groups:

$$
\begin{aligned}
\text{i-i coupling} &\leftrightarrow SO_{2(2i+1)}{}^{(i)} &&= SO_8{}^{(i)} \\
\text{k-k coupling} &\leftrightarrow Sp_{2(2k+1)}{}^{(k)} &&= Sp_6{}^{(k)} \\
\text{SDI} &\leftrightarrow Sp_{(2i+1)(2k+1)}{}^{(j)} &&= Sp_{12}{}^{(j)}
\end{aligned}
$$

The antisymmetrized L = 2 wavefunctions according to the three models are:

$$\psi^{ii} = \sqrt{0.267} \, (31) - \sqrt{0.067} \, (51) + \sqrt{0.013} \, (33) - \sqrt{0.373} \, (53) - \sqrt{0.280} \, (55)$$
$$\psi^{kk} = \sqrt{0.033} \, (31) - \sqrt{0.300} \, (51) - \sqrt{0.106} \, (33) + \sqrt{0.420} \, (53) - \sqrt{0.140} \, (55)$$
$$\psi^{SDI} = \sqrt{0.234} \, (31) + \sqrt{0.350} \, (51) + \sqrt{0.117} \, (33) - \sqrt{0.100} \, (53) - \sqrt{0.200} \, (55)$$

where, for example, (31) denotes the antisymmetric combination of j = 3/2 and j' = 1/2.

Note that the kk and ii coupling models lead to quite different amplitude coefficients and that the SDI results are in between for 3 out of 5 coefficients.

III. ASYMPTOTIC RESULTS FOR LARGE j',j >> L

It is interesting to consider the results for the amplitude coefficients in the semi-classical limit where all the single particle angular momenta are large compared to the resultant L of the Fermion pair. First we study the SDI results in the limit of large j and j'.

$$c_{jj'L} = \hat{j}\,\hat{j}' \begin{pmatrix} j & j' & L \\ 1/2 & -1/2 & 0 \end{pmatrix} \xrightarrow[j\to\infty]{} d^{(L)}_{j'-j,0}(\pi/2)$$

where $d^{(L)}_{\alpha,\beta}(\theta)$ is the Wigner d function.

In particular, for j = j', we have:

$$c_{jjL} = (2j+1) \begin{pmatrix} j & j & L \\ 1/2 & -1/2 & 0 \end{pmatrix} \xrightarrow[j\to\infty]{} d^{(L)}_{0,0}(\pi/2) = \hat{j}\,P_L(0) = \hat{j}$$

The results for the two dynamic symmetry models are quite similar:

For i,j >> L and j'-j << j,

$$d^{ii}_{jj'L} = (2j+1) \begin{Bmatrix} i & j' & k \\ j & i & L \end{Bmatrix} \xrightarrow[i,j,j'\to\infty]{} d^{(L)}_{j'-j,0}(\theta_{ij})$$

where

$$\cos\theta_{ij} = \{k(k+1)-i(i+1)-j(j+1)\}/2\sqrt{\{i(i+1)j(j+1)\}}$$

and

$$d^{kk}_{jj'L} = (2j+1) \begin{Bmatrix} k & j' & i \\ j & k & L \end{Bmatrix} \xrightarrow[k,j,j'\to\infty]{} d^{(L)}_{j'-j,0}(\theta_{kj})$$

$$\cos\theta_{kj} = \{i(i+1)-k(k+1)-j(j+1)\}/2\sqrt{\{k(k+1)j(j+1)\}}$$

For j = j',

$$d^{ii}_{jjL} = (2j+1) \begin{Bmatrix} i & j & k \\ j & i & L \end{Bmatrix} \xrightarrow[i,j,\to\infty]{} \hat{j}\,P_L(\cos\theta_{ij})$$

$$d^{kk}_{jjL} = (2j+1) \begin{Bmatrix} k & j & i \\ j & k & L \end{Bmatrix} \xrightarrow[k,j,\to\infty]{} \hat{j}\,P_L(\cos\theta_{kj})$$

Thus the results for all three coupling schemes are given by the same kind of expression involving the Wigner function, differing only in the angle.
Note that if i ≈ k and both are >> j,j', then we have $\theta_{ij} = \theta_{kj} = \pi/2$, and all three coupling schemes give the same values for the coefficients.
However, if i,j >> k, then we have the following values for cos θ_{ij}:

	SDI	i-i coupling	k-k coupling
i,j >> k	0	-1	-(j-i)/i
i = k	0	-j/i	-j/i

In the former case and if $j = j'$, then we have no j dependence for the SDI and for i-i coupling. However, there is some j dependence for k-k coupling. For $j \neq j'$, there is, of course, some j dependence since the index in the d function is $j'-j$.

IV. CONCLUSIONS

The results obtained here exhibit the connection between the surface delta interaction coupling schemes and dynamical symmetries. It is a challenging problem to see to what extent these symmetries are preserved in nuclei when the relevant single particle orbits are not degenerate.

ACKNOWLEDGMENTS

The author is grateful to D. H. Feng and C. L. Wu for suggesting this investigation and to J. N. Ginocchio for a helpful discussion. This work was supported in part by NSF grant #Phy-84-20619.

REFERENCES

[GIN80] J. N. Ginocchio, Ann. Phys. 126,234(1980).

[ARI69] A. Arima, M. Harvey and K. Shimizu, Phys. Lett. B30, 517 (1969).

[HEC69] K.T. Hecht and A. Adler, Nucl. Phys. A137, 129, (1969).

[WU85] C.L. Wu, D.H. Feng, X-G Chen, J-Q Chen and M. Guidry, Prepr., Mar.85.

[PLA66] A. Plastino, R. Arvieu, and S.A. Moszkowski, Phys. Rev. 145,837(1966)

RECEIVED May 13, 1986

The Importance of an Accurate Determination of Interacting Boson Model-2 Parameters

B. R. Barrett[1,4], I. Morrison[2], and J. G. Zabolitzky[3]

[1]Physics Division, National Science Foundation, Washington, DC 20550
[2]School of Physics, University of Melbourne, Parkville, Victoria 3052, Australia
[3]Institute for Theoretical Physics, University of Köln, 5 Köln, 41, Federal Republic of Germany

First, a brief description of the neutron-proton Interacting Boson Model (IBM-2) is given. Next, this model is applied to experimental data in order to determine its empirical parameters. Finally, we discuss why an accurate determination of these parameters is so important.

1. Introduction

The Interacting Boson Model (IBM) of Arima and Iachello [ARI76, ARI81, BAR81] has been highly successful in correlating and describing a wide variety of experimental data regarding the collective properties of medium-to-heavy mass nuclei. As originally formulated [ARI76], it is a purely phenomenological model, whereby the properties of nuclei are described in terms of interacting s (J=0) and d (J=2) bosons, such that the number of bosons is conserved. This original version did not distinguish between proton bosons and neutron bosons and is commonly referred to as the IBM-1. Later, the IBM-2 [ARI77] was developed to treat the neutron-proton interaction [ARI77] and to allow a connection to be established with the microscopic nuclear shell model. We will first outline the IBM-2 formalism and will then discuss problems related to an accurate determination of the IBM-2 parameters.

2. The IBM-2 Formalism

The principal idea of the IBM-2 is to exploit the observation of Talmi [TAL82] and others [FED79] that it is the interaction between active protons and neutrons which is mainly responsible for causing nuclei with several valence nucleons to deform. The lowest rank proton-neutron interaction which can produce this effect is a quadrupole-quadrupole interaction [TAL71, TAL82, ARI77]. For this reason, and in order to keep the number of variable parameters small, the usual form taken for the IBM-2 Hamiltonian is [ARI77, IAC79]

where

$$H^{IBM-2} = \varepsilon(\hat{n}_{d_\pi} + \hat{n}_{d_\nu}) + \kappa Q_\pi \cdot Q_\nu + M_{\pi\nu} + V_{\pi\pi} + V_{\nu\nu} \tag{1}$$

$$Q_\rho = (d^+xs + s^+x\tilde{d})^{(2)}_\rho + \chi_\rho(d^+x\tilde{d})^{(2)}_\rho, \ \rho = \pi, \nu \tag{2}$$

$$M_{\pi\nu} \equiv \text{Majorana term} = \xi_2(s^+_\pi xd^+_\nu - d^+_\pi xs^+_\nu)^{(2)} \cdot (s_\nu x\tilde{d}_\pi - \tilde{d}_\nu xs_\pi)^{(2)}$$

$$+ \sum_{k=1,3} \xi_k(d^+_\nu xd^+_\pi)^{(k)} \cdot (\tilde{d}_\nu x\tilde{d}_\pi)^{(k)} \tag{3}$$

[4]Permanent address: Department of Physics, University of Arizona, Tucson, AZ 85721

0097-6156/86/0324-0065$06.00/0

$$V_{\rho\rho} = \sum_{L=0,2,4} \tfrac{1}{2}(2L+1)^{1/2} C_{L\rho}[(d^+ \times d^+)_\rho^{(L)} \times (\tilde{d} \times \tilde{d})_\rho^{(L)}]^{(0)}, \quad \rho = \pi, \nu \tag{4}$$

$$\hat{n}_{d\rho} = (\sum_m d_m^+ d_m)_\rho, \quad \rho = \pi, \nu \tag{5}$$

$$\tilde{d}_m = (-1)^m d_{-m}. \tag{6}$$

The "•" represent scalar products and the "x" represent tensor products. The purpose of the Majorana term was to remove states which are non-symmetric under interchange of the proton and neutron degrees of freedom by shifting them up in energy. The non-symmetric neutron-proton states are now of some physical interest [BOH 84]. For simplicity, and to decrease the number of variable parameters, we have taken $\epsilon_\pi = \epsilon_\nu = \epsilon$. In general, this will not be true.

3. Application of the IBM-2 Formalism

Since the ξ_i are usually held constant, there could be ten variable parameters to be determined for each nucleus studied. This number is usually reduced to six by assuming that only $V_{\nu\nu}(V_{\pi\pi})$ contributes to relative splittings in isotopes (isotones) and that the contribution of $C_{4\rho}$ is negligible. The remaining six parameters are ϵ, κ, χ_ν, χ_π, $C_{0\rho}$, $C_{2\rho}$ ($\rho = \pi$ or ν). After the first isotope (isotone) is described, $\chi_\pi(\chi_\nu)$ is determined and is assumed to be the <u>same</u> for <u>all</u> remaining isotopes (isotones), leaving five parameters per nucleus.

The goal is to determine empirically the values of these six parameters which yield the best description of the low-lying spectra of medium-to-heavy nuclei and which at the same time vary <u>smoothly</u> with changes in the neutron- and proton-boson numbers. It is important that the set of parameters for one set of isotopes be quite similar to the set of parameters for the neighboring series of isotopes. The values of these parameters should not vary in a random manner, if the IBM-2 is a truly meaningful description of the properties of nuclei in this mass region. So, it is definitely of interest to determine empirically these six parameters as well as possible for medium-to-heavy mass nuclei. Work along this line has been carried out (e.g., [BAR81, SCH80]), but this entire mass region has so far not been investigated within the IBM-2 approach.

There is the difficulty that most IBM-2 investigations carried out to date have been done by fitting the model to the data by "eye." That is, one chooses values for the IBM-2 parameters and uses them to obtain eigenenergies and eigenvectors. These eigenenergies are compared with the experimental excitation energies for a given nucleus. New values of the parameters are selected and the process repeated until (i) a "reasonable" fit has been obtained to the experimental energies and (ii) a set of parameters has been obtained which vary <u>smoothly</u> with N and Z for neighboring isotopes and isotones. This procedure generally yields a satisfactory fit to the experimental data after a small number of iterations (less than 4 or 5). The advantage of this procedure is that (i) it uses less computer time than a least-squares fit to the data and (ii) it allows the person doing the fit to check on the smoothness of the parameter variations with N and Z. This latter feature is guided by our microscopic understanding of the IBM-2 in terms of the nuclear shell model [ARI77, BAR 81, TAL82].

To improve on this situation, we have used the Glasgow shell model code, rewritten to treat bosons [MOR80], to perform a least-squares fit to the excitation energies of a single nucleus using the IBM-2 Hamiltonian (1) <u>without</u> the Majorana term. In general, the model is applicable to only 10 to 12 experimentally known energy levels per nucleus. Since the IBM-2 Hamiltonian used contains five or six variable parameters, there are only two or less pieces of data per parameter per nucleus in the least-squares fit. What this means is that it is possible to find <u>several</u> sets of parameter values which fit the experimental data equally well (i.e., have essentially the same χ^2 value). These parameter sets can be rather different, <u>especially</u> in their values for χ_π and χ_ν. When the six parameters ε, κ, χ_ν, χ_π, C_{0_ν}, and $\overline{C_{2_\nu}}$ were all allowed to vary in the least-squares-fit procedure, a "minimum" was found in all cases studied by changes only in ε and κ, with χ_ν, χ_π, C_{0_ν}, and C_{2_ν} essentially unchanged. Typical results are shown in Table 1.

Table 1. IBM-2 parameter sets producing similar least-squares fits to the excitation energies for $^{196}_{78}$Pt$_{118}$

Case	ε(MeV)	κ(MeV)	χ_π	χ_ν	C_{0_ν}(MeV)	C_{2_ν}(MeV)	χ^2
Initial guesses	0.580	-0.180	-0.800	1.050	0.600	0.020	--
Vary all 6	0.568	-0.209	-0.800	1.052	0.600	0.018	3.83
Vary only χ_ν	0.580	-0.180	-0.800	1.083	0.600	0.020	4.94
Vary all 6, χ_ν(initial)=1.083	0.572	-0.209	-0.800	1.084	0.600	0.019	3.69
Vary only χ_π and χ_ν	0.580	-0.180	-0.853	1.160	0.600	0.020	2.86

Source: Reproduced with permission from [BAR84]. Copyright 1984 World Scientific Publishing Co.

When the eigenvectors from these different fits for a given nucleus were then used to calculate electromagnetic properties, such as transition rates and quadrupole moments, the results were found to agree with one another within 10%.

Hence, it was evident that we needed to increase the amount of data included in a given IBM-2 fit in order to "tie down" the empirical values of the model parameters. We did this by making a least-squares fit of the IBM-2 Hamiltonian simultaneously to the excitation energies of several neighboring nuclei. For our fit we chose the isotopes of Xe, Ba, Ce, Nd, and Sm with $66 \leq N \leq 80$. For these nuclei we selected 171 excitation energies, which we felt were reasonable levels to be described by the IBM-2 [LED78, KAR84], and used the Glasgow boson code to determine the values of the IBM-2 parameters which yield the best fit to these energies. In our calculations we varied the following six parameters until a minimum was obtained: ε, κ, χ_{0_π}, χ_{0_ν}, $\Delta\chi_\pi$, and $\Delta\chi_\nu$, where $\chi_\rho = \chi_{0_\rho} + \Delta\chi_\rho N_\rho (\rho=\pi$ or $\nu)$. That is, while ε and κ were taken to be the <u>same</u> for all nuclei considered, χ_π and χ_ν were allowed to vary linearly with their respective boson numbers. We performed several different fits to the experimental data, depending upon our initial guesses for the above six parameters, as shown in Table 2.

From Table 2 we observed that we were not able to obtain a particularly good fit to the experimental data. Even in the best case, the average error per level was 188 keV. From the present investigations we were able to conclude: (1) that is it not possible to obtain a reasonable fit to the experimental data with a large value of ε (i.e., a value near the closed-shell value); (2) that reversing the signs of χ_π and χ_ν leaves the final energy spectrum unchanged, due to an arbitary phase between the s boson and the d boson; (3) that energy spectra in the IBM-2 are mainly determined by ε and κ and are

Table 2. Initial and minimal IBM-2 parameter sets for the Xe, Ba, Ce, Nd and Sm isotopes

Case	ϵ(MeV)	κ(MeV)	χ_{0_π}	$\Delta\chi_\pi$	χ_{0_ν}	$\Delta\chi_\nu$	Average error per level(keV)
1. Standard choice: Initial	0.700	-0.160	-1.000	+0.125	+1.000	-0.125	--
Standard choice: at minimum	0.638	-0.226	-1.001	+0.124	+1.003	-0.113	188
2. Large fixed ϵ choice: initial	1.350	-0.160	-1.000	+0.125	+1.000	-0.125	--
Large fixed ϵ at minimum	1.350 (fixed)	-0.283	-1.000	+0.125	+1.000	-0.126	964
3. χ_π, χ_ν same sign: initial	0.700	-0.160	+1.000	-0.125	+1.000	-0.125	--
χ_π, χ_ν same sign: at minimum	0.737	-0.164	+0.996	-0.137	+0.996	-0.141	209
4. Reverse signs χ_π, χ_ν: initial	0.700	-0.160	+1.000	-0.125	-1.000	+0.125	--
Reverse signs χ_π, χ_ν: at minimum	0.638	-0.226	+1.001	-0.124	-1.003	+0.113	188

fairly insensitive to the values of χ_π and χ_ν; and (4) that the "standard" parameter choice, based on trends predicted by the nuclear shell model, produces the best overall fit to the data.   However, another fit based on the opposite trend for χ_π produced results only 10% worse. Hence, further investigations are needed and are under way in Cologne.   We will now fit the experimental data based on the $N_\pi \cdot N_\nu$ systematics, discussed by Rick Casten in an earlier talk in this session, and will also include the Majorana parameters ξ_i as variables in the fit.   These results should be available in the near future.

Acknowledgments
This work was supported in part by the National Science Foundation, Grant Nos. PHY81-00141 and INT82-11657.

References
[ARI76]   A. Arima and F. Iachello, Ann. Phys. (N.Y.) 99, 253 (1976); 111, 201 (1978); 123, 468 (1979).
[ARI77]   A. Arima, T. Otsuka, F. Iachello, and I. Talmi, Phys. Lett. 66B, 205 (1977); 76B, 139 (1978).
[ARI81]   A. Arima and F. Iachello, Annu. Rev. Nucl. Part. Sci. 31, 75 (1981).
[BAR81]   B. R. Barrett, Rev. Mex. Fis. 27, 533 (1981); Proc. Int. Sum. School N-N Interaction and Nucl. Many-Body Problems, eds. S. S. Wu and T. T. S. Kuo (World Scientific, Singapore, 1984), p. 415.

[BOH84]   D. Bohle, et al, Phys. Lett. 137B, 27 (1984).
[FED79]   P. Federman and S. Pittel, Phys. Rev. C 20, 820 (1979).
[IAC79]   F. Iachello, in Interacting Bosons in Nuclear Physics, ed. F. Iachello (Plenum, New York, 1979), p. 1.
[KAR84]   Nuclear Data Tabulations, Kernforschungszentrum, Karlsruhe, West Germany, private communication (1984).
[LED78]   Table of Isotopes, 7th edition, eds. C. M. Lederer and V. S. Shirley (Wiley, New York, 1978).
[MOR80]   I. Morrison and R. Smith, Nucl. Phys. A350, 89 (1980).
[SCH80]   O. Scholten, Ph. D. Thesis, University of Groningen (1980).
[TAL71]   I. Talmi, Nucl. Phys. A172, 1 (1971).
[TAL82]   I. Talmi, in Contemporary Research in Nuclear Physics, ed. D. H. Feng et al. (Plenum, New York, 1982); and Comments Nucl. Part. Phys. 11, 241 (1983).
[BAR84]   B. R. Barrett, in Interacting Boson-Boson and Boson-Fermion Systems, ed. O. Scholten (World Scientific, 1984), P. 25.

RECEIVED May 30, 1986

11

Shell-Model Calculations near ^{132}Sn Using a Realistic, Effective Interaction

C. A. Stone[1], W. B. Walters[1], S. D. Bloom[2], and G. J. Mathews[2]

[1]Department of Chemistry, University of Maryland, College Park, MD 20742
[2]Lawrence Livermore National Laboratory, University of California, Livermore, CA 94550

We have performed shell model calculations on positive parity states of the one, two, and three quasiparticle nuclei near ^{132}Sn using the Kallio-Kolltveit two-body interaction. A weak quadrupole and a weak pairing potential were added to the two-body interaction as core polarization corrections. We have found that the addition of a quadrupole potential improves the level spacing in ^{130}Sn and ^{134}Te. For ^{133,134}Te and ^{135}I (two and three proton systems) a pairing potential has to be included while ^{129,130}Sn and ^{131}Sb (zero and one proton systems) can not tolerate this addition. There is evidence that the $h_{11/2}$ interactions are overestimated and lead to some level misordering in ^{129,130}Sn and ^{131}Sb.

I. INTRODUCTION

There has been some success in developing effective interactions for finite nuclei using realistic two-body interacti·ns. Much of this work has been done on the lighter nuclei. Kuo and Brown [KUO66] developed a g-matrix interaction for the sd and fp shells. They used the Hamada-Johnson nucleon-nucleon interaction in their calculations with a core polarization correction. This work was extended to the region near ^{40}Ca and ^{56}Ni [KUO68] and near ^{208}Pb [KUO72]. Baldridge and Vary [BAL76] have also performed calculations in the ^{208}Pb region. The spectra of ^{204,206}Pb were calculated using the Reid soft-core potential with core polarization corrections. Lane [LAN79] used the Petrovich, McManus, and Madsen two-body interaction in a limited study of calculations near ^{132}Sn. Calculations in the heavy nuclei unfortunately been limited by the large model spaces needed and also by the quality of the available experimental data.

The ^{132}Sn region has recently shown itself to be a very good region in which to develop an effective interaction. This is a region with a strong double-shell closure, stronger than all shell closures beyond ^{16}O. There is also a large set of experi-

0097-6156/86/0324-0070$06.00/0

mental data available on the nuclei near ^{132}Sn. All five of
the neutron-hole states in ^{131}Sn have been seen [FOG84] and four
of the five proton states in ^{133}Sb have been found [BLO83] (only
the 1/2$^+$ state in ^{133}Sb is missing). Our knowledge of the two-
quasiparticle nuclides, ^{130}Sn [FOG81], ^{132}Sb [STO86a], and ^{134}Te
[KER72] is not as complete as the with one-quasiparticle nuclides
but much of the low-lying structure has been identified. The
four 3-quasiparticle nuclides, ^{129}Sn [DEG80], ^{131}Sb [STO86b],
^{133}Te [LAN80], and ^{135}I [SAM85], have been studied extensively: a
large number of levels has been seen in each nucleus.

Our goal is to develop a semi-phenomenological effective
interaction for use near ^{132}Sn using a systematic approach. We
began the study by using the single-quasiparticle nuclides to fix
the one-body portion of the Hamiltonian. This is done empirically by fitting the excitation states in ^{131}Sn and ^{133}Sb.
Single particle energies (SPE's) are determined by fitting the
experimental separation of the 7/2$^+$ state from other excitation
states. As corrections are made to a given two-body interaction
SPE's are redetermined.

After fixing the one-body Hamiltonian, the next step is to
test the two-body Hamiltonian in the two-quasiparticle systems,
^{130}Sn and ^{134}Te. As our trial two-body interaction we have used
the Kallio-Kolltveit (KK) interaction [KAL64]. The KK interaction is a G-matrix interaction based on the Scott-Moskowski cut-
off procedure. Only even components are explicitly included and
they have the form of an exponential with a hard-core. G-matrix
interactions do not take into account core polarization but
guidance for including such corrections has been given by Brown
and Kuo [BRO67]. These corrections have the form of a quadrupole
potential and a pairing potential. We have added these potentials as corrections to the KK interaction, using the two-
quasiparticle nuclides to fix their strengths.

II. Calculations

Calculations were performed at LLNL using the vectorized
shell model code, VLADIMIR [HAU76]. The model space for these
calculations included five orbits: $1g_{7/2}$, $2d_{5/2}$, $2d_{3/2}$, $3s_{1/2}$,
$1h_{11/2}$. There are 64 single particle orbitals in this
model space which poses some problems for even the simplest of
the calculations. The VLADIMIR shell model code uses an internal
occupation number representation to describe a nuclear configuration. The string of bits within a word defines a Slater
determinant. When a bit is set to one the single particle orbital which it represents is occupied and the single particle orbital is unoccupied when that bit is zero. The Cray computer has
a word size of 64 bits. Since 1 bit is reserved as a sign bit,
only 63 single particle orbitals can be represented in a single
word. VLADIMIR does have the ability of representing a Slater
determinant in multiple words but this portion of the code is
currently written in Fortran, not Assembly. Two-word calculations are presently much slower than the one-word calculations.

    We have made further truncations to our model space in order
to limit the representation to one word. For calculations on
129,130,131Sn the $\pi h_{11/2}$ orbitals were removed. This had no
effect on the Sn isotopes. When calculations were performed on
133Sb, 133,134Te, or 135I excitations were restricted from the
$1g_{7/2}$ orbitals. This effectively placed most of the
$1g_{7/2}$ neutrons into the core. These truncations had the effect
of breaking isospin conservation. Low-lying excitation states
(below ≈3 MeV) were not affected significantly but higher excitation states did not have a well-defined energy.

III. Results

    A. One-quasiparticle nuclides

    Figure one shows the results of calculations of 133Sb and
131Sn, one-quasiparticle nuclides. The calculated levels agree
very well with experimental excitation states. Figure 1 also
shows results using the bare KK interaction and with the addition
of a weak QQ potential determined in the two-quasiparticle nucli-

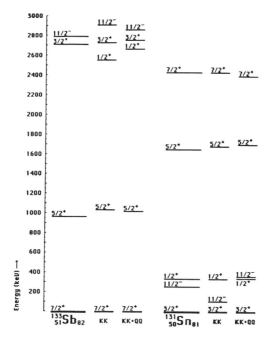

Figure 1. Results for calculations on the one quasiparticle
nuclides. Experimental states are labeled with the nucleus.
Calculated states are shown to the right, labeled with the
effective interaction that was used.

des. The poorest fit was the 11/2⁻ state and the total deviation
was only about 200 keV. Table 1 gives our best SPE's from these
studies. SPE's for ^{133}Sb and ^{131}Sn are not very different and
suggest that a single set of SPE's can be used for both proton
systems and neutron-hole systems. The SPE's for the
KK+QQ+Pairing interaction are the same as those with the KK+QQ
interaction. This is easily understood since the pairing potential will pull down the lowest 0⁺ configuration in an even-even
nucleus. In an odd-A nucleus, all states with the odd particle
(hole) coupled to this 0⁺ state will be pulled down in energy.

Table 1. Single particle energies determined
for the three effective interactions.

Orbit	Single Particle Energies			
	^{131}Sn (MeV)	^{133}Sb (MeV)	Average (MeV)	Deviation (MeV)
KK Potential				
7/2⁺	-4.5000	-4.5000	-4.5000	0.0000
5/2⁺	-3.9208	-4.0097	-3.9652	0.0444
11/2⁻	-2.9674	-3.2153	-3.0914	0.1240
1/2⁺	-2.6703	---	-2.6703	0.0000
3/2⁺	-2.2358	-2.2672	-2.2515	0.0157
KK+QQ Potentials				
7/2⁺	-4.5000	-4.5000	-4.5000	0.0000
5/2⁺	-3.9506	-4.0097	-3.9802	0.0296
11/2⁻	-2.9821	-3.2153	-3.0987	0.1166
1/2⁺	-2.7799	---	-2.7799	0.0000
3/2⁺	-2.3656	-2.2672	-2.3164	0.0492
KK+QQ+Pairing Potentials				
7/2⁺	-4.5000	-4.5000	-4.5000	0.0000
5/2⁺	-3.9506	-4.0097	-3.9802	0.0296
11/2⁻	-2.9821	-3.2153	-3.0987	0.1166
1/2⁺	-2.7799	---	-2.7799	0.0000
3/2⁺	-2.3656	-2.2672	-2.3164	0.0492

B.   Two-quasiparticle systems

The bare KK interaction gives fairly good results in calculations on ^{134}Te excitation states. The levels are somewhat
compressed and appear to lie at too low an energy. The best fit
occurs when the KK interaction has a QQ correction of
V_{QQ}=-0.00041 MeV. A T=1 pairing potential can then be used to
depress the ground state, relative to the other excitation states. The optimum strength for the pairing potential is
V_{pair}=-0.049 MeV. These results are summarized in figure 2a.

Results for calculations on ^{130}Sn were similar to those of
^{134}Te. The bare KK interaction gives a level ordering which is
approximately correct. Addition of a weak QQ potential does give

an improvement, but only because the separation of the second
2^+ and the 4^+ is becoming smaller. Eventually the fit to the
experimental 6^+, 8^+, and 10^+ states degrades giving a shallow
χ^2 minimum at about the same value as found in ^{134}Te. For
^{130}Sn, the addition of a pairing potential can not be tolerated.
These results are summarized in figure 2b.

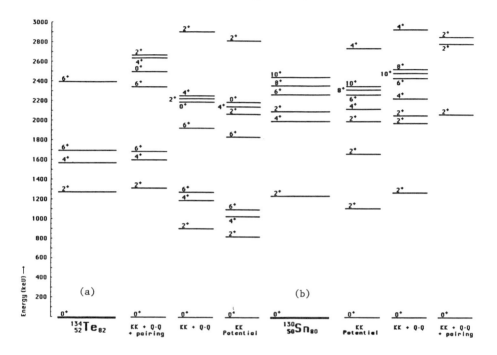

Figure 2a,b. Calculations on ^{134}Te and ^{130}Sn showing
results with the bare KK potential and the KK with core
polarization corrections. Only the positive parity states
with even angular momentum are shown.

C. Three-quasiparticle systems

We have used the three-quasiparticle nuclei to test whether
our effective interaction, derived from the two-quasiparticle
nuclei, is sufficient or whether a larger core polarization
correction is required. We also were interested in determining if
the pairing potential must be varied as the number of the protons
was increased. In the two-neutron hole ^{130}Sn we saw that a
pairing potential could not be tolerated but in the two-proton
^{134}Te a pairing potential was needed. The three-quasiparticle
nuclei would provide us a wider range of nuclei to test this
behavior.

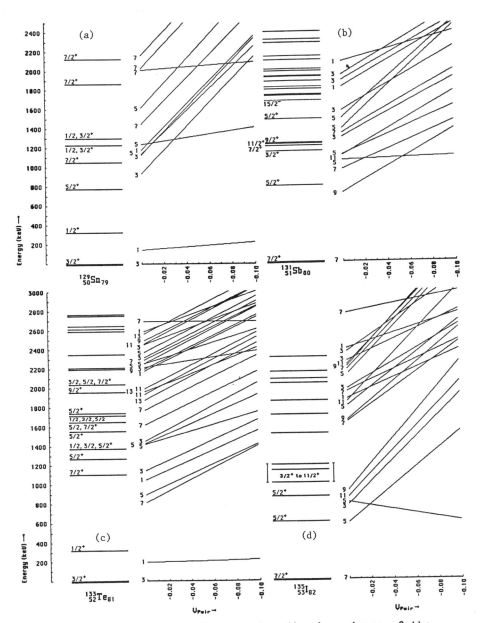

Figures 3a-3d. These figures show the dependence of the
calculated states on the pairing strength for the three quasiparticle nuclei. Only experimentally accessible states are shown.

It is not clear whether the strength of the QQ potential needs to be varied as we move from a three neutron-hole nucleus to the three proton nucleus. There are a large number of possible excitation states, with similar spins, and this makes comparison with experimental data difficult. Figures 3a through 3d show the results of the calculations on the three-quasiparticle nuclei. One feature that is apparent in the results is the need for a pairing potential in ^{133}Te and ^{135}I. In ^{129}Sn and ^{131}Sb the pairing potential does not seem to be necessary. Comparison of the results from calculations on ^{133}Te and ^{135}I with those of ^{129}Sn and ^{131}Sb shows another interesting feature. The level-ordering is good in ^{133}Te and ^{135}I but in ^{129}Sn and ^{131}Sb the level-ordering is not correct. The primary difference between the two sets of nuclei is that in ^{129}Sn and ^{131}Sb the low-lying states will have a large $h_{11/2}$ character to them. In ^{133}Te and ^{135}I the $h_{11/2}$ character will be low. Another interesting point is that some states have markedly different slopes as the pairing potential is increased. These are predominantly single particle states, single particle excitations above ground state.

IV. Conclusions/Discussion

We have found that the nuclear structure near ^{132}Sn can be reasonably well described by an ab initio effective interaction with core polarization corrections. Addition of a small amount of a quadrupole force improves the separation of the calculated states in ^{130}Sn and ^{134}Te. We found that the pairing potential was needed when there are active protons. The pairing correction may be an indication of the role of the $1g_{9/2}$ proton excitations. For the Sn isotopes it does not appear that the exclusion of the $1g_{9/2}$ protons has any affect on the excitation states. In ^{131}Sb the optimum pairing potential strength may be somewhat larger but this could not be determined. Pairing is important in the two-proton and three-proton systems. It is satisfying to note that on the whole the pairing correction to the KK interaction can account for much of the truncation effects.

Results also seem to indicate that the KK potential overestimates two-body matrix elements involving $h_{11/2}$ orbitals. This may have been seen in the one-quasiparticle nuclides; the largest deviation, through only 200 keV, was with the separation of the 11/2$^-$ state from the 7/2$^+$ state. In ^{134}Te we saw the results were quite good. For ^{130}Sn, however, the fit was not as satisfying: there was some level misordering and a QQ correction did not account for this. These states will be largely $h_{11/2}$ in character. The problems with level-ordering were most severe in ^{129}Sn and ^{131}Sb. Again, many of the low-lying states will have a large $h_{11/2}$ character to them.

Further developments of a general effective interaction for the ^{132}Sn region will have to focus on improving the characterization of the $h_{11/2}$ interactions. This will be essential if we are to attempt to calculate the negative parity states and the isomerism within this region. We will also have to consider other corrections to the two-body interaction such as the addition of density-dependent potentials, tensor potentials and odd-components to the two-body potential.

References

[BAL76]   W.J. Baldridge and J.P. Vary, Phys. Rev. C14 2246 (1976).
[BLO83]   J. Blomqvist, A. Kerek, and B. Fogelberg, Z. Phys. A314 199 (1983).
[BRO67]   G.E. Brown and T.T.S. Kuo, Nucl. Phys. A92 (1967) 481.
[DEG80]   L.-E. de Greer and G.B. Holm, Phys. Rev. C22 2163 (1980).
[FOG81]   B. Fogelberg, K. Heyde and J. San, Nucl. Phys. A352 157 (1981).
[FOG84]   B. Fogelberg and J. Blomqvist, Phys. Lett., 137B 20 (1984).
[HAU76]   R.F. Hausman, Ph.D. Thesis, University of California, Davis (1976).
[KAL64]   A. Kallio, and K. Kolltveit, Nucl. Phys. 53 (1964) 87.
[KER72]   A. Kerek, G.B. Holm, S. Borg., L.-E. de Greer, Nucl. Phys. A195, 177 (1972).
[KUO66]   T.T.S. Kuo and G.E. Brown, Nucl. Phys. 85 40 (1966).
[KUO68]   T.T.S. Kuo and G.E. Brown, Nucl. Phys. A114 241 (1968).
[KUO72]   T.T.S. Kuo and G. Herling, Nucl. Phys. A181 113 (1972).
[LAN79]   S.M. Lane, Ph.D. Thesis, Dept. of Applied Sciences, Univ. of Calif., UCRL-52825 (1979).
[LAN80]   S.M. Lane, E.A. Henry, R.A. Meyer, Univ. of Calif., UCRL-85211, 1980.
[SAM85]   M. Samri, G.J. Costa, G. Klotz, et. a. Z. Phys. A321 (1985) 255.
[STO86a] C.A. Stone, W.B. Walters, (to be published).
[STO86b] C.A. Stone, W.B. Walters, (to be published).

RECEIVED August 12, 1986

12

Shell-Model Calculations of ^{90,88}Zr and ^{90,88}Y

J. A. Becker[1], S. D. Bloom[1], and E. K. Warburton[2]

[1]Lawrence Livermore National Laboratory, University of California, Livermore, CA 94550
[2]Brookhaven National Laboratory, Upton, NY 11973

Conventional spherical shell model calculations have been undertaken to describe ^{90,88}Zr and ^{90,88}Y. In these large scale calculations valence orbitals included $1f_{5/2}, 2p_{3/2}, 2p_{1/2}$, and $1g_{9/2}$. The $d_{5/2}$ orbital was included for ^{90}Y and for high-spin calculations in ^{90}Zr. Restrictions were placed on orbital occupancy so that the basis set amounted to less than 25,000 Slater determinants. Calculations were done with a local, state independent, two-body interaction with single Yukawa form factor. Predicted excitation energies and electromagnetic transition rates are compared with recent experimental results.

I. INTRODUCTION

Gamma-ray data from the fusion-evaporation reactions[74,76] Ge + [18]O have been analyzed. Measurements consisted of excitation functions, angular distributions, $\gamma - \gamma$ coincidences, linear polarization and nuclear lifetimes. As one result, the detailed nuclear spectroscopy of ^{90}Zr and ^{88}Zr was extended to $E_x \sim 10$ MeV and $J \sim 20$ [WAR85]. Also, Yrast decay schemes were obtained for ^{90}Y and ^{88}Y, extending to $E_x = 4.5$ MeV, J = [12] for ^{90}Y, and $E_x = 5.5$ MeV, J = [15] for ^{88}Y [WAR86]. (Speculative assignments are enclosed in [ ] and uncertain assignments are enclosed in parenthesis.) The variety of measurements facilitated assignments of level spin and parity in these nuclei. The data for ^{90,88}Y are less complete than for ^{90,88}Zr because the Y isotopes are produced in weaker reaction channels.

II. CALCULATION AND RESULTS

Spherical shell-model calculations were undertaken of these nuclei, in order to gain information on their nuclear structure. The calculations reported here were done with a local, state independent two-body interaction with a single Yukawa form factor,

$$V_{1,2} = f(r)[V_0 + V_\sigma \vec{\sigma}_1 \cdot \vec{\sigma}_2 + V_\tau \vec{\tau}_1 \cdot \vec{\tau}_2 +$$

$$V_{\sigma\tau}(\vec{\sigma}_1 \cdot \vec{\sigma}_2) (\vec{\tau}_1 \cdot \vec{\tau}_2)] \tag{1}$$

Parameters based on a "realistic" finite range potential [PET69] are given in Table 1. Single particle energies, adjusted within model space so that single particle levels in neighboring odd-A nuclei are described, are given in Table 2. Effective charges and magnetic moments were used in the transition rate calculations. E2 rates were done with and additional nucleon

0097-6156/86/0324-0078$06.00/0

charge δ = 1. M1 rates were calculated with the free nucleon spin g factors [$g_s(\pi)$ = 5.59, $g_s(\nu)$ = -3.83] and an orbital g factor $g_\ell = g_\ell$(free) + δg_ℓ, where $\delta g_\ell(\pi)$ = 0.10 and $\delta g_\ell(\nu)$ = -0.05. All calculations were done with the code description in [HAU76].

Table 1. Parameters (in MeV) of the two-body potential. Range = 1.0 fm

V_0	V_σ	V_τ	$V_{\sigma\tau}$
-36.2	6.23	17.8	12.1

Table 2. Single particle energies (MeV) relative to the ^{56}Ni core

$\epsilon f_{5/2}$	$\epsilon p_{3/2}$	$\epsilon p_{1/2}$	$\epsilon g_{9/2}$	$\epsilon d_{5/2}$
2.30	3.31	5.13	2.98	6.77

^{90}Zr

^{90}Zr has a full $1f_{7/2}$, $1f_{5/2}$, $2p_{3/2}$ and $2p_{1/2}$ proton shell and a full $1g_{9/2}$ neutron shell. Low-lying excitations with positive parity are due to proton excitations from the fp shell to the $g_{9/2}$ orbital. In the model space $\pi[p_{1/2}^{-2}g_{9/2}^{+2}]$ excitations are restricted to J $\leq$ 8, and earlier calculations [GLO74] account for energy levels and transition rates within these restrictions. We have chosen to expand the model space in following ways:

(I) $\pi[f_{5/2}, p_{3/2}, p_{1/2})^{-2} g_{9/2}^2]$ for which J_{max} = 12

(II) $\pi[f_{5/2}, p_{3/2}, p_{1/2})^{-4} g_{9/2}^4]$ + I which has J_{max} = 18 and

(III) $\pi[f_{5/2}, p_{3/2}, p_{1/2})^{-2} g_{9/2}^{+2}] \nu[g_{9/2}^{-2}d_{5/2}^2]$ + I

which was J_{max} = 24. Computational restrictions were such that J_{min} = 8 and 12 for (II) and (III), respectively. Odd-parity states were described in the model space $\pi[f_5, p_3, p_1)^{-n} g_{9/2}^{+n}]$, n = 1 or 3.

Calculated even- and odd-parity levels are compared with experiment in Becker, et al. [BEC84]. An adequate description of states with J > 10 requires contributions from Models II and III. Technically, Model II and III calculations require a shift in main frame computers and are not complete. The agreement (within the model space I restriction) between experimental energy levels is reasonable for the even-parity states. The model predicts the level spin sequence correctly. Selected electromagnetic moments and transition rates are presented in Tables III and IV, respectively. Agreement is good for the first 8_1^+ state for both quadrupole and magnetic moments. The electromagnetic transition strengths predicted by Model I are in accord with the γ-ray branching ratios observed for the 10_1^+ and 9_1^+ states; agreement is not good for the decay of the 8_2^+ state. Experimentally the $10_1^+ \rightarrow 8_1^+$ and the $9_1^+ \rightarrow 8_1^+$ γ-ray branches are 100%. The possible $10_1^+ \rightarrow 8_2^+$ and $9_1^+ \rightarrow 8_2^+$ branches are not observed.

This is in accord with the transition branching listed in Table IV. The experimental γ-ray branching of the 8_2^+ state is [B.R.($8_2^+ \rightarrow 8_1^+$)]/[B.R.($8_2^+ \rightarrow 6_1^+$)] = 43.67, while the calculation predicts 27/1.

Table III. Quadrupole (e fm^2) and magnetic moments (μ_o) for ^{90,88}Zr.

$J\#\pi$	A	Q(E2)		μ	
		Theory	Exp.	Theory	Exp.
8_1^+	90	-53.4	I51(6)I[a]	+12.31	+10.85(5)[b]
8_2^+	90	+18.8		+0.31	
8_1^+	88	+25.6	I51(6)I[a]	-3.09	-1.808(4)[c]
8_2^+	88	-72.9		+12.0	

a P. Raghavan, private communication.
b O. Haüsser, et al., Nucl. Phys. A293, 248 (1977).
c T. Faesterman, et al., Hyperfine Interactions 4, 196 (1978).

Table IV. Branching Ratio (BR) Comparison for ^{90,88}Zr.

A	J_i	J_{f1}	J_{f2}	EXP.		THEORY $\frac{BR_1}{BR_2}$
				BR_1	BR_2	
90	10_1^+	8_1^+	8_2^+	100	N.O.[a]	15650/1
90	9_1^+	8_1^+	8_2^+	100	N.O.	26570/1
90	8_2^+	8_1^+	6_1^+	43(6)	57(6)	27/1
88	10_1^+	8_1^+	8_2^+	1.8(3)	98.2(3)	1/50
88	9_1^+	10_1^+	8_2^+	1.8(5)	94.4(9)	1/23
88	9_1^+	8_1^+	8_2^+	3.8(8)	94.4(9)	1/411
88	8_2^+	8_1^+	6_1^+	100	N.O.	134/1

a N.O. denotes not observed experimentally.

^{88}Zr.

^{88}Zr has a full $1f_{7/2}$, $1f_{5/2}$, $2p_{3/2}$, $2p_{1/2}$ shell for protons and 8 neutrons in the $g_{9/2}$ orbital. We describe the low lying even parity states of ^{88}Zr in the model space $\pi\nu[(f_{5/2}, p_{3/2}, p_{1/2},)^{-2} g_{9/2}^{10}]$, which has $J_{max} = 20$. Odd-parity states are described for

$J > 0$ in the model space I $\pi\nu[(f_{5/2}, p_{3/2}, p_{1/2})^{-1} g_{9/2}{}^9]$, while for $J \geq 12$ the model space II was $\pi\nu[(f_{5/2}, p_{3/2}, p_{1/2})^{-3} g_{9/2}{}^9]$. These models have $J_{max} = 15$ and 24, respectively. Results for excitation energies are given in Becker, et al. [BEC84]. The positive parity energy level spectrum is in good agreement with experiment. Note $E_x(9_1{}^+) > E_x(10_1{}^+)$ as is observed experimentally. The spectrum of negative parity states is compressed relative to experiment.

Selected electromagnetic moments are presented in Tables III. Agreement for the electric quadrupole moment and magnetic moment for the $8_1{}^+$ state is not very good for $^{88}Zr(8_1{}^+)$. There are several electromagnetic transitions rates that can be compared: The calculated value B(E2; $8_1{}^+ \rightarrow 6_1{}^+$) = 0.86 Wu it too small by a factor 6. A strong M1 transition $12_2{}^+ \rightarrow 12_1{}^+$ is measured; the model predicts B(M1) = 3.10 Wu while the experimental value is 0.95 Wu. The model also predicts B(E2; $10_1{}^+ \rightarrow 8_2{}^+$) = 7.56 Wu; the experimental observation is B(E2) > 15.4 Wu. The $8_2{}^+ \rightarrow 8_1{}^+$ transition has B(M1) = 8.1(6) x 10^{-3}; the calculation predicts B(M1) = 6.3 x 10^{-5}. Some γ-ray branching can also be compared with calculated values (see Table IV). The $10_1{}^+ \rightarrow 8_1{}^+$ decay is observed with a 1.8(3)% branch. The model predicts a 2% branch relative to the $10_1{}^+ \rightarrow 8_2{}^+$ decay. The $9_1{}^+$ state is observed to decay to the $10_1{}^+$, $8_2{}^+$, and $8_1{}^+$ states with branching ratios 1.8(5), 94.4(9), and 3.8(8)%. The model predicts that the ratio of branching ratios, [B.R.($9_1{}^+ \rightarrow 10_1{}^+$)]/[B.R.($9_1{}^+ \rightarrow 8_2{}^+$)] = 1/23 and [B.R.($9_1{}^+ \rightarrow 8_1{}^+$)] /[B.R.($9_1{}^+ \rightarrow 8_2{}^+$)] = 1/411. The $8_2{}^+$ state is observed to decay 100% $8_2{}^+ \rightarrow 8_1{}^+$ with no reported $8_2{}^+ \rightarrow 6_1{}^+$ branch. The calculation predicts [B.R.($8_2{}^+ \rightarrow 8_1{}^+$)]/[B.R.($8_2{}^+ \rightarrow 6_1{}^+$)] = 134/1.

^{90}Y

^{90}Y has a full $g_{9/2}$ shell for neutrons and one neutron in the $d_{5/2}$ shell, and one proton-hole in the f-p shell. We describe ^{90}Y in the model space

$$\{\pi[(f_{5/2}p)^{-(n+1)}(g_{9/2})^n]\nu[(d_{5/2})^1]; n=0,2\}$$

for odd-parity and

$$\{\pi[(f_{5/2}p)^{-(n+2)}(g_{9/2})^{n+1}]\nu[(d_{5/2})^1]; n=0,2\}$$

for even-parity levels. The predicted energy levels are compared with the experimental results in Fig. 1. Sufficient levels in the right region of excitation are predicted, however, the energy scale for the positive parity levels appears to be too expanded. The large gap between th $J^\pi = 7^+$ and 8^+ levels is reproduced by the calculation.

^{88}Y

The model space for ^{88}Y was

$$\{\pi\nu[(f_{5/2}p)^{-(n+1)}(g_{9/2})^{9+n}; n=0,2]\}$$

for odd-parity states and

$$\{\pi\nu[(f_{5/2}p)^{-(n+2)}(g_{9/2})^{10+n}; n=0,2]\}$$

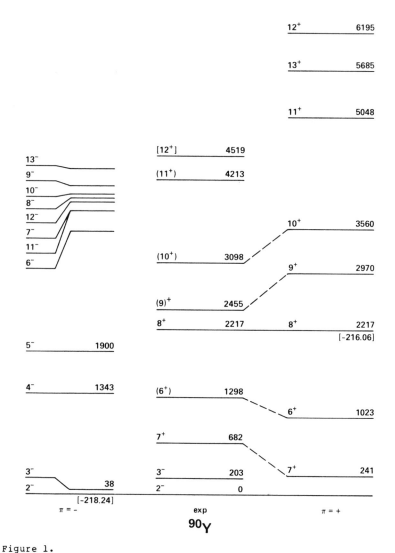

Figure 1.

Experimental and calculated ^{90}Y level schemes. Only the lowest-
lying (or several lowest-lying) levels of a given spin and parity are
included in the calculated spectra. The experimental and calculated
spectra are matched at the $8^+(\pi=+)$ and $3^-(\pi=-)$ levels. The
calculated binding energies of these two levels are given in
brackets. Only calculated levels are shown with $J \geq 6(\pi=+)$ and
$J \geq 2(\pi=-)$.

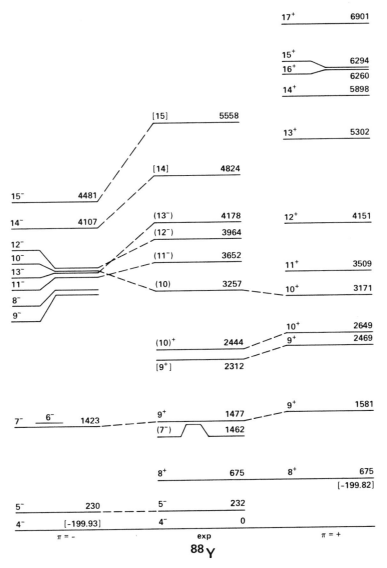

Figure 2.
Experimental and calculated [88]Y level schemes. Only the lowest-lying (or several lowest-lying) levels of a given spin and parity are included in the calculated spectra. The experimental and calculated spectra are matched at the $8^+(\pi=+)$ and $4^-(\pi=-)$ levels and only levels with spins greater than these are shown. The calculated binding energies (in Mev) of these two states are given in brackets.

for even parity states. The results are compared to experiment in Fig. 2. Again, sufficient energy levels are generated at the right excitation energy. The calculations make the apparent switch between even and odd parity for the Yrast states at $E_x = 3.5$ MeV in ^{88}Y seem quite plausable.

III. SUMMARY

A shell model calculation of ^{90,88}Zr and ^{90,88}Y has been done with a "realistic" two-body interaction. A large model space which however does not allow El γ-ray transitions was used. For ^{90,88}Zr good accounts of the observed electromagnetic moments and transitions of the lower lying positive parity levels are obtained, in particular the decay of the $10_1{}^+$ states for both nuclei. For ^{88}Zr, the strong $12_2{}^+ \rightarrow 12_1{}^+$ is identified correctly. Our calculation predicts that the model space required for ^{90}Zr high spin even parity states ($J \geq 11$) requires the breaking of the $g_{9/2}$ neutron shell. For ^{90,88}Y, the model space employed generates sufficient energy levels as approximately the correct excitation. The competition between odd- and even-parity Yrast levels is predicted, and for ^{90}Y, the large energy gap between the $J^\pi = 7^+$ and 8^+ level is reproduced.

IV.   ACKNOWLEDGMENTS

This work was performed in part under the auspices of the U.S. Department of Energy by the Lawrence Livermore National Laboratory under contract No. W-7405-ENG-48, and in part by the U.S. Department of Energy, Division of Basic Energy Sciences, under contract No. DE-AC02-76CH00016.

V. REFERENCES

[BEC84]   J. A. Becker, S. D. Bloom, and E. K. Warburton, in Proceedings of
          the International Symposium on IN-BEAM NUCLEAR SPECTROSCOPY,
          Debrecen, Hungary (1984).

[GLO74]   D. H. Gloeckner and F. J. D. Serduke, Nucl. Phys. A220, 477 (1974).

[HAU76]   R. F. Hausman, Jr., UCRL-52178, (1976).

[PET69]   F. Petrovich, H. McManus, V. A. Madsen, and J. Atkinson, Phys. Rev.
          Lett. 22, 895 (1969).

[WAR85]   E. K. Warburton, J. W. Olness, C. J. Lister, R. W. Zurmühle, and J.
          A. Becker Phys. Rev. C31, 1184 (1985).

[WAR86]   E. K. Warburton, J. W. Olness, C. J. Lister, J. A. Becker, and S. D.
          Bloom, to be published (1986).

RECEIVED May 8, 1986

Dynamic Deformation Model

Krishna Kumar

Tennessee Technological University, Cookeville, TN 38505

Some recent developments of the model are reviewed: (1) Collective bands of the γ-soft nucleus ^{124}Te are compared with experiment and with the O(6) version of the IBM-2 model. (2) The model, combined with the HFB-treatment of the density-dependent Gogny force and with the Coupled Channel Method of Tamura, is employed for α-scattering from ^{24}Mg and ^{28}Si. (3) The model, with major extensions to nuclear fission, leads to surprising predictions for the lifetimes of superheavy nuclei. Suggestions for heavy-ion synthesis are presented.

The currently popular nuclear models range all the way from completely microscopic to completely phenomenological. A prime example of the former type of models is the Hartree-Fock-Bogolyubov (HFB) model where a density-dependent-finite range nuclear force is employed [DEC80]. A prime example of the later type is the Droplet Model of Myers and Swiatecky [MYE74]. The Dynamic Deformation Model (DDM) lies in between these two extremes. This model is not completely microscopic since the single-particle properties are not derived in a self-consistent manner from a nucleon-nucleon force. However, all nucleons are taken into account and no inert core is assumed.

The DDM is not expected to lead to the "final" theory of the nucleus. (That would probably be closer to the HFB model mentioned above, or the recently developed Relativistic Fermi Liquid model [ANA83], or something very different.) Hence, no effort is made to find the "best" parameters for each nuclear region, as is done in most of the currently popular models. Instead, all model parameters are given fixed strengths and Z-A-dependences. *There are no local parameters.* This provides a perfect excuse for not obtaining perfect agreement with the experimental data!

Because of limitations of computer resources and of research personnel, calculations were previously performed for only about 25 nuclei ranging from ^{12}C to ^{240}Pu. It was shown [KUM84, and previous references cited there] that the major trends of the low-energy spectra and electromagnetic moments could be reproduced for the first time without any local parameters. More recently, the model has been extended in several directions. Because of space and time limitations, only three of these are discussed below.

Before discussing the recent developments of the model, let me remind you of the main components of the DDM: (1) The starting point is the spherical shell model of Mayer and Jensen, where the single-particle level energies are taken from the experimental spectra of odd-A nuclei with one particle (or hole) outside a closed shell. There is a *single* level scheme for all nuclei. Our version can be found in Table I of [KUM77]. (2) Quadrupole deformations (axial as well as non-axial) are introduced by employing the Rainwater method, where the average single-particle field oscillates at different frequencies in different directions. The three oscillator frequencies are related to ω_0 (an overall scaling factor for the

0097-6156/86/0324-0085$06.00/0

energies), and to two shape variables (β, γ). (3) Residual interactions of
the pairing type are taken into account. Improved BCS method is employed
where the particle-hole channel is treated on an equal footing with the
particle-particle channel. (4) The shell-correction method of calculating
the potential energy of deformation is employed, where the smooth part due
to the unreliable levels (far from the Fermi surface) is replaced by the
more reliable analytic functions of the Droplet Model [MYE74]. (5) Time-
dependent (cranking type) treatment of nine collective variables (five
quadrupole, neutron and proton energy gaps, neutron and proton Fermi
energies) is employed to calculate three rotational moments of inertia and
two vibrational mass parameters for each shape of the nucleus. (6) The
complete dynamics of rotation-vibration coupling (or anharmonicity or band-
mixing) is taken into account by employing the Kumar-Baranger treatment of
- the Bohr Hamiltonian. No assumption is made about the K-structure or the
vibrational nature of the collective bands. Instead, the collective
Hamiltonian-Matrix is diagonalized for each angular momentum of interest.
Then, the calculated levels are grouped into different bands (according to
the decay characteristics) only for the sake of elucidating possible deviations from the expected trends. The same Hamiltonian, operators, and parameters are employed for all even-even nuclei.

One of the recent developments concerns the solution of some technical problems so that the DDM codes can be run by the graduate students.
This goal has been achieved recently at the University of Sussex at
Brighton, England. The codes have been employed to calculate the spectra
of isotopes, ^{72,74,76,78,82,84}Se, ^{124,126,128,130,132,134}Te. These results
have been compared with experiment and with the IBM-2 [PAR85]. As an
example, a few results for ^{124}Te are given here. Fig. 1 gives the contour
plot of the DDM-calculated potential energy of ^{124}Te. This is the most
perfect example of a γ-soft nucleus that we have come across. The calculated potential has two well-deformed minima, one prolate ($\beta=0.2$, $\gamma = 0$),
one oblate ($\beta = 0.2$, $\gamma = 60°$). Both minima are 8.0 MeV below the spherical
maximum, that is the prolate-oblate difference is zero (within accuracy of
0.1 MeV). However, this does not lead to two sets of bands because the two
minima are separated in the γ-direction by a barrier of only 0.8 MeV. In
fact, this barrier is well-below the ground state which is raised by the
Zero-Point-Motion above the minima by 1.9 MeV. The resulting spectrum
(see Fig. 2) is close to that expected for a γ - unstable [or O(6) in the
IBM language] nucleus. However, Nature is often more interesting than our
simple models. Although, the potential and the spectrum [also, the B(E2)
values and the X(E0/E2) values, to be discussed in a longer paper] suggest
a perfect γ - unstable nucleus, the quadrupole moment of the first 2^+ state
surprises us once again. The measured value [NAQ77] is 67(15)% of the
rotational value. The corresponding DDM value is 56%. Since the DDM potential shows no prolate-oblate preference, how can it give such a large quadrupole moment? This interesting "quirk of nature" arises because the
inertial functions (which represent "fine structure" effects of the single-
particle levels) are not γ-independent. The IBM-2 calculation also yields
a substantial quadrupole moment, but at the cost of having to add an extra
term (a Q·Q term) to the Hamiltonian.

The nuclear dynamics treatment part of the DDM has recently been
combined with the HFB model (mentioned above) and with an extension of the

Coupled Channel Method. The calculation of reliable β-γ-dependent wave functions for various collective band members allows the inclusion of a large number of reaction channels. Furthermore, this marriage of nuclear structure and nuclear reaction models allows for a unified treatment of spherical and deformed nuclei, of γ-rigid and γ-soft nuclei, of different shapes in different reaction channels, and of the possibilities of substantial differences in the shapes of target and residual nuclei. This model has been employed for α-scattering as well as for neutron cross-sections [KUM85]. Two examples are reviewed here. Fig. 3, 4 give the HFB-calculated potential functions of ^{24}Mg and ^{28}Si. As expected on the basis of previous studies, ^{24}Mg is strongly prolate while ^{28}Si is strongly oblate. These shape changes lead to dramatic changes in the angular distributions of α-scattering cross-sections (see Fig. 5, 6) from the 0^+, 2^+, 4^+ members of the ground state rotational band. This shows once again the enormous sensitivity of the form factors to structural changes.

The third major development concerns the extension of the DDM to

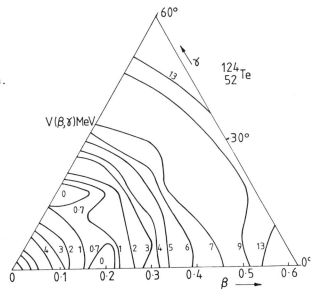

Fig. 1.   Contour Plot of V(β, γ) of ^{124}Te Calculated in the DDM (see [PAR85]).

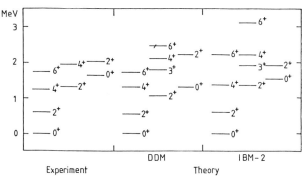

Fig. 2.   Collective Bands of ^{124}Te (see [PAR85]).

the problems of fission barriers, fission isomers, and superheavy nuclei. While the shell-correction method of calculating the potential energy function is similar to that employed in previous theories [FIS72, RAN74], there are several important differences: (1) The classical Mayer-Jensen method is used for the spherical single-particle levels. No adjustments are made to simply fit the data. (2) Energy of Zero-Point-Motion is calculated for each nucleus rather than estimated or ignored. (3) Elaborate shape definitions are replaced by a matching procedure where the fragment interaction has the correct asymptotic form. (4) Microscopically calculated mass parameter functions are employed in two-dimensional action integrals. Mass asymmetry as well as charge asymmetry are fully taken into account.

The extended model has been subjected to a number of tests. Without any parameter adjustments, the model reproduces the energies and B(E2) values of the lowest 2^+ states of doubly magic nuclei. It gives reasonable fission barriers, fission isomer energies, and fission isomer lifetimes [KUM86]. Agreement with the fission lifetimes is also quite reasonable (see Fig. 7). This did require the introduction of two additional parameters (the strength and the $A_1 A_2$-dependence) for the nuclear part of the fragment-fragment interaction. But, we believe that our model and single-particle levels are quite reasonable. The experiments done so far also support our main result that 114 is not the next magic number for the protons. Our predicted fission halflife of $^{298}[114]$ is 10^{-17} y, 36 orders of magnitude lower

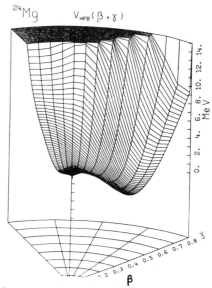

Figure 3. Reproduced with permission from [KUM85]. Copyright 1985 American Physical Society.

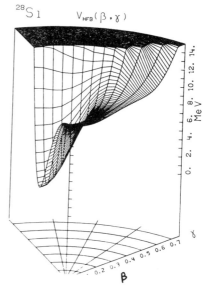

Figure 4. Reproduced with permission from [KUM85]. Copyright 1985 American Physical Society.

than the prediction of [FIS72] and 25 orders lower than [RAN74]. We find ^{310}[126] to be quite stable against fission (halflife > 10^{23} y). But the α-decay lowers the total halflife (see Fig. 8) to only 4ms. The extra stability against α-decay at lower Z values is expected to shift the most stable superheavy from ^{310}[126] to ^{306}[122]. The corresponding total lifetime is estimated to be 2.6 s.

In view of experimentalists' disappointing experiences with the predictions for superheavy nuclei, we hesitate to suggest new experiments. But our main results are based on general principles, rather than on a new set of parameters. Hence, we would like to encourage experimentalists to try once again. Our suggested heavy-ion experiments are indicated in Fig. 9. Note that the suggested projectile energy is the laboratory energy for the lighter nucleus (which is expected to be less susceptible to Coulomb-induced fission before fusion because of its semi-magic nature). The upper energy is not the absolute limit. It would be desirable to increase it in small steps and search for the optimum energy.

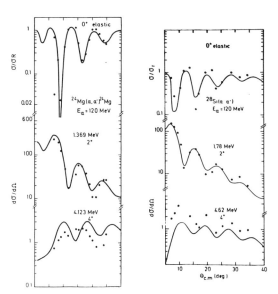

Figure 5 (left). Reproduced with permission from [KUM85]. Copyright 1985 American Physical Society.
Figure 6 (right). Reproduced with permission from [KUM85]. Copyright 1985 American Physical Society.

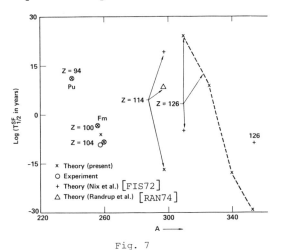

Fig. 7

Fig. 9 also gives the α-decay energies and lifetimes, obtained by combining the atomic masses from [MYE77] with the Viola-Seaborg formula for α-decay lifetimes [VIO66]. Since these energies are quite large, they should help in making a reliable identification of the superheavy nucleus and of the corresponding SuperHeavy Element (SHE).

Hopefully, our results (some of which have been presented previously [KUM83]) for the superheavy nuclei will also stimulate other theorists to recalculate fission lifetimes, and, especially, fusion cross-sections for the suggested target-projectile combinations. These cross-sections are expected to be small and the corresponding experiments are expected to be quite difficult. But the experimentalists have a chance of clearly distinguishing between different models whose predictions differ not by factors of 2 but by many orders of magnitude. Also, they might discover many new isotopes with unusual physical and chemical properties.

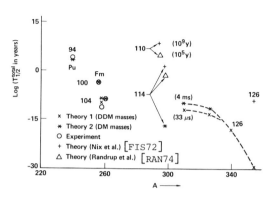

Fig. 8

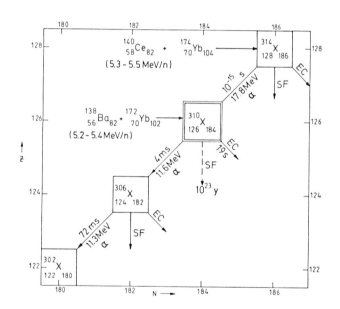

Fig. 9

References

[ANA83] M. R. Anastasio. L. S. Celenza, W. S. Pong, and C. M. Shakin, Phys. Repts. $\underline{100}$ 327 (1983).

[DEC80] J. Decharge' and D. Gogny, Phys. Rev. C. $\underline{21}$ 1568 (1980).

[FIS72] E. O. Fiset and J. R. Nix, Nucl. Phys. $\underline{A193}$ 647 (1972).

[KUM77] K. Kumar, B. Remaud, P. Aguer, J. S. Vaagen, A. C. Rester, R. Foucher, and J. H. Hamilton, Phys. Rev. C $\underline{16}$ 1235 (1977).

[KUM83] K. Kumar and M. G. Mustafa, in Proc. Int. Conf. on Nuclear Physics, Florence, Italy, August 29-Sept. 3, 1983, Vol. 1, p. 626.

[KUM84] K. Kumar, Nuclear Models and the Search for Unity in Nuclear Physics (Columbia University Press, New York, 1984).

[KUM85] K. Kumar, Ch. Lagrange, M. Girod, and B. Grammaticos, Phys. Rev. C $\underline{31}$ 762 (1985).

[KUM86] K. Kumar and M. G. Mustafa, to be published.

[MYE74] W. D. Myers and W. J. Swiatecki, Ann. Phys. (N.Y.) $\underline{84}$ 186 (1974).

[MYE77] W. D. Myers, Droplet Model of Atomic Nuclei (Plenum, New York, 1977).

[NAQ77] I. M. Naqib, A. Christy, I. Hall, M. F. Nolan, and D. J. Thomas, J. Phys. G:Nucl. Phys. $\underline{3}$ 507 (1977).

[PAR85] P. Park, A. R. H. Subber, W. D. Hamilton, J. P. Elliott, and K. Kumar, to be published.

[RAN74] J. Randrup, S. E. Larsson, P. Möller, A. Sobiczewski, and A. Lukasiak, Physica Scripta $\underline{10A}$ 60 (1974).

[VIO66] V. E. Viola and G. T. Seaborg, J. Inorg. Nucl. Chem. $\underline{28}$ 741 (1966).

RECEIVED August 11, 1986

Section I: Motivation for Further Exploration
Part 2: Nuclear Data and Systematics
Chapters 14–18

14

Nuclear Decay Data

An Important Interface Between Basic and Applied Science

C. W. Reich

Idaho National Engineering Laboratory, EG&G Idaho, Inc., Idaho Falls, ID 83415

In this paper, we discuss several categories of decay data which have
contributed to low-energy nuclear physics, indicate some of the ways
they are useful in solving problems in other areas and identify needs
for further measurements. Illustrations include half-life and
emission-probability data of actinide nuclides important for reactor
technology and useful as reference standards for nuclear-data
measurements. Decay data of highly neutron-rich fission-product
nuclides are important in such diverse areas as astrophysics and
reactor-safety research. Some of these data needs and experimental
approaches suitable for satisfying them are presented.

1. Introduction

Data on the decay properties (e.g., total and partial half-lives, total
decay energies, energies and emission probabilities of the emitted radiations)
have in the past made major contributions to our knowledge of nuclear
properties at low excitation energies and of the various modes of motion
that give rise to them. Evidence that these data continue to have a
significant influence on this field, and indeed of the continuing vitality
of the field itself, is provided by the large number and breadth of content
of the papers being presented at this symposium. Beyond this highly
interesting, but relatively restricted application, however, nuclear decay
data play an important role in many other areas of basic and applied science.

In any discussion of decay data and their applications, the following
should be kept in mind. In spite of their wide applicability, the primary
motivation for their measurement has largely been the information which they
provide for low-energy nuclear physics. The extensive measurement activity
in this area over the past three decades or so has produced a large amount
of fairly detailed data on the decay properties of unstable nuclei.
However, because of the interests and objectives of most of the measurers of
these data, the existence of this extensive body of information does not
necessarily mean that the decay-data needs of any given area are satisfied.
Such needs frequently are highly specialized and require additional
characteristics not usually considered by an experimenter whose interests
lie in other areas. To satisfy such needs, specialized measurements,
occasionally utilizing techniques specifically tailored to the particular
application, are required. As a result, the field of decay-data measurement,
although by many standards a mature one, continues to present exciting new
challenges as experimental capabilities develop and additional needs in
other areas of basic and applied science continue to be identified.

0097–6156/86/0324–0094$06.00/0
© 1986 American Chemical Society

To adequately treat these many applications and to illustrate the
specific ways in which decay data make useful, if not crucial, contributions
to them is a task that lies beyond the space and time limitations of this
paper. We have thus chosen to limit the scope of this presentation to the
discussion of several selected examples, drawn mostly from the area of
fission-reactor physics. These include the results of recent significant
developments in actinide-nuclide decay data and, in the spirit of this
symposium, decay data of fission-product nuclides off the line of β
stability and some of the problems and challenges they present to both
experimental capabilities and nuclear theory.

2. An International Coordinated Program of Actinide-Nuclide Decay-Data Measurement and Evaluation

In 1975, the IAEA convened an Advisory Group Meeting on Transactinium
Isotope Nuclear Data at Karlsruhe, F.R.G. [TND76] to survey the requirements
for and status of transactinium isotope nuclear data relevant to fission-
reactor research and technology. It was found that the accuracy of many of
the decay data was not sufficient to satisfy a number of the needs for such
applications as materials safeguards, fuel assay, sample-mass determination
and standards preparation. One of the recommendations drawn up at this
meeting called for an internationally coordinated program of decay-data
measurement and evaluation to address this situation. Subsequently, in
1977, the IAEA organized a Coordinated Research Program (CRP) to measure and
evaluate decay data for selected transactinium nuclides. Research groups
from five laboratories around the world agreed to participate in this CRP;
and work got underway in the Spring of 1978.

The work of this CRP has recently concluded and a report summarizing the
results of this task has been prepared. New, highly precise decay data on
total and/or partial half-lives for 7 nuclides, α-particle emission
probabilities ($P\alpha$) for 7 nuclides, and γ-ray emission probabilities
($P\gamma$) for 21 nuclides have been produced. The final list of recommended
decay data includes these results, as well as those from other recent
measurements, together with carefully evaluated information on a number of
other nuclides. In all, this list includes half-life values for 125
nuclides and, for the more prominent transitions only, $P\alpha$ values for 30
nuclides and $P\gamma$ values for 47 nuclides. [Some of these data represent
simply the results of previously published evaluations.] These data will be
included in the final report of the CRP, to be published as one of the IAEA
Technical Reports Series.

One example of the improvement in the quality of the data resulting from
this measurement program is provided by the $P\gamma$ values for ^{235}U. These
data were measured by three of the CRP participants. When combined with another
recent measurement, these led to a recommended $P\gamma$ value of 57.2 ± 0.5
photons/100 decays for the prominent 185.7-keV γ ray. The previously accepted
value [NDS77, TOI78] for this quantity was 54., with no uncertainty given.

3. Actinide Half-Lives as Standards for Nuclear Data Measurements

A knowledge of the half-life is required for any application in which
quantitative assay of material for radionuclide content is desired. In
fission-reactor research and technology, for example, accurate half-life

data are required for many areas related to the safeguarding of special
nuclear materials. In the precise measurement of neutron-induced reaction
cross sections important for studies of fast-reactor fuel cycles, such
information is necessary for the accurate mass assay of, and the correction
for nuclide decay in, the samples utilized.

In recognition of these needs, a list of recommended half-life values
for a number of actinide nuclides useful as standards for nuclear-data
measurements was prepared several years ago [VAN83]. This list included
both total and partial half-life data for 11 such nuclides - the important
isotopes of U, Np and Pu, as well as ^{252}Cf. Subsequently, these data were
revised [REI85a] to incorporate the results of new measurements, some of
them from the work of the CRP. These data are given in a recent paper
[REI85b], as well as in the final report of the IAEA CRP.

Important comments regarding some of these values need to be made. For
example, the presently recommended half-life values for ^{237}Np and ^{252}Cf
are, respectively, $(2.14\pm0.01)\times10^6$ y and (2.645 ± 0.008) y. Although the
quoted precision of the former is quite good ($\sim0.5\%$), this value is based
on the results of only one measurement. Some evidence that it may be
significantly in error has been given by [MEA83], who finds that his
measured ratio of the ^{237}Np fission cross section to that of ^{235}U in the
MeV region of neutron energies differs from those of other experimenters who
used different techniques. One potential source of this discrepancy is the
mass assay of the ^{237}Np samples, which depends ultimately on the value
used for the ^{237}Np half-life. A satisfactory resolution of this
discrepancy would help support the use of ^{237}Np as a neutron flux standard
in the MeV region. Regarding the ^{252}Cf half-life, [SMI83] reports in a
recent analysis that the present situation is confused. The results of ten
measurements reported with high precisions (ranging from $\sim0.4\%$ to
$\sim0.04\%$) over the past twenty years, tend to cluster around two distinct
values, viz. 2.638 y and 2.651 y. Evidence exists which provides support
for each of these, and it is thus not clear which (if indeed either) of them
represents the "correct" ^{252}Cf half-life.

The status of these values emphasizes the importance of the observation
that simply because data exist whose quoted precision is quite adequate for
certain (indeed many) purposes it cannot be assumed that there is no need
for further measurements. While it is highly unlikely that any changes in
these two half-life values which result from the eventual resolution of
these above-mentioned discrepancies will significantly affect basic nuclear
physics, these discrepancies nonetheless present serious problems for many
reactor-related nuclear-measurement efforts.

4. Short-Lived Fission-Product Decay Data

Because of their intimate link with energy production in nuclear
reactors, fission products and their nuclear data have long occupied an
important position in reactor technology. In recent years, interest in
short-lived fission-product decay data has increased markedly, as their
relevance to different areas of research and technology has become recognized.
In addition to their importance for estimation of the fission-product decay-
heat source term in nuclear reactors, the increasing attention being focused
on the assessment of the hazards associated with the release, transport and

deposition of fission products following postulated reactor accidents has produced a need for additional information on their energy spectra, especially for the isotopes of the volatile elements. In astrophysics there is a need [MAT84] for data on the highly neutron-rich fission-product nuclides to help refine and further constrain present models of elemental synthesis in stellar environments. In the analysis of recent experiments at nuclear reactors [BOE84] to search for neutrino oscillations, a knowledge of the reactor antineutrino spectrum, especially the high-energy region (which depends sensitively on the decay properties of the short-lived fission products), is needed.

In addition to the difficulties in producing good samples, the study of short-lived fission-product decay schemes presents special problems. Because of the complexity of the spectrum of the emitted radiation and the large number of energy levels that can be populated, the observation of all the γ-ray transitions and the correct placement of them within the daughter-nucleus level scheme is a difficult and extremely time-consuming process, if it is indeed possible at all. Consequently, in actual practice, it is almost never done. For most purposes in low-energy nuclear physics this is not a serious problem, since it is usually only the level properties in the first MeV or two of excitation that are of interest, and these can be obtained with considerable confidence from even a relatively incomplete decay scheme. In cases where an accurate knowledge of how the emitted energy is distributed, however, the lack of a complete decay scheme can be a fundamental limitation. A simple illustration of this, discussed in [REI85b], is provided by the ^{87}Br decay ($T_{1/2}$ = 55.7 s). A recent very detailed study [RAM83] has led to a deduced distribution of the β intensity feeding the levels in the ^{87}Kr daughter that is quite different from that in [NDS79]. [RAM83] also report Pγ values ∿30% smaller than those in [NDS79]. These differences in the proposed decay scheme lead to significantly different values for the average energies per decay (important for reactor decay-heat assessment), as shown in Table 1.

Table 1
Comparison of average decay-energy values (in keV) for ^{87}Br deduced from two different sets of data

Quantity	[RAM83]	[NDS79]
$<E_\beta>$	1643.	1861.
$<E_\nu>$	2122.	2358.
$<E_\gamma>$	3204.	4113.
sum*	6969.	8332.

For a "perfect" intensity balance, the three average-energy values should sum to Q_β(=6830 $\pm$ 120).

The ^{87}Br decay has a number of simplifying features [REI85b], which make its study simpler than those of most of the short-lived fission products. Since, even for this "simple" case (which even so took several years to complete), a detailed study produced such drastic changes in many of the deduced decay properties, caution should be taken in using presently available data on such nuclides for many applications. It seems likely, [REI85b], for example, that the somewhat poorer prediction of the β- and γ-ray components of the decay-heat source term using the ENDF/B-V decay data than was obtained using ENDF/B-IV resulted from the inclusion in the former of decay data for a number of short-lived fission products that were obtained from just such incompletely determined decay schemes.

Many of the applications, particularly the more challenging ones, of
fission-product decay data have involved the fission products as a group.
Because of their large number (>700) and the fact that almost no data have
been available for many of them, it has always been necessary for such
applications to utilize calculated values for the unmeasured quantities. To
do this, a number of different approaches have been taken to describe the
β-decay properties (β-strength functions) of these nuclides [TAK72,
MAN82, KLA83]. The most comprehensive of these approaches has been that of
Klapdor and his co-workers [KLA83] who, using a specific model for the
β-strength function, provide predictions for a wide variety of
fission-product-related phenomena in astrophysics and nuclear physics.

These different approaches give different predictions for many of the
phenomena of interest; and if one is to apply them with some degree of
confidence to the solution of a wide range of problems, it is necessary to
establish which approaches, or which aspects of the different approaches,
are reliable and which are not. This is not always a simple matter. Part
of the problem in assessing the various theoretical approaches results from
the complexity of the situation. In different applications, one is really
testing different aggregates of various types of data (fission yields, Q_β
values, average decay energies, etc.); and it is difficult to isolate the
effect of any one individual component. To permit reliable predictions of
phenomena related to the decay properties of fission products as a group, a
more systematic approach is required.

Based on past experience, the outlines of such an approach are clear.
Both theory and experiment, in close collaboration, are needed to refine the
models used to calculate the β-decay properties of the individual nuclides.
On the theoretical side, a number of interesting features are beginning to
emerge. Using a model of the β-strength function recently developed by
[KRU84], [KRA85] has calculated the decay properties of a number of nuclides.
From these results, he reaches the following conclusions. Because of nuclear-
structure effects, there is no simple, global β-strength function
systematics that is sensitive enough to permit reliable extrapolations. In
general, deformation effects in the far unstable nuclei tend to shift the
β strength down, resulting in larger $\langle E_\beta \rangle$ values and smaller $\langle E_\gamma \rangle$
values as compared to spherical nuclides with the same Q_β value. Perhaps
most important, no shell model appears capable of providing reliable β-
strength predictions over the whole mass range with only one parameter set.

Experimentally, the measurements must be tailored specifically to
produce that information most sensitive to refining the theoretical
concepts. Because it is not feasible to measure all the nuclides, only
those whose information is the most important should be studied.
Conventional decay-schemes studies do not seem appropriate, because of the
complexity of the decay schemes, the errors to which they are subject (cf.
the discussion of ^{87}Br above) and the large amount of time needed to carry
them out. Direct measurement of the β-strength functions themselves,
utilizing total-absorption γ spectrometry and, where relevant,
delayed-neutron-gamma coincidence techniques, promises to provide a means of
producing the necessary information in a reasonable time.

Acknowledgments

This work was carried out under the auspices of the U.S. Department of
Energy under DOE Contract No. DE-AC07-76ID01570.

References

[BOE84].   See, for example, F. Boehm and P. Vogel, Ann. Rev. Nucl. Part. Sci. 34, 125 (1984).

[KLA83].   H. V. Klapdor, Prog. Part. Nucl. Phys. 10, 131 (1983) and references contained therein.

[KRA85].   K.-L. Kratz, private communication (1985); see also Nucl. Phys. A417, 447(1984).

[KRU84].   J. Krumlinde and P. Möller, Nucl. Phys. A417, 419 (1984).

[MAN82].   F. M. Mann, C. Dunn and R. E. Schenter, Phys. Rev. C 25, 524 (1982).

[MAT84].   See, for example, G. J. Mathews, in Proceedings of the NEANDC Specialists Meeting on Yields and Decay Data of Fission Product Nuclides, Brookhaven National Laboratory, October 24-27, 1983, U.S. DOE Report BNL-51778 (1984), pp. 485-502.

[MEA83].   J. W. Meadows, Nucl. Sci. Engr. 85, 271 (1983).

[NDS77].   Nuclear Data Sheets 21, 91 (1977).

[NDS79].   Nuclear Data Sheets 27, 389 (1979).

[RAM83].   S. Raman, B. Fogelberg, J. A. Harvey, R. L. Macklin, P. H. Stelson, A. Schröder and K.-L. Kratz, Phys. Rev. C 28, 602 (1983).

[REI85a].   C. W. Reich, in Proceedings of the IAEA Advisory Group Meeting on Nuclear Standard Reference Data, Geel, Belgium, 12-16 November, 1984, IAEA-TECDOC-335 (IAEA, Vienna, 1985) pp. 390-396.

[REI85b].   C. W. Reich, "Nuclear Decay Data: Some Applications and Needs", invited paper included in the Proceedings of the International Conference on Nuclear Data for Basic and Applied Science, Santa Fe, NM, 13-17 May, 1985 (to be published).

[SMI83].   J. R. Smith, U.S. DOE Report EGG-PBS-6406 (September, 1983).

[TAK72].   K. Takahashi, Progr. Theor. Phys. 47, 1500 (1972) and references contained therein.

[TND76].   Proceedings of the IAEA Advisory Group Meeting on Transactinium Isotope Nuclear Data, Karlsruhe, F.R.G., Nov. 3-7, 1985, IAEA Report IAEA-186, Vols. I-III, (IAEA, Vienna, 1976.)

[TOI78].   Table of Isotopes, 7th Edition, C. M. Lederer and V. S. Shirley, eds., (John Wiley, New York, 1978).

[VAN83].   R. Vaninbroukx and A. Lorenz, in Nuclear Data Standards for Nuclear Measurements, IAEA Technical Reports Series, No. 227 (IAEA, Vienna, 1983) pp. 69-70.

RECEIVED August 11, 1986

15

The Importance of Level Structure in Nuclear Reaction Cross-Section Calculations

M. A. Gardner and D. G. Gardner

Lawrence Livermore National Laboratory, University of California, Livermore, CA 94550

We continue to observe the necessity of describing nuclei in the first few MeV above their ground states with discrete levels in order to obtain accurate cross-section and isomer-ratio calculations. There appears to be no adequate way in which one can use a level-density expression to represent or substitute for the level structure. The level information should not only consist of those levels that are obtained from experiment, but must be supplemented with the level structure theoretically known to be present by model calculations. For the calculation of isomer ratios, gamma-ray branchings among the discrete levels are also required. Discrete-level sets are not only important in describing nuclei at low excitation energies, but are also valuable in estimating several calculational parameters, particularly for nuclei off the line of stability.

Introduction

In the course of carrying out nuclear reaction cross-section calculations of the statistical-model Hauser-Feshbach type, we continue to observe the necessity of describing nuclei in the first few MeV above their ground states with discrete levels in order to obtain accurate results[GAR85a]. There is no adequate way in which one can use a level-density expression to represent or substitute for the level structure. Furthermore, all levels must be included; the level information obtained from experiment must be supplemented with that known to be present theoretically. This is true both for spherical nuclei, where typically tens of levels are found in the first 1 to 4 MeV of nuclear excitation, and for deformed nuclei, where perhaps a thousand levels are present in the first 1.5 MeV. We have found that discrete-level sets are not only important in the description of nuclei at low excitation energies but are also very valuable in the prediction of a number of calculational parameters, particularly for nuclei off the line of stability. Discrete-level information can be used to obtain improvements on global parameterizations of level densities; to predict unknown values of D_{ob}, the average s-wave neutron resonance spacing; to compute correct radiation widths; and to deduce absolute E1 and M1 gamma-ray strength functions. If correct gamma-ray branchings among the discrete levels are provided, reliable isomer ratios can be computed. In addition, complete discrete-level sets allow one to analyze primary gamma-ray spectra.

Discrete-Level Descriptions of Nuclei at Low Excitation Energies

The accurate calculation of a reaction cross section near the threshold requires that the product nucleus be described with a sufficient number of discrete levels. Figure 1 illustrates this with the calculation of the ^{169}Tm(n,3n) excitation function for incident neutron energies in the first few MeV above the reaction threshold of 14.96 MeV[GAR84a]. We obtained good agreement with the experimental data[BAY75,NET72,NET76,VEE77] when we used 63 modeled discrete levels for ^{167}Tm in its first 1.4 MeV of nuclear excitation[HOF84]. However, when we described ^{167}Tm down to its ground state with the Gilbert-Cameron level-density expression and the Cook-modified parameters[COO67], we computed an excitation function that was about 60% lower in the first 3 MeV above the threshold.

Figure 2 shows another example of the need for discrete-level descriptions of nuclei. Two computations of the ^{89}Y(n,γ) excitation function[GAR84b] were made. In the first, the ^{89}Y and ^{90}Y nuclei were described above the ground state with an additional 24 levels provided by E. A. Henry[HEN77]; in the second, the additonal levels were replaced with the Gilbert-Cameron level-density formulae and the Cook-modified parameters. Since the first level above the ground state in ^{89}Y lies at 0.9 MeV, no inelastic

0097-6156/86/0324-0100$06.00/0

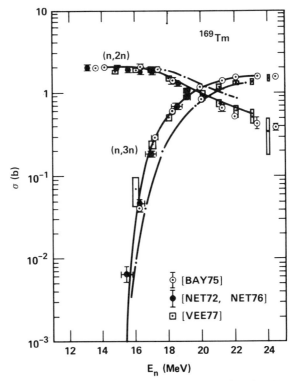

Figure 1. Calculated (n,2n) and (n,3n) excitation functions for
^{169}Tm [GAR84a], with discrete-level descriptions of ^{167}Tm that
included 63 levels (solid curves) and only the ground-state
level (broken curves), compared with experimental data from
[BAY75,NET72,NET76,VEE77].

scattering can occur below this energy. Yet, the calculation that used only level-density expressions predicted that the (n,n') cross section was already 400 mb at 0.9 MeV! This overestimation of the inelastic cross section leads to a significant lowering of the (n,γ) cross section.

Another effect of substituting a level-density description for discrete levels can also be seen in this figure. Note that the calculation done using level densities immediately above the ground state is still lower than that done using 25 levels for ^{90}Y in the low-keV energy range. This cannot be due to incorrectly computed inelastic scattering competition. It comes from the fact that the description of ^{90}Y with only level-density formulae leads to partial radiation widths that are calculated to all be about the same. In Table 1 we see that with 25 levels in ^{90}Y, the p-wave width and the f-wave width are calculated to be twice as large as that for s-wave neutrons (in agreement with experiment[BOL77]), and the larger capture cross section is accounted for.

The role of a good discrete-level description of the target nucleus in the calculation of a neutron capture excitation function is shown again in Fig. 3, where our initial computation of the ^{209}Bi(n,γ) cross section vs neutron energy is plotted. In this calculation, the ^{209}Bi is described by 36 discrete levels up to 3.5 MeV in nuclear excitation; these levels were provided by R. A. Meyer[MEY85]. Note that for incident energies between 0.5 and 3 MeV, the capture cross section actually increases, in agreement with

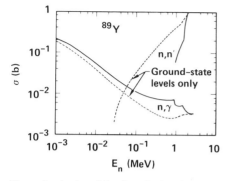

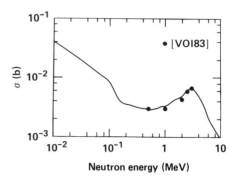

Figure 2. (n,γ) and (n,n') excitation functions for ^{89}Y, calculated with discrete-level descriptions for ^{89}Y and ^{90}Y that included 25 levels (solid curves), and only the ground-state level (dashed curves).

Figure 3. Our initial calculation of the ^{209}Bi(n,γ) excitation function compared with measurements from [VOI83].

Table 1. Calculated and measured partial radiation widths for ^{90}Y.

Neutron partial wave	Calculated width (meV)		Measured width (meV) [BOL77]
	25 levels in ^{90}Y	1 level in ^{90}Y	
s	183	183	115
p	342	170	307
d	186	152	—
f	251	132	—
g	179	113	—

measurements[VOI83], because of the lack of inelastic scattering competition. This is to be expected, since the first three levels above the ground state in ^{209}Bi are widely spaced at 0.896, 1.608, and 2.443 MeV, respectively. A level-density description of ^{209}Bi above its ground state would surely overestimate the inelastic scattering competition in this region, leading to an incorrect energy dependence for the capture cross section.

Discrete Levels and Level-Density Parameterizations

Most modern statistical-model codes allow the matching of the low-energy portion of a level-density expression to an available discrete-level set. Since we use the Gilbert-Cameron level-density formalism in our version of the STAPRE code[UHL70], we accomplish this by adjusting the values of the energy constant, E_0, in the constant-temperature portion and of the pairing energy, δ, in the Fermi-gas portion in such a way that the two functions and first derivatives remain continuous. We find this procedure to be quite valid for spherical nuclei, provided that the level set is sufficiently complete. From such a set, information can be extracted about the spin cut-off parameter as well as about other parameters, enabling one to obtain good agreement with known D_{ob} values or to predict such quantities for nuclei off the line of stability. In this way, the level structure can be used to improve on global parameterizations of the level density.

For example, in calculations that we made involving a series of strontium and yttrium isotopes in the mass range of 86 to 90, we used the discrete-level schemes provided by Henry[HEN77,HEN85]; each set was composed of 25 to 60 levels. We found that almost every isotope required a different value of k, the constant in the mean-square spin projection expression $\langle m^2 \rangle = kA^{2/3}$, which is related to the energy-dependent spin cut-off parameter $\sigma^2(E)$ by the formula $\sigma^2(E) = (6/\pi^2)\langle m^2 \rangle(a[E - \delta])^{1/2}$. Although it is common practice to use average values of $k = 0.146$[GIL65] or 0.24[REF80], we obtained a range of k values, from 0.146 to 0.29, when we attempted to extract them from the discrete-level sets. Table 2 shows our findings. Such fluctuations are in agreement with the observations of Reffo[REF80].

Table 3 summarizes the two ways in which σ^2 can be deduced from discrete-level sets. The first is the maximum likelihood estimator expression, where there are N levels of various J_i spins. This method can be more sensitive to small differences in σ^2 unless the levels exhibit a nonrepresentative distribution of mainly high or low spins. The second expression allows one to extract σ^2 from the spin distribution and is useful unless several spins are missing. Figures 4(a) and (c) show the results of the estimator method for ^{89}Y, described with 60 discrete levels, and for ^{88}Y, described with 25 levels. Difficulties are encountered in both cases since each nucleus has low-lying, high-spin states that introduce disproportionate weightings because of the $(J + 1/2)^2$ term. Note that for ^{89}Y in Fig. 4(a), if one had information on the first 15 levels only, up to an excitation of 3.1 MeV, the choice of $k = 0.24$ would be made. In Fig. 4(b), although the spin distribution method indicates that $k = 0.146$ is reasonable for ^{89}Y, the estimator method indicates that k should have a somewhat larger value. Finally, in Fig. 4(d), the spin distribution in ^{88}Y clearly indicates that the global k value of 0.146 is not suitable and that a k of about 0.29 is preferred.

The following are examples of what happens when one value of k is chosen over another. The use of a k value of 0.24 instead of 0.146, with no other changes, leads to about a 30% difference in the Fermi-gas level density. In the calculation of the ^{89}Y$(n,\gamma)^{90}$Y cross section, the choice of $k = 0.17$ instead 0.146 for ^{90}Y, based on the discrete levels, led to a D_{ob} change of 15% and a lowering of the capture cross section by about 10% in the E_n range of a few hundred keV.

It has been our experience that one frequently can use discrete-level information to predict unknown values of D_{ob}. For example, we have recently begun cross-section calculations for a suite of bismuth isotopes with target masses ranging from 203 to 212 and are using preliminary discrete-level sets provided by Meyer[MEY85]. These schemes presently consist of 30 to 40 levels each. In a number of cases, the parity distributions are poor, with most levels having the same parity; Meyer is now carrying out further evaluations for these sets. Since D_{ob} is only known experimentally for ^{210}Bi, we used Meyer's preliminary level schemes for the various bismuth isotopes to adjust our level densities and thereby obtain initial estimates of all the D_{ob} values. These are summarized in Table 4. As one sees, the values decrease for the lighter-mass isotopes, spanning more than two orders of magnitude, as one moves away from the closed neutron shell. For the compound nucleus ^{210}Bi, the calculated value of 5.21 keV is in reasonable agreement

Table 2. **Values of k, the constant in the mean-square spin projection expression, that were extracted from the discrete-level sets for isotopes of yttrium and strontium.**

Isotope	k	Isotope	k
^{90}Y	0.169	—	—
^{89}Y	0.146	^{89}Sr	0.146
^{88}Y	0.292	^{88}Sr	0.146
^{87}Y	0.171	^{87}Sr	0.146
^{86}Y	0.240	^{86}Sr	0.240

Table 3. **The two methods of deducing the energy-dependent spin cut-off parameter, $\sigma^2(E) = (6/\pi^2)(kA^{2/3})(a[E - \delta])^{1/2}$, from discrete-level sets.**

(1) Use of the maximum likelihood estimator expression:

$$\sigma^2(E) = \frac{1}{2N_f(E)} \sum_{i=1}^{N_f(E)} (J_i + 1/2)^2$$

(2) Use of the spin distribution of the levels up to energy E:

$$P(J) = N_f(E)\left(\frac{2J + 1}{2\sigma^2(E)}\right)\exp\left[-\frac{(J + 1/2)^2}{2\sigma^2(E)}\right]$$

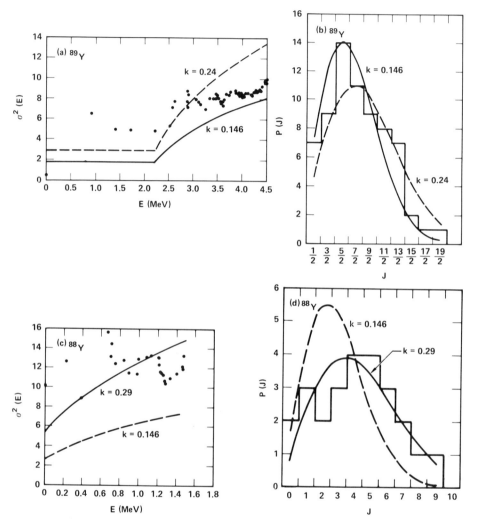

Figure 4.   (a,b) The ^{89}Y spin cut-off parameter, σ^2, vs E, and the spin distribution, $P(J)$, obtained from 60 discrete levels and from calculations with k values of 0.146 and 0.24. At energies below the pairing energy, the value of σ^2 is arbitrarily held constant and equal to $(6/\pi^2)$ $(kA^{2/3})$.   (c,d) The ^{88}Y spin cut-off parameter and the spin distribution obtained from 25 discrete levels and from calculations with k values of 0.146 and 0.29.

with the experimental value of 4.5 ± 0.6 keV[MUG84]. Figure 5 shows plots of preliminary calculations of the neutron capture excitation functions for the ^{204}Bi–^{210}Bi targets. At an incident energy of 10 keV, we see that the cross section can vary by a factor of about 40. Besides the use of the discrete-level information to improve our level densities and to predict D_{ob} values, the position of the first unbound resonance in each neutron-capture excitation function can be estimated roughly by dividing the D_{ob} value by 2. The result should correspond to the lowest incident energy at which we can expect the statistical-model calculation, averaging through the separated and overlapping resonance regions, to be valid.

Further comments on the size and extent of this bismuth cross-section calculational effort should be made. Table 5 summarizes the number of target states that we will need to consider (the 21 states include both ground and isomeric states), the number of separate reaction excitation functions that must be included, as well as estimates of the number of computer runs and the CDC 7600 CPU computer time that will be required. In our calculations, we typically use energy-bin sizes of 10 to 250 keV, depending on the reaction type and the energy ranges of concern.

Table 4. Initial estimates of values of D_{ob}, the average s-wave neutron resonance spacing, for isotopes of bismuth.

Compound nucleus	D_{ob} (MeV)
^{204}Bi	3.07×10^{-5}
^{205}Bi	1.40×10^{-5}
^{206}Bi	1.16×10^{-4}
^{207}Bi	1.33×10^{-4}
^{208}Bi	7.21×10^{-4}
^{209}Bi	3.42×10^{-3}
^{210}Bi	5.21×10^{-3}
^{211}Bi	1.26×10^{-3}

Table 5. Summary of target states, type and number of reactions, and estimated computer time required to produce the complete set of calculated neutron cross sections for a suite of bismuth isotopes for incident energies up to 20 MeV.

Target	(n,γ)	(n,n')	(n,2n)	(n,3n)	(n,4n)
^{203}Bi	X		X		
^{204}Bi	X	X	X		
^{205}Bi	X	X	X	X	
^{206}Bi	X	X	X	X	
^{207}Bi	X	X	X	X	
^{208}Bi	X	X	X	X	
^{209}Bi	X	X	X	X	
^{210}Bi	X	X	X	X	X
^{211}Bi	X		X	X	X
^{212}Bi			X	X	X
Number of target states			21		
Number of reactions					
(n,γ)			45		
(n,n')			32		
(n,2n)			45		
(n,3n)			39		
(n,4n)			13		
Total number of reactions			174		
Number of computer runs			80		
Estimated CDC 7600 time needed			800 min		

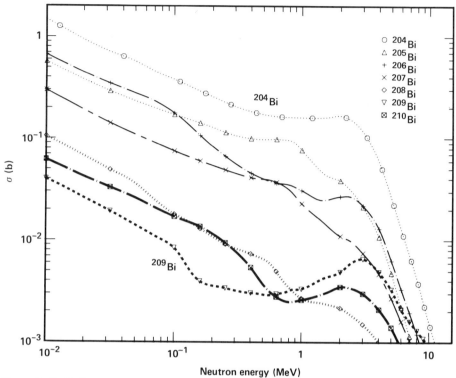

Figure 5. Preliminary calculations of the neutron capture excitation functions for the ^{204}Bi–^{210}Bi targets.

Discrete Levels and Isomer Ratios

So far we have discussed the importance of discrete-level descriptions of nuclei at low excitations and the way in which such information can be used to optimize level densities. If valid information on the gamma-ray branchings among the discrete levels is available, one can compute reliable isomer ratios. In the mass-90 region, for example, we have been making cross-section calculations for neutron reactions on the unstable isotopes ^{87}Y and ^{88}Y with the inclusion of a number of isomeric states. As mentioned earlier, in these calculations we used the discrete-level sets and gamma-ray branchings provided by Henry[HEN77,HEN85]. We were able to assess the reliability of this input, as well as the other calculational parameters employed in the computations for incident neutrons, by making companion calculations of proton-induced reaction cross sections on strontium isotopes and comparing the results with recent measurements[BAR83]. Figure 6 shows an example of the good agreement we obtained between the calculations and the measurements for the (p,xn) cross sections on ^{88}Sr. The calculations are shown as bands, the bounds corresponding to the limiting cases of the isospin being completely damped or conserved. No use of these proton data was made to modify any of the calculational parameters. No ad hoc parameter adjustments were allowed. The observed agreement of the cross-section results for the production of the ground and isomeric states of ^{87}Y validates both the gamma-ray branching information among the discrete states and the level-density parameters deduced from the levels. We found that our calculations for all of the reactions on the targets ^{87}Sr and ^{88}Sr were in quite satisfactory agreement with the data.

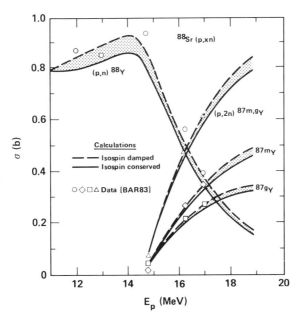

Figure 6.   Calculated excitation functions for the (p,xn) reactions on ^{88}Sr, compared with measurements from [BAR83].

For the deformed odd-odd nuclei in the lanthanide and actinide regions, we have been using the modeled level structures of Hoff et al.[HOF85a]. Typically, this method provides several hundred (sometimes up to 1000) discrete levels in the first 1.5 MeV of nuclear excitation, as well as gamma-ray branching ratios. These large numbers of discrete levels at low energies are expected because of rotational enhancement effects in these mass regions. Since the energy dependence of the rotational enhancement is still an open question[HAN83,KAT78], however, it is not yet clear to us just how much level-density parameter information can be extracted from these discrete levels. But we have made a number of isomer-ratio calculations with these sets for (n,γ) and (n,2n) reactions on ^{175}Lu, ^{237}Np, ^{241}Am, and ^{243}Am. In each case, we have found that the hundreds of discrete levels and the gamma-ray branchings provided by the modeling are necessary to achieve agreement with experiment, that many rotational bands must be included in order to obtain a sufficiently representative selection of K quantum numbers, and that the levels of each band must be extended to appropriately high values of angular momentum.

We made all of the isomer-ratio calculations with our version of the STAPRE code, with no fission competition. Gamma-ray cascades leading to the ground- and isomeric-state products were modeled using our E1 and M1 strength-function systematics among the continuum bins[GAR84b]. The discrete levels were depopulated according to Hoff's method, where the gamma-ray transitions are assumed to take place within a rotational band, with no interband crossing except from the band heads. These de-excite by E1 and M1 transitions that have an E^3 energy dependence and are constrained by the selection rule $\delta K = 0, \pm 1$. M1 transitions are intrinsically faster than E1 transitions by a factor of 6. For the ^{175}Lu(n,γ) and ^{237}Np(n,2n) reactions, when we tried using the same E1 and M1 strength functions among the discrete levels as we do in the continuum, we obtained calculated isomer ratios equal to those obtained using the constant E1/M1 ratio of 0.167.

Table 6 summarizes the characteristics of the modeled and experimental level sets for the nuclei that we have studied so far; the modeled sets provided by Hoff are designated as Set A. Note, as an example, that the available experimental level information for ^{236}Np consists of only 5 levels (including 3 rotational bands) up to 0.22 MeV in nuclear excitation, whereas the modeled Set A consists of 998 levels in the first 1.48 MeV and includes 94 rotational bands. Sets B through D were obtained by truncating the 998-level set just below selected band heads. In Fig. 7(a), curve A shows our calculated g/m ratio for ^{236}Np production

Table 6. Comparison of modeled and experimental level sets.

Level-set designation		Energy range (MeV)	Number of levels	Number of rotational bands	Range of quantum Nos.	
					K	I
^{174}Lu	A	0–1.50	433	82	0–9	0–12
	Expt	0–0.97	57	10	0–7	0–7
^{176}Lu	A	0–1.50	291	62	0–9	0–12
	Expt	0–0.99	38	7	0–7	0–10
^{236}Np	A	0–1.48	998	94	0–6	0–20
	B	0–0.91	453	71	0–6	0–15
	C	0–0.36	50	15	0–6	0–11
	D	0–0.20	14	5	1–6	0–9
	Expt	0–0.22	5	3	1–6	1–6
^{242}Am	A	0–1.50	788	107	0–7	0–13
	Expt	0–1.10	30	9	0–5	0–7
^{244}Am	A	0–1.40	769	101	0–8	0–15
	Expt	0–0.78	44	15	0–6	0–6

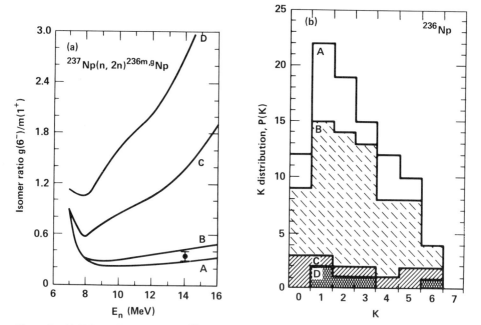

Figure 7.   (a) Values of the calculated ^{237}Np(n,2n) isomer ratio, g/m, obtained using ^{236}Np level sets A–D. The experimental g/m data point is from [MYE75].   (b) Values of the distribution of the rotational quantum number P(K) for ^{236}Np level sets A–D.

via the ^{237}Np(n,2n) reaction from threshold up to 16 MeV using the ^{236}Np Set A. The short-lived, 22.5-h, ^{236}Np (assigned to be ^{236m}Np by the level modeling) beta decays 48% to ^{236}Pu, which is of concern in reactor shielding. Curve B was computed with the level set of 453 levels obtained by truncating Set A just below the 72nd band head. Both sets yield an isomer ratio in agreement with the single measured value at 14 MeV[MYE75]. Figure 7(b) gives the distributions of *K* quantum numbers, *P(K)*, for the level sets associated with the curves. Curves C and D were computed with Sets C and D, which do not have sufficiently representative samplings of *K* values and which therefore lead to isomer ratios that are too high.

Role of Discrete Levels in Deriving Absolute Dipole Strength Functions

Accurate calculations of neutron or proton capture cross sections, gamma-ray production spectra, etc., require that the magnitude and energy dependence of the gamma-ray transmission coefficients be well characterized. We have found that the parameterization of the gamma-ray strength function is a reliable way to predict gamma-ray transmission coefficients[GAR84b]. Recently, we were able to derive absolute dipole strength function information for ^{176}Lu using Hoff's set of 291 modeled discrete levels shown in Table 6 and two types of experimental data: an average resonance capture (ARC) study of ^{175}Lu using 2-keV neutrons[HOF85b], and neutron capture cross-section measurements[MAC78,BEE80]. First, as seen in Fig. 8, we calculated a ^{175}Lu(n,γ) excitation function that agreed well with the measurements, using the total radiation width calculated from our initial dipole systematics, along with other calculational input[GAR85b]. Then we calculated the relative primary gamma-ray intensities to all of the 291 discrete levels in ^{176}Lu following the capture of 2-keV neutrons and compared the results with the ARC measurements. Figure 9 shows the final results. The bands are the calculated values of the reduced dipole transition intensities to the modeled set of ^{176}Lu levels, together with a one standard deviation error limit

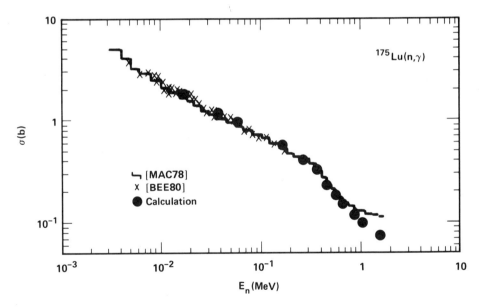

Figure 8. ^{175}Lu(n,γ) excitation function, calculated using our original dipole strength function systematics, compared with experimental data from [MAC78,BEE80].

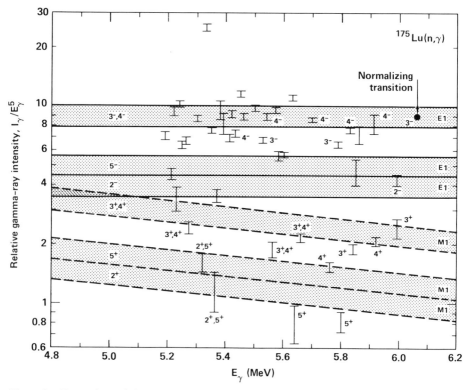

Figure 9.   Comparison of the calculated and measured [HOF85b] reduced transition intensities for primary gamma rays following 2-keV neutron capture by ^{175}Lu.

obtained from the estimated number of radiating states. The bars indicate experimental values and their error limits, with some showing spin and parity assignments. All measurements and calculations are normalized to the single indicated transition.

Originally, the comparison showed that our calculated M1/E1 ratio was low by a factor of 3. In order to preserve the total radiation width value that had given agreement with experiment, and also match the M1/E1 ratio from the ARC measurements, we had to increase our M1 strength function by a factor of 3 and decrease our E1 strength in the region around 5 to 6 MeV. Figure 10 compares the newly derived strength functions (the solid curves) with those predicted by our original systematics (the dashed curves). We revised the E1 strength function downward around 5 MeV, while still maintaining continuity with the photonuclear region, by changing the value of the one free parameter in our systematics, E_x, from 5 to 11 MeV, as indicated by the arrows in the figure. This parameter, E_x, is the energy at which our energy-dependent Breit-Wigner line shape[GAR84b] has the same value as that for the Lorentzian line shape (the dotted curve), both defined with the same giant dipole resonance parameters.

As seen in Fig. 11, when we applied the new E1 systematics in the mass-90 region, we obtained excellent agreement with experimental measurements for ^{89}Y[AXE70] and ^{90}Zr[SZE79]. Our original systematics had predicted E1 strengths that were about 30% higher. We are continuing this work in other mass regions to verify and further improve our systematics.

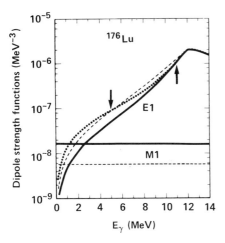

Figure 10. Recently derived absolute dipole strength functions for ^{176}Lu (solid curves), compared with those predicted by our original systematics (dashed curves). The arrows indicate the value assigned to the one free parameter, E_x. In our recent derivation, $E_x = 11$ MeV, and in our original systematics, $E_x = 5$ MeV.

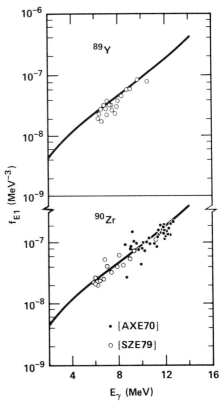

Figure 11. E1 strength functions for ^{89}Y and ^{90}Zr, calculated using our revised systematics, compared with measurements from [AXE70, SZE79].

Use of Discrete Levels in Analysis of Primary Gamma-Ray Spectra

Another interesting utilization of Hoff's modeled discrete levels for ^{176}Lu is presented in Fig. 12, where we show calculated primary dipole transitions to the ^{176}Lu modeled levels with energies in the range of 0.40 to 0.82 MeV following 2-keV neutron capture by ^{175}Lu. Also shown are the positions of the accessible levels and their spins and parities. The spectrum was constructed by smearing the calculated transition intensities with a unit Gaussian line shape with a 6-keV resolution. Note these features of the plot: (a) for gamma rays from 5.5 to 5.9 MeV we see 5 doublets, 1 triplet, and 1 quadruplet, all unresolved, more than expected on statistical grounds[HOF85b]; (b) the 2^- + 5^- doublet just below 0.44 MeV excitation appears to have the same shape and intensity as the singlet 4^- peak at 0.466 MeV, while the triplet, quadruplet, and doublet peaks at 0.66, 0.75, and 0.77 MeV, respectively, also show a close resemblance in shape and intensity; (c) the true singlet E1 peaks, due to transitions to 4^- levels at 0.82, 0.60, and 0.47 MeV, show a pronounced increase in intensity because of the energy dependence of the E1 strength function itself, in addition to the expected E_γ^3 energy dependence. This kind of plot should be quite useful in interpreting the experimental data, particularly where unresolved multiple peaks occur.

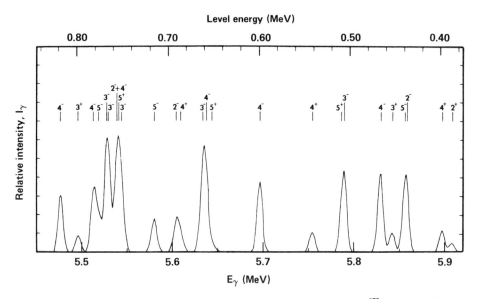

Figure 12. A portion of the calculated primary gamma-ray spectrum for the ^{175}Lu(n,γ) reaction.

Conclusion

It is evident from the above examples that the use of complete sets of discrete levels is an essential key in making accurate statistical-model Hauser-Feshbach calculations. Since these level sets must include both the observed and modeled information for completeness, additional personnel and expertise are required to produce them. It should also be pointed out that the use of large discrete-level sets in calculations usually requires significant increases in computer time.

We see that many problems still need to be solved in order to obtain accurate results in Hauser-Feshbach calculations. Some examples are the energy dependence of rotational enhancement of levels in deformed nuclei, the energy and mass dependence of M1 gamma-ray transitions, the importance of E2 transitions, and better estimates of fission barriers. Work in each of these areas will benefit greatly from a better understanding of the discrete levels, particularly in nuclei away from stability.

Acknowledgments

The authors are grateful to M. N. Namboodiri of the Lawrence Livermore National Laboratory for presenting this paper at the Symposium. This work was performed under the auspices of the U.S. Department of Energy by Lawrence Livermore National Laboratory under contract No. W-7405-ENG-48.

References

[AXE70] P. Axel et al., *Phys. Rev. C* **2**, 689 (1970).

[BAR83] D. W. Barr, S. A. Beatty, M. M. Fowler, J. S. Gilmore, R. J. Prestwood, E. N. Treher, and J. B. Wilhelmy, "(p,xn) Measurements on Strontium Isotopes," *Nuclear Chemistry Division Annual Report FY 82*, Los Alamos National Laboratory, Los Alamos, NM, LA-9797-PR (1983).

[BAY75] B. P. Bayhurst et al., *Phys. Rev. C* **12**, 451 (1975).

[BEE80] H. Beer, F. Kappeler, and K. Wisshak, "The Neutron Capture Cross Sections of Natural Yb, ^{170}Yb, ^{175}Lu, and ^{184}W in the Energy Range from 5 to 200 keV for the ^{176}Lu-Chronometer," *Proc. Int. Conf. Nuclear Cross Sections for Technology, Knoxville, TN, 1979*, National Bureau of Standards, Gaithersburg, MD, NBS Special Publication 594 (1980), p. 340.

[BOL77] J. W. Boldeman et al., *Nucl. Sci. Eng.* **64**, 744 (1977).

[COO67] J. L. Cook, H. Ferguson, and A. R. Musgrove, *Nuclear Level Densities in Intermediate and Heavy Nuclei*, Australian Atomic Energy Commission, AAEC/TM-392 (1967).

[GAR84a] M. A. Gardner, "^{169}Tm(n,3n)^{167}Tm Cross-Section Calculations Near Threshold," *Nuclear Chemistry Division FY 84 Annual Report*, Lawrence Livermore National Laboratory, Livermore, CA, UCAR-10062-84/1 (1984).

[GAR84b] D. G. Gardner, "Methods for Calculating Neutron Capture Cross Sections and Gamma-Ray Energy Spectra," *Neutron Physics and Nuclear Data in Science and Technology, Volume 3, Neutron Radiative Capture* (Pergamon Press, New York, 1984), Chapter III.

[GAR85a] M. A. Gardner, *Calculational Tools for the Evaluation of Nuclear Cross-Section and Spectra Data*, Lawrence Livermore National Laboratory, Livermore, CA, UCRL-91947 (1985); to be published in *Proc. Int. Conf. Nuclear Data for Basic and Applied Science, Santa Fe, NM, May 13–17, 1985*.

[GAR85b] D. G. Gardner, M. A. Gardner, and R. W. Hoff, "Absolute Dipole Gamma-Ray Strength Functions for ^{176}Lu," *Proc. Conf. Capture Gamma-Ray Spectroscopy and Related Topics, Knoxville, TN*, American Institute of Physics, New York, NY, AIP Conf. Proc. #125 (1985), p. 513.

[GIL65] A. Gilbert and A. G. W. Cameron, *Can. J. Phys.* **43**, 1446 (1965).

[HAN83] G. Hansen and A. S. Jensen, "Energy Dependence of the Rotational Enhancement Factor in the Level Density," *IAEA Advisory Group Meeting on Basic and Applied Problems of Nuclear Level Densities*, Brookhaven National Laboratory, Upton, NY, BNL-NCS-51694 (1983), p. 161.

[HEN77] E. A. Henry, Lawrence Livermore National Laboratory, Livermore, CA, private communication (1977).

[HEN85] E. A. Henry, Lawrence Livermore National Laboratory, Livermore, CA, private communication (1985).

[HOF84] R. W. Hoff and W. R. Willis, "Level-Structure Modeling of Odd-Mass Deformed Nuclei," *Nuclear Chemistry Division FY 84 Annual Report*, Lawrence Livermore National Laboratory, Livermore, CA, UCAR 10062-84/1 (1984).

[HOF85a] R. W. Hoff, J. Kern, R. Piepenbring, and J. B. Boisson, "Modeling Level Structures of Odd-Odd Deformed Nuclei," *Proc. Conf. on Capture Gamma-Ray Spectroscopy and Related Topics, Knoxville, TN*, American Institute of Physics, New York, NY, AIP Conf. Proc. #125 (1985), p. 274.

[HOF85b] R. W. Hoff et al., *Nucl. Phys.* **A437**, 285 (1985).

[KAT78] S. Kataria et al., *Phys. Rev. C* **18**, 549 (1978).

[MAC78] R. L. Macklin, D. M. Drake, and J. J. Malanify, *Fast Neutron Capture Cross Sections of ^{169}Tm, ^{191}Ir, ^{193}Ir, and ^{175}Lu for 3 < E_{11} < 2000 keV*, Los Alamos National Laboratory, Los Alamos, NM, LA-7479-MS (1978).

[MEY85] R. A. Meyer, Lawrence Livermore National Laboratory, Livermore, CA, private communication (1985).

[MUG84] S. F. Mughabghab, *Neutron Cross Sections*, Vol. 1, Part B (Academic Press, New York, 1984).

[MYE75] W. Myers et al., *J. Inorg. Nucl. Chem.* **37**, 637 (1975).

[NET72] D. R. Nethaway, *Nucl. Phys.* **A190**, 635 (1972); also, unpublished (n,3n) cross-section measurements, Lawrence Livermore National Laboratory, Livermore, CA (1972).

[NET76] D. R. Nethaway, unpublished (n,2n) cross-section measurements, Lawrence Livermore National Laboratory, Livermore, CA (1976).

[REF80] G. Reffo, "Phenomenological Approach to Nuclear Level Densities," *Theory and Applications of Moment Methods in Many-Fermion Systems* (Plenum Press, New York, 1980).

[SZE79]   G. Szeflinska et al., *Nucl. Phys.* **A323**, 253 (1979).

[UHL70]   M. Uhl, *Acta Phys. Austriaca* **31**, 245 (1970).

[VEE77]   L. R. Veeser et al., *Phys. Rev. C* **16**, 1792 (1977).

[VOI83]   J. Voignier, S. Joly, and G. Grenier, "Mesure De La Section Efficace De Capture Radiative Du Lanthane, Du Bismuth, Du Cuivre Naturel et De Ses Isotopes Pour Des Neutrons D'Energie Comprise Entre 0,5 et 3 MeV," *Proc. Int. Conf. Nuclear Data for Science and Technology, Antwerp, Belgium* (1983), p. 759.

RECEIVED May 8, 1986

Distribution of Electromagnetic Transition Amplitudes

P. J. Brussaard [1] and J. J. M. Verbaarschot [2]

[1] Fysisch Laboratorium, Rijksuniversiteit te Utrecht, P.O. Box 80.000, 3508 TA Utrecht, the Netherlands
[2] Max-Planck-Institut für Kernphysik, Postfach 103980, D-6900 Heidelberg, Federal Republic of Germany

The present-day shell-model computer programmes of nuclear physics can handle very large model spaces, and often in practice the upper bounds on dimensionalities are set by the capacity of the computer. However, still larger configuration spaces can be treated successfully by the methods of statistical spectroscopy to obtain average properties of many-particle systems. These methods are based on the use of distribution functions with parameters that can be evaluated without the explicit calculation and diagonalization of a high-dimensional hamiltonian matrix. The basic philosophy behind this method is that the average properties depend on the particular model at hand, whereas the fluctuations about the average are universal. The latter remark means that the fluctuations can be described by the gaussian orthogonal ensemble. Of course, the results are less detailed than those one would obtain, if possible, from the explicit diagonalization of the energy matrix and the use of the resulting many-dimensional wave functions.

The most familiar example is the energy spectrum of a nuclear hamiltonian that can be described by a gaussian distribution. It must be remarked that for a system of many, say N, noninteracting particles the gaussian distribution of the eigenvalues is a direct consequence of the central limit theorem for large values of N.

For interacting particles, to the contrary, not every hamiltonian will lead to a gaussian energy spectrum. For example, the square of a traceless one-body operator as a hamiltonian will produce a χ^2-distribution. Proper ensemble averaging of the energy density, however, will then generate the gaussian distribution of eigenvalues, as it is also obtained by many large-scale shell-model calculations that have been performed with a wide variety of realistic interactions. In a proper ensemble the influence of the unwanted hamiltonians is essentially eliminated.

Let the k-body hamiltonian be given by

$$H = \sum_{\alpha\beta} H_{\alpha\beta} \, a_\alpha^\dagger \, a_\beta \tag{1}$$

where $a_\alpha^\dagger |0\rangle = |\alpha\rangle$ represents the k-particle state $_\blacktriangle|\alpha\rangle$. In the corresponding gaussian orthogonal (i.e., showing orthogonal invariance) ensemble the distinct matrix elements $H_{\alpha\beta}$ are independently distributed as gaussian variables according to the relations

$$\langle H_{\alpha\beta}\rangle = 0 \text{ and } \langle(H_{\alpha\beta})^2\rangle = (1+\delta_{\alpha\beta})v^2 \tag{2}$$

where the acute brackets denote ensemble averaging [BRO81; POR65].

When these matrices are applied in an N-particle configuration space (N>>k) the moments $\langle H^p\rangle$ of the hamiltonian behave on ensemble averaging in the limit of large dimensionality as in the case of noninteracting particles without averaging. This reflects the dominance of binary correlations in the operator products H^p when the ensemble averaging is performed in the 'dilute' system (k << N).

For a space of eigenvectors of matrices of the gaussian orthogonal ensemble (k = N) the distribution of values of matrix elements of electromagnetic transition operators is gaussian, as follows from the central limit theorem. The ensemble averaging of hamiltonians guarantees that no correlations exist between the hamiltonian structure and the particular transition operator that is considered.

However, we are concerned with a many-body system that is described by a particular one- and two-body interaction. For each eigenvalue E_i the many-particle eigenfunction ψ_i, obeying the Schrödinger equation

$$H\psi_i = E_i\psi_i \tag{3}$$

can be expanded in a set of basis states ϕ_α as

$$\psi_i = \sum_\alpha c_{i\alpha}\,\phi_\alpha \tag{4}$$

or inversely

$$\phi_\alpha = \sum_i c_{i\alpha}^*\psi_i \tag{5}$$

where the basis state ϕ_α is expanded in terms of the eigenstates ψ_i.

The values of $|c_{i\alpha}|^2$ for a given state ψ_i turn out to show a secular variation as a function of the diagonal matrix elements $\langle\phi_\alpha|H|\phi_\alpha\rangle$ [VER79]. The secular behaviour itself depends on the energy E_i which reflects a correlation between the hamiltonian and the transition operator. Thus one should expect the distribution of the electromagnetic-transition amplitudes to be gaussian only for a small energy range of the energies W and W' of the initial and final many-particle states, respectively. The square of the absolute values of the transition amplitudes for the single-particle operator 0, $|\langle W'||0||W\rangle|^2$, between spaces of fixed angular momentum, isospin, isospin projection and parity shows a secular variation as a function of W and W' [DRA77]. The secular behaviour is obtained after local averaging over a small interval around (W,W'). The absolute values of the transition amplitudes are taken to show fluctuations about the secular value according to the Porter-Thomas assumption [POR56].

This means that the amplitudes themselves, i.e. the reduced matrix elements $x = \langle W'||0||W\rangle$ are locally distributed as a gaussian, $G(x;0,\sigma^2)$, with zero mean and variance given by the local average, i.e.

$$P(\langle W'||0||W\rangle = x) \, dx = G(x;0,\sigma^2)dx \tag{6}$$

with

$$\sigma^2(W',W) = \overline{|\langle W'||0||W\rangle|^2} \tag{7}$$

Here the horizontal bar denotes local averaging over a small interval around (W,W').

The secular behaviour of $|\langle W'||0||W\rangle|^2$, i.e. $\overline{|\langle W'||0||W\rangle|^2}$, can be evaluated if one knows for given initial (or final) state the strength distribution and the density of final (or initial) states.

The moments of the strength distribution for given initial state $|W\rangle$, i.e. $\rho_W'(W')$, can be rewritten as

$$\sum_{W'} W'^{p} \overline{|\langle W'||0||W\rangle|^2} = \frac{1}{d''} \sum_{W''\approx W} \langle 0W''||H^p||0W''\rangle \tag{8}$$

where d'' denotes the number of states in a small interval around W (in the right-hand member of eq. (8) it is assumed that proper angular-momentum coupling is taken care of). It is seen that the moments (8) are averages over the spectral distribution of states $0|W''\rangle$ and thus are expected to define gaussian distribution functions [FRE71]

$$\rho_W'(W') = \frac{n_W'}{\sigma_W'\sqrt{2\pi}} \exp\left[-\frac{(W'-E_W')^2}{2\sigma_W'^2}\right] \tag{9}$$

with the parameters being the moments for $p = 0,1,2$ defined in eq. (8).

Similarly the many-body eigenvalue distribution for a two-body interaction is very close to gaussian [FRE71; MON75]. The distribution of final states $|W'\rangle$ is given by

$$\rho_0'(W') = \frac{d'}{\sigma_0'\sqrt{2\pi}} \exp\left[-\frac{(W'-E_0')^2}{2\sigma_0'^2}\right] \tag{10}$$

where the parameters are derived from the moments (p = 0,1,2)

$$\int_{-\infty}^{+\infty} W'^P \rho_0'(W')dW' = \frac{1}{d'} \sum_{\alpha} \langle \phi_\alpha | H^P | \phi_\alpha \rangle = \frac{1}{d'} Tr(H^P) \tag{11}$$

The secular behaviour of $\sigma^2(W',W)$, defined in eq. (7), can now be expressed in terms of the distributions (9) and (10) as

$$\sigma^2(W',W) = \overline{|\langle W'||0||W\rangle|^2} = \frac{\rho_W'(W')}{\rho_0'(W')} \tag{12}$$

In order to obtain the frequency function of the amplitudes $|\langle W'||0||W\rangle|$ for a fixed value W one now has to integrate over W' the product of the secular behaviour of $|\langle W'||0||W\rangle|$ and the Porter-Thomas distribution. The distribution function of the amplitudes $|\langle W'||0||W\rangle|$ for arbitrary values of initial and final energies then follows after a second integration [VER79]

$$f(x = |\langle W'||0||W\rangle|) =$$

$$2 \int_{-\infty}^{\infty} dW \rho_0(W) \int_{-\infty}^{\infty} dW' \rho_0'(W') \frac{1}{\sigma(W',W)\sqrt{2\pi}} \exp\left[-\frac{x^2}{2\sigma^2(W',W)}\right] \tag{14}$$

In fig. 1a-c the distribution function (14) is compared for isoscalar M1-, E2- and E4-electromagnetic transition amplitudes with shell-model results for ^{20}Ne(T=0,T_z=0), calculated in the sd-configuration space. A surface delta interaction was used with strength parameters [BRU77] $A_0=A_1=$ 1 MeV. The single-particle energies with respect to the ^{16}O-core are $\varepsilon(0d5/2) = -4.15$ MeV, $\varepsilon(1s1/2) = -3.28$ MeV and $\varepsilon(0d3/2) = 0.93$ MeV. The curves with the shorter tail represent eq. (14). The curves with the longer tail result when for fixed initial (final) state correlations between final (initial) states are taken into account.

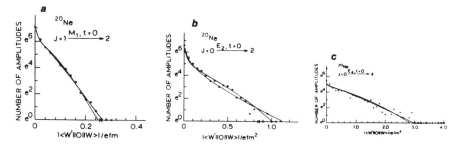

Figure 1a-c. Reproduced with permission from VER85. Copyright 1985 Springer-Verlag.

In fig. 2a-f the results are given for isoscalar and isovector transition amplitudes for ^{21}Ne($T=\frac{1}{2}$, $T_z=\frac{1}{2}$). (The values of the isovector amplitudes have been multiplied by a factor $\sqrt{3}$, compared to the conventional definition [BRU77].)

For further details the reader is referred to [VER85].

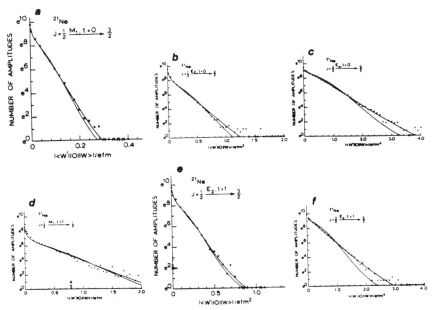

Figure 2a-f. Reproduced with permission from VER85. Copyright 1985 Springer-Verlag.

References

BRO81   T.A. Brody, J. Flores, J.B. French, P.A. Mello, A. Pandey and S.S.M. Wong, Revs. Mod. Phys. <u>53</u> 385 (1981)

BRU77   P.J. Brussaard and P.W.M. Glaudemans, 'Shell-model applications in nuclear spectroscopy', North-Holland Publishing Company, Amsterdam 1977.

DRA77   J.P. Draayer, J.B. French and S.S.M. Wong, Ann. Phys. (N.Y.) <u>106</u> 472, 503 (1977)

FRE71   J.B. French and S.S.M. Wong, Phys. Lett. <u>35B</u> 5 (1971)

MON75   K.K. Mon and J.B. French, Ann. Phys. (N.Y.) <u>95</u> 90 (1975)

POR56   C.E. Porter and R.G. Thomas, Phys. Rev. <u>104</u> 483 (1956)

POR65   C.E. Porter and N. Rosenzweig in 'Statistical theories of spectra: fluctuations' (C.E. Porter, ed.), Academic Press, New York 1965, p.235

VER79   J.J.M. Verbaarschot and P.J. Brussaard, Phys. Lett. <u>87B</u> 155 (1979)

VER85   J.J.M. Verbaarschot and P.J. Brussaard, Z. Phys. <u>A321</u> 125 (1985)

RECEIVED June 13, 1986

17

$N_p \cdot N_n$ Systematics and Their Implications

R. F. Casten[1]

Institut für Kernphysik, University of Köln, Federal Republic of Germany

A substantial simplification of the systematics in nuclear phase transition regions is obtained if the data
are plotted against the product, $N_p \cdot N_n$, of the number of valence protons and neutrons instead of against
N, Z, or A as is usually done. Such a scheme leads to a
unified view of nuclear transition regions and to a
simplified scheme for collective model calculations.

It is generally considered that the residual proton–neutron interaction is responsible for the onset of deformation [TAL62, FED77, CAS81] in
nuclei. Since the total p–n strength should be simply related [CAS85,
HAM65] to the number of interacting pairs of valence protons and neutrons,
that is, to the product $N_p \cdot N_n$ (counted as valence holes past midshell),
one might expect nuclear systematics to be much simpler if plotted against
such a parameter. This is indeed the case, as seen for the A=130 region in
Fig. 1 (top). To exploit this approach when significant shell or subshell
closures are present requires careful counting of valence particles. Thus,
for example, near A=150, the Z=64 gap is assumed to be active and thus the
proton shell is Z=50-64 for N<90 but Z=50-82 thereafter. Likewise, near
A=100 a realistic [FED77] proton shell is Z=38-50 for N<60 and Z=28-50 for
N≥60. $N_p \cdot N_n$ plots, subject to these definitions, are given in Figs.
1-2. They reveal that, again, a remarkable simplification results. Only
the N=90 points in Fig. 1 deviate from a smooth curve: this simply reflects
the fact that the Z=64 gap is still partly intact for N=90. Indeed, one can
exploit the otherwise smooth systematics by shifting these N=90 points to
this smooth curve and extracting effective valence proton numbers that
reflect the evolving proton subshell structure.

$N_p \cdot N_n$ plots are useful in another way as well. Fig. 3 compares plots for **different** regions and reveals their nearly identical

[1]Permanent address: Brookhaven National Laboratory, Upton, NY 11973

0097-6156/86/0324-0120$06.00/0

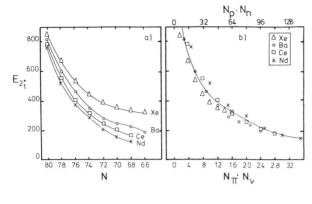

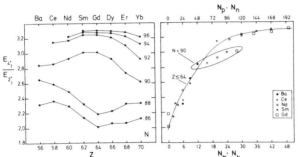

Fig. 1.
$N_p N_n$ and con-
ventional plots
for the A=130 and
150 regions. For
the latter region
only the nuclei
with Z≤64 are
shown in the
$N_p N_n$ plot.
The Z≥66 nuclei
lie on a sepa-
rate, parallel
curve (see Fig.
3).

structure. In fact, three regions in Fig. 5 (A=100, A=150, Z<64, and A=150, Z>66) have virtually the same slopes and the other two are only slightly different. The principal difference in the curves lies in their horizontal location. These findings are at first sight surprising since these regions were long thought to be widely different. However, simple estimates [CAS85, SOR84] of the p-n interaction in the highly overlapping orbits whose filling is crucial to the onset of deformation show that it is indeed roughly equal in the three regions of steeper slope in Fig. 3 and somewhat less for the other two regions. The horizontal shifts in Fig. 3 are also easily under- stood. In each region, a certain number of more or less inert orbits must be filled before the crucial, highly overlapping ones begin to fill. The horizontal position of a phase transition in an $N_p \cdot N_n$ plot reflects the number and degeneracy of such inert orbits.

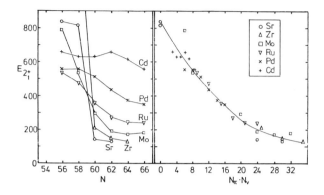

Fig. 2.
Examples of the simplification brought about in the $N_p N_n$ scheme in the A=100 region.

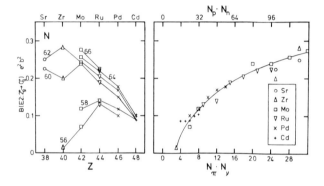

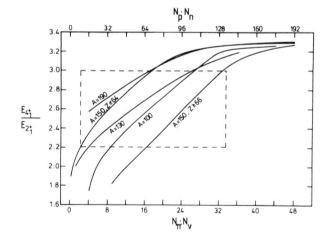

Fig. 3.
Comparison of smooth curves through the $E_4^+{}_1/E_2^+{}_1$ plots for five transition regions from A=100-200.

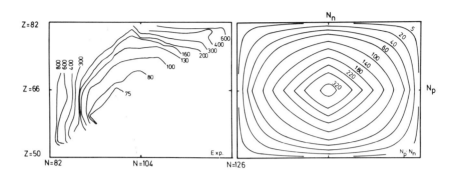

Fig. 4.   Curves in the N-Z plane representing isodeformation contours:
left, empirical contours of $E_{2^+_1}$ energies; right, contours of constant
$N_p \cdot N_n$.

The fact that different $N_p \cdot N_n$ plots are so similar has two
immediate consequences besides a unified interpretation of diverse regions.
One is that such a plot gives greater confidence in extrapolation far off
stability.   This can be of use either in estimating the structure of such
nuclei or in choosing particularly crucial nuclides for study.   This point
is highlighted in Fig. 4 which shows isodeformation contours [CAS85] for the
A=150 region.   The theoretical curves are those of constant $N_p \cdot N_n$ which
define nuclei with the same structure. In an N-Z plot, then, such isodeformation curves will be hyperbolas which will have nearly vertical or horizontal contours at the edges of major shell regions.   This behavior is in
contrast to many typical collective model calculations (e.g., RAG74) which
lead to roughly circular contours. It is remarkable that the data display
just the features expected in the $N_p \cdot N_n$ scheme.   However, it is also
clear that this could be tested much more decisively by the addition of even
the first 2^+ energy of a few neutron rich isotopes of Xe-Sm near A=150.
        The other consequence is that $N_p \cdot N_n$ systematics offer a way to
greatly simplify collective model calculations. Normally, such calculations
are parameterized for each nucleus individually or for a set of isotopes.
The $N_p \cdot N_n$ curves suggest that an entire region can be treated as a unit
in which the collective parameters are taken as smooth functions of
$N_p \cdot N_n$ only.   Moreover, Fig. 3 suggests that the same set of parameters

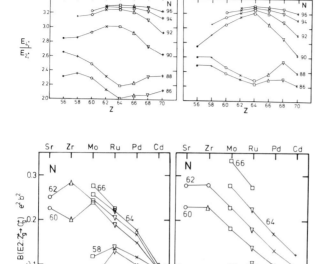

Fig. 5
Examples of empirical (left) and calculated (right) results for the A=150 and 100 regions.

could even be used for _several_ regions of similar slope. This idea can be illustrated with an IBA calculation of ≈100 nuclei in the three vibrator → rotor regions of similar slope in Fig. 3 using the IBA-1 Hamiltonian

$$H = \varepsilon n_d - \kappa Q \cdot Q \qquad : \qquad \varepsilon = \varepsilon_0 \cdot e^{-\theta(N_\pi N_\nu - N_0)} = \varepsilon_0 e^{\theta N_0} e^{-\theta N_\pi N_\nu} \qquad (1)$$

where N_π, N_ν are boson numbers and where θ is related to the slope and N_0 to the horizontal displacement in an $N_p \cdot N_n$ plot. In Eq. (1), a vibrator to rotor transition is obtained by allowing ε to decrease as a function of $N_\pi \cdot N_\nu$. Eq. (1) contains five parameters: ε_0, θ, κ, χ (an internal parameter in Q), and N_0. To achieve the utmost economy of parameters ε, κ, θ, and χ are held constant for all 100 nuclei, and N_0 has a separate value for each of the three regions. This gives 7 parameters for ≈100 nuclei or, in fact, 6 parameters if one uses the second form for ε in Eq. (1) where ε_0

and N_0 are combined. Examples of the results [CAS85] are shown in Fig. 5. Given the highly constrained parameterization, the predictions are in remarkable agreement with the data. The overall trajectories of each transition region are closely reproduced and so are many of the detailed systematics. The reason for this agreement lies less in the specific features of the IBA as in the exploitation of the $N_p \cdot N_n$ scheme: therefore, one expects that similar simplifications might apply to other collective models.

Summary

1) The $N_p \cdot N_n$ scheme provides a major simplification of nuclear systematics, particularly in transition regions.

2) It provides information on evolving shell structure.

3) $N_p \cdot N_n$ plots show a remarkable similarity in transition regions previously thought to behave quite differently. $N_p \cdot N_n$ is an appropriate unit to measure the rapidity of different phase transitions. Moreover, this rapidity is primarily a function of the p-n interaction in certain crucial highly overlapping orbits.

4) $N_p \cdot N_n$ systematics are very useful for extrapolation far off stability and can be exploited to simplify collective model calculations.

Acknowledgments

I would like to acknowledge my collaborators on much of this work, P. von Brentano, A. Gelberg, W. Frank, and H. Harter, as well as I. Talmi, S. Pittel, K. Heyde, B. Barrett, O. Scholten, P. Van Isacker, P. Lipas, and A. Aprahamian for many useful comments. Work supported by contract DE-AC02-76CH00016 with the USDOE, the BMFT, and the Von Humboldt Foundation.

References

[CAS81] R.F. Casten et al., Phys. Rev. Lett. <u>47</u> 1433 (1981).

[CAS85] R.F. Casten, Phys. Lett. <u>152B</u> 145 (1985), Phys. Rev. Lett. <u>54</u> 1991 (1985), Nucl. Phys. <u>A443</u> 1 (1985), R.F. Casten et al., Nucl. Phys. (in press), and R. F. Casten et al., to be published.

[FED77] P. Federman and S. Pittel, Phys. Lett. <u>69B</u> 385 (1977), and Phys. Lett. <u>77B</u> 29 (1978).

[HAM65] I. Hamamoto, Nucl. Phys. <u>73</u> 225 (1965).

[RAG74] Ragnarsson et al., Nucl. Phys. <u>A233</u> 329 (1974).

[SOR84] R.A. Sorensen, Nucl. Phys. <u>A420</u> 221 (1984).

[TAL62] I. Talmi, Rev. Mod. Phys. <u>34</u> 704 (1962).

RECEIVED August 25, 1986

18

Nuclear Masses Far from Stability
Interplay of Theory and Experiment

P. E. Haustein

Chemistry Department, Brookhaven National Laboratory, Upton, NY 11973

Mass models seek, by a variety of theoretical
approaches, to reproduce the measured mass surface and
to predict unmeasured masses beyond it. Subsequent
measurements of these predicted nuclear masses permit
an assessment of the quality of the mass predictions
from the various models. Since the last comprehensive
revision of the mass predictions (in the mid-to-late
1970's) over 300 new masses have been reported. Global
analyses of these data have been performed by several
numerical and graphical methods. These have identified
both the strengths and weaknesses of the models. In
some cases failures in individual models are distinctly
apparent when the new mass data are plotted as
functions of one or more selected physical parameters.
Several examples will be given. Future theoretical
efforts will also be discussed.

I. Introduction

A continuing effort among experimentalists who study nuclei far from
beta stability is the measurement of the atomic mass surface. As a
manifestation of the nuclear force and the nuclear many body system, atomic
masses signal important features of nuclear structure on both a macroscopic
and microscopic scale. It has thus been a challenge to nuclear theorists
to devise models which can reproduce the measured mass surface and to
predict successfully the masses of new isotopes. Both the measured mass
surface and that beyond it which can be predicted by these models serve as
important input to a variety of fundamental and applied problems, e.g.,
nucleosynthesis calculations, predictions of decay modes of exotic nuclei
far from stability, nuclear de-excitation by particle evaporation, decay
heat simulations, etc.

Well determined masses of nuclei which lie far from beta stability can
provide very sensitive tests of atomic mass models. While a single new
mass measurement from one previously uncharacterized isotope carries with
it only limited information about the quality of mass predictions from the
models, important trends frequently become evident across isotopic
sequences or when global comparisons of many new masses are made against
the various mass models. It is in this context that a comprehensive and
critical assessment of the predictive properties of atomic mass models is
presented with the aim of identifying both the successes and failures in
the models. A summary of a portion of this effort has been published
earlier [HAU84].

0097-6156/86/0324-0126$06.00/0

II. New Masses, Analysis Methods, and Global Comparisons

The last comprehensive update of the atomic mass predictions from nine
different models was published in 1976 [MAR76]. Additional predictions
from other models appeared in the late 1970's and early 1980's [MON78],
[MOL81], [UNO82]. In each case, one of the atomic mass evaluations
periodically provided by Wapstra [MAR76], [WAP77], [WAP84], served as the
experimentally determined mass data base on which the adjustable parameters
of the models were determined. Since the 1975 Wapstra evaluation (which
was used in the formulation of many of the models published in 1976) over
300 new mass measurements have been made. An examination of where these
new measurements occur in the Chart of Nuclides reveals that they are
distributed among almost all the elements at their most neutron-rich or
neutron-deficient isotopes. Especially long isotopic sequences of new
masses occur in the Na, Rb, and Cs nuclei and in alpha decay chains which
originate from ^{176}Hg and ^{178}Hg.

It is quite instructive to compare these new measurements (which lie
outside the data bases available at the time the various mass models were
formulated) with predictions from the models. For such comparisons it is
convenient to define Δ = Predicted Mass − Measured Mass. $\Delta > 0$ thus
denotes cases where the binding energy has been predicted to be too low and
conversely, $\Delta < 0$ corresponds to a prediction of too much nuclear binding.
Table 1 summarizes average and root−mean−square deviations for twelve
models.

Table 1. Average and Root-Mean-Square Deviations (all energies in keV)

Model[+]	Data Base Used	Old Masses[++] $\langle \Delta \rangle$	RMS-Δ	New Masses	New Masses $\langle \Delta \rangle$	RMS-Δ	RMS Ratio
M	1971	209	1327	270	−551	1566	1.18
GHT	1975	20	718	276	−478	1096	1.53
SH	1971	−6	718	257	−195	954	1.33
MN	1977	−4	835	213	279	970	1.16
B	1971	−459	1506	121	−768	1772	1.18
BLM	1971	1984	2747	146	1991	3125	1.14
LZ	1975	7	276	268	87	589	2.13
UY	1975	0	393	219	110	1100	2.80
CK	1975	5	312	258	186	1314	4.21
JGK	1975	6	212	271	219	1361	6.42
MS	1975	−7	159	267	−6	695	4.37
JE	1975	0	363	239	24	952	2.62

+ M = Myers, GHT = Groote et al., SH = Seeger & Howard, MN = Möller & Nix,
   B = Bauer, BLM = Beiner et al., LZ = Liran & Zeldes, UY = Uno & Yamada
   (linear shells), CK = Comay & Kelson, JGK = Jänecke, Garvey-Kelson,
   MS = Monahan & Serduke, JE = Jänecke & Eynon.
++ Relative to the 1975 Wapstra masses.

Several significant trends are apparent. $\langle \Delta \rangle$ values for the 1975
masses are quite small, typically a few kilovolts except for those models

which used the 1971 Wapstra masses. RMS-Δ values range from 159 to
2747 keV. This spread is a reflection of the degree of conformation of the
calculated mass surface to the measured one afforded, in those models with
smaller RMS-Δ values, by the use of increasingly larger numbers of
adjustable parameters. When global comparisons are made for all the models
to the new masses (up to 276 nuclei) reported since 1975-77 one notes that
approximately half of the models display net positive <Δ> values and the
remainder net negative values. RMS-Δ values for the new masses reveal that
all models show poorer fits to these masses than to the 1975 or '77 data
base. The last column lists the ratio of the rms-deviations of the new
masses to the old ones. The more "fundamental" models (e.g., liquid drop,
droplet, simple shell) have larger RMS-Δ deviations when compared to the
old masses than the models based on mass relations or complicated shell
corrections. However the comparison also reveals that these simpler
approaches show substantially smaller enlargement of rms deviations; models
based on mass relations (CK, JGK, and MS) exhibit larger RMS ratios, up to
factors of 6.4, which result from progressively poorer predictions for
nuclei especially far from stability.

III. Analysis of Selected Individual Models

A. Seeger and Howard: Figure 1 displays Δ values for new masses as a
function of neutron number for this model (semiempirical liquid drop plus
shell corrections). The solid lines pass through sets of points where Δ

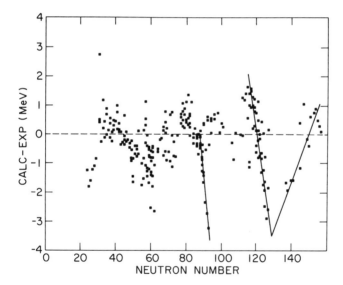

Fig. 1   Delta values of new masses as a function of neutron number for the
         model for Seeger and Howard.

values are linear with neutron number N. The more obvious of these is the
rapidly falling trend which begins near N = 114 and continues to N = 126,
followed by a reversal that extends to N = 150. Another trend starts for
points with N = 82 and extends for approximately ten neutron numbers. The
correlation of these effects with neutron shell closures at N = 82 and 126
may be understood by examination of the treatment of the microscopic shell
corrections in the Seeger and Howard model [SEE75]. Without invoking an
ad hoc enlargement of the N = 126 shell gap by 3 MeV (which tapered
smoothly and symmetrically to zero at N = 108 and N = 144) it was not
possible to obtain simultaneously the proper single particle level ordering
and a good fit to the known (1971) mass surface. The N = 82 gap is
affected to second order by this prescription. While an optimized fit to
the 1971 masses was obtained in this way, the trend in the predictions of
masses of nearly all new isostopes (since 1971) with N = 114 to 150 and
some new masses for isotopes with N = 82 to 92 is clear evidence that the
procedure described above will not work satisfactorily for these nuclei
which lie further from stability.

B. Jänecke, Garvey-Kelson: The value of 6.42 in the RMS Ratio column in
Table 1 for this model is a result of a small number of very poorly
predicted masses for nuclei quite far from stability. Closer to stability
this model (and similar ones, e.g., CK and MS) provide excellent
predictions with Δ values usually within ±1.5 MeV. A useful way to
illustrate this feature in these models is by plotting Δ values as a
function of how far each isotope is from the valley of beta stability. For
this purpose, the quantity $N - Z - (0.4A^2)/(200 + A)$ gives the difference
in neutron numbers of the isotope of interest and that of the isotope
nearest the stability line of the same element. Positive values of this
quantity correspond to neutron-rich nuclei, negative values to proton-rich
nuclei, and values near zero represent nuclei close to the stability line.
Figure 2 displays Δ values versus number of neutrons from stability for
this model. Many points cluster about the dashed horizontal Δ = 0 line but
there is a clear trend that shows that proton-rich nuclei are not bound
enough and neutron-rich nuclei are too well bound. Use of this model (and
the others of similar type) for calculations of r-process nucleosynthesis
will therefore introduce a strong bias that results from the prediction of
the location of the neutron drip line too close to stability. As shown
here the trend is approxiimately proportional to T_z^3 and is a reflection
of need for correction terms [JAN84] in the transverse Garvey-Kelson mass
relationship on which these models are based.

C. Jänecke and Eynon: This model, which involves the solution of
inhomogeneous third order partial difference equations, represents one
approach that aims to correct the deficiencies of type noted above in
models that employ (homogeneous) mass relations. In particular, the
introduction of an inhomogeneous term is meant to account for variations in
the effective neutron-proton residual interaction as a function of nucleon
number and neutron excess. It is therefore instructive to compare the
quality of mass predictions from this model to those which derive their
mass predictions from solutions of homogeneous partial difference
equations. Figure 3 shows a plot of Δ values versus neutrons from
stability for the Jänecke and Eynon model. Two features are immediately
apparent: (1) the largest Δ values are considerably reduced -- note the

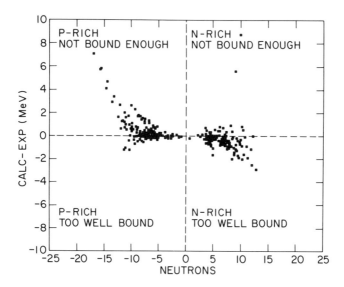

Fig. 2   Delta values for new masses as a function of neutrons from
         stability for the model of Jänecke, Garvey–Kelson.

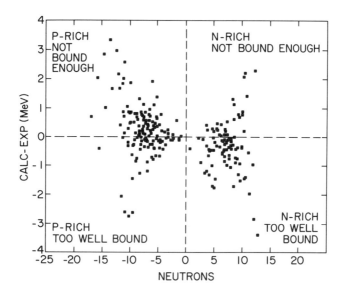

Fig. 3   Delta values for new masses as a function of neutrons from
         stability for the model of Jänecke and Eynon.

change of a factor of 2.5 in the vertical scale relative to Figure 2; and
(2) points scatter more uniformly into the four quadrants of the plot,
suggesting that the isospin dependence has been more successfully treated.
One expects, therefore, that this model would be more suitable for use as
input for nucleosynthesis calculations or in other applications requiring
more reliable predictions far off the stability line.

IV.   Summary and Future Directions

The analysis methods described here have highlighted some of the
systematic features in the predictive properties of several of the commonly
used atomic mass models. Additional understanding of these features and
the availability of many new atomic masses for isotopes far from the
stability line will serve as a basis for improving the models. The need
clearly exists for a comprehensive revision and update of the mass
predictions. A project, coordinated by the author, has been started to
accomplish this. It is expected that new sets of mass predictions from a
number of groups may be available late in 1986.

Acknowledgments

This research was carried out at Brookhaven National Laboratory under
contract DE-AC02-76CH00016 with the U. S. Department of Energy and
supported by its Office of High Energy and Nuclear Physics.

References

[HAU84] P. Haustein, Proceedings of the 7th International Conference on
         Atomic Masses and Fundamental Constants (AMCO-7), 3-7 September
         1984, Darmstadt-Seeheim, FRG, O. Klepper, ed., Technische
         Hochschule Darmstadt, 413 (1984).
[JAN84] J. Jänecke, Proceedings of the 7th International Conference on
         Atomic Masses and Fundamental Constants (AMCO-7), 3-7 September
         1984, Darmstadt-Seeheim, FRG, O. Klepper, ed., Technische
         Hochschule Darmstadt, 420 (1984).
[MAR76] S. Maripuu, At. Data Nuc. Data Tables 17 411 (1976).
[MOL81] P. Möller and J. R. Nix, At. Data Nucl. Data Tables 26 165 (1981).
[MON78] J. E. Monahan and F. J. D. Serduke, Phys. Rev. C17 1196 (1978).
[SEE75] P. A. Seeger and W. M. Howard, Nucl. Phys. A238 491 (1975).
[UNO82] M. Uno and M. Yamada, Report INS-NUMA-40, Waseda University, Tokyo,
         Japan (1982).
[WAP77] A. H. Wapstra and K. Bos, At. Data Nucl. Data Tables 19 175 (1977).
[WAP84] A. H. Wapstra and G. Audi, Nucl. Phys. A432 1 (1984).

RECEIVED May 8, 1986

Section I: Motivation for Further Exploration

Part 3: Astrophysics

Unit A: Astrophysical Models
Chapters 19-21
Unit B: Nuclear Data and Theory Related to Astrophysics
Chapters 22-26

19

Nuclear Astrophysics Away from Stability

G. J. Mathews, W. M. Howard, K. Takahashi, and R. A. Ward

Lawrence Livermore National Laboratory, University of California, Livermore, CA 94550

Explosive astrophysical environments invariably lead to the production of nuclei away from stability. An understanding of the dynamics and nucleosynthesis in such environments is inextricably coupled to an understanding of the properties of the synthesized nuclei. In this talk a review is presented of the basic explosive nucleosynthesis mechanisms (s-process, r-process, n-process, p-process, and rp-process). Specific stellar model calculations are discussed and a summary of the pertinent nuclear data is presented. Possible experiments and nuclear-model calculations are suggested that could facilitate a better understanding of the astrophysical scenarios.

Introduction

Even after several decades of research [BUR57] into the mechanisms by which the elements are synthesized in stars, it is still often true that the degree to which an astrophysical environment can be understood is limited by the degree to which the underlying microscopic input nuclear physics data have been measured and understood. As new and more exotic high-temperature astronomical environments have been discovered and modeled (and as observations and models for more familiar objects have been refined) the needs for more and better data for nuclei away from stability have increased. In this brief overview, we discuss a few of the explosive astrophysical environments which are currently of interest and some of their required input nuclear data.

We begin with a discussion of the poorly understood mechanisms for heavy-element nucleosynthesis and some of our efforts to understand these environments. Then we turn to a discussion of the exotic environments for hot hydrogen burning and some of our experimental and theoretical efforts to obtain the associated nuclear data.

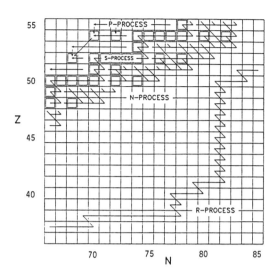

Fig. 1   The mechanisms for heavy-element nucleosynthesis drawn
as lines representing the dominant isotopes produced
during the processes [MAT85].

Heavy-Element Nucleosynthesis

Most nuclides with $A \gtrsim 70$ are synthesized by neutron
capture. Figure 1 is from a recent review article [MAT85] on the
various stellar processes for neutron capture nucleosynthesis. For the
most part, these processes correspond to different time scales. The
s-process describes neutron capture on a slow time scale compared with
typical beta-decay lifetimes near the line of stability, and thus leads
to the formation of a continuous chain of stable heavy elements from
the iron group to ^{209}Bi. The r-process, on the other hand,
corresponds to neutron capture on a time scale which is rapid compared
with beta-decay lifetimes. In the limit of high neutron density, this
process appears as a chain of isotopes for each element which
represents the point of $(n,\gamma) \underset{\leftarrow}{\overset{\rightarrow}{}} (\gamma,n)$ equilibrium far from
stability. The n-process is what actually seems to be predicted by
most stellar models [MAT85] for either the r-process or s-process, i.e.
a competition between beta decay and neutron capture which can not be
treated with the same mathematical simplicity as in the classical

s-process and r-process [MAT85, SEE65]. The p-process is a somewhat
less frequent process (as evidenced in abundances) which probably is
the result of photodisintegration reactions in high-temperature [WOO78]
or high-electron-density [HAR78] regimes.

The s-Process

Figure 2 is from some of our recent studies [MAT84a, MAT84b,
HOW85] of the details of s-process nucleosynthesis in the framework of
stellar models which produce neutrons by a $^{22}Ne(\alpha,n)^{25}Mg$ reaction
during sequential thermal pulses in a helium-burning shell [IBE77].
What is plotted is the σN curve (neutron capture cross section times
abundance) for stable isotopes along the chain of elements which are
produced in the s-process. This calculation corresponds roughly to the
conditions thought to exist in the interior of a 7 $M_\odot$ red-giant star
on the asymptotic giant branch. Stars in this phase of evolution are
unstable to a thermonuclear runaway of the helium-burning shell. In
the explosive environment of one of these thermal pulses the neutron
densities can become so high that successive neutron captures can occur
out to nuclei well away from stability. Furthermore, the temperatures
become so high (T $\sim$ 3×10^8K) that the Boltzmann population of
nuclear excited states can lead to dramatic changes of the
neutron-capture and beta-decay rates.

In a classical s-process [MAT85] (low neutron density) Fig. 2
would be a smooth curve. However, as variations in the neutron
density and temperature are taken into account, a structure emerges due
to branch points, i.e., unstable nuclei with competing neutron-capture
and beta-decay rates.

Thus, in order to understand such environments it is necessary to
calculate complete network of the competitions between neutron capture
and beta decay as well as their corrections for the thermal population
of excited states. With regard to this latter correction it is
particularly important to know the low-energy level structure of nuclei
away from stability. This structure will affect the beta decay
properties differently from the neutron capture properties. In a
separate contribution to this conference, [TAK85] we will discuss the
corrections for beta decay. Basically this becomes important if a
low-lying excited state can undergo a Gamow-Teller allowed decay. The

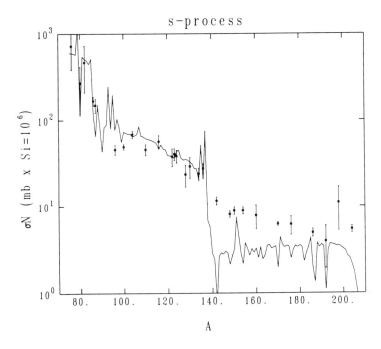

Figure 2   The σN curve calculated [MAT84ab, HOW85] in the
          framework of stellar models [IBE77] for the s-process in
          thermally pulsing red-giant stars.

corrections to the cross sections become the greatest for low-lying
states with considerably different spins.

   Figure 3 is an illustration of some calculated [HOM76] thermal
correction factors for ground-state neutron-capture cross sections for
a number of isotopes near the line of stability. From this figure it
is clear that these correction factors can be significant. For the
benefit of anyone who might like to attack this problem. Table I
summarizes some of what we consider to be the most important quantities
to better refine as input to the s-process.

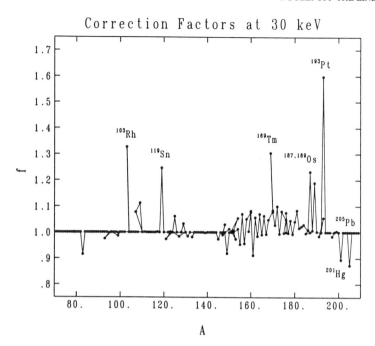

Figure 3   Correction factors, f, for the neutron-capture cross
           section for various isotopes due to the thermal
           population of excited states at stellar temperatures
           (from [HOW76]). Isotopes with f > 10% are labeled.

TABLE I   Some nuclear data which would be most useful for
s-process nucleosynthesis

1. Stable nuclei for which the neutron capture cross section has
not been measured; ^{66}Zn, ^{72,73}Ge, ^{77}Se, ^{99}Ru.

2. Important unstable nuclei for which the neutron capture cross
section has not been measured; ^{79}Se, ^{85}Kr,
^{107}Pd, ^{147}Pm, ^{151}Sm, ^{166}Ho, ^{186}Re, ^{192}Ir, ^{205}Pb.

3. Nuclei with large thermal correction factors for the
ground-state neutron capture cross sections; ^{103}Rh,
^{119}Sn, ^{169}Tm, ^{187,189}Os, ^{193}Pt, ^{201}Hg, ^{205}Pb.

4. Branch points with large corrections for
thermally-enhanced beta decay; ^{60}Co, ^{63}Ni, ^{79}Se, ^{93}Zr,
^{99}Tc, ^{107}Pd, ^{134}Cs, ^{151}Sm, ^{152,154,155}Eu, ^{160}Tb, ^{163}Dy,
^{163}Ho, ^{176}Lu, ^{181,182}Hf, ^{187}Re, ^{187}Os, ^{204,205}Tl.

The r-process

The required nuclear input data for the r-process are summarized in detail elsewhere [MAT83a, MAT85]. The astrophysical site for the r-process is still not known, although a number of promising possibilities have been proposed [MAT85, SEE65, BLA81, THI79, COW82]. Basically there are two fundamental time scales which must be determined from better nuclear input data before the ambiguity surrounding the r-process site can be resolved. One is just how high the neutron density must be to reproduce the r-process abundances. This quantity depends on the ratios of neutron-capture to beta-decay rates. For some scenarios [BLA81, COW82, CAM83] it is possible to reproduce the observed r-process abundances without reaching (n,γ) equilibrium. Then the abundance peaks will largely be determined by the neutron capture cross sections away from stability as demonstrated in [CAM83].

The cross sections away from stability have for the most part been estimated from global Hauser-Feshbach calculations [HOM76] although near neutron closed shells direct radiative capture may be more appropriate [MAT83b]. Along this line we are currently investigating the viability of a new statistical formalism [VER84] based on a random matrix approach to describe the statistical fluctuations. This formalism has the advantage that it goes to the correct limit when the number of channels becomes small. In any of these calculations it becomes extremely important to know the level structure for low-lying states (which may have significant gamma channels) and states near 1-3 MeV excitation which may have resonance contributions for nuclei away from stability. Some progress in this latter regard has been made [KRA83, WIE84] by utilizing data from delayed neutron emission to identify some of the states populated in the inverse neutron-capture reaction.

If (n,γ) equilibrium is achieved, then the abundances will be determined by the beta-decay rates of unmeasured nuclei far from stability. Although some progress has been made [KLA81, TAK85] in shell model calculations of these rates, further studies are warranted.

The second quantity which must be determined for the r-process is the dynamical time scale during which the neutron density must remain high in order to produce the actinides (which can not be produced by

the s-process due to alpha particle decay at ^{210}Bi). This quantity
depends on the sum of the neutron-capture and beta-decay lifetimes as
one moves away from stability. Essentially, the r-process must live
long enough for nuclei to capture out to far from stability and then
beta decay up to the mass numbers of the actinides. This places a
severe constraint on dynamical processes such as shock-driven explosive
helium burning during a supernova [BLA81, KLA81]. At present, it
appears that the calculated beta-decay rates may be too slow [COW85].
Clearly, more refined determinations of neutron-capture cross sections
and beta-decay rates are desired before the astrophysical site for the
r-process can be determined.

p-Process

The p-process is responsible for the production of a number of
neutron deficient nuclei. The astrophysical site for the p-process is
not well established, but the most viable models at the present time
[WOO78, HAR78] attribute this process in one way or another to
photodisintegration reactions of heavier more abundant species. In
[WOO78] the photodisintegration is thought to occur in the high
temperature regions after passage of a supernova shock. In [HAR78] it
has been suggested that that energetic photons ($\sim$19MeV) from the
^{3}H(p,γ)^{4}He reaction may induce photonuclear reactions before the
photons are thermalized. In both of these cases it is desirable to
know photonuclear (γ,n), (γ,p), and (γ,α) rates neutron
deficient nuclei.

Hot Hydrogen Burning

For a host of astrophysical environments (e.g. novae, supernovae,
supermassive stars, accreting neutron stars, and dense inhomogeneous
cosmologies [WAL81]) hydrogen burning may occur at temperatures far in
excess of the temperatures (10^{6}<T<10^{8}K) associated with normal
main-sequence stellar evolution (see Fig. 4). In such environments,
charged-particle reaction data for unstable nuclei become important.

Almost any time there is thermonuclear hydrogen burning, there is
a possibility for proton reactions on unstable nuclei. Well known
examples are the ^{7}Be(p,γ)^{8}B reaction in the sun (which is a
particularly important link in the chain of reactions leading to the

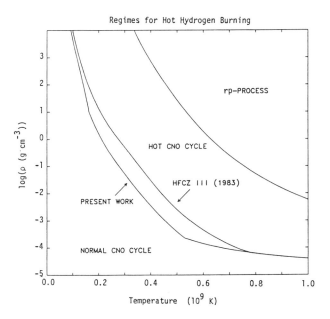

Figure 4   Regions of the density vs. temperature plane in which
the various hydrogen-burning processes are dominant
[MAT84c].   The normal CNO cycle occurs in stars slightly
larger than the sun.   The hot (beta-limited) CNO cycle
is particularly important in supermassive stars.   The
rp-process is important during the thermonuclear
runaways on accreting neutron stars which may be the
source of X-ray bursts.

production of solar neutrinos detectable in the ^{37}Cl experiment
[BAH78]) the ^{22}Na$(p,\gamma)^{23}$Mg reaction in the Ne-Na cycle, and the
reactions of ^{26}Al in the Mg-Al cycle.   When the temperatures are
high, a few other reaction rates also become important.   For
$T \gtrsim 2\times10^8$K the waiting point for the normal hydrogen-burning
CNO cycle shifts [MAT84c] from ^{14}N to ^{13}N, and then, via the
^{13}N$(p,\gamma)^{14}$O reaction, shifts to the production of ^{14}O and
^{15}O.   This is the hot (beta-limited) CNO cycle [WAL81], which is
particularly significant in the evolution of supermassive (M >
10^4M$_\odot$) stars [FUL85].   This hot hydrogen-burning scenario may also

come into play on accreting white dwarfs [WAL81] and in the formation
of x-ray bursts from accreting neutron stars [AYA82, WOO83].

As the temperature and density continue to increase, the
$^{15}O(\alpha,\gamma)^{19}Ne$ and $^{15}O(\alpha,p)^{18}F$ reactions lead to break out
from the CNO cycle to a process of rapid proton capture (rp-process)
which involves sequential proton captures out to the proton drip line
or until the Coulomb barrier becomes too large. Each of these
transitions to higher-temperature reactions lead to orders-of-magnitude
increases in the rates of energy production. Thus, in addition to
effects on nucleosynthesis, the dynamics of the various high
temperature environments are intimately coupled to the cross sections
for proton and alpha-particle capture reactions on unstable nuclei. In
a few cases [WAL81] even the question of whether the next proton or
aphha capture leads to a bound nuclear state can have a dramatic effect
on the evolution of the environment.

Essentially two different approaches have been attempted to supply
the necessary data. The most straightforward approach (which we will
call the finesse approach) is to utilize nuclear data obtained by
conventional means combined with a model for the nuclear structure and
reaction mechanisms to derive the desired input datum. This has been
done for the $^{13}N(p,\gamma)^{14}O$ reaction [MAT84c, CHU85, LAN85], the
^{14}O, ^{15}O, and ^{19}Ne reactions [WIE85], and heavier nuclei
[SCH84]. This approach has been quite productive since many of the
reactions of interest are probably dominated by one or a few resonances
whose radiative and particle widths can be inferred indirectly. There
is still quite a bit that can be done with this approach, for example
to better identify the energies and widths of the resonances of
interest, particularly for nuclei far on the proton-rich side of
stability which tend to become the waiting points for this process.

The other approach to obtain these data (which we will call the
brute-force approach) is to produce a beam of radioactive heavy ions to
be focused onto a target of hydrogen or ^{4}He (or in a few cases to do
measurements on a radioactive target [FIL83]). The radioactive
ion-beam brute-force approach is much more difficult but may provide
more information. We have been involved in a modest effort [HAI83,
MAT84d] to develop this technology along with a number of other labs
[BOY80, NIT84, AUR85].

Acknowledgment

Work performed under the auspices of the U.S. Department of Energy by the Lawrence Livermore National Laboratory under contract number W-7405-ENG-48. Work supported in part by the Lawrence Livermore National Laboratory Institute for Geophysics and Planetary Physics.

References

[AUR85] J. D'Auria, (ISOL TRIUMF proposal; 1985).

[AYA82] S. Ayasli and P. C. Joss, Ap. J., 256, 637 (1982).

[BAH78] J. N. Bahcall, Rev. Mod. Phys., 50, 881 (1978).

[BLA81] J. B. Blake, S. E. Woosley, T. A. Weaver, and D. N. Schramm, Ap. J., 248, 315 (1981).

[BOY80] R. N. Boyd, in "Proc. Workshop on Radioactive Ion Beams and Small Cross Sections", R. N. Boyd (ed.) (Burr Oak, Ohio; 1981), Ohio State Univ. Preprint, p. 30.

[BUR57] G. R. Burbidge, E. M. Burbidge, W. A. Fowler, and F. Hoyle, Rev. Mod. Phys., 29, 547 (1957).

[CAM83] A. G. W. Cameron, J. J. Cowan, and J. W. Truran, Ap. Sp. Sci., 91, 235 (1983).

[CHU85] E. L. Chupp, et al. Phys. Rev. C21, 1031 (1985).

[COW82] J. J. Cowan, A. G. W. Cameron, and J. W. Truran, Ap. J., 252, 348 (1982).

[COW85] J. J. Cowan (Priv. Comm.).

[FIL83] B. W. Fillipone, et al., Phys. Rev. C28, 2222 (1983).

[FUL85] G. M. Fuller, S. E. Woosley, and T. A. Weaver, Ap. J. 1985 (in press).

[HAI83] R. C. Haight, G. J. Mathews, R. M. White, L. A. Aviles, and S. E. Woodard, Nucl. Inst. Meth., 212, 245 (1983).

[HAR78] T. G. Harrison, Ap. J., 36, 199 (1978).

[HOW85] W. M. Howard, G. J. Mathews, K. Takahashi, and R. A. Ward, (Submitted to Ap. J. 1985).

[HOM76] J. A. Holmes, S. E. Woosley, W. A. Fowler, and B. A. Zimmerman, Atomic and Nuclear Data Tables, 18, 305 (1976).

[IBE77] I. Iben, Ap. J., 217, 788 (1977).

[KLA81] H. V. Klapdor, T. Oda, J. Mentzinger, W. Hillebrandt, and F.-K. Thielemann, Z. Phys. A., 299, 213 (1981).

[KRA83]   K.-L. Kratz, W. Ziegert, W. Hillebrandt, and F.-K.
           Thielemann, Astron. Ap., 125, 381 (1983).
[LAN85]   K. H. Langanke and O. S. van Roosmalen, and W. A. Fowler,
           (Submitted Nucl. Phys. A).
[MAT83a]   G. J. Mathews, in "Proc. Specialists Meeting on Yields and
           Decay Data of Fission Product Nuclides", (Brookhaven National
           Laboratory, 1983).
[MAT83b]   G. J. Mathews, A. Mengoni, F.-K. Thielemann, and W. A.
           Fowler, Ap. J., 270, 740 (1983).
[MAT84a]   G. J. Mathews, W. M. Howard, K. Takahashi, and R. A. Ward, in
           "Neutron-Nucleus Collisions as a Probe of Nuclear Structure",
           J. Rapaport, R. W. Finlay, S. M. Grimes, and F. S. Dietrich
           (eds.), (Burr Oak, Ohio 1984) (American Institute of Physics,
           New York), p. 511.
[MAT84b]   G. J. Mathews, W. M. Howard, K. Takahashi, and R. A. Ward, in
           "Capture Gamma-Ray Spectroscopy and Related Topics-1984", S.
           Raman (ed.), (Knoxville, Tenn.) (American Institute of
           Physics, New York), p. 766.
[MAT84c]   G. J. Mathews and F. S. Dietrich, Ap. J., 287, 969 (1984).
[MAT84d]   G. J. Mathews, R. C. Haight, and R. W. Bauer, "Proceedings of
           the Workshop on the Prospects for Research with Radioactive
           Beams from Heavy-Ion Accelerators", J. M. Nitschke (ed.)
           (Washington, DC; 1984).
[MAT85]   G. J. Mathews and R. A. Ward, Rep. Prog. Phys. (in press)
           1985.
[NIT84]   J. M. Nitschke, "Proceedings of the Workshop on the Prospects
           for Research with Radioactive Beams from Heavy-Ion
           Accelerators", J. M. Nitschke (ed.) (Washington, DC; 1984).
[SCH84]   P. Schmalbrock, et al., in "Capture Gamma-Ray Spectroscopy
           and Related Topics-1984", S. Raman (ed.), (Knoxville, Tenn.)
           (American Institute of Physics, New York), p. 785.
[SEE65]   P. A. Seeger, W. A. Fowler, and D. D. Clayton, Ap. J. Suppl.,
           97, 121 (1965).
[TAK84]   K. Takahashi and K. Yokoi (to be published in Atomic and
           Nuclear Data Tables).
[TAK85]   K. Takahashi, B. S. Meyer, G. J. Mathews, W. M. Howard, S. D.
           Bloom, and P. M\oller, (this conference proceedings).

RECEIVED September 2, 1986

Shell-Model Calculations of β-Decay Rates for s- and r-Process Nucleosyntheses

K. Takahashi, G. J. Mathews, and S. D. Bloom

Lawrence Livermore National Laboratory, University of California, Livermore, CA 94550

Examples of large-basis shell-model calculations of Gamow-Teller β-decay properties of specific interest in the astrophysical s- and r- processes are presented. Numerical results are given for: i) the GT-matrix elements for the excited state decays of the unstable s-process nucleus ^{99}Tc; and ii) the GT-strength function for the neutron-rich nucleus ^{130}Cd, which lies on the r-process path. The results are discussed in conjunction with the astrophysics problems.

1. Introduction

For a given astrophysical scenario for the s- or r- process, the corresponding elemental and isotopic abundances can be reliably computed only if the relevant nuclear (and, in some cases, atomic) physics input data are available. In return, careful studies of required nuclear properties and comparisons of the calculated and observed abundances often give a hint as to the astrophysical conditions appropriate for the s- or r- process site (see [MAT85a] for a review).

Some of the most important input data in such analyses are often the nuclear β-decay rates. Unfortunately, evaluating β-decay transition rates under astrophysical circumstances is not always straightforward because of i) the thermal population of excited states at high temperatures, requiring a knowledge of unobserved β transitions from the excited states and ii) the ionization which necessitates an introduction of atomic physics into the calculations, especially in s-process studies. For example, the importance of bound-state β^- decay in certain nuclei has been demonstrated [TAK83]. Reliable predictions of unknown β transition rates in heavy nuclei can be extremely difficult even for rather well-known nuclei. In what follows, we present some examples of large-basis shell-model calculations for such unknown β decays.

2. ^{99}Tc problem

The discovery of Tc (most probably ^{99}Tc) in red-giant stars is one of the strongest pieces of evidence that the s-process is indeed occurring in stellar interiors. On the other hand, the thermal population [CAM59] of the low-lying $7/2^+$ (140 keV) and $5/2^+$ (181 keV) states in ^{99}Tc induces Gamow-Teller allowed transitions (Fig. 1) and thus reduces the effective half-life of ^{99}Tc at high temperature relative to the terrestrial value of 2.1×10^5 yr. At a typical s-process temperature of 3×10^8 K, the half-life could be as short as a few years [COS84], suggesting that ^{99}Tc might not survive the s-process environment. If most ^{99}Tc is expected to decay to ^{99}Ru, then this expectation is in contradiction with the observations of substantial ^{99}Tc at the surface of at least some red-giant stars. This seeming conflict is the "^{99}Tc problem". Toward a solution of the problem, we have calculated the ^{99}Tc stellar β-decay rate and applied this rate to an s-process model.

0097-6156/86/0324-0145$06.00/0

Using the Lanczos method ([WHI80], [HAU76]), we have performed a large-basis shell-model calculation of the log-ft values for the unknown Gamow-Teller transitions shown in Fig. 1 [TAK85a]. With the use of a realistic two-body effective interaction [KAL64] and a model space consisting of low-seniority excitations in the (1g,2d) shell, the low-lying positive-parity states in ^{99}Tc-^{99}Ru and in an analogous isotonic pair ^{97}Nb-^{97}Mo are reproduced reasonably well. The calculated log-ft values are shown in Fig. 1 in brackets. They have been normalized to a known GT decay from ^{97}Nb. These transition rates imply an effective ^{99}Tc half-life of ~ 20 yr at a temperature of 3×10^8 K. The fact that this half life is longer by at least a factor of five compared with the previous values from systematics (e.g. [COS84], [YOK85]) is encouraging, but this alone does not solve the "problem" without referring to a specific s-process model.

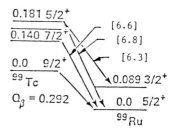

Fig. 1. ^{99}Tc β^- decays of astrophysical interest. Calculated log-ft values for the unknown GT decays are given in brackets. The energies are in MeV.

Probably the most promising astrophysical site for the s-process is the recurrent thermal-pulse and third dredge-up phase in the He-burning shell of intermediate mass stars [IBE77], where neutrons are produced by the ^{22}Ne$(\alpha,n)^{25}$Mg reaction. Within an analytic version of this model, it was shown in network calculations [MAT85b] that ^{99}Tc indeed survives the s-process: Because of the drastic increase of the neutron production rate for temperatures above ~ 3×10^8 K [FOW75], the net production of ^{99}Tc compensates its β-decay destruction.

Since that work, more quantitative s-process calculations have been done in conjunction with the detailed numerical thermal-pulse model calculations by Becker [BEK81], leading to the same conclusion [TAK85b]. It is worth noting that the observed Tc abundance relative to Zr, Nb, Mo and Ru [SMI83] can be well accounted for even if the ^{99}Tc β-decay half-life is as short as a few years at 3×10^8 K.

3. ^{130}Cd decay

The astrophysical site that was responsible for the bulk of the solar system r-process material is not yet known ([SCH83],[MAT85a]). Whatever the scenario is, however, the r-process path is expected to pass at least near the neutron-magic nucleus ^{130}Cd. Otherwise, the abundance peak near mass number A=130 may be difficult to explain. In a classical r-process [MAT85a], the relative heights of the abundance peaks near A=80, 130 and 194 are largely determined by the relatively slow β-decay rates of such neutron-magic (N=50, 82 and 126) nuclei with these mass numbers.

We have again utilized the Lanczos algorithm to compute the β-strength function for the ^{130}Cd(0^+) → ^{130}In(1^+) Gamow-Teller transitions. Low-seniority excitations are allowed within the (1g,2d,3s,1h) shell, which are mixed by the Kallio-Kolltveit [KAL64] two-body effective interaction. The single-particle energies are taken from a detailed analysis [STO85] of one-, two-, and three- quasi-particle nuclei near the Z=50, N=82 closed shells with the same two-body force.

The resultant β-strength functions are displayed in Fig. 2 as a sequence in the number of Lanczos iterations. As for the absolute values, we have assumed a quenching factor of 0.5 for the GT sum-rule. It can be understood that the half-life itself converges very quickly. For the predicted ground-state β-decay Q-value of 6.4 MeV ([LIR76], [JAN76], [COM76]), the calculated ^{130}Cd half-life is 0.08 sec as shown in Fig. 3 in comparison with the TDA calculations with a GT residual interaction [KLA84].

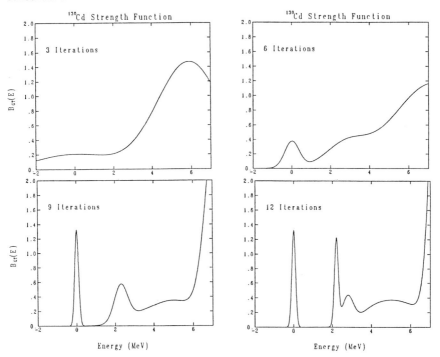

Fig. 2. Calculated GT-strength function for the ^{130}Cd ground-state (0^+) decays to the low-lying 1^+ (T=16) states in ^{130}In as a sequence of the number of Lanczos iterations. The abscissa is the excitation energy in ^{130}In. For convenience of drawing, the minimum width is chosen to be 100 keV.

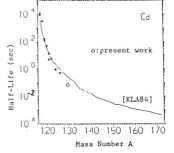

Fig. 3. Calculated ^{130}Cd half-life in comparison with the TDA predictions for Cd isotopes by Klapdor et al. [KLA84]. The squares represent known experimental data.

As a test of our prediction for ^{13}CCd, we have calculated the β-decay rates for the GT transitions from the lowest 1^+ state of ^{130}In to the ground (0^+) and first-excited (2^+) states of ^{130}Sn. The resultant ^{130}In total half-life is 0.26 sec for the assumed Q_β=10.2 MeV, which is comparable with the reported value (0.33 sec) for the low-spin $(1,3)^+$ state [RUD85].

More systematic calculations with the present method will certainly hel to clarify our understanding of the β-decay properties of spherical nuclei fa off the line of stability, which are needed in r-process studies. In particular, a study of the effects of the β-decay of low-lying states thermally populated in the high temperature r-process environment is due. Suc effects have not been included in any r-process model yet attempted. Finally we mention that a different approach (i.e. RPA) is probably called for in order to deal with deformed nuclei effectively [BRA85].

Acknowledgments

Work performed under auspices of the U.S. Department of Energy by the Lawrence Livermore National Laboratory under contract number W-7405-ENG-48, and supported in part by the LLNL Institute for Geophysics and Planetary Science.

References

[BEK81] S.A.Becker, in Physical Processes in Red Giants, eds. I.Iben, Jr. and A.Renzini (Reidel, 1981), p.141
[BRA85] B.S.Meyer, W.M.Howard, G.J.Mathews, P.Möller and K.Takahashi, this meeting
[CAM59] A.G.W.Cameron, Astrophys.J. 130 452 (1959)
[COM76] E.Comay and I.Kelson, Atm.Nucl.Dat.Tables 17 463 (1976)
[COS84] K.R.Cosner, K.H.Despain and J.W.Truran, Astrophys.J. 283 313 (1984)
[FOW75] W.A.Fowler, G.R.Caughlan and B.A.Zimmerman, Ann.Rev.Astron.Astrophys. 13 69 (1975)
[HAU76] R.F.Hausman, Jr., Ph.D. thesis, University of California Radiation Laboratory Rept. UCRL-52178 (1976)
[IBE77] I.Iben, Jr., Astrophys.J. 217 788 (1977)
[JAN76] J.Jänecke, Atm.Nucl.Dat.Tables 17 455 (1976)
[KAL64] A.Kallio and K.Kolltveit, Nucl.Phys. 53 87 (1964)
[KLA84] H.V.Klapdor, J.Metzinger and T.Oda, Atm.Nucl.Dat.Tables 31 81 (1984)
[LIR76] S.Liran and N.Zeldes, Atm.Nucl.Dat.Tables 17 431 (1976)
[MAT85a] G.J.Mathews and R.A.Ward, Rept.Prog.Phys., in press
[MAT85b] G.J.Mathews, K.Takahashi, R.A.Ward and W.M.Howard, Astrophys.J., in press
[RUD85] G.Rudstam, P.Aagaard and H.-U. Zwicky, The Studsvik Science Research Laboratory Rept. NFL-42 (1985)
[SCH83] D.N.Schramm, in Essays in Nuclear Astrophysics, eds. C.A.Barnes, D.D.Clayton and D.N.Schramm (Cambridge Univ., 1983) p.325
[SMI83] V.V.Smith and G.Wallerstein, Astrophys.J. 273 742 (1983)
[STO85] C.A.Stone, W.B.Walters, S.D.Bloom and G.J.Mathews, this meeting
[TAK83] K.Takahashi and K.Yokoi, Nucl.Phys. A404 578 (1983)
[TAK85a] K.Takahashi, G.J.Mathews and S.D.Bloom, Phys.Rev. C, submitted
[TAK85b] K.Takahashi, G.J.Mathews, R.A.Ward and S.A.Becker, in Proc.5th Moriond Meeting on Nucleosynthesis and its Implications to Nuclear and Particle Physics, Les Arcs, 1985 (Reidel) in press
[WHI80] R.R.Whitehead, in Moment Method in Many Fermion System, eds. B.J. Dalton, S.S.Grimes, J.D.Vary and S.A.Williams (Plenum, 1980), p.235
[YOK85] K.Yokoi and K.Takahashi, Kernforschungszentrum Karlsruhe Rept. KfK-3849 (1985)

RECEIVED May 16, 1986

β-Delayed Fission Calculations for the Astrophysical r-Process

B. S. Meyer, W. M. Howard, G. J. Mathews, P. Möller, and K. Takahashi

Lawrence Livermore National Laboratory, University of California, Livermore, CA 94550

We discuss RPA calculations of the Gamow-Teller properties of neutron-rich nuclei to study the effect of β-delayed fission and neutron emission on the production of Th, U and Pu chronometric nuclei in the astrophysical r-process. We find significant differences in the amount of β-delayed fission when compared with the recent calculations of Thielemann et al. (1983). In the simplest case of a constant abundance along the r-process path, however, the inferred production ratios in both calculations are similar.

The study of the decay of nuclei from the r-process path back to the β-stability line is an area of astrophysics which requires knowledge of β-strength functions off the line of beta stability. This knowledge is especially important for the determination of the abundances of the progenitors of the Th-U-Pu chronometers since β-delayed fission and neutron emission during decay back to the stability line may significantly affect the final abundance distribution of these nuclei ([BER69], [WEN75], [KOD75], [KRU81]). The β-strength function for nuclei along the decay back paths [coupled with neutron separation energies (S_n), fission barrier heights (B_f) and β-decay Q-values (Q_β)] determines the amount of β-delayed fission and neutron emission that occurs during the cascade back to the β-stability line.

A recent analysis by Thielemann et al. [THI83] of the effects of β-delayed processes on the progenitors of the Th-U-Pu chronometers showed that these processes (delayed fission in particular) did indeed significantly influence the final abundances of the chronometer progenitors. This leads to a long age for the Galaxy. In view of the importance of this result, it is useful to re-examine the calculation with a nuclear model that includes the effects of nuclear deformation on the β-decay rates, fission barriers, and neutron separation energies self-consistently.

Since many of the nuclei involved are presumably highly-deformed, we have used the Nilsson RPA code of Krumlinde and Möller [KRU84], which calculates β-strength functions, using an infinite-range residual (Gamow-Teller) interaction with a strength χ_{GT} = 23/A MeV. With this code, we have calculated β-strength functions for 118 nuclei lying in the mass range from A=232 to A=255 and from the line of β-stability to the r-process path given by Thielemann et al. [THI83] (see Fig.1). These 118 nuclei satisfied the criteria for possible β-delayed fission (Q_β of precursor $\gtrsim B_f$ of emitter) and/or possible β-delayed neutron emission (Q_β of precursor $\gtrsim S_n$ of emitter) as determined from the values given in [HOW80].

The required input values are the deformation parameters ε_2 and ε_4 and the numbers κ and μ which determine the size of the $\mathbf{l} \cdot \mathbf{s}$ force and $\mathbf{l}^2$ terms for the harmonic oscillator. We adopted the values κ_p = 0.0577 and μ_p = 0.650 for protons and κ_n = 0.0635 and μ_n = 0.325 for neutrons over the entire range studied, in accordance with [HOW80]. The values of ε_2 and ε_4 for each nucleus were taken from the same source.

0097–6156/86/0324–0149$06.00/0

From our β-strength distributions, we then calculated rates as a function of excitation energy in the daughter nucleus. These rates and the Q_β, B_f, and S_n values [HOW80] give the amount of β-delayed fission and neutron emission for each daughter nucleus. Our results are shown in Fig. 1. The numbers in the squares in Fig. 1(a) give the percentage of decays resulting in a daughter nucleus excitation energy greater than or equal to the fission barrier height. The numbers in the squares in Fig. 1(b) give the percentage of decays resulting in an daughter nucleus excitation energy greater than or equal to the neutron separation energy but less than the fission barrier height. In order to determine the maximum possible effect of delayed fission, we take the number from (a) as the amount of β-delayed fission and the number from (b) as the amount of β-delayed neutron emission in the daughter nucleus. This undoubtedly leads to an overestimate of delayed fission probabilities in some cases.

For the purpose of simple comparison with [THI83], we assume constant abundances (1.0 per isobar) along the r-process path shown in Fig.1. After β-delayed fission and neutron emission during decay back, the final abundances are those shown in Table 1. Clearly β-delayed fission plays a large role in decay back at A ≲ 250. On the other hand, β-delayed neutron emission does not affect the abundances too much, since loss from chain A is more or less compensated for by gain from chain A+1, except in a few cases (A = 234, 236, 239, 244, 248, and 249). Of course, a more significant effect of delayed neutron emission will be seen if we take more realistic initial abundances along the r-process path, which usually exhibit a strong even-odd effect.

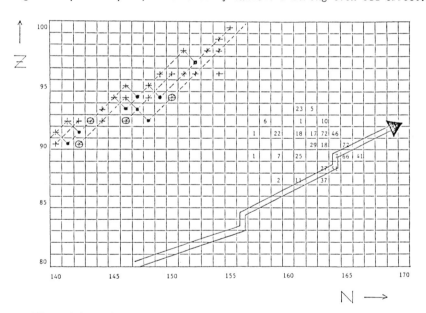

$N \longrightarrow$

Fig. 1(a).   Calculated maximum possible values of β-delayed fission probabilities (in %) shown for the precursor nuclei. The arrow follows the r-process path given in [THI83]. The crosses indicate β-stable nuclei, and the dashed lines indicate α decays. The r-process nuclear cosmo-chronometers are circled.

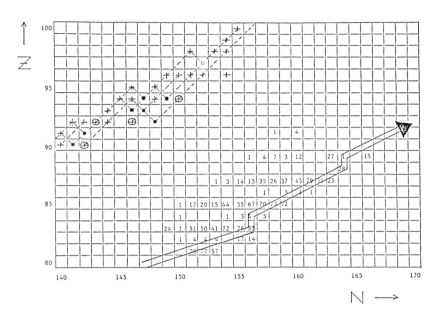

$N \longrightarrow$

Fig. 1(b). Calculated minimum possible values of β-delayed neutron emission probabilities (in %). See the caption to Fig.1(a).

Table 1. Isobaric abundance changes during the cascade from the initial value of 100 % at the r-process path shown in Fig. 1 down to the β-stable nuclei marked by crosses.

A	%	A	%
232	78	244	47
233	102	245	112
234	70	246	110
235	118	247	124
236	133	248	64
237	92	249	64
238	97	250	47
239	67	251	32
240	115	252	67
241	111	253	18
242	103	254	11
243	94	255	6

From the results in Table 1, we can find the r-process production ratios of interest to nuclear cosmo-chronology: $^{232}Th/^{238}U$, $^{235}U/^{238}U$, and $^{244}Pu/^{238}U$. These are obtained by summing the α-decay progenitors for each isotope and considering the leak due to spontaneous fission. Our results are $^{232}Th/^{238}U = 1.60$ $^{235}U/^{238}U = 1.42$, $^{244}Pu/^{238}U = 0.57$. Our value for $^{232}Th/^{238}U$ ratio in this approximation of constant r-process abundances would imply a lower limit on the Galaxy's age of 8.8 Gyr. However, it is important to note that the final production ratios will be dependent on the explicit r-process calculation employed [FOW85]. (See also [MEY85], [THI83] and [YOK83] for further details on nuclear cosmo-chronology.)

For comparison, we give the values for the chronometer production ratios determined by Thielemann et al. [THI83] for constant abundances along the r-process path: $^{232}Th/^{238}U$ = 1.63, $^{235}U/^{238}U$ = 1.21, and $^{244}Pu/^{238}U$ = 0.13. To our surprise, our $^{232}Th/^{238}U$ ratio agrees well with [THI83], despite the large difference between the two calculations in the amount of β-delayed fission (compare Fig. 1(a) with Fig. 2 of [THI83]). This result suggests that β-delayed processes do indeed significantly affect the abundances of the progenitors of the r-process chronometers. The other two sets of ratios differ rather strongly, however. These differences are not surprising because of the different β-strength functions and adopted mass formulae (for Q_β and S_n), although both sets of calculations used the same fission barrier heights. It is worth emphasizing that the virtue of our calculations are that all input data were taken from a single source [HOW80], i.e., the effects of nuclear deformations on the fission barriers and β-strength functions were treated self-consistently. We understand that many uncertainties still exist in this calculation and, therefore, that further work is still required. We hope to pursue this study in the near future and will be aided by improvements in the code used for calculating the β-strength functions (e.g., the inclusion of the first-forbidden decay) and in the set of Q_β, B_f and S_n values. We will also include a more realistic determination of the competition between delayed fission and delayed neutron emission. These improvements should lead to a better understanding of decay back to the β-stability line and help in the search for the astrophysical site(s) for the r-process.

Acknowledgments

We express our thanks to Prof. William A.Fowler for useful discussions and encouragement.

Work performed under auspices of the U.S.Department of Energy by the Lawrence Livermore National Laboratory under contract number W-7405-ENG-48, and supported in part by the LLNL Institute for Geophysics and Planetary Science. One of us (BSM) acknowledges the support of National Science Foundation Graduate Fellowship.

References

[BER69] E.Yu.Berlovich and Yu.N.Novikov, Phys.Lett. 29B 155 (1969)
[FOW85] W.A.Fowler and C.C.Meisel, in Proc.Symp. on Cosmogonical Processes, Boulder, Colorado, 1985, in press
[HOW80] W.M.Howard and P.Möller, Atm.Nucl.Dat.Tables 25 219 (1980)
[KOD75] T.Kodama and K.Takahashi, Nucl.Phys. A239 489 (1975)
[KRU81] J.Krumlinde, P.Möller, C.-O.Wene and W.M.Howard, in Proc.4th Int.Conf. on NUclei Far From Stability, Helsingør, 1981; CERN Rpt. 81-09, p.260
[KRU84] J.Krumlinde and P.Möller, Nucl.Phys. A417 419 (1984)
[MEY85] B.S.Meyer and D.N.Schramm, in Proc.5th Moriond Astrophysics Meeting on Nucleosynthesis and its Implications on Nuclear and Particle Physics, Les Arcs, 1985 (Reidel) in press
[THI83] F.-K.Thielemann, J.Metzinger and H.V.Klapdor, Astron.Astrophys. 123 162 (1983)
[WEN75] C.-O.Wene, Astron.Astrophys. 44 233 (1975)
[YOK83] K.Yokoi, K.Takahashi and M.Arnould, Astron.Astrophys. 118 6 (1983)

RECEIVED April 11, 1986

Shell-Model Plus Pairing Calculations of β-Delayed Neutron Emission Properties at $A \cong 90$-100

Shaheen Rab and Adnan Shihab-Eldin

Kuwait Institute for Scientific Research, P.O. Box 24885 Safat, Kuwait

Expressions are derived for reduced transition probabilities of allowed Gamow-Teller beta decay using Seniority Truncated Exact Diagonalization (STED) method within the framework of shell-model plus pairing. When expressed in terms of fractional occupation probability parameters U and V, the derived formulae are identical to those of BCS method. These expressions are used to calculate beta-delayed neutron emission probability and beta decay half life of Br and Rb isotopes. The results are found to be sensitive to pairing strength parameter G and optimum agreement with experimental values are obtained for G(n) = 0.25. The calculated values for these quantities are in agreement with the overall trend of experimental results. Moreover, they show better agreement with experimental values than similar calculation using BCS pairing method.

Introduction

The beta delayed neutron emission properties of neutron rich nuclei have attracted a lot of interest recently because of advancements in experimental techniques. The information on neutron spectrum, beta decay half-life $t_{1/2}$, delayed neutron emission probability p_n for nuclei far from the line of stability are important in nuclear reactors, nuclear physics and astrophysics. These properties, for nuclei around mass $A \simeq 90 - 100$, have been studied previously by statistical theory [HAR78] [GJO78] and Gross theory [TAK69] [YAM70]. Application of shell model (SM) plus pairing was first suggested in the mid seventies [SHI77]. The SM plus pairing model [OLI80] gave better overall agreement with experimental results for $t_{1/2}$ and p_n. The pairing forces are usually treated in BCS quasi particle theory [KIS63] [LAN64]. Inclusion of Gamow Teller force in SM plus pairing (BCS) improved the fit better [OLI80]. The deviation from the experimental curve is still large compared to uncertainties and the odd-even fluctuation is the main problem we are concerned with here. The problem of particle non-conservation in BCS introduces an averaging effect on the properties of neighbouring isotopes. Furthermore, the blocking effect is not taken into account in these calculations. To overcome these BCS limitations, we propose to handle pairing force in SM using Seniority Truncated Exact Diagonalization (STED) scheme. Expressions are derived for reduced transition probability B in STED to calculate beta strength function, S, $t_{1/2}$, and p_n. The overall agreement with the experimental results of $t_{1/2}$, p_n in STED for $^{92\text{-}102}$ Rb and $^{87\text{-}92}$ Br is found to be better than the BCS results.

Seniority Truncated Exact Diagonalization (STED) Scheme

Only Gamow Teller (GT) allowed β-decay is studied in this work as it plays the dominant role in the specified mass range. The experimentally studied nuclei are assumed to be

0097-6156/86/0324-0153$06.00/0
© 1986 American Chemical Society

spherical and hence we use spherical shell model basis. The number conserving and conceptually simpler STED scheme developed by John [JOH68] is used to deal with the pairing force. The usual fermion creation and destruction operators C^+, C are used to define pair creation (destruction) operator a^+ (a).

$$a_i^+ = \frac{1}{\sqrt{2(2i+1)}} \sum_{m>0} (-1)^{i-m} C_{im}^+ C_{i-m}^+ \tag{1}$$

The Hamiltonian is written as:

$$H = \sum_j e_j n_j - \sum_{ij} G_{ij} \sqrt{(\Omega_i \Omega_j)} a_i^+ a_j \tag{2}$$

where e_j is the single particle energy, n_j is the occupation number, G is the pairing force constant, $\Omega_i = 1/2 (2i + 1)$, the pair degeneracy. The truncated basis wave functions are:

$$\text{Seniority Zero State} \equiv |p_i\rangle = \frac{1}{\eta_i (p)} (a_i^+)^p |0\rangle \tag{3}$$

with η_i (p) being the normalization constant.

$$\text{Seniority One State} \equiv |\bar{p}_j\rangle = \frac{1}{\bar{\eta}_j(p)} (a_j^+)^p C_{jm}^+ |0\rangle \tag{4}$$

with p pairs + 1 particle, again, $\bar{\eta}_j(p)$ is the normalization constant.

$$\text{Seniority Two State} \equiv |\bar{\bar{p}}_j\rangle = \frac{1}{\bar{\bar{\eta}}_j(p)} (a_j^+)^p \frac{1}{\sqrt{2}} a_{jj}^+ (JM) |0\rangle , J \neq 0 \tag{5}$$

with p pairs + 2 odd particles in orbital j, $\bar{\bar{\eta}}_j(p)$ the normalization constant. The anticommutation rules of C^+, C and the commutation of a^+, a together with Wick's theorem, are repeatedly used in diagonalizing the Hamiltonian.

Reduced Transition Probability for β Decay—B

The β-decay strength function S_β is proportional to the square of the matrix elements of β-operator between final and initial states. So is B and is defined by:

$$B(GT; I_i \rightarrow I_f) = \frac{1}{2I_i + 1} \sum_{M_i M_f \mu} |\langle I_f M_f | T_\mu^1 | I_i M_i \rangle|^2 = \frac{1}{2I_i + 1} |\langle I_f ||T^1|| I_i \rangle|^2 \tag{6}$$

$$T_\mu^1 = \sum_k t_-(k) \sigma_\mu(k) \text{ , is the allowed GT beta operator.} \tag{7}$$

Once B is calculated, we can compute S, $t_{1/2}$ and p_n. In second quantized notation, T can be expressed as:

$$T^1 = \sum_{\alpha\beta m_\alpha m_\beta} \langle \alpha\, m_\alpha \,|\, T_\mu^1 \,|\, \beta\, m_\beta \rangle \, C_{\alpha m_\alpha}^+(\pi) \, C_{\beta m_\beta}(\nu) \tag{8}$$

By substituting T^1 from (7) into (8), we calculate all possible expressions for B. The possible transitions are classified as: I. even-even→odd-odd, II. odd-even → even-odd, III. even-odd→ odd-even, IV. odd-odd→ even-even. The space is truncated beyond seniority 2 for protons and neutrons. The resulting expressions are written in terms of V^2 and U^2, probability of occupancy and emptiness, to compare them with those of BCS theory. We only give the final forms.

I.
$$\left| \pi(p_i)\,\upsilon(q_j) \right\rangle \to \left| \pi(\bar{p}_i)\,\upsilon(\overline{q-1})_i \right\rangle, \qquad B = (T^1)_{ij}^2 \left(1 - \frac{p}{\Omega_i}\right)^{U_{p_o}^2} \left(\frac{q}{\Omega_j}\right)^{V_{n_o}^2} \tag{9}$$

II.

(i) $\left| \pi(\bar{p}_i)\,\upsilon(q_j) \right\rangle \to \left| \pi(p+1)_i\,\upsilon(\overline{q-1})_j \right\rangle$, $\quad B = \dfrac{(T_{ij}^1)^2}{2i_{p_o}+1} \left(\dfrac{p+1}{\Omega_i}\right)^{V_p^2} \left(\dfrac{q}{\Omega_j}\right)^{V_{n_o}^2}$ (10)

(ii) $\left| \pi(\bar{p}_i)\,\upsilon(q_j) \right\rangle \to \left| \pi(\bar{\bar{p}}_i)\,\upsilon(\overline{q-1})_j \right\rangle$, $\quad B = (T_{ij}^1)^2 \left(1 - \dfrac{p}{\Omega_i-1}\right)^{U_{p_o}^2} \left(\dfrac{q}{\Omega_j}\right)^{V_{n_o}^2}$ (11)

III.

(i) $\left| \pi(p_i)\,\upsilon(\bar{q}_j) \right\rangle \to \left| \pi(\bar{p}_i)\,\upsilon(q_j) \right\rangle$, $\quad B = \dfrac{(T_{ij}^1)^2}{2j_{n_o}+1} \left(1 - \dfrac{p}{\Omega_i}\right)^{U_{p_o}^2} \left(1 - \dfrac{q}{\Omega_j}\right)^{U_n^2}$ (12)

(ii) $\left| \pi(p_i)\,\upsilon(\bar{q}_j) \right\rangle \to \left| \pi(\bar{p}_i)\,\upsilon(\overline{\overline{q-1}})_j \right\rangle$, $\quad B = (T_{ij}^1)^2 \left(1 - \dfrac{p}{\Omega_i}\right)^{U_{p_o}^2} \left(\dfrac{q}{\Omega_j-1}\right)^{V_{n_o}^2}$ (13)

IV.

(i) $\left| \pi(\bar{p}_i)\,\upsilon(\bar{q}_j) \right\rangle \to \left| \pi(p+1)_i\,\upsilon(q_j) \right\rangle$, $B = \dfrac{1}{2l_i+1} (T_{ij}^1)^2 \left(\dfrac{p+1}{\Omega_i}\right)^{V_p^2} \left(1 - \dfrac{q}{\Omega_j}\right)^{U_n^2} \delta_{l_i 1}$ (14)

(ii) $\left[\left| \pi(\bar{p}_i)\,\chi\upsilon(\bar{q}_j) \right]_{M_i}^{l_i} \right\rangle \to \left| \pi(\bar{\bar{p}}_i)\,\upsilon(q_j) \right\rangle$, $\quad B = \dfrac{(T_{ij}^1)^2}{2j_{n_o}+1} \left(1 - \dfrac{p}{\Omega_i-1}\right)^{U_{p_o}^2} \left(1 - \dfrac{q}{\Omega_j}\right)^{U_n^2}$ (15)

(iii) $\left| \pi(\bar{p}_i)\,\upsilon(\bar{q}_j) \right\rangle \to \left| \pi(p+1)_i\,\upsilon(\overline{q-1})_j \right\rangle$, $B = \dfrac{(T_{ij}^1)^2}{2i_{p_o}+1} \left(\dfrac{p+1}{\Omega_i}\right)^{V_p^2} \left(\dfrac{q}{\Omega_j-1}\right)^{V_{n_o}^2}$ (16)

(iv) $\left[\left| \pi(\bar{p}_i)\,\chi\upsilon(\bar{q}_j) \right]_{M_i}^{l_i} \right\rangle \to \left| \pi(\bar{\bar{p}}_i)\,\upsilon(\overline{q-1})_j \right\rangle$, $B = (T_{ij}^1)^2 \left(1 - \dfrac{p}{\Omega_i-1}\right)^{U_{p_o}^2} \left(\dfrac{q}{\Omega_j-1}\right)^{V_{n_o}^2}$ (17)

All these expressions for B agree with those of earlier BCS calculations [SAK64] [RAN72] [OLI80].

Applications to $^{87\text{-}92}$ Br and $^{92\text{-}102}$ Rb Precursors

The model calculations are carried out for $_{35}\text{Br} \xrightarrow{\beta^-} {}_{36}\text{Kr} \rightarrow {}_{36}\text{Kr+n}$ and $_{37}\text{Rb} \xrightarrow{\beta^-} {}_{38}\text{Sr} \rightarrow {}_{38}\text{Sr+n}$ transitions assuming the core with $Z = 28$ and $N = 50$. The single particle energies are taken from experiments when available, otherwise from systematics. The β-decay half life of the parent is given by:

$$t_{1/2}^{-1} = \sum_{E_i \geq 0}^{Q_\beta} S_\beta (E_i) f(Q_\beta - E_i) \tag{18}$$

where

$$S_\beta (E_i) = \frac{1}{D(g_v/g_A)^2} B(GT, E_i) \tag{19}$$

with $D = 6260 \pm 60$ sec., g_V g_A, are the vector and axial vector constants of γ-current. The delayed neutron emission probability p_n is defined as the fraction of β-decays to intermediate levels in the emitter which finally leads to neutron emission. Assuming no γ-ray emission above B_n, p_n is defined as:

$$p_n \% = \frac{\displaystyle\sum_{i, B_n \leq E_i \leq Q_\beta} S_\beta(E_i) f_i(Q_\beta - E_i)}{\displaystyle\sum_{E_i \geq 0}^{Q_\beta} S_\beta (E_i) f(Q_\beta - E_i)} \times 100\% \tag{20}$$

Results and Discussions

The calculated half-lives $(t_{1/2})$ for Br and Rb isotopes are shown in Figs. 1 and 2. We also show in the same figures the calculated half-lives using the BCS method [OLI80] and the experimental values. Similarly in Figs. 3 and 4, we show the delayed neutron branching ratios, p_n and compare them again with those obtained using BCS method and the experimental values. It can be concluded from these figures that the STED method gives a better overall agreement with the experimental results than the BCS method, for both $t_{1/2}$ and p_n. The noticeable odd-even fluctuation in the $t_{1/2}$ values for the BCS method is not present in the STED results, which follow very closely the experimental trends. The calculations for Br isotopes were repeated for different values of $G(n)$ and were found sensitive to it. The best fit was at $G(n) = 0.25$, $G(p) = 0.3$ for both $t_{1/2}$ and p_n. Preliminary analysis of the energy distributions and strengths of the configurations responsible for the GT-beta decay indicate that this difference in the two sets of calculations is due to the fact that STED takes account of the blocking effect while BCS ignores it completely. As a result of these encouraging results, we are currently extending this work to include a GT-type residual interaction

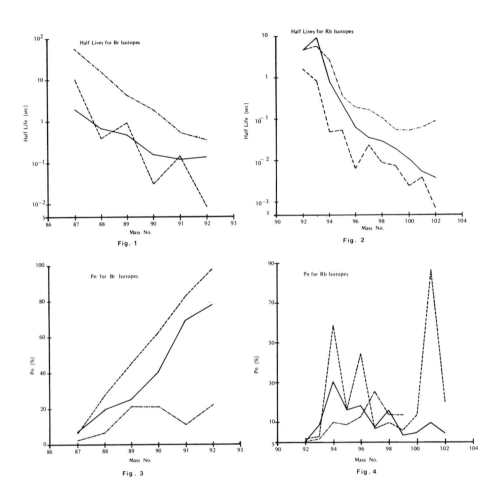

Fig. 1

Fig. 2

Fig. 3

Fig. 4

Legend
——— STED Values
- - - - - BCS Results
-·-·-·- Experimental Values

within the STED framework, which is expected to give yet better agreement with experimental results. This may be of value when using the STED method to predict the $t_{1/2}$ and p_n values for very short-lived delayed neutron precursors that are very far from the stability line and are not yet accessible to experimental measurements.

References

[CAM80] X. Campi and M. Epherre, *Phys. Rev.* C 22 (1980) 2605.

[GJO78] O.K. GjoHernd, P. Hoff, A.C. Pappas, *Nucl. Phys.* A 303 (1978).

[HAR78] J.C. Hardy, B. Jonson, P.G. Hansen, *Nucl. Phys.* A 305 (1978) 15.

[JOH68] O.D. John, Ph.D. Thesis, University of California, Berkeley, (1968).

[KAR82] K.L. Kratz, et. al., *Z. Phys. A—Atoms and Nuclei* 306, 239(1982).

[KIS63] L.S. Kisslinger and R.A. Sorensen, *Reviews of Modern Physics* 35 (1963) 853.

[LAN64] A.M. Lane, *Nuclear Theory,* New York, Benjamin Cummings (1964).

[OLI80] Z.M. Oliveria, Ph.D. Thesis, University of California, Berkeley (1980).

[RAN72] J. Randrup, Ph.D. Thesis, University of Aarhus, Denmark (1972).

[REE83] P.L. Reeder and R.A. Warner, *Phys. Rev.* C 28, 4 (1983) 1740.

[SAK64] M. Sakai and S. Yoshida, *Nucl. Phys.* 50 (1964) 494.

[SHI77] A.A. Shihab-Eldin, F.K. Nuh, W. Halverson, S.G. Prussin, W. Rudolph, H. Ohm and K.L. Kratz, *Phys. Lett.* 69 B (1977) 143.

[TAK69] K. Takahaski and M. Yamada, *Prog. Theor. Phys.* 41 (1969) 1470.

[THI81] C. Thibault et. al., *Phys. Rev.* C 23 (1981) 2720.

[YAM70] M. Yamada and K. Takahaski, Leysin Conf. (1970) CERN (70-30) 397.

RECEIVED May 19, 1986

Random Phase Approximation Calculations of Gamow-Teller β-Strength Functions in the A = 80-100 Region with Woods-Saxon Wave Functions

K.-L. Kratz[1], J. Krumlinde[2], G. A. Leander[3], and P. Möller[2]

[1]Institut für Kernchemie, Universität Mainz, D-6500 Mainz, Federal Republic of Germany
[2]Department of Mathematical Physics, Lund Institute of Technology, Lund University, Box 118, S-22100 Lund, Sweden
[3]UNISOR, Oak Ridge Associated Universities, Box 117, Oak Ridge, TN 37831

We discuss some features of a model for calculation of β-strength functions, in particular some recent improvements. An essential feature of the model is that it takes the microscopic structure of the nucleus into account. The initial version of the model used Nilsson model wave functions as the starting point for determining the wave functions of the mother and daughter nuclei, and added a pairing interaction treated in the BCS approximation and a residual GT interaction treated in the RPA-approximation. We have developed a version of the code that uses Woods-Saxon wave functions as input. We have also improved the treatment of the odd-A Δν=0 transitions, so that the singularities that occured in the old theory are now avoided.

The calculation of the β-strength function involves evaluating the matrix element of the β transition operator between the initial wave function ψ_i of the mother nucleus and the wave functions ψ_f of the final states in the daughter nucleus. A model for the β-strength function is essentially equivalent to developing a model for the wave functions ψ. We shall here give a brief outline of one such model developed by [KRU84] and discuss in somewhat greater detail some new features we recently added to the model.

The model developed by [KRU84] uses Nilsson wave functions as the starting point for constructing the wave function ψ_i of the ground state of the mother nucleus and the wave functions ψ_f of the ground and excited states of the daughter nucleus. It is instructive to consider the effect of the various improvements, which have been added to the model beyond the use of Nilsson wave functions, on some particular transition, for which also an experimental value is available. In ref. [BOH75] transitions in the rare earth region are discussed and ref. [KRU84] selected the transition $[523 \frac{7}{2}]_p - [523 \frac{5}{2}]_n$ for ^{170}Yb. The log ft value for this transition is 4.8. The connection between the transition matrix element $<K+1|\sigma\pm|K>$ and the ft value is given by the formula [BOH75]

$$<K+1|\sigma_\pm|K>^2 = \frac{1}{4} \frac{8.3 \times 10^3}{ft(sec)} \frac{1}{<I_i = \Omega_i \ \Omega_i \ 1K|I_f \ \Omega_f = \Omega_i + K>^2}$$

Thus the experimental value for the above matrix may be obtained. We find that for the above transition

$$<K + 1 \ |\sigma_\pm|K>^2_{exp} \cong 0.04$$

It is pointed out in ref. [BOH75] that a pure deformed oscillator model, assuming good asymptotic quantum numbers, would yield the value 1 for the above matrix element. With Nilsson wave functions we obtain 0.86. The small reduction is due to the mixing of orbitals due to the $\bar{\ell} \cdot \bar{s}$ and ℓ^2 terms in the single-particle potential and a small amount of ε_4 distortions. Now, with the addition of pairing, the increasing complexity of the wave functions reduces the calculated value of the transition matrix element to 0.37. According to [BOH75] one can expect a reduction by about a factor of 4 due to pairing for levels close to the Fermi surface, because the reduction is $u_p u_n$ or $v_p v_n$ and u and v are $\frac{1}{\sqrt{2}}$ at the Fermi surface. That we obtained a smaller reduction in this particular case is due to the fact that the entering pairing factors are larger than $\frac{1}{\sqrt{2}}$. An additional reduction of $<K + 1|\sigma_{\pm}|K>^2$ in the model developed by [KRU84] comes about from the addition of a residual interaction, specific to GT decay, $V_{GT} = :\beta^{1-} \cdot \beta^{1+}:$, to the Hamiltonian, which was treated in the RPA approximation. With this interaction included, the square of the matrix element is reduced to

$$<K+1|\sigma_{\pm}|K>^2 = 0.09$$

The V_{GT} residual interaction is introduced to account for the retardation of low-energy GT decay rates, and as we saw in our example above, the strength was reduced by about a factor of four. The calculated strength is still about a factor of 2 larger than the observed strength. The calculated strength could be further reduced by increasing further the strength χ of the V_{GT} interaction, which is set at $\chi = 23/A$ MeV by [KRU84] as is also done by most other investigators. However, the strength χ is determined by the requirement that the calculated position of the giant Gamow-Teller resonance agrees with the experimental results. Thus, the mechanism behind the remaining factor-of-two discrepancy between the experimental and calculated strength is thought to be of a different origin. Two mechanisms that are being investigated are couplings to 2p2h states [BER82] and the $\Delta(1232)$ isobar [BOH81]. To account for the missing strength in half-life calculations we divide the calculated strength by 2 for such applications. The calculated strengths shown in this contribution are not divided by 2 however.

We have added some new features to the model described above, which was developed by [KRU84]. In particular we have observed that the perturbation expressions for the $\Delta v = 0$ transitions for odd-mass and odd nuclei used by [KRU84] (eqs. (43) - (47) in that paper) break down occasionally. Similar expressions have also been used earlier by [HAL67] and [RAN73]. When the expressions break down, a single $\Delta v = 0$ transition may have a strength that is many times the sum rule $S_\beta^- - S_\beta^+ = 3(N-Z)$ for the $\Delta v = 2$ transitions. However, the equations (43) - (47) of ref. [KRU84] can be modified somewhat to remove this difficulty. The equations for the $\Delta v = 0$ strength contain sums over terms with amplitudes $A_p(n\omega) = 1/(E_p - E_n - \omega)$. The quantities ω are the roots of the RPA equations and are "close" to the asymptotes $E_p + E_n$, but not close enough to cause any singularity in the $\Delta v = 2$ transition strengths. By "accident" they may be so close to the quantity $E_p - E_n$ that the perturbation expansion breaks down and a singularity occurs. This singularity can be removed by introducing a width d, and moving the pole E_p - E_n into the complex plane. Thus, we replace $A_p(n\omega)$ =

$$\frac{1}{E_p - E_n - \omega}$$

by

$$A_p(n\omega) = \frac{1}{2}\left[\frac{1}{E_p - E_n + id - \omega} + \frac{1}{E_p - E_n - id - \omega}\right] = \frac{E_p - E_n - \omega}{(E_p - E_n - \omega)^2 + d^2}$$

In figs. 1a - 1b we show the results for the original model (a) which exhibits a pronounced singularity and for d = 0.1 MeV and d = 1.0 MeV (b). In fig. 1b we have indicated the d = 0.1 MeV results with dashed lines. We see that differences between the d = 0.1 MeV and d = 1.0 MeV results are minor and that the results therefore are insensitive to the width d, as should be. We have selected d = 0.1 MeV for our calculations.

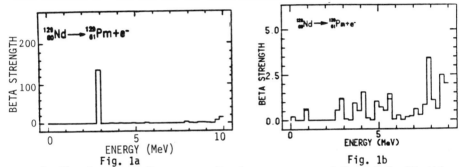

Fig. 1a   Fig. 1b

In fig. 2 we exhibit some results for a sequence of Rb nuclei with this modification included. We have also corrected an error that occured in an earlier version of the computer code for some $\Delta v = 0$ transitions in the spherical case. The error was usually small. Compared to [KRU84] fig. 2 contains only these two modifications and the differences relative to the earlier results are small. However the introduction of the width d is very important in some other cases as can be seen in fig. 1.

The calculated results are discussed extensively in [KRU84], to which we refer for a more complete discussion. Here, let us just note that for the top four spectra in fig. 2 we used the parameter set "A = 100" and for the lower four figures we used the parameter set "N = 60". The N = 60 set reproduces best the large increase in strength of the low-energy peak that is seen experimentally in ^{95}Rb relative to ^{93}Rb. The N = 60 single-particle parameter set was adjusted to reproduce results in the vicinity of the N = 56 subshell [AZU78], so it is to be expected that this set gives the better description of some features in this region.

One major difficulty in the Nilsson model is the determination of the single-particle parameters κ and μ for various regions of nuclei. Usually these parameters are determined by adjusting calculated single-particle levels to experimental data. Since κ and μ can vary rather unpredictably from region to region, the model is therefore somewhat unsuitable for extrapolations to unknown regions of nuclei. With the aim of being able to calculate properties for such nuclei more reliably we are now developing the code to accept wave

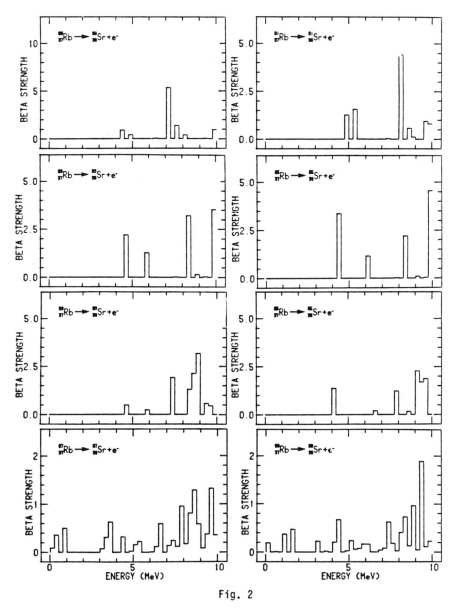

Fig. 2

functions from both Woods-Saxon and folded-Yukawa single-particle potentials
as input. The parameters of such models vary smoothly with Z and N and are
therefore expected to extrapolate more reliably to new and to unknown regions

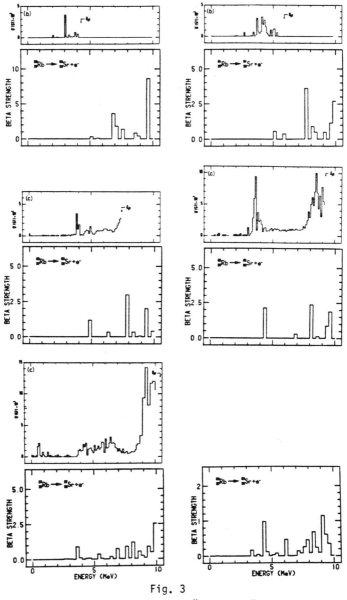

Fig. 3

of nuclei. See for example the studies [MOL81a], [MOL81b] and [BEN84] of the folded-Yukawa potential, which covers the entire periodic system. In this initial study we shall use Woods-Saxon wave functions, mainly because such a code has been made available to us by [NAZ84] in a form that runs on VAX and NORSK DATA computers. The folded-Yukawa code currently runs only on CDC computers. The Woods-Saxon model is described in [DUD82].

Figure 3 shows β-strength functions calculated with Woods-Saxon wave functions for a sequence of Rb nuclei. Above all but the last of the calculated strength functions, which have the words BETA STRENGTH along the vertical axis, there are plots of experimental results from [KRA83] and [KRA81]. The experimental results have the label B'(GT) along the vertical axis. The results with Woods-Saxon wave-functions are similar to the results obtained with the oscillator model, and also agree fairly well with experiment.

Here we shall just comment on two aspects of the results. First, the large increase in strength seen experimentally when going from ^{93}Rb to ^{95}Rb is slightly less well reproduced with the Woods-Saxon wave-functions, than with the Nilsson wave functions with the N = 60 set. However, the set N = 60 was optimized to reproduce the N = 56 subshell conditions. The Woods-Saxon calculation was performed with a "universal" parameter set [DUD82], which should in general be more reliable for extrapolations far from stability. Second, for the deformed ^{97}Rb and ^{99}Rb results there is very little strength at low energy, in contrast to the experimental situation and the modified oscillator results. However, the Woods-Saxon calculation was run for a deformation β_2 that was somewhat smaller than the appropriate value for ^{97}Rb and ^{99}Rb.

Due to circumstances beyond our control, we have only been able to present very few initial results with the Woods-Saxon wave functions here. We hope, however, to explore the model more fully in the near future, in particular to run ^{97}Rb and ^{99}Rb with a more appropriate value of β_2 and to explore additional regions of nuclei far from stability, for which applications the Woods-Saxon single-particle model should be advantageous compared to the Nilsson model, which is less reliable for extrapolations.

Acknowledgments

We are grateful to J. Dudek and W. Nazarewicz for making a copy of their Woods-Saxon code available to us.

References

[AZU78] R.E. Azuma, G. L. Borchert, L.C. Carraz, P.G. Hansen, B. Jonson, S. Mattson, O.B. Nielsen, G. Nyman, I. Ragnarsson and H.L. Ravn, Phys. Lett. 80B 4 (1978).
[BEN84] R. Bengtsson, P. Möller, and J.R. Nix, Phys. Scr. 29 402 (1984).
[BER82] G.F. Bertsch, and I. Hamamoto, Phys. Rev. C26 1323 (1982).
[BOH75] A. Bohr, and B.R. Mottelson, Nuclear Structure, vol. II (Benjamin, New York, 1975) pp. 306, 307.
[BOH81] A. Bohr and B. R. Mottelson, Phys. Lett. 100B 10 (1981).
[DUD82] J. Dudek, Z. Szymanski, T. Werner, A. Faessler, and C. Lima, Phys. Rev. C26 1712 (1982).
[HAL67] J.A. Halbleib, Sr., and R.A. Sorensen, Nucl. Phys. A98 542 (1967).
[KRA81] K.L. Kratz, H. Ohm, A. Schröder, H. Gabelmann, W. Ziegert, H. V. Klapdor, J. Metzinger, T. Oda, B. Pfeiffer, G. Jung, L. Alquist and G.I. Crawford, Proc. 4th Int. Conf. on nuclei far from stability, Helsinger, 1981 (CERN 82-09, Geneva, 1981) p. 317.
[KRA83] K.L. Kratz, Priv. comm. (1983).
[KRU84] J. Krumlinde, and P. Möller, Nucl. Phys. A417 419 (1984).
[MOL81a] P. Möller, and J.R. Nix, Nucl. Phys. A361 117 (1981).
[MOL81b] P. Möller, and J. R. Nix, At. Data Nucl. Data Tables 26 165 (1981).
[NAZ84] W. Nazarewicz, and J. Dudek, Priv. com. (1984).
[RAN73] J. Randrup, Nucl. Phys. A207 209 (1973).

RECEIVED August 20, 1986

Identifying Nilsson States in $A \cong 100$ Nuclei by Investigating Gross β-Decay Properties

K.-L. Kratz[1], H. Gabelmann[1], W. Ziegert[1], V. Harms[1], J. Krumlinde[2], and P. Möller[2]

[1]Institut für Kernchemie, Universität Mainz, D-6500 Mainz, Federal Republic of Germany
[2]Department of Mathematical Physics, Lund Institute of Technology, Lund University, Box 118, S-22100 Lund, Sweden

Since most nuclei in the region of deformation at A≈100 can only be produced with rather low yields which makes detailed spectroscopic studies difficult, we have examined possibilities of extracting nuclear structure information from easily measurable gross β-decay properties. As examples, comparisons of recent experimental results on ^{99}Rb-Y and ^{101}Rb-Y to RPA shell model calculations using Nilsson-model wave functions are presented and discussed.

The existence of strong deformation for neutron-rich A≅100 nuclei was already extablished 15 years ago [CHE70], and data continue to collect on this region up to now. They revealed a rather complex picture indicating rapid nuclear shape changes and shape coexistence [BEN84,MEY85] which can be related to the occurrance of several (quasi-) spherical and deformed subshells for both Z and N (see Fig.1). While in the past years spectroscopic studies were mainly restricted to even-even nuclei, more recent investigations have turned towards odd-mass nuclei which allowed first conclusions on individual Nilsson orbitals as well as the determination of the Nilsson Hamiltonian in the A≈100 mass region (see e.g. [SIS84]).

Usually, characterization of nuclear structures in transitional nuclei requires rather sophisticated spectroscopic investigations, e.g. measurements of level life-times, γ-transition multipolarities and magnetic moments. Unfortunately, most of the nuclei of interest in the A≈100 mass region are fission products far from β-stability which - at present - can only be produced with rather low yields. This makes detailed spectroscopic studies very time-consuming, and in a number of cases even impossible. Therefore, we have examined possibilities of determining shape transitions and Nilsson states in very-neutron-rich nuclei via correlated changes in easily measurable **gross** β-decay properties, such as the half-life ($T_{1/2}$), the delayed neutron emission probability (P_n), or the gross β-branching pattern in terms of the β-strength function (S_β). To be more specific, we thoroughly compare existing experimental data on these quantities with shell model [KRU84] expectations for different nuclear structures, and from agreement, respectively disagreement between measurements and predictions information on Nilsson parameters (x, μ, odd-particle states) and deformation is deduced.

The trigger for this attempt was the observation of strong variations in the shape of S_β of neutron-rich odd-mass Rb isotopes [KRA81,83,84] which could be related to the spherical N=56 subshell closure and to the sudden onset of strong deformation at the deformed N=60 shell gap (see Fig.1). In

0097-6156/86/0324-0165$06.00/0
© 1986 American Chemical Society

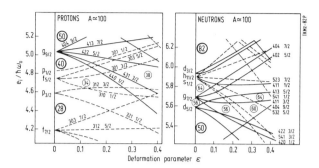

Fig.1. Single-particle levels for protons and neutrons in the
A=100 region for the modified oscillator potential of the
Nilsson model as a function of prolate deformation.

the case of the N=60 isotone ^{97}Rb, the shape of S_β lead to the π[431 3/2]
assignment for the ground state (g.s.) of this nucleus, which constituted
some of the first evidence for establishing deformation in medium-mass odd-
nucleon isotopes. To give an example for the possible sensitivity of shell
structure on gross β-decay properties, in Fig.2 the situation for ^{97}Rb is
shown in a simplified, schematic way. As is discussed in some detail in
[KRA83,84], depending on the valence-proton Nilsson orbital, quite different
shapes of S_β will occur. With the π[301 3/2] configuration, S_β will have a
distribution similar to that shown in the upper part of Fig.2, with the
lowest Gamow-Teller (GT) transition ($\nu g_{7/2} \rightarrow \pi g_{9/2}$) at medium excitation
energy. This will result in a 'long' $T_{1/2}$ and a 'high' P_n-value. For the
connection of both quantities to S_β, see [KRA84]. On the other hand, when
assuming the odd proton in the [431 3/2] Nilsson orbital, part of the
$\nu g_{7/2} \rightarrow \pi g_{9/2}$ strength will be shifted down to (near) the g.s. of ^{97}Sr,
resulting in an S_β distribution similar to that in the lower part of Fig.2.
This will give a 'short' $T_{1/2}$ and a 'low' P_n. It is reasonable to assume
that such differences in β-decay properties will occur also for other odd-
particle nuclei in the A≈100 mass region, thus providing a new and simple
method for an identification of Nilsson states.

Fig. 2.   Schematic model-S_β's
for the decay of a far-unstable
nucleus, as e.g. ^{97}Rb, and their
influence on $T_{1/2}$ and P_n. For
discussion, see text.

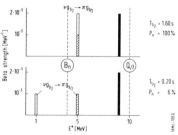

A good example for testing the applicability of our approach is the
β-decay of the odd-proton (Z=37) nucleus ^{99}Rb to the odd-neutron (N=61)
daughter ^{99}Sr. As was shown in [PFE84], even from a rather time-consuming
spectroscopic investigation of ^{99}Rb decay at OSTIS, resulting in g.s. band
properties considerably improved over those reported in an earlier

publication [WOH83], an unambiguous N=61 Nilsson orbital assignment was not possible since both calculated $(g_K-g_R)/Q_0$ curves for ν [411 3/2] and ν[541 3/2] overlapped with the experimental value for deformations of $\varepsilon \geqslant 0.3$. However, we were able to demonstrate that, when including β-intensity arguments for the decays of mother and daughter nucleus (in particular log ft-values for the Rb→Sr and Sr→Y g.s. to g.s. transitions), and with additional comparison to shell model predictions using the N=60 Nilsson parameter set [KRU84], the ν[541 3/2] orbital for the ^{99}Sr ground band could be ruled out. In this way, also two excited rotational bands built on the ν[413 5/2] and ν[422 3/2] Nilsson configurations were identified (see Fig.1 of [PFE84]). It should be pointed out in this context, that good agreement between experimental energies and log ft-values for the different band heads with shell model predictions can be obtained, **provided an appropriate choice of Nilsson parameters and deformation is made.** As is demonstrated in [KRU84] for neutron-rich Rb isotopes and as will be more generally discussed for the A = 80-110 region in [KRA85], Nilsson parameters fitting experimental data of spherical nuclei near stability, like the standard A=100 set of [LAR76], tend to become inappropriate for far-unstable deformed nuclei. This may, for example, be the reason for some incorrect proton orbital assignments to band heads in ^{99}Y [WOH85] which were based on BCS+pairing predictions using Nilsson parameters adjusted to single-particle levels of near-stable Y isotopes.

In principle, the same conclusion on the Nilsson orbital of the ^{99}Sr g.s. as obtained by the above mentioned γ-spectroscopic work can already be drawn from a comparison of easily measurable **gross** β-decay features (requiring ~5 h measuring time for β- and n-multiscaling, compared to ~3 weeks for γ-spectroscopy) to predictions of our RPA shell model. In the

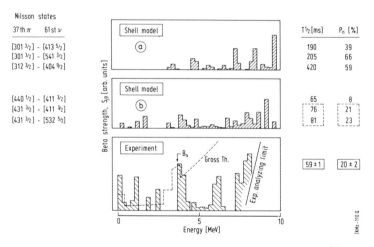

Fig.3.   Comparison of the experimental S_β distribution of ^{99}Rb decay with RPA shell model predictions (middle part) involving different odd-particle Nilsson states lying near the Fermi level (left part). In the right part, the corresponding $T_{1/2}$ and P_n values are listed.

middle part of Fig.3, the β-strength pattern for ^{99}Rb decay observed in experiment is compared to (schematical) S_β predictions involving the odd-particle Nilsson states shown in the left part of the figure. Two different types of strength distributions are calculated: (a) **without** low-lying GT-strength, and (b) **with** g.s. to (near) g.s. GT-transitions. In the right part of Fig.3, the corresponding model-$T_{1/2}$ and P_n are listed together with our experimental values. Clearly, from both the shape of S_β as well as from $T_{1/2}$ and P_n, the Nilsson orbital combinations for type (a) decay patterns can be excluded. Furthermore, for type (b) strength functions the π[440 1/2] - ν[411 3/2] combination can be ruled out - although having the correct S_β shape - because the estimated P_n value is a factor 2.5 too low. With this, the [431 3/2] orbital can be assigned to the 37th proton in ^{99}Rb, while for the 61st neutron no distinction between the Nilsson states [411 3/2] and [532 5/2] is possible from the gross β-decay properties of this isotope. However, a similar comparison of theoretical $T_{1/2}$ and P_n with experimental data for ^{99}Sr decay (see Tab.1) allows the unambiguous assignment of the ν[411 3/2] orbital to the g.s. of ^{99}Sr and, moreover, confirms the earlier assignment of the π[422 5/2] orbital for the ^{99}Y ground band [MON82].

Tab.1. Comparison of experimental $T_{1/2}$ and P_n for ^{99}Sr to shell model predictions for different odd-particle Nilsson orbitals near the Fermi level

Nilsson states		$T_{1/2}$ [ms]	P_n [%]
61st ν	39th π		
[413 5/2]	- [431 3/2]	550	0.29
[411 3/2]	- [431 3/2]	177	0.03
[411 3/2]	- [301 3/2]	190	0.03
[541 3/2]	- [431 3/2]	1065	0.34
[532 5/2]	- [422 5/2]	1172	1.21
[411 3/2]	- [422 5/2]	275	0.20
	Experiment:	285±15	0.19±0.10

The second example is related to the possible identification of Nilsson states in odd-nucleon A=101 isotopes. In the case of ^{101}Rb, the 37th proton may occupy the [301 3/2], [431 3/2] or [312 3/2] Nilsson orbital, and similarly the 63rd neutron in the daughter nucleus ^{101}Sr may be in different states near the Fermi surface (see Fig.1). The corresponding odd-particle Nilsson orbital combinations may again yield different S_β distributions and, consequently, different $T_{1/2}$ and P_n (see Fig.4). From a very recent experiment performed at CERN-ISOLDE, a rather short β-decay half-life of $T_{1/2}$=(32±5) ms and a relatively low P_n-value of (24±5) % were obtained. Comparison of these (preliminary) results with our shell model predictions favours an S_β of type (b) and thus suggests the π[431 3/2] g.s. configuration for the precursor ^{101}Rb, but it cannot determine the N=63 Nilsson assignment for ^{101}Sr. However, as in the A=99 case, this ambiguity can be removed by considering gross β-decay properties of the latter nucleus to levels in ^{101}Y (see Fig.5). Apart from an older measurement at ISOLDE which yielded a $T_{1/2}$ of about 180 ms, the only other information on ^{101}Sr decay comes from recent studies at TRISTAN. Wohn et al. [WOH83] have established two rotational bands in ^{101}Y, and Reeder et al. [REE85] have measured $T_{1/2}$ and P_n for the Sr precursor. As can be seen from Fig.5, a comparison of these experimental data with our shell model predictions for different Nilsson orbital combinations for the 63rd neutron and the 39th proton suggests the [411 3/2] assignment for the ^{101}Sr g.s. and confirms the π[422 5/2] configuration for the ^{101}Y ground band [WOH83]. Furthermore, deformations of ε=0.35±0.03 can be deduced for both isotopes.

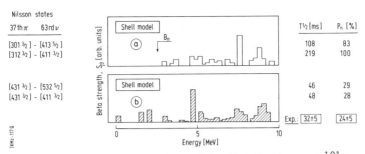

Fig. 4. Shell model predictions of S_β distributions of ^{101}Rb decay (middle part) involving different odd-particle Nilsson orbitals (left part). In the right part, the corresponding model-$T_{1/2}$ and P_n are listed and compared to experimental data.

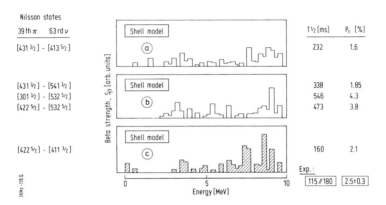

Fig. 5. Shell model predictions of S_β distributions of ^{101}Sr decay (middle part) involving different odd-particle Nilsson states near the Fermi surface (left part). In the right part, the corresponding model-$T_{1/2}$ and P_n are compared to existing experimental data.

So far, we have considered β-decay mother-daughter pairs with (almost) identical g.s. deformations, since our shell model implicitly contains this constraint [KRU84]. Consequently, in case of different g.s. deformations for mother and daughter, or in case of shape coexistence, our method is less conclusive. Nevertheless, by calculating S_β with both deformations and taking in a sense the 'average' of both predictions, we can at least deduce some information on the effective structure properties of such nuclei. Following this line, we can, for example, demonstrate that the general shape of S_β as well as $T_{1/2}$ and P_n of ^{97}Rb can only be understood when assuming a g.s. deformation of $\epsilon \simeq 0.3$ for the precursor and $\epsilon \ll 0.2$ for the g.s. of the daughter. Moreover, in order to reproduce the magnitude of the experimental S_β, we have to assume a change in deformation to $\epsilon \simeq 0.3$ for levels above 0.5 MeV in ^{97}Sr. This observation is consistent with the interpretation of shape coexistence in this N=59 isotone [PFE81,KRA83,MEY85] and determines the g.s. configuration of ^{97}Sr to be [411 3/2], in contrast to our earlier assignment of $J^{\pi}=1/2^+$ [PFE81].

In summary, we have demonstrated that the comparison of present generation shell model predictions for very-neutron-rich odd-nucleon isotopes in the A ≈ 100 region with experimental **gross** β-decay properties may serve to identify, within limits, nuclear shape changes and Nilsson parameters, and to determine nuclear deformation. However, in order to deduce reliable information care must be excercised when selecting the single-particle parameters in the models used in the calculation of β-decay properties. With this, theoretical estimates of $T_{1/2}$ and decay patterns may serve as a valuable guide to the selection of experiments for investigating exotic isotopes of unknown properties, and - more generally - may result in more reliable extrapolations up to the neutron drip line. For example, with regard to astrophysical application, our shell model predictions of $T_{1/2}$, being on the average a factor of two longer than those estimated by Klapdor et al. [KLA84], might (again) raise questions about the plausibility of r-process nucleosynthesis in He-burning environments.

Acknowledgment

Thanks are due to the organizers of this symposium and to the ACS. Without their support, K.-L. Kratz and P. Möller could not have participated. This work was supported by the German BMFT and by the Swedish NSRC.

References

[BEN84] R. Bengtsson et al., Phys. Scripta 29 402 (1984).
[CHE70] E. Cheifetz et al., Phys. Rev. Lett. 25 38 (1970).
[KLA84] H.V. Klapdor et al., At. Data Nucl. Data Tables 31 81 (1984).
[KRA81] K.-L. Kratz et al., Proc. 4th Int. Conf. on Nuclei Far From Stabi-
         lity, Helsingør, Denmark, CERN 81-09 317 (1981).
[KRA83] K.-L. Kratz et al., Z. Physik A312 43 (1983).
[KRA84] K.-L. Kratz, Nucl. Phys. A417 447 (1984).
[KRA85] K.-L. Kratz et al., 'Beta-Strength Function Systematics of A = 100
         Nuclei', to be published.
[KRU84] J. Krumlinde and P. Möller, Nucl. Phys. A417 419 (1984).
[LAR76] S.E. Larsson et al., Nucl. Phys. A261 77 (1976).
[MEY85] R.A. Meyer, Hyperfine Interactions 22 385 (1985), and Refs. therein.
[MON82] E. Monnand et al., Z. Physik A306 183 (1982).
[PFE81] B. Pfeiffer et al., Proc. 4th Int. Conf. on Nuclei Far From Stabi-
         lity, Helsingør, Denmark, CERN 81-09 423 (1981).
[PFE84] B. Pfeiffer et al., Z. Physik A317 123 (1984).
[REE85] P.L. Reeder et al., these proceedings.
[SIS84] K. Sistemich et al., Proc. Int. Symp. on In-Beam Nuclear Spectros-
         copy, Debrecen, Hungary, Vol. I 51 (1984), and Refs. therein.
[WOH83] F.K. Wohn et al., Phys. Rev. Lett. 51 873 (1983).
[WOH85] F.K. Wohn et al., Phys. Rev. C31 621 and 634 (1985), and these
         proceedings.

RECEIVED August 20, 1986

New Delayed-Neutron Precursors from TRISTAN

P. L. Reeder [1], R. A. Warner [1], M. D. Edmiston [2], R. L. Gill [3], and A. Piotrowski [3]

[1]Pacific Northwest Laboratory, Richland, WA 99352
[2]Bluffton College, Bluffton, OH 45817
[3]Brookhaven National Laboratory, Upton, NY 11973

Recent development of reliable, high intensity ion sources for the TRISTAN on-line mass separator facility has greatly expanded the number of very neutron-rich fission products available. Half-lives and delayed neutron emission probabilities are being measured for these nuclides with a high efficiency neutron counter. During the past year, previously unknown P_n values have been measured for 11 precursors; ^{75}Cu, ^{100}Rb, $^{100-102}Sr$, $^{100-102}Y$, ^{148}Cs, ^{149}Ba, and ^{149}La. For ^{100}Rb, combining our P_n value with the previously measured ratio of P_{2n}/P_n gives a P_{2n} of 0.14 ± 0.04 %.

The TRISTAN on-line mass separator facility at Brookhaven National Laboratory provides mass separated beams of delayed neutron precursors at over 52 mass numbers. Many of these mass numbers include several isobars which are precursors. In addition, some of the nuclides have two or more beta decaying isomers which may or may not be precursors. We have engaged in a series of experiments to measure the half-lives and delayed neutron emission probabilities (P_n) of as many of these precursors as possible [REE83-1,REE83-2]. Because delayed neutron emission can be detected with high sensitivity in the presence of large intensities of beta and gamma radiations, we have used neutron counting to search for new isotopes among the very neutron-rich fission products [REE83-1,REE85].

We report here our recent half-life and P_n measurements among the low-yield fission products. The P_n values for precursors around mass 100 are unusually low, which may be a manifestation of ground state deformation in this vicinity. We also describe our observations on the three new isotopes ^{75}Cu, ^{124}Ag, and ^{149}Ba.

Mass separated beams of fission products from TRISTAN were deposited on an aluminized Mylar tape in the center of a high efficiency neutron counter. The neutron counter contained 40 ^3He filled counter tubes embedded in polyethylene and surrounded by Cd, boral, and polyethylene shields. Experiments were done with two alternative arrangements of the counter tubes. For higher efficiency ($\approx 50\%$) and flat neutron energy response, the tubes were arranged in three concentric rings. This arrangement gave a neutron residence half-time of 25 μs. For beta-neutron coincidence experiments, it is desirable to have an even shorter residence time. This was achieved by arranging the counter tubes in four quadrants and packing them as close as possible to the beam deposition point. The residence half-time was 11 μs with this configuration but the neutron counting efficiency was $\approx 34\%$.

0097-6156/86/0324-0171$06.00/0
© 1986 American Chemical Society

Beta particles at the deposition point passed through the Mylar tape and a 0.0075 cm thick Al vacuum window and were counted with a 450 mm^2 totally depleted surface barrier Si detector 1000 μm thick. The geometric solid angle for a point source under these conditions was estimated to be 17%; however, the measured efficiency was about 14%.

The experiments consisted of simultaneous multiscaler measurements of growth and decay curves for the beta counting rate, neutron counting rate, beta-neutron coincidence rate, and accidental coincidence rates. Data were stored in four 128 channel multiscalers. The ion beam was switched by upstream electrostatic deflection. Cycle times were chosen to give about 2 half-lives of beam-on and 14 half-lives of beam-off. Each cycle included a 1 s period during which the tape was moved about 2 cm. This distance was sufficient to completely remove long-lived beta activity from the beta counter. However, if there were long-lived neutron activities in a short cycle time experiment, these neutrons were still counted in successive cycles. These neutrons were corrected for in the data analysis. To simplify the analysis of background components, the multiscalers were turned on for several channels before the start of each beam-on pulse.

The residence time was determined for our neutron counter by measuring the time intervals between beta start signals and neutron stop signals. With a residence half-time of 11 μs and a coincidence resolving time of 40 μs, 92% of the true coincidence events were included. The fraction of true events not detected does not influence the present results because we normalize the P_n measurements to a known P_n value measured under identical conditions. The coincidence rate was measured by a simple overlap coincidence module where the beta pulse input was stretched to 40 μs by a gate and delay generator. To measure the accidental coincidence rate, the same beta pulse was sent to a second coincidence module and overlapped with neutron pulses which had been delayed 45 μs. After correcting each coincidence rate for deadtime effects, the difference was the true coincidence rate.

The beta singles, neutron singles, and coincidence growth and decay curves were analyzed by the least square fitting program MASH [WOH78]. This program explicitly includes parent, daughter, and granddaughter relationships. Delayed neutron branching to the one-unit-lower mass chain as well as branching to isomeric states can be included in the beta decay curve analysis. The program outputs the saturation counting rate of each component present in the sample. Half-lives can be varied by an iterative procedure to find the best fits.

The determination of P_n values is based on the beta saturation counting rate ($C^\beta{}_{sat}$), the neutron saturation counting rate ($C^n{}_{sat}$), the beta-neutron coincidence saturation counting rate ($C^{\beta n}{}_{sat}$), the beta counting efficiency (ϵ_β), and the neutron counting efficiency (ϵ_n). The usual relation for the delayed neutron emission probability is

$$P_n = \frac{C^n_{sat} / \epsilon_n}{\sigma^\beta_{sat} / \epsilon_\beta} \qquad (1)$$

By measuring the coincidence rate, we can obtain P_n from

$$P_n = \frac{\sigma^{\beta n}_{sat}}{\sigma^\beta_{sat} \cdot \epsilon_n} \qquad (2)$$

If one could reproduce the neutron and beta efficiencies from one mass to the next, one could use a known P_n value to determine the ratio of beta efficiency to neutron efficiency and then calculate all other P_n values using Eq. 1. With our present apparatus, the neutron efficiency is insensitive to changes from one mass number to the next, but the beta efficiency depends more critically on how the beam is tuned. We thus use Eq. 2 to calculate P_n because the beta efficiency does not appear in the expression. In some mass chains with several precursors (either isomers or isobars) it is possible to determine the beta efficiency for one of the prominent precursors from the ratio of the coincidence counting rate to the neutron counting rate. We then calculate P_n values for the other precursors in that mass chain by use of Eq. 1. In all cases we use the neutron counting efficiency determined for ^{98}Rb by use of Eq. 2 and the P_n value of 13.6 ± 0.9 %.

The half-lives and P_n values are presented in Table I. These results differ slightly from a previous report of this work [WAR85] due to additional experiments performed at masses 81-83, 98-100, 131-132, and 147-149. The latest experiments used the ring configuration for the neutron counter, whereas the previous experiments used the quadrant configuration. We have assumed equal neutron counting efficiencies for all precursors. The uncertainties shown on the P_n values include all the statistical uncertainties plus a systematic uncertainty of 10% on the neutron efficiency for data taken with the ring configuration or a systematic uncertainty of 20% for data taken with the quadrant configuration. These systematic uncertainties account for the energy dependence of the neutron counter efficiency and for our lack of knowledge of the neutron energy spectra of these precursors. Upper limits are based on twice the uncertainty when the error on the P_n exceeded the value.

Table I also shows recommended P_n values from a recent compilation [MAN84]. The present results include 11 precursors for which no P_n values have been reported previously. The $^{100-102}$Y P_n values are significant in that these precursors were expected to contribute a large fraction of the unmeasured total delayed neutron yield in microscopic summation calculations [ENG83, REE84]. If one compares the experimental P_n values to estimated values using the prescription given in [MAN84], one observes a number of low ratios in the vicinity of mass 100. In Fig. 1 where the ratio of experimental P_n to calculated P_n is plotted versus neutron number, one sees a group of unusually low P_n values for Rb, Sr, and Y precursors with neutron number just above the shell closure at 60. These low P_n values may be another manifestation of the region of deformed nuclides around mass 100 which has been identified by gamma spectroscopy experiments.

Table I. Half-lives and delayed neutron emission probabilities

Precursor	Half-life (s)	P_n (%) [a]	P_n (%) (Mann) [b]
75 Cu	1.3 ± 0.1	3.5 ± 0.6	
79 Ga	2.85 ± 0.01	0.055 ± 0.012	0.10 ± 0.01
80 Ga	1.69 ± 0.01	0.69 ± 0.16	0.86 ± 0.08
81 Ga	1.218 ± 0.004	11.7 ± 1.2	12.2 ± 1.2
82 Ga	0.609 ± 0.003	20.9 ± 2.2	21.9 ± 2.6
83 Ga	0.308 ± 0.004	62.8 ± 6.3	44. ± 8.
95 Rb	0.377 ± 0.001	9.0 ± 1.1	8.59 ± 0.60
97 Rb	0.169 ± 0.001	26.1 ± 5.4	26.9 ± 1.9
98 Rb	0.106 ± 0.001	13.6 ± 0.9 [c]	13.4 ± 0.9
99 Rb	0.059 ± 0.001	20.7 ± 2.3	13.1 ± 1.8
100 Rb	0.059 ± 0.010	5.0 ± 1.0	
97 Sr	0.429 ± 0.005	‹0.05	0.006 ± 0.003
98 Sr	0.653 ± 0.002	0.23 ± 0.05	0.32 ± 0.13
99 Sr	0.269 ± 0.001	0.093 ± 0.012	0.33 ± 0.13
100 Sr	0.201 ± 0.001	0.75 ± 0.08	
101 Sr	0.115 ± 0.001	2.49 ± 0.25	
102 Sr	0.069 ± 0.015	4.8 ± 2.3	
97 Y(m)	1.18 ± 0.04	‹0.08	0.11 ± 0.04
97 Y(g)	3.76 ± 0.02	0.054 ± 0.012	0.059 ± 0.008
98 Y	0.548 ± 0.001	0.23 ± 0.05	0.23 ± 0.05
99 Y	1.470 ± 0.007	1.09 ± 0.11	2.6 ± 2.2
100 Y	0.735 ± 0.004	0.85 ± 0.09	
101 Y	0.431 ± 0.007	2.07 ± 0.21	
102 Y	0.44 ± 0.06	6.0 ± 1.7	
127 In(m)	3.70 ± 0.04	0.54 ± 0.11	0.70 ± 0.08
128 In	0.800 ± 0.001	0.030 ± 0.007	0.060 ± 0.015
129 In(m)	1.18 ± 0.03	2.52 ± 0.52	3.0 ± 0.5
129 In(g)	0.61 ± 0.01	0.13 ± 0.03	0.26 ± 0.05
130 In(m)	0.532 ± 0.006	1.72 ± 0.18	4.4 ± 1.6
130 In(g)	0.278 ± 0.007	0.91 ± 0.10	1.43 ± 0.11
131 In	0.278 ± 0.003	1.70 ± 0.18	1.76 ± 0.25
132 In	0.204 ± 0.006	6.8 ± 1.4	4.3 ± 0.9
147 Cs	0.229 ± 0.001	26.4 ± 2.9	26.4 ± 3.7
148 Cs	0.146 ± 0.003	25.1 ± 2.5	
147 Ba	0.892 ± 0.001	0.021 ± 0.018	0.031 ± 0.22
148 Ba	0.653 ± 0.002	0.057 ± 0.020	0.055 ± 0.013
149 Ba	0.356 ± 0.008	0.58 ± 0.08	
147 La	4.02 ± 0.01	0.041 ± 0.017	0.033 ± 0.008
148 La	1.38 ± 0.02	0.143 ± 0.015	0.13 ± 0.02
149 La	1.10 ± 0.03	1.07 ± 0.13	

[a] This work   [b] Recommended values from [MAN 84]   [c] Used for determination of neutron counting efficiency

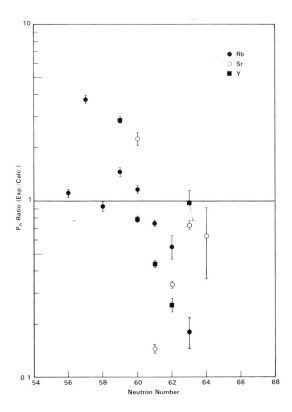

Fig. 1. Ratio of experimental delayed neutron emission probability to calculated emission probability plotted versus neutron number of the precursor nuclide. Calculated values are based on Liran and Zeldes mass formula [LIR76].

Both ^{98}Rb and ^{100}Rb are known to be beta-delayed two-neutron precursors. The P_{2n} for ^{98}Rb was determined in [REE81]. For ^{100}Rb only the ratio of P_{2n}/P_n was measured [JON81]. Combining this ratio with the P_n measured here gives the P_{2n} of 0.14 ± 0.04 % for ^{100}Rb.

We have previously published our results on the new isotopes ^{75}Cu and ^{124}Ag. Both these nuclides were produced by the FEBIAD ion source which has a high ionization efficiency for any element which diffuses from the target matrix. For ^{75}Cu, the half-life is 1.3 ± 0.1 s and the P_n is 3.5 ± 0.6 % [REE85]. For ^{124}Ag only the half-life of 0.54 ± 0.08 s was measured [REE83-1]. The High Temperature Thermal ion source produces particularly good yields of Sr, Y, Ba, and La nuclides. This source was used to measure the new nuclide ^{149}Ba as shown in Table I.

References

[ENG83] T.R. England, W.B. Wilson, R.E. Schenter, and F.M. Mann, Nucl. Sci. Eng. 85 139 (1983).

[JON81] B. Jonson, H.A. Gustafsson, P.G. Hansen, P. Hoff, P.O. Larsson, S. Mattsson, G. Nyman, H.L. Ravn, and D. Schardt, Proceedings IVth Int. Conf. Nuclei Far From Stability, Helsingor, June 7-13, 1981, CERN, Geneva, 1981. CERN 81-09.

[LIR76] S. Liran and N. Zeldes, At. Data Nucl. Data Tables 17 431 (1976).

[MAN84] F.M. Mann, M. Schreiber, R.E. Schenter, and T.R. England, Nucl. Sci. Eng. 87 418 (1984).

[REE81] P.L. Reeder, R.A. Warner, T.R. Yeh, R.E. Chrien, R.L. Gill, M. Shmid, H.I. Liou, and M.L. Stelts, Phys. Rev. Lett. 47 483 (1981).

[REE83-1] P.L. Reeder, R.A. Warner, and R.L. Gill, Phys. Rev. C 27 3002 (1983).

[REE83-2] P.L. Reeder and R.A. Warner, Phys. Rev. C 28 1740 (1983).

[REE84] P.L. Reeder and R.A. Warner, Nucl. Sci. Eng. 87 181 (1984).

[REE85] P.L. Reeder, R.A. Warner, R.M. Liebsch, R.L. Gill, and A. Piotrowski, Phys. Rev. C 31 1029 (1985).

[WAR85] R.A. Warner and P.L. Reeder, Int. Conf. Nuclear Data for Basic and Applied Science, Santa Fe, NM, May 13-17, 1985. PNL-SA 12858.

[WOH78] F.K. Wohn, K.D. Wunsch, H. Wollnik, R. Decker, G. Jung, E. Koglin, and G. Siegert, Phys. Rev. C 17 2185 (1978).

RECEIVED May 9, 1986

Level Densities near the Neutron Separation Energy in Sr-93→97

David D. Clark[1], Robert D. McElroy[1], R. L. Gill[2], and A. Piotrowski[2]

[1]Ward Laboratory, Cornell University, Ithaca, NY 14853
[2]Brookhaven National Laboratory, Upton, NY 11973

Experimental studies of beta-delayed neutron spectra
are a unique source of information on level parameters
of neutron-rich nuclides far from the valley of stability
that are of astrophysical and theoretical interest. Spectra
from mass-separated fission products have been investigated
at the TRISTAN facility at Brookhaven to develop systematics
of neutron resonances in the 1-100 keV range as an experimental
base for testing theoretical extrapolations of level parameters
and neutron cross sections from stable to very unstable
nuclides. Neutron energy is determined by time-of-flight
and associated beta energy by a counter telescope. The
high resolution, clean response function, and beta-energy
sensitivity are exploited to deduce level energies and
densities above the neutron separation energy in the neutron-
emitting nuclide. Experiments to date on five Rb precursors
reveal many newly resolved peaks below 100 keV. Level
parameters from analyses nearing completion are presented
for the neutron-emitters Sr-93-->97 (precursors Rb-93-->97).
Results are in reasonable agreement with values obtained
with the formulas of Gilbert and Cameron.

A frequent decay mode of very neutron-rich nuclides such as fission
products is the two-stage cascade in which a precursor nuclide (PR) beta-
decays to an excited state of a neutron-emitter nuclide (NE) with enough
excitation energy to form a grandchild nuclide (GC) in ground or excited
states. This phenomenon affords unique empirical clues to neutron cross
sections of far-unstable nuclides. The key point is to exploit the inverse
relationship between neutron emission and absorption by compound-nuclear
levels above the neutron binding energy. Data of sufficient quality
and quantity to allow construction of empirically based cross sections
for very neutron-rich nuclides would be of interest for astrophysics
and for nuclear theory. Such cross sections would in some cases be dominated
by individual isolated resonances and in others would be viewed as averages
over a number of closely spaced resonances. In either case, delayed
neutron spectra are a unique source of data for nuclides that are completely
inaccessible to direct measurement. The validity of this approach has
been demonstrated recently by Kratz et al. [KRA83a] who applied it in
the case of two precursors, Br-87 and K-50, and has been discussed further
by Wiescher et al. [WIE85].

0097-6156/86/0324-0177$06.00/0
© 1986 American Chemical Society

Over the past several years we have conducted a program to develop a high resolution time-of-flight (TOF) system for delayed neutron spectra below 100 keV and to explore its application to mass-separated fission product precursors at the TRISTAN facility at the High Flux Beam Reactor at Brookhaven National Laboratory. The original goal was a system with sufficient resolution, efficiency, and background rejection to obtain natural line widths of a statistically useful number of levels; interpretation of such data to obtain level densities and cross sections would require a minimum of additional assumptions. Although that goal proved to be unfeasible, spectra have been obtained with very considerably improved resolution (below 100 keV) over that in earlier ones taken with Cuttler-Shalev ion chambers (Kratz et al. [KRA83b], Reeder et al. [REE80], Rudstam [RUD79]) and with proton recoil counters (Greenwood and Caffrey [GRE83]). The TOF spectra therefore exhibit many more resolved peaks and make possible more reliable deductions of level densities. We report here nearly final results for neutron-emitters Sr-93 through Sr-97, which have half-lives of seconds or less and have (respectively) four to eight more neutrons than the highest mass Sr isotope that can be studied by neutron capture on stable Sr isotopes.

Experimental Method

The basic experimental method has been described in earlier reports [CLA83, MCE85]. Briefly, the TOF apparatus consists of a plastic scintillator telescope to detect betas that feed neutron-emitting levels and a Li-6 glass scintillator to detect the emitted neutrons. A fast coincidence between the two beta detectors provides the stop pulse for a time-to-amplitude converter (TAC); the start pulse is from the neutron detector. The observed FWHM of the TOF peak due to betas in the telescope and gammas in the neutron detector was typically about 2.9 ns under running conditions. The slow signals from the neutron and thick beta detectors are digitized and recorded in three-parameter event mode. The neutron detector pulse height is used to eliminate many of the pulses due to gammas. The thick beta detector pulse height can be used to obtain the spectrum of betas that are in coincidence with selected neutron time (energy) gates and thereby aid in deciding which parts of the neutron spectrum feed different levels in the final (GC) nuclide. The neutron scintillator is of type NE912 glass, 0.95 mm thick and 127 mm diameter, shielded on its face by 6.7 mm of Boral and 14 mm of lead and on its sides by 6.4-mm boron-loaded plastic. Path lengths of 50, 71, and 120 cm have been used, and in later runs as many as four different neutron detectors were used at the same time, in effect running four quasi-independent simultaneous experiments. Time dependent and equilibrium gamma singles spectra of the active deposit were also recorded in most runs.

The most important parameter of the detection system is its response function. We have studied this extensively in Monte Carlo and other calculations. The calculated time-spectrum response to monoenergetic neutrons is composed of a Gaussian timing curve (2.97-ns FWHM), a trapezoidal contribution from detector thickness and non-axial paths, and an exponential tail, calculated by Monte Carlo, from multiple scattering in the neutron scintillator. (Spectrum distortion due to neutrons multiply scattered by structural and other parts of the apparatus and arriving at the neutron detector with incorrect flight times was also established by Monte Carlo and found to be negligible.) The time-spectrum response is a very clean, nearly symmetric peak with a FWHM resolution in the energy spectrum of 2% to 4% for the flight paths that were used. The only feasible check of this calculated response function is measuring the spectrum of Br-87 and comparing the analyzed results with an expected spectrum derived from the precisely known resonance energies and widths found by Raman et al. [RAM83] in cross section measurements. We plan such a calibration of our system when the appropriate ion source is available to us at TRISTAN.

Data Reduction and Analysis

The three-parameter event mode data were scanned with various software windows and the results were processed in standard fashion. A brief outline follows. The raw TOF spectra showed a delayed-neutron region, a narrow peak due to true beta-gamma coincidences, and on the other side of the peak a "non-physical" region due to "events" in which the stop signal occurred before the start signal; this region was useful for correctly subtracting random coincidences. The neutron pulse height spectrum had the characteristic appearance for Li-6 glass -- a broad neutron peak on a sloping gamma platform. The gamma-only portions were used as software gates in scanning the TOF spectra to determine a realistic shape for the random coincidence platform lying under the delayed neutrons. A large number of delayed-neutron peaks were discernible in the TOF spectra gated by the neutron portion; many of these were fitted to a convolution of the calculated system response function with Lorentzian shapes of various widths. However, only the two prominent peaks in Rb-95 showed statistically significant natural line widths [MCE85]. Therefore the use of a level width distribution as a missing-level indicator was not possible and an intensity distribution approach was adopted.

The intensity method requires a careful analysis of the entire TOF spectrum because it compares the summed intensities in resolved peaks with the total intensity including the unresolved continuum. Li-6 glass exhibits a number of scattering resonances which add exponential tails of varying sizes and decay lengths to the spectral peaks while reducing their primary size. Monte Carlo calculations were carried out following a method of Kinney [KIN76] to determine the reduction and tail parameters as a function of incident neutron energy. Using these, a partial spectrum stripping was performed to remove the tails and correct the spectrum shape. These corrections can be significant to as low a neutron energy as 70 keV. As part of the same unfolding operation the Li-6 (n,alpha) cross section was used to provide the system efficiency.

The fluctuations in neutron peak intensities arise from the Porter-Thomas distributed beta decay widths to levels in the NE nuclide. In the simplest case only a single state in the GC nuclide can be fed and only one neutron partial wave is significant. The observed levels will be a subset of levels in the NE nuclide and will be distributed in energy following a Wigner distribution. In a typical GC nuclide, however, there will be a number of accessible final states and the delayed neutron spectrum will be a superposition of transitions from several parts of the NE nuclide level structure.

In the case where a single final state is accessible in the GC nuclide
the level spacing may be estimated rather simply. A suitable energy
interval is chosen by considering the system resolution and the spacing
of observed peaks. The ratio of the summed intensity in discrete peaks
to the total neutron intensity is simply related through the Porter-Thomas
distribution to the average level spacing in the energy interval. For
each of the Sr nuclides in this study more than one state is available
and neutron peaks must be assigned to particular final states and partial
neutron branching ratios must be determined. This is the most difficult
part of the analysis. Each isotope is treated separately, but in general
since the level density is an exponentially increasing function of excitation
energy the average intensity of peaks will drop sharply as higher final
states in the GC nuclide are considered. To obtain a total level density
for all spin states requires spin and parity assignments for the PR and
GC nuclides and to deduce a level density parameter a the neutron binding
energy B_n in the NE nuclide must be known. These and other values are
given in Table 1.

Results and Conclusions

The results of these studies are presented in Table 2 and compared
with values calculated with the formulas of Gilbert and Cameron [GIL65].
In Rb-93 and Rb-94 it was assumed that any transition strong enough to

Table 1.   General Data

PR nuclide			NE nuclide		Neutron spectra		GC nuclide		
A	Spin	P_n	B_n	Spins	KE range	No. of discrete peaks	Spin	P_n^i	
		%	MeV		keV			%	
93	$5/2^-$	1.4	5.46	$3/2^-$	36-156	9	0^+	85.2	[a]
94	3^-	10.2	6.83	$2,3,4^-$	12-100	16	$7/2^+$	73	[a]
95	$5/2^-$	8.6	4.36	$3/2^-$ $3/2,5/2,7/2^-$	9-100	10 16	0^+ 2^+	42 ± 4 49 ± 4	* *
96	2^-	14.2	5.89	$1/2^+$	--	--	$1/2^+$	24	[a]
97	$3/2^+$	26.9	4.01	$1/2^+$ $3/2,5/2^+$ $1/2^+$	3-40	0 10	0^+ 2^+ 0^+	0 ± 20 50 ± 9	* *[a]

*This work
[a] Reference KRA83b

Section II: Experimental Exploration of Current Issues
Part 1: Intruder States in Nuclei
Chapters 27-39

Issues currently undergoing experimental investigation appear in this section. Relative numbers of papers in each part (intruder states in nuclei, octupole modes in nuclei, high-spin-state investigations using heavy ions, and nuclear moments) roughly represent the current division of interest in nuclear structure research. These experiments are sources of new information that can be used to constrain and test nuclear models. Because each generation of nuclear models tends to be "fitted" in some sense to the then-current data base, the further generation of experimental data in unexplored areas is very important for two reasons: predictive powers of models are tested for performance in new regions of the nuclear landscape, and carefully designed experiments can sometimes serve to critically test the basis for a theoretical model.

In addition to many new results, many new and sophisticated advances in experimental techniques are described in this section. The rich variety of experimental innovation promises to yield an enormous wealth of new data, especially for nuclei far off the line of stability—those that are critical for testing models of nuclear structure and in astrophysics. The chapters in this section, therefore, confirm that a lively and clever group of experimenters is prepared to provide data necessary for future development of nuclear theories and models.

Toward a Shell-Model Description of Intruder States and the Onset of Deformation

K. Heyde[1], P. Van Isacker[1], R. F. Casten[2], and J. L. Wood[3]

[1]Institute for Nuclear Physics, Proeftuinstraat 86, B-9000 Gent, Belgium
[2]Brookhaven National Laboratory, Upton, NY 11973
[3]School of Physics, Georgia Institute of Technology, Atlanta, GA 30332

Basing on the nuclear shell-model and concentrating on the monopole,pairing and quadrupole corrections originating from the nucleon-nucleon force,both the appearance of low-lying 0^+ intruder states near major closed shells(Z=50, 82)and sub-shell regions (Z=40,64) can be described.Moreover,a number of new facets related to the study of intruder states are presented.

In a nucleus,the major part of the nucleonic motion is determined by the average single-particle field.This field approximation implies the existence of magic numbers i.e. 20,28,50,82,126 ,determining extra stability for the particular nuclei having proton and/or neutron number equal to such a magic number.The energy separation between the last filled single-particle orbitals and the lowest,unfilled orbitals varies from ≈6 MeV(light nuclei) to≈3.5 MeV(Z=82 gap).Besides,a number of subshell closures have by now been established i.e. Z=40,64.In these cases,a much smaller energy gap of 1 and 2.5 MeV,respectively results.It is precisely the existence of particular configurations that makes the study of nuclear structure approximate tractable.Most nucleons can be considered to contribute to the "core" nucleus and thus to the average field leaving only a small number of nucleons,called valence nucleons(particles or holes) outside closed shells(see fig.1).In region I,few valence nucleons are present

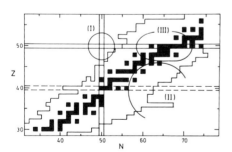

Figure 1. Schematic divisions of nuclei in three major regions: (I) region near doubly closed shells; (II) region of strongly deformed nuclei; (III) region of intruder states.

and the nuclear shell-model techniques[SHA63] allow for a detailed study of low-lying nuclear excited states.In region I,the pairing properties between identical nucleons are primordial and strongly collective motion does not develop until many valence protons and neutrons (region II [BOH75]) moving in identical single-particle (or spin-orbit partner) orbitals[SHA53,FED79] are present.In this case,the strong proton-neutron interaction is the dominant contribution and leads to permanently deformed nuclei.

The separation in two regions I,II very much relies on the existence of a set of well defined closed shell configurations.Since the value of the shell gaps can be shown to be A dependent(see sect.2 [GU077,SOR84]),a more detailed analysis of the variation of shell gaps as a function of A will have to be studied.

There now also exists an idealized region III(see fig.1) where one type of nucleons is near a closed shell configuration(very few valence nucleons) and the other has a maximal number of valence nucleons(mid-shell situation).It has been shown that in such nuclei,the closed shell can be excited and low-lying particle-hole (p-h)excitations result(1p-2h or 2p-1h in odd-A nuclei having originally 1h or 1p outside a closed shell).These "intruder" states for which one would expect,at first,a larger excitation energy,have been studied in detail for odd-A nuclei in [HEY83],where most closed-shell regions are covered in detail.In the present discussion(specific for even-even nuclei),we present a shell-model approach for describing <u>both</u> the small excitation energy and the particular A-dependence of such intruder states throughout region III nuclei.

2.Shell-model description of intruder states

In a schematic presentation(see fig.2) of a single-closed shell nucleus, we take as the lowest intruder state (0^+ state) a proton (π) 2p-2h configuration

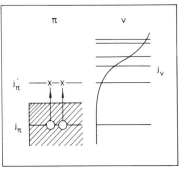

Figure 2. Schematic representation of a proton (π) intruder 2p-2n 0^+ configuration where j_π denotes the regular orbital, j'_π the intruder orbital and j_ν the neutron orbitals filled in a BCS way (ν).

with the valence neutrons distributed over the available single-particle orbi-
tals.The unperturbed energy then becomes

$$E_{intr.}^{unp.}(0^+) = 2(\varepsilon_{j_\pi'} - \varepsilon_{j_\pi}) \tag{1}$$

We now shortly discuss the different energy corrections due to both the strong
pairing correlations in the 2p and 2h configurations and the proton-neutron interaction affecting both the proton single-particle energies and the binding
energy of these intruder 2p-2h configurations due to core polarization effects
of the underlying 0^+ core.

2.1 Monopole field correction

The proton single-particle energies become modified due to the proton-
neutron interaction in the following way

$$\tilde{\varepsilon}_{j_\pi} = \varepsilon_{j_\pi} + \sum_{j_\nu} v_{j_\nu}^2 (2j_\nu +1)E(j_\pi,j_\nu) \tag{2}$$

with $E(j_\pi,j_\nu)$ the average proton-neutron interaction energy[TAL63] and $v_{j_\nu}^2$ the
BCS occupation probabilities of the j_ν neutron orbitals.Because of these
single-particle energy variations,a first correction to the unperturbed proton
intruder configurations results as

$$\Delta E_M = 2(\tilde{\varepsilon}_{j_\pi'} - \tilde{\varepsilon}_{j_\pi}) - 2(\varepsilon_{j_\pi'} - \varepsilon_{j_\pi})$$

$$= 2\sum_{j} v_{j_\nu}^2 (2j_\nu +1) \left[E(j_\pi',j_\nu) - E(j_\pi,j_\nu)\right] \tag{3}$$

This monopole correction has been studied in some detail for both the large
shell Z=50,82 and the subshell Z=40,64 regions[HEY85].it thereby becomes clear
that for the large spherical single-particle shell gaps(50,82) only a small
modification of the original gap results.On the contrary,for the subshell regions,when N increases,an almost complete eradication of the subshell gaps
(for Z=40 near N=58,60;for Z=64 near N=88,90) results and thus,in the latter
cases,we end up with nuclei similar to the region II type of fig.1,thus giving
rise to strongly deformed nuclei which is also experimentally the case.

2.2Pairing field correction

When exciting a 2p-2h configuration,we form 0^+ coupled configurations as
the lowest-lying states therby obtaining a large pairing binding energy gain.
This latter value can be calculated using the residual nucleon-nucleon force
or estimated,starting from proton separation energies as

$$\Delta E_p = \Delta E_{pairing}(part.) + \Delta E_{pairing}(holes) \tag{4}$$

$$\Delta E_{pairing}(holes) = 2S_p(Z,N) - S_{2p}(Z,N) \tag{5}$$

(similar expression for $\Delta E_{pairing}(part.)$),where $\Delta E_{pairing}$ is determined at(if
experimentally known) or near the doubly-closed shell configurations i.e.

^{208}Pb, ^{146}Gd, ^{132}Sn, ^{90}Zr,..).This pairing correction results in an important lowering of the intruder state energy,which is of the order of $E_p \approx 4\text{-}5$ MeV at Z= 50,N=82;$E_p \approx 2.5$ MeV at Z=82,N=126 region,but largely independent of the mass number A.

2.3 Quadrupole proton-neutron energy correction

Up to now,we always assumed a 0^+ coupled pair distribution to give a good description of both the proton 2p-2h excitation and of the neutron distribution over the valence model space.The quadrupole component of the proton-neutron force will induce $0^+ \to 2^+$ pair breaking for both protons and neutrons which we call the core polarization effect.Thus,0^+ ground state and intruder wave functions are changed into

$$|0^+_\pi \otimes 0^+_\nu> \to |0^+_\pi \otimes 0^+_\nu> + \alpha|2^+_\pi \otimes 2^+_\nu> + \dots . \qquad (6)$$

Using perturbation theory and the lowest seniority shell-model description of the 0^+ pair distribution,one calculates a quadrupole polarization energy gain

$$\Delta E_Q = \kappa^2_{sm} N_\nu (\Omega_\nu - N_\nu).F \qquad (7)$$

where κ_{sm} is the strength of the quadrupole proton-neutron force component ($\kappa_{sm} Q_\pi . Q_\nu$ with $Q_\rho \equiv (r\sqrt{\frac{m\omega}{\hbar}})^2 Y_2(\hat{r}_\rho)$) and N_ν the number of neutron pairs (F contains the particular j'_π ,j_π and j_ν orbital information).If many valence nucleons are present,many 2^+ proton-neutron pairs can become admixed and perturbation theory breaks down.In the latter case,using an SU(3) wave function for describing the repartition of 0^+ and 2^+ coupled proton and neutron pairs, one obtains the expression

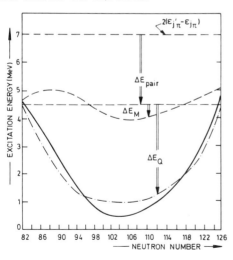

Figure 3. The total energy correction (full thick line) due to both the monopole (ΔE_M ; dashed line), the pairing (ΔE_p ; straight dashed line) and the quadrupole correlations (ΔE_Q ; dot-dashed line), for the Pb(Z = 82) region. The unperturbed energy (upper straight dashed line) is also shown.

$$\Delta E_Q = 2\kappa_0 N_\pi N_\nu (\Omega_\pi - N_\pi)^{1/2} \cdot (\Omega_\nu - N_\nu)^{1/2} \tag{8}$$

(κ is the strength of the quadrupole force in IBM-2 calculations[ARI84] in a sd-boson model space with $\kappa = \kappa_0 (\Omega_\pi - N_\pi)^{1/2} (\Omega_\nu - N_\nu)^{1/2}$ [SCH80]).In the quadrupole energy gain,for given N_π near closed shells ($N_\pi \ll \Omega_\pi$),a very particular N_ν dependence results,being maximal at mid-shell configurations,conform with the known experimental facts on intruder states in even-even nuclei near or at closed-shells(Sn,Pb,..).

Collecting all parts ,one obtains the final expression,

$$E_{intr.}(0^+) = 2(\varepsilon_{j_i} - \varepsilon_{j_\pi}) + \Delta E_M + \Delta E_P + \Delta E_Q \tag{9}$$

(relation which is illustrated in fig.3 for the Pb region).The expression (9) for $E_{intr.}(0^+)$,is obtained in a separable form pointing out clearly the different contributions from the nucleon-nucleon interaction.This is of course an approximation to the more general result one would obtain when carrying out a full HFB calculation in determining

-the proton(neutron)single-particle energies and orbitals in a self-consistent way for each Z,N,A

-diagonalizing in the appropriate basis,for describing the lowest excited states,the residual nucleon-nucleon interaction [FED79]

We also point out that,when the spherical shell-gap,giving rise to the possibility of defining 2p-2h excitations,disappears due to the monopole correction ΔE_M,the above procedure of separating the different parts of the nucleon-nucleon interaction(single-particle field,pairing correlations and quadrupole correlations)will break down. This is signaled by the intruder 0^+ state crossing the regular 0^+ ground state of the nuclei we are discussing.This is, for instance, the case in the Zr nuclei (near $N \approx 60$) and the Gd nuclei ($N \approx 90$).

3. New facets of intruder states

Here,we shortly mention the possibilities of other intruder state properties:

-"Scaling " property for the intruder excitation energy[VAN85] .Recently,scaling properties have been discussed at length by R.F.Casten [CAS85,85a] when studying low-lying regular collective excitations($2^+,4^+$ energies,B(E2) values)

-intruder states in odd-odd nuclei[MAL82,VAJ83,HUY85,NES82]

-one-broken pair intruder states:these are situations where we consider the $(j_{p_1} j_{p_2})J_{max}$. $(j_h^{-2})0^+;J_{max}$. configurations.These should occur at an energy, roughly $\Delta E_{pairing}$(part.) above the lowest 0^+ intruder state.Possibilities for such states can exist in the Pb region[NES83,HES85]

-intruder states near neutron closed shells[HEY83]

Acknowledgments

The authors are most grateful to S. Pittel, I. Talmi, A. Aprahamian, M. Huyse, R. A. Meyer, J. Sau, P. Van Duppen, and D. Warner for many stimulating discussions. This research was performed in part under contract DE-AC02-76CH00016 and DE-AS05-80ER10599 with the United States Department of Energy and the NATO grant RG 0565/82/D1.

References

[ARI84] A.Arima and F.Iachello,in Advances in Nuclear Physics,vol.13,139(1984)

[BOH75] A.Bohr and B.Mottelson,Nuclear Structure,vol.2(ed.W.A.Benjamin Inc., Reading,1975),ch.4

[CAS85] R.F.Casten,Phys.Rev.Lett.54,1991(1985)

[CAS85a R.F.Casten,W.Frank and P.von Brentano,Nucl.Phys.A,to be publ.

[FED79] P.Federman and S.Pittel,Phys.Rev.C20,820(1979)

[GOO77] A.L.Goodman,Nucl.Phys.A287,(1977)

[HES85] W.Hesselink,proceedings this conference and priv.comm.

[HEY83] K.Heyde,P.Van Isacker,M.Waroquier,J.L.Wood and R.A.Meyer,Phys.Repts.102 291(1983) and refs.therein

[HEY85] K.Heyde,P.Van Isacker,R.F.Casten and J.L.Wood,Phys.Lett.155B,303(1985)

[HUY85] M.Huyse et al.,proceedings this conference and priv.comm.

[MAL82] J.Van Maldeghem,J.Sau and K.Heyde,Phys.Lett.116B,387(1982)

[NES82] P.Van Nes et al.,Nucl.Phys.A379,35(1982)

[NES83] P.Van Nes et al.,Phys.Rev.C27,1342(1983)

[SCH80] O.Scholten,Ph.D Thesis,University of Groningen,The Netherlands,1980

[SHA53] A.de Shalit and M.Goldhaber,Phys.Rev.C92,1211(1953)

[SHA63] A.de Shalit and I.Talmi,Nuclear Shell Theory,(Academic Press,New York, 1963),ch22

[SOR84] R.A.Sorensen,Nucl.Phys.A420,221(1984)

[VAJ83] S.Vajda et al.,Phys.Rev.C27,2995(1983)

[VAN85] P.Van Duppen,M.Huyse,J.L.Wood,K.Heyde and P.Van Isacker, to be publ.

RECEIVED May 8, 1986

28

In-Beam Spectroscopy Using the (t,p) Reaction
Recent Results near $A = 100$

E. A. Henry[1], R. J. Estep[2], Richard A. Meyer[1], J. Kantele[3], D. J. Decman[1], L. G. Mann[1], R. K. Sheline[2], W. Stöffl[1], and L. E. Ussery[4]

[1]Lawrence Livermore National Laboratory, University of California, Livermore, CA 94550
[2]Department of Chemistry, Florida State University, Tallahassee, FL 32306
[3]Department of Physics, University of Jyväskylä, SF-40100, Jyväskylä, Finland
[4]Los Alamos National Laboratory, Los Alamos, NM 87545

Charged particle spectroscopy using the (t,p) reaction has been employed for more than two decades to study the low-energy structure of nuclei. This reaction has contributed significantly to the elucidation of single-particle and collective phenomena for neutron rich nuclei in virtually every mass region. We have begun to use the (t,p) reaction in conjunction with in-beam γ-ray and conversion-electron spectroscopy to bring additional understanding to low-energy nuclear structure. In this report we briefly discuss the experimental considerations in using this reaction for in-beam spectroscopy, and present some results for nuclei with mass near 100.

EXPERIMENTAL METHODS

Until now the only methods available for studying the beta unstable nuclei with a mass near 100 were the prompt γ-ray decay and beta decay of fission products, charged-particle spectroscopy using two-neutron transfer reactions, and, to a limited extent, in-beam spectroscopy using reactions like ($^{18}O, ^{16}O\gamma$). In-beam spectroscopy using the (t,pγ) reaction has several features that make it an attractive technique to complement these methods. 1) Even-even nuclei that have two neutrons more than the heaviest target can be studied by the (t,pγ) reaction with useful cross sections. 2) The levels in the product nucleus are populated by both direct and compound nuclear reactions. Thus the set of levels that are identified at low excitation energies can be quite complete. 3) The spin distribution of levels populated is broader than is usually the case in beta decay. The ground state band is often populated up to the 8^+ member. In the same experiment 0^+ states can also be populated, probably by the direct reaction mechanism. 4) The (t,p) reaction has a unique signature, an energetic proton that identifies that particular channel.

The disadvantages of this reaction place some real constraints on its use. 1) The (t,p) cross section is only about 5 percent of the total cross section. 2) The dominant reaction, usually (t,2n), produces abundant prompt γ rays. 3) Reactions such as (t,n) and (t,d) [as well as (t,p)] often result in short-lived beta decaying products. 4) The usual in-beam techniques such as angular distributions are complicated by the necessity to use the outgoing proton to identify the reaction. As a result of the first three disadvantages, much of the γ-ray and electron count rates are not from the (t,p) reaction and thus experiments of reasonable duration have limited statistics.

We have developed γ-ray and conversion-electron spectroscopy techiques that take advantage of the energetic proton as an indicator of the (t,p)

0097-6156/86/0324-0190$06.00/0

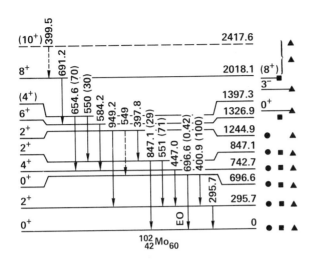

Fig. 1. Level scheme of ^{102}Mo determined from (t,pγγ) studies. Previous studies are beta decay (circles), (^{18}O, ^{16}Oγ) (squares), and (t,p) triangles.

reaction. Our conversion-electron spectrometer and its use with the (t,p) reaction have been described in the literature [DEC84, STO84]. For γ-ray spectroscopy, a 1-mm thick cylindrical plastic scintillator detects the protons to gate germanium detectors. A thin tapered cylindrical aluminum absorber inserted into the scintillator prevents reaction deuterons and scattered tritons from reaching the scintillator, while allowing energetic protons to do so. The geometric solid angle for the scintillator is about 30% of 4π. Typically, when the γ-ray singles rate is 10,000 cps, the pγ coincidence rate is 20-80 cps, and the pγγ coincidence rate is 1-5 cps. The γ rays associated with the (t,p) reaction are the dominant ones in the spectrum that is gated by the proton detector, with those from the (t,2n) reaction being attenuated by a factor of fifty or more.

NUCLEAR STRUCTURE STUDIES NEAR A=100

In 1970 Cheifitz et al. [CHE70] presented experimental evidence of rotation-like nuclear structure for ^{102}Zr and several nearby nuclei. They found that as N=50 or Z=50 closed shells were approached, indicators of deformation (e.g. $E_4{}^+/E_2{}^+$) varied farther from the rotational limits, but were also significantly different from the values for spherical nuclei. Experimentally, many of the nuclei in this mass region display complex low energy structure. Thus the experimental knowledge of the structure of nuclei surrounding the deformed region near A=100 must be as complete as possible to confidently apply a theoretical description. We report here on preliminary results from four in-beam studies of nuclei in this mass region, and discuss the results briefly in the context of the nuclear structure of the region.

^{102}Mo. The (^{18}O,^{16}Oγ) reaction has been used by Koenig et al. [KOE81] in an attempt to establish the yrast levels in ^{102}Mo. Our pγγ coincidence data confirm the $4_1{}^+$ and $6_1{}^+$ levels established in that study (See Fig. 1). However, it is clear from our data that the $8_1{}^+ \rightarrow 6_1{}^+$ transition is at 691 keV, establishing the $8_1{}^+$ level at 2018 keV. The 655-keV γ ray, assigned by Koenig et al. as

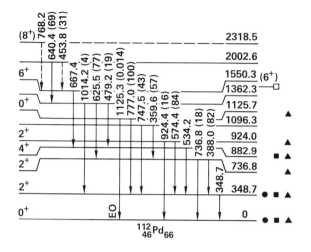

Fig. 2. Level scheme of ^{112}Pd determined from (t, pγγ) studies. Previous studies are beta decay (circles), fission fragment decay (squares), and (t,p) triangles.

the $8^+_1 \rightarrow 6^+_1$ transition, actually feeds the 4^+_1 level directly from a new level at 1397 keV. We suggest that this new level has a J^π value of 4^+.

Gamma rays with energies of 398 and 401 keV are known from previous studies and are observed as a multiplet in the (t,pγ) data. These γ rays depopulate the 2^+_2 and 0^+_1 levels, respectively. However, a gate on these γ rays reveals coincidences with all the members of the yrast band, not just the $2^+_1 \rightarrow 0^+_1$ transition as expected. A narrower gate set at about 399 keV reveals weak coincidences with the $8^+_1 \rightarrow 6^+_1$, $6^+_1 \rightarrow 4^+_1$, and $4^+_1 \rightarrow 2^+_1$ transitions. Though the statistics are poor, the intensities for these transitions in the coincidence spectrum are consistent with the 399-keV γ ray feeding the 8^+_1 level. We suggest that this transition may be the $10^+_1 \rightarrow 8^+_1$ yrast transition. If this assignment is confirmed, it will be the first experimentally observed backbend in the neutron rich nuclei near A=100. A backbend at the 10^+_1 level would agree with the calculation by Tripathi et al. [TRI84] of a pronounced backbend at the 10^+_1 level in ^{102}Mo.

^{112}Pd. Little has been known experimentally about the level structure of ^{112}Pd. Previous studies are in agreement only for the J^π assignment of the 2^+_1 level at 349 keV [CAS72,CHE70]. We have established five new levels in ^{112}Pd (see Fig. 2), and have been able to propose J^π assignments for some of the ^{112}Pd levels from decay patterns. The coincidence results established a ground state transition depopulating the 736 keV level, indicating that it is a 2^+ level. The level at 924 keV has a transition to the 2^+_1 level, and a γ ray seen in the proton gated singles is probably the ground state transition, suggesting a J^π value of 2^+ for that level also. Our data confirm that the 4^+_1 level occurs at 882 keV, in agreement with the results of the fission fragment decay studies, but that the previous tentative assignment of the 6^+_1 level was incorrect. Instead, we establish the 6^+_1 level at 1550 keV, and have evidence that suggests that the 8^+_1 level occurs at 2319 keV. New levels at 1096, 1362, and 2002 keV are based on the coincidence results.

Stachel et al. [STA82] have interpreted the structure of Ru and Pd nuclei in terms of the transition between the SU(5) (vibrational) and O(6) (γ-unstable) limits of the Interacting Boson Model (IBM). The energy ratios $E_{4_1}^+/E_{2_1}^+$ and $E_{6_1}^+/E_{2_1}^+$ are consistent with those of the O(6) limit for the heavier Pd and Ru nuclei. Stachel et al. point out that the energy ratio $E_{2_2}^+/E_{2_1}^+$ is not well reproduced in their calculation. Experimentally, that ratio is closer to the SU(5) limit than the O(6) limit, even for ^{112}Pd. However, if the suggested 2_3^+ level at 924 keV in ^{112}Pd is instead the 2_2^+ level within the model space, the $E_{2_2}^+/E_{2_1}^+$ ratio is just above that of the O(6) limit. In ^{112}Pd the $E_{0_2}^+/E_{2_1}^+$ energy ratio is higher than for the lighter Pd nuclei, but still only midway between the SU(5) and the O(6) limits. Stachel et al. suggest that the 0_2^+ level is an intruder level similar to those known in nearby Cd nuclei. Taken together, this evidence indicates that ^{112}Pd cannot be described by the simple O(6) limit of IBM alone. Recent calculations show that complete structures resulting from intruder states must be included with a large degree of mixing occurring between the two configurations [KUS85].

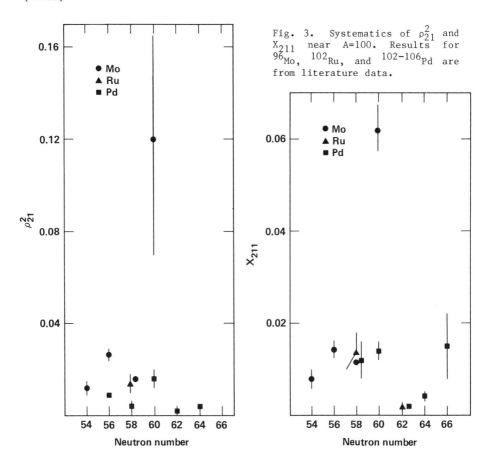

Fig. 3. Systematics of ρ_{21}^2 and X_{211} near A=100. Results for ^{96}Mo, ^{102}Ru, and $^{102-106}$Pd are from literature data.

EO Transitions. We have measured EO transitions in the neutron rich nuclei ^{100,102}Mo, ^{106}Ru, and $^{108-112}$Pd using the (t,p) reactions and coincidence techniques. Listed in Table 1 are the values for the EO(K) branching ratios from our measurements. Where half-life information and E2 branching ratios are available from our own measurements or from the literature, ρ^2 and X values have been determined (see Fig. 3). The ρ^2_{21} and X_{211} values for ^{102}Mo are larger than in our previous report [DEC83] because of the additional γ-ray recently observed at 399 keV in that nucleus.

^{102}Mo has been described as a transitional nucleus which is quite close to being rotational, as its neighbor ^{104}Mo is. If we use Rasmussen's model [RAS60] for the X value ($X_{\beta11} \propto \beta^2$), we find that the deformation parameter β for ^{102}Mo is about 2.3 times that of ^{100}Mo. This suggests that the 0^+_2 level in ^{102}Mo might be characterized as the beta-vibrational level in a rotating nucleus. However, while the yrast levels of ^{102}Mo appear to be evolving toward a rotational character, the lower energy levels might be better characterized by a model that mixes vibrational and γ-unstable limits of the IBM-2 [SAM82].

The measured ρ^2_{21} and X_{211} values for ^{106}Ru and $^{108-112}$Pd are quite small. This would be consistent with the suggestion that these nuclei are evolving

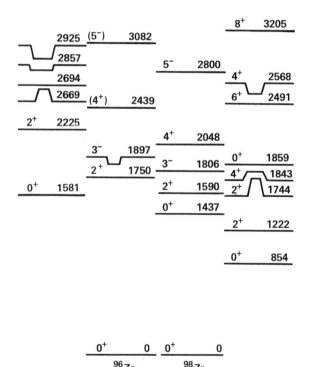

Fig. 4. Levels of ^{96}Zr and ^{98}Zr from (t,p$\gamma\gamma$) studies. Levels are grouped according to their decay properties.

Table 1. EO(K) branching ratios for 0^+_2 levels.

Nucl.	$\dfrac{I_{EO(K)}}{I_{total}}$. (X10^6)		
^{100}Mo	71200	$+$	3600
^{102}Mo	4780	$\mp$	780
^{106}Ru	30	$\mp$	12
^{108}Pd	43	$\mp$	19
^{110}Pd	122	$\mp$	29
^{112}Pd	126	$\mp$	57

from the SU(5) and the O(6) limit of the IBM as neutrons are added [STA82], since the $0_2^+ \to 0_1^+$ transition is forbidden in both limits. As pointed out earlier however, Stachel et al. suggest that the 0_2^+ levels in Pd nuclei may be due to intruder states since their energies are much too low compared to a calculation using the IBM. It may be that the somewhat higher values of ρ_{21}^2 and X_{211} for ^{112}Pd reflect this influence.

^{96}Zr and ^{98}Zr. The preliminary results of our studies for the levels in ^{96}Zr and ^{98}Zr are shown in Fig. 4. The levels for each nucleus are segregated into two groups. Levels within a group γ decay with their largest reduced transition rates to other levels within the group. For ^{98}Zr, the strongest E0 transitions are between the two groups, rather within them, indicating possible shape differences between the 0^+ levels in the two groups. In ^{96}Zr, the (4^+) and (5^-) assignments indicated are not definite, but these levels seem to be the lowest that could have these J^π assignments. For some levels the group preference based on the γ decay criterion is modest. However, on the whole, a coherent pattern is developed.

Bengtsson et al. [BEN84] show that there is extra stability toward a spherical shape in the ^{96}Zr ground state brought on by the $2d_{5/2}$ neutron subshell closure. Focusing on the group that includes the ground state in each nucleus, we see the repeated pattern of levels with J^π of 2^+, 3^-, and, possibly, 4^+ and 5^- at energies which are reasonably close in the two nuclei. The comparison in Fig. 5 suggests that, insofar as excited states in ^{98}Zr are concerned, there may be some additional stability toward spherical shapes afforded by filling the $3s_{1/2}$ neutron orbital. The closeness of the energies of the two ground state groups should not be taken too seriously. The two accompanying "bands" must undoubtedly mix with the respective ground state groups to bring them closer together in energy. Nevertheless, this comparison graphically demonstrates the complex interplay between possible models for nuclear structure in this region.

ACKNOWLEDGMENTS

    A portion of this work is part of the doctoral thesis of one of the authors (RJE). This work was performed under the auspices of the U.S. Department of Energy by Lawrence Livermore National Laboratory under contract No. W-7405-Eng-48.

REFERENCES

BEN84   R. Bengtsson, et al., Phys. Scr. 24, 402 (1984).
CAS72   R. F. Casten, et al., Nucl. Phys. A184, 357 (1972).
CHE70   E. Cheifitz, et al., Phys. Rev. Lett. 25, 38 (1970).
DEC83   D. J. Decman, et al., Nuclear Chemistry Division Annual Report, FY83,
        Lawrence Livermore National Laboratory, Livermore, CA, UCAR 100062-83/1
        pg 181 (1983).
DEC84   D. J. Decman, et al., Nucl. Inst. Meth. 219, 523 (1984).
KOE81   J. Koenig, et al., Phys. Rev. C24, 2076 (1981).
KUS85   D. F. Kusnezov and R. A. Meyer, Nuclear Chemistry Annual Report, FY85,
        Lawrence Livermore National Laboratory, Livermore, CA, UCAR 10062-85/1
        (1985).
RAS60   J. O. Rasmussen, Nucl. Phys. 19, 85 (1960).
SAM82   M. Sambataro and G. Molnar, Nucl. Phys. A376, 201 (1982).
STA82   J. Stachel, P. Van Isaker, and K. Heyde, Phys. Rev. C25, 650 (1982).
STO84   W. Stöffl and E. A. Henry, Nucl. Inst. Meth. 227, 77 (1984).
TRI84   P. Tripathi, S. Sharma, and S. Khosa, Phys. Rev. C29, 1951 (1984).

RECEIVED July 1, 1986

29

Possible Evidence for α-Clustering in the Double Subshell Closure Nucleus ^{96}Zr

G. Molnár[1,2], B. Fazekas[1], T. Belgya[1], Á. Veres[1], Steven W. Yates[2], E. W. Kleppinger[2], and Richard A. Meyer[3]

[1]Institute of Isotopes, Hungarian Academy of Sciences, Budapest H-1525, Hungary
[2]Department of Chemistry, University of Kentucky, Lexington, KY 40506-0055
[3]Nuclear Chemistry Division, Lawrence Livermore National Laboratory, University of California, Livermore, CA 94550

Evidence, based on recent (n,n´γ) reaction experiments and previous particle transfer studies is presented in support of a coexisting four-particle, four-hole band built on the 1581-keV first excited 0^+ state in the doubly closed subshell nucleus ^{96}Zr. An alternative explanation for this band in terms of alpha-clustering appears reasonable.

Shape coexistence arguments were first invoked to explain the first excited 0^+ state and the associated rotational band in doubly closed shell ^{16}O [MOR56]. Since that time the common occurrence of coexisting deformed states near closed shells due to particle-hole pair excitations has been established both in even-even and odd-mass nuclei [HEY83]. This phenomenon has not been thoroughly investigated, however, in closed subshell regions [HEY83,WOO84]. In this paper we present evidence for shape coexistence at double subshell closure, in ^{96}Zr.

Near closed major shells, such as N=50, nuclear shapes are spherical. A gradual transition from spherical to deformed shape is expected to take place as neutron number increases, with maximum deformation at mid-shell, as observed in both even-even [CHE70] and odd-mass [MON82] nuclei near A~100. However, the Z=40 proton subshell closure in this region leads to the appearance of intruder configurations which coexist with the ground-state configuration. The influence of this configuration mixing on the level structure of transitional Mo nuclei was described by recent neutron-proton interacting boson model (IBM-II) calculations which took into account explicitly the two-particle, two-hole proton excitations across the Z=40 subshell [SAM82].

A special situation occurs at ^{96}Zr (Z=40), where the neutron subshell closure (N=56) gives this nucleus a double subshell closure. Thus, particle-hole pair excitations across both subshell gaps are possible. This could produce situations much like those in doubly closed shell ^{16}O, in which the lowest-lying intruder deformed band has been shown to arise due to such excitations [BRO64].

The first excited state of ^{96}Zr is a 0^+ state, as in doubly closed shell ^{16}O or closed shell/subshell ^{90}Zr. Support for the identification of the 0_2^+ state in ^{96}Zr as an intruder deformed state comes from particle transfer

0097-6156/86/0324-0196$06.00/0

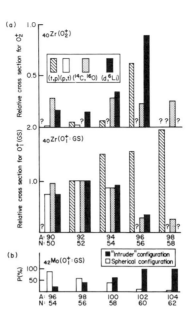

Fig. 1. a) Relative cross sections for transfer reactions populating the first two 0^+ states of Zr nuclei. All cross sections are normalized to the ground state cross section of ^{96}Zr. b) Percentage composition of the ground states of the target Mo nuclei. Unshaded bars: configuration with one proton boson; shaded bars: "intruder" configuration with three proton bosons.

data. Figure 1(a) summarizes the results of two-proton [MAY82], two-neutron [BAL71] and alpha-particle [VAN84] transfer reaction studies; Figure 1(b) shows the admixtures of the basic spherical and the "intruder" deformed boson configuration for the ground state wavefunctions [SAM82] for the Mo targets used in some of these studies. The most striking feature is the unusually high alpha-pickup strength to the 0_2^+ state of ^{96}Zr [VAN84]. Even though some of the two-particle transfer cross sections do not correlate exactly with the alpha-transfer cross sections, which may be due to interference effects, the high value of the cross section ratio in favor of the 0_2^+ state is a common feature. Apparently, this state differs markedly in nature from the supposedly spherical ground state; we suggest that this is due to its character as a four-particle, four-hole deformed state, like the 0_2^+ state of doubly closed shell ^{16}O [BRO64].

In order to elucidate the nature of the 0^+ excitations in ^{96}Zr and to search for associated structures, we have performed $(n,n'\gamma)$ reaction studies. Gamma-ray singles and angular distribution measurements were first carried out at the reactor neutron beam facility of the Institute of Isotopes in Budapest. These measurements led to the identification of several hitherto unknown transitions from previously unplaced levels [FAZ84]. Excitation function measurements performed at the University of Kentucky pulsed neutron beam facility have shown that a 644.2-keV transition, also identified in the Budapest experiments, populates the 0_2^+ state. Parallel beta decay studies [MEY85] indicate that the correct energy of the latter is 1581.4 $\pm$ 0.5 keV, hence the 644-keV transition deexcites the well known 2225-keV state. These results have also been confirmed by parallel in-beam (t,pγ-γ) studies [HEN85]. The angular distributions of the 2225-keV and 644-keV gamma rays deexciting this level to the first and second 0^+ states, respectively, have been measured at the Kentucky facility. As Fig. 2 shows, they are identical and give a unique 2^+ spin-parity assignment, in contrast with the previously suggested value of 3^- [FLY70,SAD75].

We associate the 2225-keV 2_3^+ state with the third largest peak observed in the (d,^{6}Li)^{96}Zr reaction study by van den Berg et al. [VAN84]. Although these authors prefer a 5^- assignment, their angular distribution data are also consistent with spin 2^+ which gives about 60 percent alpha-pickup strength with respect to the ground state strength. We have searched for a possible doublet at this energy with the (n,n'γ) reaction which should lead to significant population of a 5^- state, but found no evidence for a second state. Hence the (d,^{6}Li) reaction leads to strong population of the 2_3^+ state as well. The measured $B(E2;2_3^+\rightarrow0_2^+)/B(E2;2_3^+\rightarrow0_1^+)$ ratio of 10^2 demonstrates a strong preferential decay to the 0_2^+ state. These two facts suggest the 2225-keV state as the 2^+ member of an intruder deformed band built on the 0_2^+ first excited state in ^{96}Zr. According to the angular distribution pattern of the 1107-keV transition depopulating the 2857-keV state (see Fig. 3) this level is an obvious candidate for the 4^+ band member. It can be associated with the 2.87-MeV peak in the (d,^{6}Li) spectrum of van der Berg et al. [VAN84] even though their particle angular distribution is best fitted with L=3. In Fig. 4 we summarize our results, emphasizing the proposed band structure.

It is important to understand what gives rise to the established coexisting intruder band in doubly closed subshell ^{96}Zr. The slight predominance of the intruder deformed configuration in the ground state wavefunction of ^{100}Mo shown in Fig. 1(b) cannot explain the tremendous difference between the alpha-pickup strengths to the two 0^+ states of ^{96}Zr clearly demonstrated by Fig. 1(a). This is not unexpected, since the simultaneous occurrence of proton and neutron subshell closures should produce a marked difference with respect to other nuclei near the middle of the 50 to 82 neutron shell where two-proton, two-hole excitations can account entirely for the observed shape coexistence phenomena.

Recently, Iachello and coworkers [IAC82,84] have developed the nuclear vibron model (NVM) which introduces bosonic degrees of freedom with $J^\pi=0^+$ and 1^- to describe dipole deformation due to clustering. This model can account for several features of nuclei such as the low-lying negative-parity states with small alpha-decay hindrance factors in ^{222}Ra and the unusually large population of groups of excited states in the actinides by the (d,^{6}Li) alpha-pickup reaction [IAC84]. Since this last feature is also exhibited by ^{96}Zr, it may be possible to interpret the observed coexisting intruder band as a dipole band associated with alpha-clustering, even though clear-cut evidence for the associated negative-parity band is still lacking.

In summary, the present (n,n'γ) results, together with previous particle transfer data, give first evidence of a coexisting four-particle, four-hole

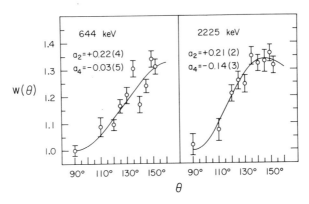

Fig. 2.   Angular distributions of the 644-keV and 2225-keV gamma rays obtained at 3.4-MeV neutron energy used in assigning the spin-parity of the 2225-keV level as 2^+.

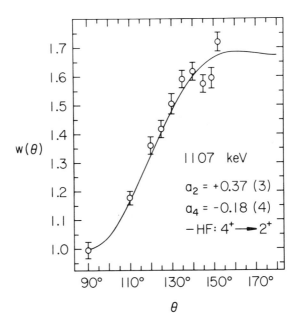

Fig. 3.   Angular distribution of the 1107-keV gamma ray measured at 3.4-MeV neutron energy. The solid line represents the angular distribution calculated with the Hauser-Feshbach theory for a streched $4^+ \to 2^+$ transition.

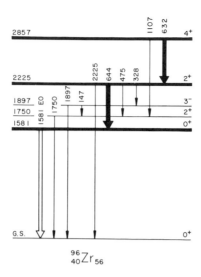

Fig. 4.   Partial level scheme for doubly closed subshell ^{96}Zr showing the 4p-4h intruder band.

intruder band built on the 1581 keV 0^+_2 state in the doubly closed subshell nucleus ^{96}Zr.   Interpretation of the band as a normal quadrupole deformed band presents some difficulties and suggests that an explanation in terms of alpha-clustering may be more reasonable.   Various experiments are under way to aid in the search for other band members and to establish the degree of collectivity of the states involved.

Acknowledgments

We wish to thank Prof. F. Iachello for enlightening discussions and Drs. H. J. Daley, E. A. Henry, K. Heyde, H. Mach, K. Sistemich, A. M. van den Berg and J. L. Wood for useful comments.   One of us (G. M.) wishes to thank members of the University of Kentucky and of the Lawrence Livermore National Laboratory for their support and hospitality.   This work was supported in part by the Hungarian Academy of Sciences, the National Science Foundation, and the U. S. Department of Energy through contract Nr. W-7405-Eng-48.

References

[BAL71]   J.B. Ball, R.L. Auble, and P.G. Roos, Phys. Rev. C4 196 (1971); E.R. Flynn, J.G. Beery, and A.G. Blair, Nucl. Phys. A218 285 (1974).
[BRO64]   G.E. Brown and A.M. Green, Nucl. Phys. 75 401 (1964); 85 87 (1966).
[CHE70]   E. Cheifetz R.C. Jared, S.G. Thompson, and J.B. Wilhelmy, Phys. Rev. Lett. 25 38 (1970).

[FAZ84]   B. Fazekas, T. Belgya, G. Molnár, and Á. Veres, in Proc. Int. Symp.
          In-Beam Nuclear Spectroscopy, Zs. Dombrádi and T. Fényes, eds.
          (Akadémiai Kiadó, Budapest, 1984).
[FLY70]   E.R. Flynn, D.D. Armstrong, and J.G. Beery, Phys. Rev. $\underline{C1}$ 703
          (1970).
[HEN85]   E.A. Henry et al., to be published.
[HEY83]   K. Heyde, P. Van Isacker, M. Waroquier, J.L. Wood, and R.A. Meyer,
          Physics Reports $\underline{102}$ 291 (1983).
[IAC82]   F. Iachello and A.D. Jackson, Phys. Lett. $\underline{108B}$ 151 (1982).
[IAC84]   F. Iachello, in Proc. Int. Conf. Clustering in Nuclei, Chester,
          England (July, 1984).
[MAY82]   W. Mayer et al., Phys. Rev. $\underline{C26}$ 500 (1982); R.S. Tickle, W.S. Gray,
          and R.D. Bent, Nucl. Phys. $\underline{A376}$ 309 (1982).
[MEY85]   R.A. Meyer et al., to be published.
[MON82]   E. Monnand et al., Z. Phys. A $\underline{306}$ 183 (1982); R.A. Meyer et al.,
          Nucl. Phys. $\underline{A439}$ 510 (1985).
[MOR56]   H. Morinaga, Phys. Rev. $\underline{101}$ 254 (1956).
[SAD75]   G. Sadler et al., Nucl. Phys. $\underline{A252}$ 365 (1975); T.A. Khan et al., Z.
          Phys. A $\underline{275}$ 289 (1975).
[SAM82]   M. Sambataro and G. Molnár, Nucl. Phys. $\underline{A376}$ 201 (1982).
[VAN84]   A.M. van den Berg, A. Saha, G.D. Jones, L.W. Put, and R.H.
          Siemssen, Nucl. Phys. $\underline{A429}$ 1 (1984); A. Saha, G.D. Jones, L.W. Put,
          and R.H. Siemssen, Phys. Lett. $\underline{82B}$ 208 (1979).
[WOO84]   J.L. Wood, in Proc. Int. Symp. In-Beam Nuclear Spectroscopy, Zs.
          Dombrádi and T. Fényes, eds. (Akadémiai Kiadó, Budapest, 1984).

RECEIVED July 15, 1986

30

Structure Transition in Heavy Y Isotopes

G. Lhersonneau[1], Richard A. Meyer[2], K. Sistemich[1], H. P. Kohl[1], H. Lawin[1], G. Menzen[1], H. Ohm[1], T. Seo[3], and H. Weiler[1]

[1]Institut für Kernphysik, Kernforschungsanlage Jülich, D-5170 Jülich, Federal Republic of Germany
[2]Lawrence Livermore National Laboratory, University of California, Livermore, CA 94550
[3]Reactor Research Institute, Kyoto University, Osaka, Japan

The structures of the neutron-rich isotopes ^{97}Y, ^{98}Y and ^{99}Y reflect with special clearness the rapid change of the nuclear shape at neutron number 60. The discovery of a new isomer in ^{97}Y has provided evidence for the shell-model character of this nucleus even at high excitation energies while ^{99}Y shows the properties of a symmetric rotor already in the ground state. The level pattern of the intermediate isotope ^{98}Y indicates shape coexistence.

The understanding of the rapid structure change of the neutron-rich nuclei with A ~ 100 is a fascinating topic for far-off-stability studies. A sudden transition from spherical to deformed nuclear shapes takes place when the neutron number raises from 58 to 60. This is especially so in the Sr and Zr isotopes where the energies of the first excited 2^+ levels decrease by more than a factor of 5 between $^{96}_{38}Sr_{58}$ and $^{98}_{38}Sr_{60}$ and between $^{98}_{40}Zr_{58}$ and $^{100}_{40}Zr_{60}$. Both ^{98}Sr and ^{100}Zr are deformed in their ground states, but probably have coexisting spherical shapes at low excitation energies.

This fact and the results of several experimental and theoretical studies suggest that the nuclei around A = 100 change their shapes rapidly but that they have complex potential energy surfaces. In particular, these nuclei are supposed to be soft with respect to γ deformations. However, recent investigations on odd-mass nuclei revealed properties of classical symmetric rotors. A good example is $^{99}_{39}Y_{60}$, the isotone of ^{98}Sr and ^{100}Zr, where an extended ground-state band and several side bands have been found [PFE81, MON82, WOH83, PET85, MEY85].

Now, new information on ^{97}Y has been obtained at the fission product separator JOSEF [LAW76] which indicates that this nucleus has shell model character even at high excitation energies. Thus, no signs of a particular softness are observed on either side of the structure transition in the Y isotopes. Shape coexistence may exist only in the intermediate nucleus $^{98}_{39}Y_{59}$ where a band with rotational properties has been known for long to exist [GRU72, SIS76] based on an excited state of 495 keV.

A new isomer in ^{97}Y

A level scheme of ^{97}Y which contains levels with spins up to 9/2 had been established [MON76, PFE81] from the study of the β⁻ decay of ^{97}Sr. Recently, an isomeric state with a half-life of 144 ms in ^{97}Y has been discovered [LHE85] which decays through a sequence of levels with higher spins into the $g_{9/2}$ state. The isomer lies at 3523 keV and is depopulated through a transition of 162 keV with a probable multipolarity of E3, α_T = 1.00(19). Its transition rate of about 2 single particle units is characteristic for a particle-to-hole decay [BL085].

The isomeric state and its depopulation can best be understood with the configurations which are proposed in Fig. 1. The isomer is a three quasiparticle state with spin and parity 27/2⁻ and the 162 keV transition basically consists of the conversion of a $h_{11/2}$ neutron particle into a $d_{5/2}$ neutron hole. The depopulation proceeds through a series of levels where the $g_{9/2}$ proton is coupled to states of the two-neutron core. Thus, the energies of the most intense γ transitions of 792, 912 and 990 keV agree well with the average energy differences of the 6⁺, 4⁺, 2⁺ and 0⁺ states of the cores ^{96}Sr and ^{98}Zr. In the lower part of the scheme there exist candidates for all but one member of the $g_{9/2} \otimes 2^+$ multiplet.

Similar core-coupled states have been observed in less neutron-rich nuclei at A ~ 100. An example is ^{93}Nb [MEY77] with two valence neutrons beyond the N = 50 shell and one proton in the $g_{9/2}$ orbital. Analoguously, it is concluded that the new levels of ^{97}Y reflect the influence of the subshell closures at Z = 40 (or 38) and N = 56 which is considerable in ^{96}Zr (or ^{94}Sr). Apparently, these shell closures are important even at high excitation in ^{97}Y, although this isotope is next neighbour to deformed nuclei.

Shape coexistence in ^{98}Y

The isotope ^{98}Y is a very remarkable nucleus. Several isomeric states with half-lives between 3 ns and 8 µs have been observed in it. A band of levels with rotational character and very little staggering is based on the 8 µs isomer at 495 keV. It had been discovered more than a decade ago [GRU72] and has regained interest since rotational bands have been observed in several odd-mass nuclei with N ⩾ 60.

In Fig. 1 is shown the part of the level scheme of ^{98}Y which is observed in the study of the decay of the µs isomers (results of β⁻ decay studies are given in [SIS76, BEC83]). Compared to earlier work the relative intensities of the γ transitions and K$_\alpha$ rays have been determined with higher precision which allows to deduce the mixing ratios δ for the ΔI = 1 transitions inside the band (which show the typical behaviour for a rotational band) and the conversion coefficients for several of the γ transitions, see Fig. 1. Additional γ lines have been observed between the levels at 170 and 495 keV, and new information on the life times of levels was achieved from γ-γ-t measurements.

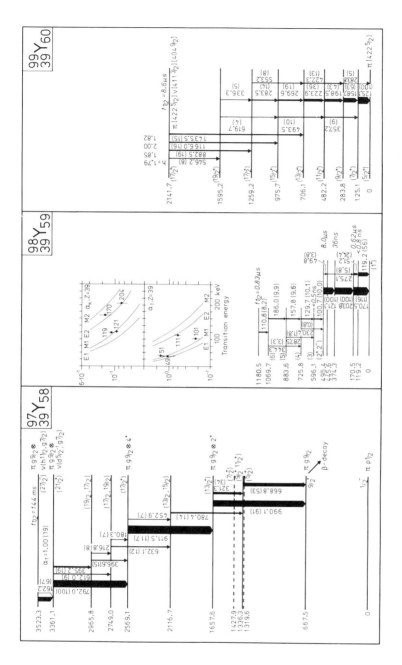

Fig. 1: Those parts of the level schemes of the Y isotones at N ∼ 60, which are observed in the decay of isomers (the hatched levels in ⁹⁷Y are populated in the β⁻ decay). The theoretical conversion coefficients are taken from [ROE78]. h = log(H)/n, where H is the hindrance of the transitions and n = Kᵢ - K_f - λ the degree of K forbiddenness [MEY85].

A reasonable fit to the energies of the members of the band can be obtained with the rotational formula $E = E_0 + A \left[I(I+1) - K^2 \right] + B \left[I(I+1) - K^2 \right]^2$ if the parameters $K = 2$, $E_0 = 455.3$ keV, $A = 17.2$ keV and $B = -22$ eV are used. A considerably better fit is, however, possible [PEK84] with the formula $E = E_0 + A \left[I(I+1) - K^2 \right] + aI$.

Although the value of $A = 17.2$ keV is small (it corresponds to a moment of inertia of about 90 % of the rigid rotor value) it is not unexpected, since a value of $A = 18.0$ keV has been deduced for the ground state band of the odd-mass neighbour $A = 99$. If $K = 1$ or $K = 3$ were used for the fit with the classical rotational formula then values of $A \sim 27$ keV and ~ 12 keV would result which are not compatible with the knowledge about other nuclei in this mass region. Hence, $K = 2$ is proposed for the band in ^{98}Y.

The parity of the members of the band cannot be determined unambiguously with the available information. Spin and parity 1^+ are assigned to the ground state of ^{98}Y since the β^- decays from ^{98}Sr into this level and its own decay into the ground state of ^{98}Zr are probably allowed [BEC83,SIS76]. At first sight the conversion coefficients of the γ transitions of 121, 204, 51 and 119 keV which connect the band head with the ground state indicate multipolarities of M1 and/or E2 and, hence, no parity change. Thus the band head should have positive parity.

But it cannot be ruled out that the 121 keV transition has a multipolarity of E1 with a small admixture of M2 instead of M1/E2. A mixing parameter of $\delta^2 = 4 \cdot 10^{-2}$ would account for both the conversion coefficient of this transition and for the half-life of the 495 keV level (with hindrance factors of $4.6 \cdot 10^7$ and 3.3 for the E1 and M2 fraction, respectively). If the 121 keV line has this character, then the parity of the band members is negative.

Calculations of the excitation energies of the band heads of odd-odd nuclei in the $A = 100$ region [HOF84, MEY84] offer candidates for both parities, namely $\{\pi[422\ 5/2]\nu[404\ 9/2]\}2^+$ and $\{\pi[303\ 5/2]\nu[404\ 9/2]\}2^-$ which both have the same neutron configuration. There is evidence [MEY84] that this intruder configuration causes also isomerism in the odd-mass isotones ^{97}Sr and ^{99}Zr, see below. Another fact of interest is, that both the proton and the neutron configuration originate from the $g_{9/2}$ single particle orbital for the 2^+ alternative with the possibility of a strong deformation - driving proton-neutron interaction.

While there is little doubt about the rotational properties of the band which is based upon the 495 keV state, the nature of the levels below the band head and of those which are only populated in the β^- decay of ^{98}Sr is not clear. These levels do not show a membership to bands and it is probable that most of them are not of rotational character. In particular, although the energy of 119 keV of the first excited state is similar to the energy of the first excited members of rotational bands of the neighbouring odd-mass nuclei it is improbable that an unperturbed ground state band exists in ^{98}Y. If so then it would be astonishing that none of the higher lying members was fed in the decay of the isomers. Moreover it is difficult to choose a convenient value of K for such a band which would explain the strong hindrance of more than 10^{10} for the direct decay from the band head at 495 keV into the ground state.

The ground and first excited states of ^{98}Y may rather be related to corresponding levels in the isotones ^{97}Sr and ^{99}Zr for which a J-2 and J-1 character of the particle-vibrator coupling type has been proposed [MEY84]. The lowest levels of these isotones are depicted in fig. 2. The striking similarity of the observed transition probabilities indicates that both nuclei have the same structure in spite of the differences in energy. The second excited states are possibly due to a low lying intruder configuration from a deformed shape.

The symmetric rotor ^{99}Y

The level scheme of ^{99}Y has been studied both through the β⁻ decay of ^{99}Sr [PFE81,PET85] and via an isomeric state at 2142 keV with high spin which is directly populated in fission [MON82,MEY85]. This nucleus contains the most extended rotational band which has been found in an odd-A nucleus at A ~ 100. In addition, several side bands have been observed.

The properties of ^{99}Y suggest that this nucleus is a classical symmetric rotor. Thus, configurations can be assigned to all the bands in accordance with the predictions of the Nilsson model for A ~ 100 and a deformation of ε ~ 0.3. Also the mixing ratios δ for the ΔI = 1 members of the bands can be accounted for in the classical picture of rotational nuclei. The half-life of the isomer at 2142 keV is obviously due to K forbiddenness.

Conclusions

The available knowledge on the three neighbouring neutron-rich isotopes of Y shows that these nuclei have very different structures:

^{97}Y has properties which are basically determined through the valence nucleons beyond the core ^{96}Zr (or ^{94}Sr). The existence of the isomer of three-quasiparticle character indicates that there is no particular softness against deformation even at high excitation energy.

^{98}Y most probably has coexisting nuclear shapes with a well developed rotational band on the 495 keV level and other levels which seem to be rather of vibrational nature.

^{99}Y is characterized through the occurence of rotational bands all over the investigated region of energies up to 2 MeV as a symmetric rotor.

These results demonstrate that a difference of only one neutron causes a considerable change of the nature of the nuclei at A ~ 100 and that the study of the isotopes with odd nucleon numbers can provide insight into the details of the shape transition. The transition in the Y isotopes seems to be even more rapid than in the Sr and Zr chains where the N = 60 isotones still have coexisting shapes and where the shell-model character of the N = 58 isotones at high excitation energies has not yet been tested.Further investigations are, however, needed in order to confirm in detail the proposed interpretation of the level schemes of the Y isotopes and to see whether similarly rapid structure changes occur in the Rb and Nb isotopes at N ~ 60.

Fig. 2: The lowest levels of the isotones of ^{98}Y. The values of $t_{1/2}$ for the first excited states and of α_K were measured at JOSEF. For details on ^{97}Sr see [PFE81, KRA83, MEY84.

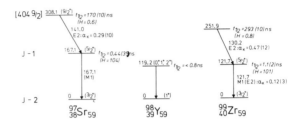

References

[BEC83] K. Becker, Thesis, Univ. Gießen, F.R. Germany (1983)

[BLO85] J. Blomqvist, FACK Stockholm, Sweden, priv. comm.

[GRU72] J.W. Grüter, Thesis, Univ. zu Köln, F.R. Germany (1972)

[HOF84] R.W. Hoff et al., Report UCRL 91143 (1984)

[KRA83] K.L. Kratz et al., Z. Phys. A312 43 (1983)

[LAW76] H. Lawin et al., Nucl. Instr. Meth. 137, 103 (1976)

[LHE85] G. Lhersonneau et al., to be published

[MEY77] R.A. Meyer and R.P. Yaffe, P.R. C15, 390 (1977)

[MEY84] R.A. Meyer and K. Sistemich, in Atomic Masses and Found. Const. 7, O. Klepper ed., Techn. Hochschule Darmstadt, (1984), p. 523

[MEY85] R.A. Meyer et al., Nucl. Phys. A439 510 (1085)

[MON76] E. Monnand et al., Proc. 3rd Int. Conf. on Nuclei far from Stability, Cargese, France, 1976, Report CERN 76-13, p. 477

[MON82] E. Monnand et al., Z. Physik A306 183 (1982)

[PEK84] L. Peker, Brookhaven Natl. Lab., Upton, New York, priv. comm.

[PET85] R.F. Petry et al., Phys. Rev. C31 621 (1985)

[PFE81] B. Pfeiffer et al., Proc. 4th Int. Conf. on Nuclei far from Stability, Helsingor, Denmark, 1981, Report CERN 81-09, p. 423

[ROE78] F. Rösel et al., At. Nucl. Data Tables 2 191 (1978)

[SIS76] K. Sistemich et al., Proc. 3rd Int. Conf. on Nucl. far from Stability, Cargese, France, 1976, Report CERN 76-13, p. 495

[WOH83] F.K. Wohn et al., Phys. Rev. Lett. 51 873 (1983)

RECEIVED July 15, 1986

31

Rotational Structure of Y Nuclei in the Deformed $A = 100$ Region

F. K. Wohn[1], John C. Hill[1], and R. F. Petry[2]

[1]Ames Laboratory, Iowa State University, Ames, IA 50011
[2]University of Oklahoma, Norman, OK 73019

Level structures of four Y isotopes with mass A=99–102 were studied via the decays of their mass–separated Sr parents using the TRISTAN isotope separator on–line to the high flux beam reactor at BNL. In ^{99,101}Y we observed 5 rotational bands and identified their band heads with the proton Nilsson orbitals 5/2[422], 5/2[303], 3/2[301], 1/2[431], and 3/2[431]. (The 3 lowest–energy bands have also been observed in ^{101,103}Nb.) Single–quasiparticle symmetric–rotor calculations of level energies, single–particle E1 or M1 and collective E2 transition rates provide an excellent description of the observed level energies and transition rates. In ^{100,102}Y we identified $K^{\pi} = 1^+$ bands (also observed in ^{102,104}Nb) that seem to be "paring free" two–quasiparticle bands with moments of inertia nearly equal to those of a rigid spheriodal nucleus.

Neutron–rich A=100 nuclei have attracted considerable interest for more than a decade since they belong to a new region of deformation. The first indications of deformation in this region were rotational energy level patterns and strongly enhanced B(E2) values consistent with equilibrium deformations β of 0.3–0.4 [CHE71]. Subsequent detailed spectroscopic studies focused primarily on deformed even–even isotopes of Sr, Zr, Mo and Ru. In the last few years, as ISOL (isotope separator on–line) facilities have made available shorter–lived and more neutron–rich nuclei, evidence has accumulated on odd–A nuclei, permitting the determination of valence Nilsson orbitals for both odd–Z and odd–N nuclei in the A=100 region. To a lesser extent, evidence on odd–odd deformed nuclei now exists.

Before discussing the four Y nuclei that we have recently studied, some of the unusual features of deformed A=100 nuclei that were revealed via spectroscopic studies are briefly discussed. One feature is the remarkably rapid onset of deformation that occurs for even–even nuclei at neutron number N=60. Fig. 1 shows the unusually sharp drop in the 2_1^+ level energy at N=60 for Sr and Zr nuclei. That the onset of deformation does not begin prior to N=60 can be attributed to subshell effects that are mutually reinforcing for the nucleon numbers Z=38,40 and N=56,58 [BEN84]. Recent considerations of the influence of subshell effects on the onset of deformation show a close analogy between the A=100 region and the rare earth region [CAS81], but the effect is more than twice as abrupt for A=100 nuclei than for the rare earths [WOH83].

0097–6156/86/0324–0208$06.00/0

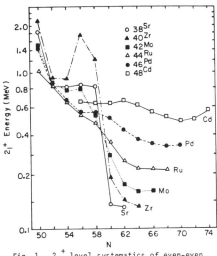

Fig. 1. 2_1^+ level systematics of even-even A=100 nuclei.

Fig. 2 further illustrates the abruptness of the shape transition. The energy ratio approaches the value of 10/3 characteristic of an axially symmetric rotor for Sr or Zr nuclei with N ≥ 60. One should thus also expect Y, which is intermediate in Z between Sr and Zr, to have a similarly sharp onset of deformation leading to a rotational structure for N ≥ 60.

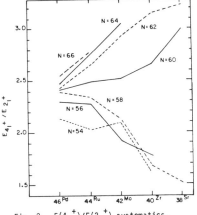

Fig. 2. $E(4_1^+)/E(2_1^+)$ systematics.

A second unusual feature of A=100 nuclei is the large values of moments of inertia $\mathscr{J}$ deduced from rotational bands. Fig. 3, which shows systematics of inertial moments, includes the four Y nuclei discussed in this paper.

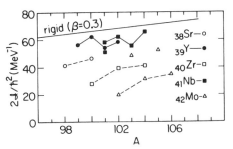

Fig. 3. Moments of inertia for deformed A=100 nuclei.

Note that several odd-A nuclei have moments of about 80–85% of the rigid value. Values exceeding 90% are obtained for the $K^\pi = 1^+$ bands discussed below. These large moments are much higher than the ≤ 50% values of the deformed rare earths, and this can be interpreted as a consequence of an unusually low strength of the pairing interaction [PET85].

A third, and quite intriguing, feature of deformed A=100 nuclei is the apparent coexistence of spherical and highly deformed shapes. All of the N=60 isotones ^{98}Sr, ^{100}Zr, ^{102}Mo, ^{104}Ru and ^{106}Pd have an "extra" 0^+ state (the 0_2^+ state) that cannot be reproduced in collective models as a simple collective excitation of the core. The very low-lying 0_2^+ level in ^{98}Sr and ^{100}Zr, which lies at 215 and 331 keV respectively, can best be understood as shape coexistent 0^+ states with the ground state being dominantly a symmetric rotor and the 0_2^+ state being dominantly a spherical, or nearly spherical, shape [WOH86]. In contrast,

the heavier Mo and Ru isotopes were interpreted as asymmetric rotors with non-zero
values of the geometric asymmetry parameter γ
[SHI83], [SUM80].

In terms of the three limiting group symmetries used in the interacting boson model
(IBM), the shape transition of Ru or Pd nuclei
can be explained as a SU(5) to O(6) transition
[STA82]. The transition of Sr or Zr nuclei
is characterized instead as SU(5) to SU(3).
For Mo nuclei, the shape transition appears to
be more complicated than the above and has not
yet been adequately described.

The unusual features of the A=100 nuclei
briefly reviewed above continue to stimulate
strong interest in the structure of these
nuclei. Theoretical studies, including potential energy and microscopic shell model calculations, as well as a more detailed review of
the features of the A=100 region than space
here permits, are discussed in [PET85].

II. DEFORMED ODD-A Y ISOTOPES

A clear gap in the study of this region
was a lack of information on single-particle
states in deformed nuclei with N ⩾ 60. Only
recently, since we began our first Y study
with ^{99}Y, has such information emerged. The
status through the end of 1984 was recently
summarized by [PET85].

From studies of the decay of an 8.6-μs
isomer in ^{99}Y [MON82] and our study [PET85]
of the decay of ^{99}Sr, we have identified 5
rotational bands in ^{99}Y. [PET85] gives the
details of our ^{99}Sr decay study and [WOH85]
gives details of a particle-rotor calculation
of the rotational band structure and transi-

tion probabilities. In the following, only
the 5 single-quasiparticle band calculations
of [WOH85] are presented. Other levels, such
as particle-vibration (3 quasiparticle) states
in ^{99}Y are discussed in [PET95].

Since so little was known about the
single particle structure of deformed A=100
nuclei, our approach in [WOH85] was to use a
textbook version of the particle-rotor model,
as outlined by [BUN71], that had been systematically used to study deformed rare earths.

The model parameters fall into two clas-
ses, an "energy" class and a "transition"
class. The energy class includes Δ_p (the
proton pairing gap), the 3 Nilsson parameters
(κ, μ, and δ), the inertial parameter $\hbar^2/2\mathcal{J}$
of the axially symmetric deformed core, and
the Coriolis attenuation parameter k. Fig. 4
shows the Nilsson parameters and Δ_p we used.

For all bands we used $\delta = 0.34$ and an inertial parameter of 21.8 keV. The bandheads
were calculated, not arbitrarily set [WOH85].

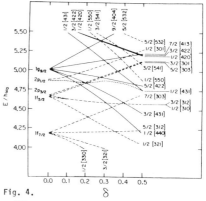

Fig. 4.

Nilsson diagram for protons with $\kappa = 0.067$
and $\mu = 0.53$; BCS model Fermi level for Y
with $\Delta_p = 0.69$ MeV is indicated by dots.

Of the energy parameters, k was determined by our ^{99}Y data. We found k = 0.77 for unique-parity and k = 1.35 for normal-parity bands.

The transition class of parameters were g' (the ratio of the spin g-factor and the free proton g-factor) and H (the enhancement factor of the ΔK=0 El Nilsson matrix elements). We found g' = 1.0 for the unique-parity bands and g' = 0.7 for the normal-parity bands. An H of 3.0 was found to give good reproduction of the relative El and Ml rates for the Coriolis-mixed 5/2[303] and 3/2[301] bands. The bands and transition rates are given in Fig. 5.

Bandhead		Calculation	Experiment
^{99}Y	5/2[303]	26 ps	–
	3/2[301]	33 ps	–
	1/2[431]	49 ps	–
^{103}Nb	5/2[303]	4.8 ns	4.7±0.5 ns
	3/2[301]	1.9 ns	2.0±0.6 ns

In order to test our calculation against absolute transition rates, we turned to ^{103}Nb, since the same 3 lowest bands were seen and El transitions from the 5/2[303] and 3/2[301] bandheads in ^{103}Nb had measurable half-lives [SE084]. We did not change the Nilsson El matrix elements but changed only the pairing factor correction (due to the different quasiparticle energies for ^{99}Y and ^{103}Nb). We obtained the lifetimes given in the table above, which clearly shows that H=3.0 works remarkably well also for ^{103}Nb.

The "adjustment" factors (k, g', H) of our "textbook" version of the particle-rotor model are much closer to unity for ^{99}Y than is typical for the rare earth nuclei. It is interesting to speculate if this can be simply a consequence of the lower density of single-particle states and lower pairing energy for the neutron-rich A=100 nuclei.

Our study of ^{101}Y levels from the decay of ^{101}Sr is not

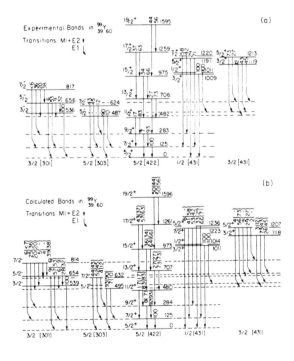

Figure 5. Nilsson bands and relative transition intensities for ^{99}Y. For ease of comparison the exp. intensities are given in (a) and (b).

yet complete, but the 4 bands shown in Fig. 6 are definite. In ^{101}Y the 3/2[431] band lies lower and the 1/2[431] band lies higher in energy than in ^{99}Y, hence identification of the latter band is more difficult for ^{101}Y.

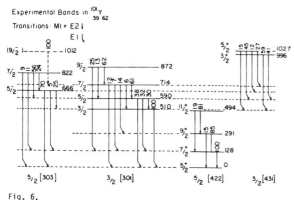

Experimental Bands in ^{101}Y 39 62
Transitions: MI + E2 ↓

Fig. 6.
Nilsson bands and relative transition intensities for ^{101}Y.

The similarity between the level structures of ^{99}Y and ^{101}Y is striking. (Also, both ^{101}Nb and ^{103}Nb have the same 3 lowest bands [SE084] with very similar level spacings.) Indeed, all known odd-A deformed A=100 nuclei appear to have very similar deformed cores.

As stated earlier, coexistence of deformed and spherical shapes is characteristic of the lighter deformed A=100 nuclei. Both ^{99}Y and ^{101}Y have anomalous states (at 599 and 890 keV respectively) that are not associated with the lower energy bands. These states decay only to the 3/2[301] bandhead, hence could be the $p_{1/2}$ state expected for the 39th proton in a spherical potential. Spherical A=100 nuclei have a 2_1^+ energy of ~800 keV, and the anomalous states are fed only by ~800 keV γ rays, which supports our speculation that they may be $p_{1/2}$ particles coupled to a spherical core.

III. DEFORMED ODD-ODD Y ISOTOPES

As mentioned above, deformed A=100 nuclei have unusually weak pairing, as indicated by moments of inertia of 70% of the moment of a rigid spheroid (see Fig. 3) for the $K^\pi = 0^+$ ground bands of e-e nuclei. With the addition of 1 quasi-particle, the deformed odd-A nuclei have moments of inertia that increase to ~85%. With 2 unpaired quasiparticles, the deformed o-o nuclei could thus have moments nearly equal to the rigid moment of inertia.

Nuclei in which the effective neutron and proton pairing gaps have vanished are referred to as "pairing-free" nuclei. The deformed odd-odd A=100 nuclei thus present a unique opportunity to observe the pairing-free nuclear phase at conditions of low angular momentum and low excitation energy.

[FED77] proposed that the unusually weak pairing for A=100 nuclei is associated with the very strong coupling between $g_{9/2}$ protons and $g_{7/2}$ neutrons. For o-o nuclei, we expect the deformed states with the strongest p-n coupling would be most likely to approach the pairing-free phase. Thus o-o nuclei having low-lying Nilsson orbitals with strong $\pi g_{9/2}$ and $\nu g_{7/2}$ components should have the required strong p-n coupling.

Odd-odd Y and Nb nuclei with N ≥ 60 are ideal candidates, since the odd proton state 5/2[422] has a large $g_{9/2}$ component and N=61 or 63 nuclei have a 3/2[411] ground band with a large $g_{7/2}$ component. These orbitals should couple to form a low-lying $K^\pi = 1^+$ band that should be strongly fed by the β decay of the e-e parent.

Figure 7 gives our proposed $K^\pi = 1^+$ bands. The lower 1^+ bands we propose to be nearly "pairing free" bands and their moments of inertia (shown in Fig. 3) range from 88% to 94% of the rigid value. Our TRISTAN data was used for ^{100,102}Y. The ^{102,104}Nb data came from [MEY84]. The log$\underline{\text{ft}}$ values and γ intensities shown in Fig. 7 are discussed in [PEK85]. The presentation by L. K. Peker in these proceedings gives additional arguments.

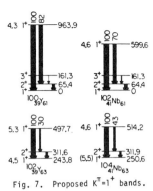

Fig. 7. Proposed $K^\pi = 1^+$ bands.

REFERENCES

[BEN84] R. Bengtsson, P. Moller, J.R. Nix and J. Zhang, Phys. Scripta 29, 402 (1984).

[BUN71] M.E. Bunker and C.W. Reich, Rev. Mod. Phys. 43, 348 (1971).

[CAS81] R.F. Casten, D.D. Warner, D.S. Brenner and R.L. Gill, Phys. Rev. Lett. 47, 1433 (1981).

[CHE71] E. Cheifetz, R.C. Jared, S.G. Thompson and J.B. Wilhelmy, Phys. Rev. C 4, 1913 (1971).

[FED77] P. Federman and S. Pittel, Phys. Lett. 69B, 385 (1977) and 77B, 29 (1978).

[MEY84] R.A. Meyer, R.W. Hoff, and K. Sistemich, FY 1984 Annual Report of Nuclear Chemistry Division, LLL, UCAR 10062-84/1, p. 41.

[MON82] E. Monnand, J.A. Pinston, F. Schussler, B. Pfeiffer, H. Lawin, G. Battistuzzi, K. Shizuma, and K. Sistemich, Z. Phys. A 306, 183 (1982).

[PEK86] L.K. Peker, F.K. Wohn, John C. Hill, and R.F. Petry (to be published).

[PET85] R.F. Petry, H. Dejbakhsh, J.C. Hill, F.K. Wohn, M. Schmid, and R.L. Gill, Phys. Rev. C 31, 621 (1985).

[SEO84] T. Seo, A. Schmitt, H. Ahrens, J.P. Bocquet, N. Kaffrell, H. Lawin, G. Lhereonneau, R.A. Meyer, K. Shizuma, K. Sistemich, G. Tittel, N. Trautmann, Z. Phys. A 315, 251 (1984).

[SHI83] K. Shizuma, H. Lawin, and K. Sistemich, Z. Phys. A 311, 71 (1983).

[STA82] J. Stachel, N. Kaffrel, E. Grosse, H. Emling, H. Folger, R. Kulessa, and D. Schwalm, Nucl. Phys. A383, 429 (1982).

[SUM80] K. Summerer, N. Kaffrell, E. Stender, N. Trautmann, K. Broden, G. Skarnemark, T. Bjornstad, I. Haldorsen, and J. A. Maruhn, Nucl. Phys. A339, 74, (1980).

[WOH83] F.K. Wohn, John C. Hill, R.F. Petry, H. Dejbakhsh, Z. Berant, and R.L. Gill, Phys. Rev. Lett. 51, 873 (1983).

[WOH85] F.K. Wohn, John C. Hill, and R.F. Petry, Phys. Rev. C 31, 634 (1985).

[WOH86] F.K. Wohn, John C. Hill, C.B. Howard, K. Sistemich, R.F. Petry, R.L. Gill, and H. Mach, Phys. Rev. C (submitted August 1985).

RECEIVED May 2, 1986

32

Coexistence in the Cd Isotopes and First Observation of a U(5) Nucleus

A. Aprahamian

Department of Chemistry, Clark University, Worcester, MA 01610

The ^{118,120}Cd nuclei were studied from the decay of Ag produced from the fission of ^{235}U. The resulting systematics of the cadmium nuclei show that the intruding configuration bandheads have in fact risen in excitation energy relative to those observed in ^{112,114}Cd and are well-separated from the two-phonon triplet of states. These results also led to the identification of a set of five states, in ^{118}Cd whose close-spacing in excitation energy and decay transition probabilities seem to qualify ^{118}Cd as the first example of a U(5) nucleus.

The presence of intruder states near the Z=50 closed shell has been well documented and tested [HEY82] for ^{112,114}Cd, whose structures showed an anomalous grouping of five states with a centroid at approximately 1.2 MeV instead of the expected vibrational triplet of states. The low-lying excitation structure of these nuclei and their E2 transition probabilities have been successfully interpreted in terms of the mixing of an intruding rotational configuration with the normally expected vibrational configuration where the intruder states are described as particle-hole excitations resulting from the excitation of a pair of protons across the Z=50 major oscillator shell gap.

Detailed descriptions of the model is given elsewhere [HEY82], [HEY83]. The basic idea, however, is to calculate two sets of states where one corresponds to the vibrational states and the other corresponds to the rotation-like intruder states which arise from the particle-hole excitations, and then to mix the two resulting configurations. The hamiltonians for both configurations are esssentially the same:

$$H_{\substack{VIB \\ ROT}} = \varepsilon_\pi\, n_{d_\pi} + \varepsilon_\nu\, n_{d_\nu} - \kappa\, Q_\pi \cdot Q_\nu$$

where ε_π and ε_ν are the proton and neutron boson energies and the $\kappa Q_\pi \cdot Q_\nu$ is the neutron-proton quadrupole interaction. The main difference between the two configurations arises from the $n(n-1)$ particle number dependence of the quadrupole interaction where n is the effective number of valence particles. The particle number dependence of the $\kappa Q_\pi \cdot Q_\nu$ term results in the prediction of a "V" shaped systematics for the excitation energies of the intruding

0097-6156/86/0324-0214$06.00/0
© 1986 American Chemical Society

configuration for a given isotopic chain where the intruder states decrease in energy towards the middle of the neutron shell and rise thereafter. Although comparisons of excitation engeries and transition branching ratios in Cd nuclei near the neutron midshell (^{112,114}Cd) show very good agreement with the above mentioned model, not enough data existed away from midshell to test the predicted rise in excitation energy for the intrudering configuration. Studies of ^{118,120}Cd were undertaken in order to extend the known Cd systematics, and to determine the evolution of the intruding configuration.

These nuclei were studied from the decay of Ag produced in the fission of enriched ^{235}U at the on-line isotope separator, TRISTAN, at Brookhaven National Laboratory by several standard spectroscopic techniques, including the measurements of angular correlations by a fixed-four detector system, γ-γ and β-γ coincidence, and lifetimes of some excited states.

The systematics of the low-lying states of the Cd nuclei are shown along with the new results from the present study for ^{118,120}Cd in Fig. 1. Note that the centroid of the quintuplet of states in ^{112,114}Cd is essentially the same as that of the 0+, 2+, 4+ triplet in ^{118,120}Cd. However, the third 0+ state has gone up in energy to 1615 KeV in ^{118}Cd [APR84] and 1745 KeV in ^{120}Cd. It is clear that the intruder configuration has risen in energy relative to the ^{114}Cd case which represents the neutron midshell. This observed increase in excitation energy of the intruder configuration bandhead confirms the coexistence and mixing interpretation for the structure of Cd nuclei. A good signature of the extent of configuration mixing is given by the B(E2) ratio R which is defined in the top portion of Fig. 1. The values of R, shown in Fig. 1, range from a maximum of 5.3×10^4 in ^{114}Cd to <3 in ^{120}Cd. This large decrease in R is due to the reduced mixing which results from the larger distances between the two configurations as one moves away from the neutron midshell.

Intruder states are certainly not unique to the Z=50 region of nuclei [HEY83]. In fact, much larger neutron-proton effects are expected for nuclei near Z=82 since the neutrons are in the 82-126 shell. The evolution of an intruding configuration in a series of Pb nuclei were reported recently [VAN85]. The effects of the neutron-proton quadrupole interaction have recently been extended [HEY85] to include not only a complete description of

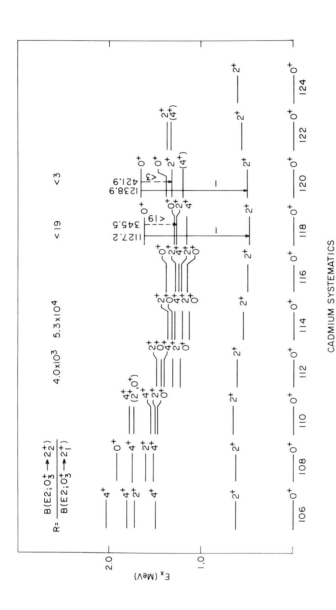

CADMIUM SYSTEMATICS

Fig. 1   Systematics of the low-lying states in cadmium nuclei with the addition of the results of the present study.

intruding configurations, but also, the onset of deformation in nuclei, where the intruding rotational configuration comes down low enough in excitation energy to become the ground state. The idea is summarized in Fig. 2 for Pb, Sm, Cd, Sn, Zr and Sr nuclei. The two determining factors for the excitation energies of intruding configurations are the shell gap and the magnitude of the neutron-proton quadruple interaction [HEY85]. In the Cd nuclei, the gap is at 5.4 MeV and the lowest occurrence of the intruding configuration is at 1.2 MeV. In the Pb nuclei, the shell gap is smaller at 3.5 MeV and there is a larger neutron-proton quadrupole effect, therefore, the intruding configuration comes below the first excited 2+ in ^{192}Pb, only rising above the second 0+ state in 200 Pb. The Sm and Zr, Sr nuclei experience subshell closures at Z=64 and Z=40, respectively. These nuclei have shell gaps of 2-2.5 MeV, the intruding configuration comes down in energy and becomes the ground state, thereby introducing deformation. One of the well known driving forces for particle excitations is the occupation of spin orbit-pairs. For example, in the Zr nuclei, the excitation of a pair of protons across the Z=40 subshell closure results in the occupation of the g9/2 orbital (Fig. 2). The neutrons are in the 50-82 shell, therefore, there is a high occupation of the g7/2 orbital. The result is increased spatial overlap of neutron-proton wavefunctions and, therefore, maximum interaction effects. It seems that the neutron-proton quadrupole interaction and its dependence on particle number provide a satisfactory explanation for the existence of intruding configurations throughout the medium and heavy nuclei, as well as, for the onset of deformation, implying that it may be a most important and unifying effect for nuclei.

As pointed out earlier, the mixing of the intruding configuration with the normal states decreases dramatically as the former goes up in excitation energy. This separation or isolation of the intruding configuration presents a unique opportunity to test the vibrational structure of these nuclei. In fact, the resulting level scheme for ^{118}Cd up to 2.1 MeV shows a special clustering of levels, as shown in Fig. 3: a 2+ state at 488 KeV, a triplet of 4+, 2+, and 0+ levels at 1165 KeV, 1270 KeV, and 1286 KeV, respectively, the intruder 0+ at 1615 KeV, and a set of five closely spaced levels with a centroid at 1989 KeV. Angular correlation measurements have allowed unique spin assignments of 2+, 0+, and 3+ for the levels at 1916 KeV, 2074 keV, and 2092 KeV, respectively, by the use of χ^2 vs. arc-tan δ plots [APR86]. The

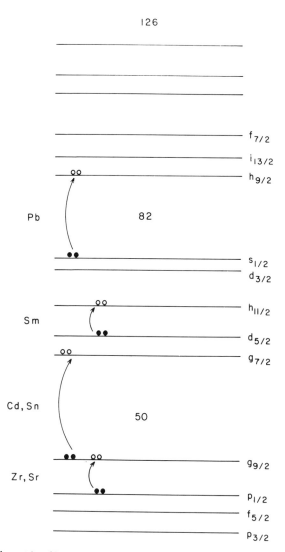

Fig. 2   A schematic diagram of the relative sizes of shell and subshell gaps
          for Pb, Sm, Cd, Sn, Zr and Sr nuclei.

remaining two levels have more than one possible spin assignment. The level
at 1929 KeV is either a 3+ or 4+, while the level at 1936 KeV is most likely a
5+ or a 6+. The B(E2) values are shown on the transition arrows of Fig. 3.
The close-spacing of these five levels, and their strong preference of decay

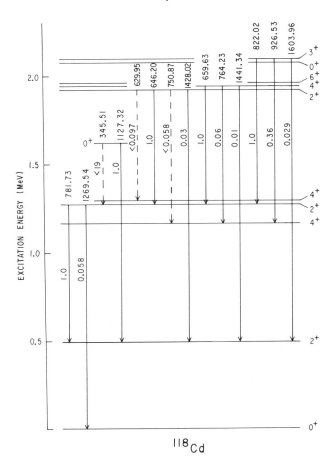

Fig. 3   Partial level scheme of ^{118}Cd.   The B(E2) values are normalized to 1.0 for the strongest transition depopulating a given level.

to members of the 0+, 2+, 4+ triplet compared to the first 2+, indicate that this quintuplet of states are the five members of the three phonon multiplet.   This is the first experimental evidence for the existence of such states.   Finally, identification of these states qualifies ^{118}Cd as the best example of the U(5) symmetry of the IBA.

Acknowledgments

The author would like to thank all her collaborators in the work reported here: D. S. Brenner, R. F. Casten, R. L.Gill, H. Mach, A. Piotrowski, D. Rehfield and J. Van Dyk.

References

[APR86]   A. Aprahamian, et al, to be published.

[APR84]   A. Aprahamian, D. S. Brenner, R. F. Casten, R. L. Gill, A. Piotrowski, and K. Heyde, Phys. Lett. 140B 22 (1984).

[HEY82]   K. Heyde, P. Van Isacker, M. Waroquier, G. Wenes, and M. Sambataro, Phys. Rev. C25 3160 (1982).

[HEY83]   K. Heyde, P. Van Isacker, M. Waroquier, J. L. Wood, and R. A. Meyer, Phys. Rpts. 102 291 (1983).

[HEY85]   K. Heyde, P. Van Isacker, R. F. Casten, and J. L. Wood, Phys. Lett. 155B 303 (1985).

[VAN84]   P. Van Duppen, E. Coenen, K. Deneffe, M. Huyse, K. Heyde, and P. Van Isacker, Phys. Rev. Lett. 52 1974 (1984).

RECEIVED May 13, 1986

E0 Transitions and Intruder States

Recent Results

J. Kantele, R. Julin, M. Luontama, and A. Passoja

Department of Physics, University of Jyväskylä, SF-40100 Jyväskylä, Finland

Results from studies of 0^+ states and E0 transitions in ^{102,104}Pd and ^{202,204}Pb are reported. Systematics of low-spin states in Pd isotopes and proton 2p-2h intruders in Pb isotopes are presented and discussed.

Among the most interesting recent topics in low-energy nuclear-structure physics have been the intruder states. As resulting from intense experimental and theoretical studies, a unified picture of them seems now to be emerging [HEY85]. One of the tools in the study of the intruders has been provided by the E0 transitions which are particularly effective in obtaining information on 0^+ states in even-even nuclei (see [KAN84] for a review of the subject). Understanding the E0's requires detailed knowledge especially of the radial parts of the initial and final-state wave functions. For this reason, these transitions are sensitive probes of the various nuclear models: For example, for the different versions of the IBA, the level energies seem to be fairly easy to predict, the B(E2)'s are more difficult, and the E0's usually really hard.

In our studies of intruding 0^+ states and associated E0 transitions, we use several newly developed special nuclear spectrometry techniques, notably in-beam conversion-electron and internal-pair spectrometry with various combinations of magnetic and semiconductor devices [KAN84]. A recent set-up based on an intermediate-image (Siegbahn-Slätis) magnetic electron transporter is sketched in Fig. 1. With a 110 mm^{2}x3 mm Si(Li) detector, the transmission of this device is about 7 % and the momentum band-width $\Delta p/p=17$ %. With a 5 mm thick Ge detector, the system is capable of detecting electron lines up to 7-8 MeV.

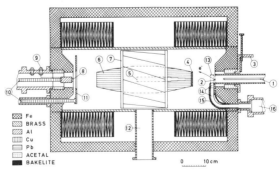

Fig. 1. A magnetic + Si(Li) combination conversion-electron spectrometer based on an "old" Siegbahn-Slätis magnet. 1) beam, 2) target, 3) target-changing system, 4) collimator and current measurement, 5) Faraday-cup, 6) Pb shield, 7) anti-positron baffle, 8) detector, 10) cold fingers, 13) cylindrical plastic scintillator 14) light guide, 16) P.M. tube.

0097–6156/86/0324–0221$06.00/0
© 1986 American Chemical Society

102,104Pd and Systematics of Low-Spin States in Pd Isotopes

From the point of view of the present understanding of the intruder states, the even Pd nuclei are of special interest since they are expected to exhibit proton intruders of the type 2p-6h. Systematic information exists already for the 2p-2h and the 2p-4h intruders in the Sn and Cd isotopes [BRO78, BÄC81, APR84], so that the Pd isotopes represent the next step towards more complex states. Such a step is important since it may show whether, when going away from closed shells, the intruder picture will get obscured by the increasing number of the particle degrees of freedom.

We have studied the low-spin states in 102,104Pd using (p,2n) and (p,p') reactions, and Coulomb excitation with 16O ions [LUO85]. Conversion-electron spectroscopy with the instrument shown in Fig. 1, as well as conventional gamma-ray and gamma-gamma coincidence spectroscopy (including delayed coincidences), were employed in obtaining the results summarized in Fig. 2.

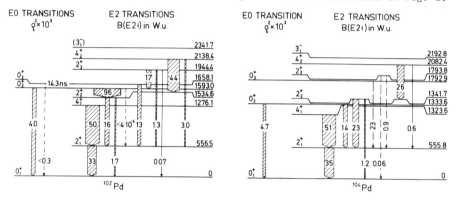

Fig. 2. Partial level schemes of 102,104Pd.

Some features in the level schemes of both 102Pd and 104Pd resemble those of the simple quadrupole phonon picture; the two-phonon triplets would consist of the 0_2^+, 2_2^+, and 4_1^+ states in 102Pd and of the 0_2^+, 2_2^+, and 4_1^+ states in 104Pd. However, a glance at Fig. 2 immediately reveals serious discrepancies from the simple phonon model. The information on the 0_3^+ and 2_3^+ levels in 104Pd is scarce, except that they both are connected with the 2_1^+ state via E2 transitions having probably moderate speeds [LUO85, NDS]. In 102Pd, however, there is a very conspicuous feature in that the isomeric 0_2^+ state lies below the more phonon-like 0_3^+ state, and only at a distance of 65 keV from it. Although there is no direct evidence for the mixing of these two states (e.g., the interconnecting E0 transition has not been observed), the small energy interval and the fact that these states are connected to completely different levels (cf. Fig. 1) seem to suggest a very weak mixing of the two 0^+ states. Another peculiar feature is the extremely strongly hindered E2($0_2^+ \rightarrow 2_1^+$) transition with B(E2) <4x10^{-4} W.u. which gives rise to the largest known value for X_{211}>400. As compared with the Cd (112,114Cd) and Sn isotopes (114,116,118Sn) [JUL80, BÄC81], the mixing between the 0_2^+ and 0_3^+ states seems to be stronger in these nuclei, as implied, e.g., by E0 transitions between the excited 0^+ states which are especially strong in the Sn nuclei. The $0_3^+ \rightarrow 2_1^+$ transition is also less hindered in Cd and Sn (again, this transition is strongest in Sn). That the 0_2^+ and 0_3^+ states are more mixed in ^{116}Sn than in ^{112}Cd is also suggested by the (^{3}He,n) results, although not fully conclusively [FIE77].

The decay characteristics of the 0_2^+ level in ^{102}Pd are very similar to those of the 0_3^+ states in ^{112,114}Cd. Therefore, similar excitation mechanisms could be responsible for these states.

According to both experimental and theoretical investigations [BRO78, BÄC84, APR84, HEY83], the 2p-2h states in the Sn nuclei and the 2p-4h states in the Cd nuclei lie lowest in the mid-shell region, i.e., near N=66. This is demonstrated in a transparent way by Aprahamian et al. [APR84] who also suggest that the ratio $R=B(E2;0_3^+-2_2^+)/B(E2;0_3^+-2_1^+)$ is largest for N=66 for the Cd nuclei. In keeping with this picture is the low-lying strongly collective band observed by Hasselgren et al. [HAS85] in ^{110}Pd with the 0_2^+ band-head at 946.7 keV (see the systematics of low-spin states in the Pd nuclei in Fig. 3 based mainly on [NDS]). However, the decay characteristics of the 0_2^+ state in ^{102}Pd and those of the 0_2^+ state in ^{110}Pd are conspicuously

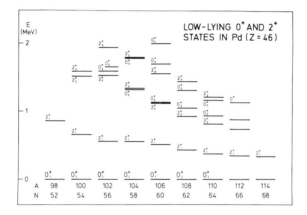

Fig. 3. Low-lying 0^+ and 2^+ states in even Pd nuclei.

different at least in that the latter state is connected to the 2_1^+ one via a collective E2 transition.

The only known state in the Pd nuclei that seems to have properties similar to those of the 0_2^+ state in ^{102}Pd is the 0_4^+ state in ^{106}Pd. According to [LUO85a], this level is characterized by the values of $X_{411} \approx 40$ and $R \approx 3,500$ which both are smaller than the corresponding values for the 0_2^+ state in ^{102}Pd but much larger than those for any other known level in the Pd nuclei. Consequently, the 0_4^+ level in ^{106}Pd may be of similar origin as the isomeric 0_2^+ state in ^{102}Pd; unfortunately, no corresponding 0^+ state has been observed in ^{104}Pd. In view of the intruder-state behaviour suggested in [APR84], neither the energies nor the R ratios of the two related states mentioned seem to fit into the picture.

The situation with the Pd intruders is thus quite unclear, and no obvious explanations of the data seem to be available. It is not clear whether the same excitation mechanism is present in the above intruder candidates in ^{102,106,110}Pd. Neither is it clear that the model presented in [WEN81] based on an interplay between quadrupole vibrations and np-mh excitations — which is a promising theoretical approach — can, e.g., explain the

extremely weak E2 transitions that seem to occur quite frequently, just to
mention some of the problems.   It is obvious   that more systematic work is
needed to elucidate the physics underlying the intruder states.

EO Transitions in ^{202,204}Pb and Intruder-State Systematics of
Even-Even Lead Isotopes.

Among the best examples of the intruder-state systematics in even
nuclei are the 0^+ states observed in the light lead isotopes $^{190-200}$Pb by Van
Duppen et al. [VAN84, VAN85].   One or two low-lying 0^+ states were found in
all of these nuclei, as well as evidence for rotational-like bands built on
0_2^+ states in ^{192,194,196}Pb.   The 0_2^+ states are interpreted as belonging to
the   deformed   2p-2h   proton   intruder   configuration   of $\{1/2^+[440]^{-2},$
$9/2^-[514]2\}$. As the proton pairing-vibrational 0^+ states, at 4100 keV in
^{206}Pb [AND77] and at 5234 keV in ^{208}Pb [JUL85], are equivalent to a proton
2p-2h configuration, the lead intruder systematics have been extended from
the closed neutron shell almost to the mid-shell region.   In this paper, we
report the discovery of the "missing" ^{202,204}Pb intruder candidates.
In our measurements, the levels in ^{202,204}Pb were excited in
^{203,205}Tl(p,2n) reactions at E_p=14.5 MeV.   Conventional gamma-ray and conversion-electron spectrometry were employed, complemented by our newly developed
centroid-shift timing technique with a lens-type electron spectrometer
[KAN82].   A typical electron spectrum obtained with a natural Tl target is
illustrated in Fig. 4 that also shows the energies of the new 0^+ states (3 in
^{202}Pb, 1 in ^{204}Pb).   The rest of our results are summarized in Table 1.

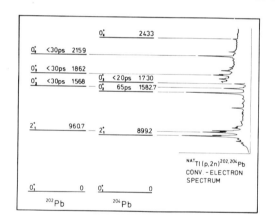

Fig. 4.   Energies of 0^+ states in ^{202,204}Pb as established in the
present work.   On the right, we show a section of an electron spectrum.

In Fig. 5, we have collected the energies of the known 0^+ states in
$^{190-208}$Pb.   The crosses for the isotopes 190-202 represent the empirical
estimates for the above-mentioned intruder configuration as suggested in
[VAN84].   The best candidates for the intruders found in our work are the 0_4^+
states in both nuclei.   This assignment is suggested on the basis of the
energy fit and the large X values associated with these states.
To see how well the picture on the proton-neutron residual interaction discussed in [HEY85] describes the Pb systematics, a fit to the experi-

mental intruder-state energies was made, and the quadrupole component of the interaction obtained. On the basis of such a fit [KAN85], the dashed curve shown in Fig. 5 was drawn and can be taken as another evidence for the underlying theoretical model.

Table 1. Results on 0^+ states of ^{202,204}Pb

Isotope	Level	$t_{1/2}$(ps)	ρ^2_{i1}	X_{i11}	$I_K(E0)/I_K(E2)$
202	0^+_2	<30	>4x10^{-3}	0.08(3)	15(6)
202	0^+_3	<30	>0.8x10^{-3}	0.06(2)	6(2)
202	0^+_4	<30	2x10^{-3}	0.5(2)	24(10)
204	0^+_2	65(20)	7.4(25)x10^{-5}	3.2(3)x10^{-3}	0.60(6)
204	0^+_3	<20	1.3x10^{-3}	>0.045	>5
204	0^+_4			>0.6	>15

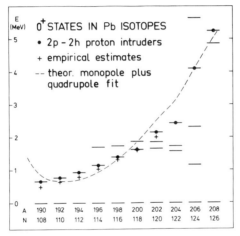

Fig. 5. 0^+ states in even-mass lead isotopes $^{190-208}$Pb.

Concluding remarks

The examples given above in two different energy regions show that, although much understanding of the intruder states in even-even nuclei has already been gained, several important questions still remain unanswered. In spite of the facts that the mere identification of an intruder state is not always unambiguous and that the energy intervals and maybe even the level ordering of the intruder band can be obscured by band mixing, it is still tempting to attempt to find some kind of fingerprints of the intruder states. In the following, we give a tentative list of possible fingerprints of proton intruder states in even-even nuclei:

1) Appearance of extra state(s) in (or below) the energy region of states which can be explained as being due to two quadrupole phonons, to two quasi-particles, or to some other familiar origin.

2) Systematic behaviour of the energy of the state(s) from isotope to isotope, with a minimum in the mid-shell region. (Is ^{102}Pd an exception?)
3) Band structure built on the top of an intruder candidate; enhanced E2 transitions within the band.
4) Strong population in two-proton transfer for 2p-2mh type of states.
5) Retarded E2 transitions to the 2^+_1 state. (This is not always the case.)
6) E0 transition to the ground state having a moderate intensity (usually of the order of 1 to 10 "milliunits" and less than 1 "s.p.u.").
7) Strong E2 transition from the intruding 0^+ state to a close-lying (lower) 2^+ state which usually is the 2^+_2 state.

Obviously, it cannot be expected that all these fingerprints are true for every intruder state (in fact, some of the proposed features may even be "orthogonal" to each other). It is likely that the above list must be revised, as our knowledge of the intruder states continues to increase.

Acknowledgments

We wish to thank J. L. Wood for many stimulating discussions.

References

[AND77]   R. E. Anderson, P. A. Catay-Csorba, R. A. Emigh, E. R. Flynn, D. A. Lind, P. A. Smith, C. D. Zafiratos, and R. M. DeVries, Phys. Rev. Lett. 39 987 (1977).

[APR84]   A. Aprahamian, D. S. Brenner, R. F. Casten, R. L. Gill, A. Piotrowski, and K. Heyde, Phys. Lett. 140B 22 (1984).

[BÄC81]   A. Bäcklin, N.-G. Jonsson, R. Julin, J. Kantele, M. Luontama, A. Passoja, and T. Poikolainen, Nucl. Phys. A351 490 (1981).

[BRO78]   J. Bron, W. H. A. Hesselink, L. K. Peker, A. Von Poelgeest, J. Uitzinger, H. Verheul, and J. Zalmstra, J. Phys. Soc. Jpn. Suppl. 44 513 (1978).

[FIE77]   H. W Fielding, R. E. Anderson, C. D. Zafiratos, D. A. Lind, F. E. Cecil, H. H. Wieman, and W. P. Alford, Nucl. Phys. A281 389 (1977).

[HAS85]   L. Hasselgren et al., to be published.

[HEY83]   K. Heyde, P. Van Isacker, M. Waroquier, J. L. Wood, and R. A. Meyer, Phys. Repts. 102 291 (1983).

[HEY85]   K. Heyde, P. Van Isacker, R. F. Casten, and J. L. Wood, Phys. Lett. 155B 303 (1985).

[JUL80]   R. Julin, J. Kantele, M. Luontama, A. Passoja, T. Poikolainen, A. Bäcklin, and N.-G. Jonsson, Z. Phys. A296 315 (1980).

[JUL85]   R. Julin, J. Kantele, M. Luontama, A. Passoja, and J. Blomqvist, to be published.

[KAN82]   J. Kantele, R. Julin, M. Luontama, and A. Passoja, Nucl. Instr. and Meth. 200 253 (1982).

[KAN84]   J. Kantele, in Heavy Ions and Nuclear Structure, ed. by B. Sikora and Z. Wilhelmi, Harwood Acad. Publishers, 1984.

[KAN85]   J. Kantele, M. Luontama, W. Trzaska, R. Julin, A. Passoja, and K. Heyde, to be published.

[LUO85]   M. Luontama, R. Julin, J. Kantele, A. Passoja, W. Trzaska, A. Bäcklin, N.G. Jonsson, and L. Westerberg, to be published.

[LUO85a] M. Luontama, R. Julin, A. Passoja, and W. Trzaska, to be published.

[NDS]   Nuclear Data Sheets.

[VAN84]   P. Van Duppen, E. Coenen, K. Deneffe, M. Huyse, K. Heyde, and P. Van Isacker, Phys. Rev. Lett. 52 1974 (1984).

[VAN85]   P. Van Duppen, E. Coenen, K. Deneffe, M. Huyse, and J. L. Wood, Phys. Lett. (in press).

[WEN81]   C. Wenes, P. Van Isacker, M. Waroquier, K. Heyde, and J. Van Maldeghem, Phys. Rev. C23 2291 (1981).

RECEIVED August 5, 1986

Recent Experimental Investigations and Interacting Boson Model Calculations of Even Te Isotopes

J. Rikovska[1], N. J. Stone[1], V. R. Green[1], and P. M. Walker[2]

[1]Clarendon Laboratory, OX1 3PU, Oxford, United Kingdom
[2]Daresbury Laboratory, WA4 4AD, Warrington, United Kingdom

The most interesting question occurring in studies of low excitations in even-even Te isotopes (Z = 52, N = 62-72) is the problem of existence of "intruder" states. These states have their origin in excitation of a pair of particles across a closed shell gap and have been observed in nuclei near to both neutron and proton closed shells (or subshells) [HEY83] . There are several main features of these states:
- considerable deformation due to residual proton-neutron interaction amongst the increased number of valence particles outside a closed shell as compared with "normal" vibrational-like configuration, which implies that the proton (neutron)intruder configuration should be lowest in energy in the middle of neutron (proton) shell, increasing rapidly as one moves away from the centre,
- a rotational band built on the intruder 0^+ state,
- enhanced E0 transition probability between intruder and normal 0^+ states and E0 admixtures in transitions between the intruder and normal states of the same spin because of the difference in shape of the two configurations,
- increased excitation cross section in specific ew nucleon transfer reactions as compared with the cross section to the ground state.

In even-even nuclei around Z = 50, the intruder configurations, although somewhat mixed with the normal ones have been identified in Cd (Z = 48) (4h – 2p states) [HEY82, MHE84] and Sn (Z = 50)(2p – 2h states) [WEN81]. Recent nuclear orientation experiments at the Daresbury DOLIS facility [SHA85a, GRE85] together with preliminary results from electron conversion measurement [GRE85], (^{3}He, n) reactions [FIE78] and thermal neutron capture reactions [ROB83] provide a possible experimental basis for identification of these states in even Te isotopes. All data available on low-lying 0^+, 2^+ and 4_1^+ and 4_2^+ state energies for even Cd and Te isotopes are shown in Fig. 1. to stress similarities and differences between these two isotopic chains, where the intruder configurations should be of related structure, proton 4h – 2p in Cd and 4p – 2h in Te.

Experimental data suggests that there are three different kinds of excited 0^+ states in Te isotopes. The second 0^+ state, simply considered as a two-phonon vibrational state, shows rapidly increasing energy with increasing neutron number, particularly between N = 68 and 70. The same change is observed in Cd between N = 60 and 62. The behaviour can be accounted for by increasing γ-softness of the nuclei [MEY77]. Intruder configuration admixture can explain the relatively low energy of the 0_2^+ state in ^{118}Te (see below) but otherwise those states are almost pure normal configuration. As a second type of 0^+ state, in Cd mixed intruder 0_3^+ states have been identified in the closed vicinity of the two-phonon triplet, however in Te they appear to be found close to the three-phonon quintuplet and are probably strongly mixed with states belonging to this vibrational excitation. The evidence for this is rather incomplete at this stage. In ^{118}Te there are two candidates for higher 0^+ states (at 1517 and 1845 keV) and one 0^+ state (at 1710 keV) has been identified in ^{120}Te in (^{3}He, n) reactions [FIE78]. We tentatively assign the 1517 keV state in ^{118}Te and the

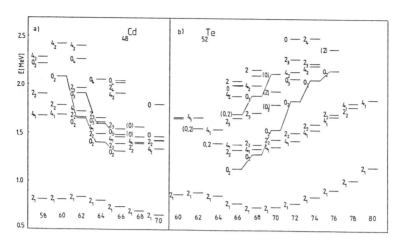

Fig. 1.   Systematics of positive parity states in even Cd and Te isotopes.
The lines connect states discussed in the text.

1710 keV state in ^{120}Te as having strong intruder admixture. For heavier
Te isotopes the intruder configuration lies probably higher in energy
($\sim$ 2 MeV in ^{122}Te and $\sim$ 3 MeV in ^{124}Te). The decay of the 1940 keV state in
^{122}Te has been shown in our recent nuclear orientation experiment to be
entirely consistent with the assignment 0^+ [SHA85b]. This state is possibly a
mixture of three phonon vibration and intruder configurations as discussed
below. In the third class of 0^+ state, the 1845 keV (^{118}Te), 1883 keV(^{124}Te)
and 1982 keV (^{128}Te) are very likely mostly of three-phonon vibrational
origin as seems to be suggested by the mode of their decay and smooth variation of their energy with neutron number.

A question which remains is the strength with which the intruder 0^+
states would be populated in the decay of ^{118,120}I ($I^\pi = 2^-$). For example,
in the similar case of ^{112}Ag ($I^\pi = 2^-$) β-decay the two 0_3^+ and 2_3^+ intruder
states in ^{112}Cd are populated with almost the same probability as the 0_2^+ and
2_2 states. It seems therefore that there is no clear reason why the intruder
states, especially 2^+, should not be populated in the decay of iodine isotopes. Candidates for 2^+ intruder states are the 2_3^+ states at 1482 keV in
^{118}Te and 1535 keV in ^{120}Te. They show similar decay patterns to those of
the 2_3^+ states, assigned as mixed intruder, in Cd isotopes, i.e. mostly to the
2_1^+ and the ground states. The simple vibrational selection rule for E2
transitions ($\Delta n = \pm 1$, n-number of vibrational phonons) rules out their interpretation as three phonon states, because of their strong decay to the one
phonon 2_1^+ state. Unfortunately, very little is known about the population of
these states in nuclear reaction. No clear evidence for a $\Delta I = 2$ rotational
band in Te isotopes built on a 4p - 2h 0^+ state was found in (^{3}He, n) [FIE78],
(α,xn) and (^{12}C, 2n) reactions [CHO82,VAN82]. Probably a reaction, which can
produce hole states directly, like Xe(d, ^{6}Li) would be more useful than two-
proton transfer in the search for 4p - 2h excitations. Our first measurements of E0 transition strength show E0 transitions between $0_2^+ \rightarrow 0_1^+$ and

$0_3^+ \to 0_2^+$ in ^{118}Te together with E0 admixtures in the $2_2^+ \to 2_1^+$ transitions in ^{118}Te and $2_3^+ \to 2_1^+$ in ^{120}Te.

Summarizing the above experimental facts concerning the existence of intruder states in even Te isotopes it is clear that there is relatively little on which to base a calculation of low-lying states of these isotopes. Nevertheless, following the successful application of the interacting boson model (IBM2) to even Cd isotopes [HEY82] similar calculations have been performed on Te. A standard IBM2 calculation, using NPBOS code was performed for $^{114-124}$Te in the space of one proton boson ($N_\pi = 1$) and corresponding number of neutron bosons N_ν. The parameters used (see Fig. 2) were chosen to fit energy levels, the electric quadrupole and magnetic dipole moments of the first 2^+ states (where known) and some ratios of reduced E2 transition probabilities, and are in line with the systematics of the IBM2 parameters in this mass region. The same calculation was performed for the intruder 4p – 2h

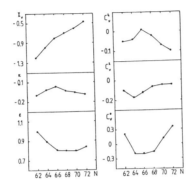

Fig. 2.   Parameters of the IBA2 Hamiltonian as a function of neutron number. $\chi_\pi = -0.8$, $\xi_1 = \xi_3 = -0.09$, $\xi_2 = 0.12$ and $C_\pi^L = 0$ (L = 0,2,4) were constant for all isotopes. All parameters are in MeV except for $\chi_{\pi,\nu}$ (dimensionless).

configuration, i.e. $N_\pi = 3$ and the same N_ν. All the parameters describing the neutron part of the IBM2 Hamiltonian were kept the same as for the normal configuration. The strength of the neutron-proton quadrupole-quadrupole interaction κ is approximately proportional to the product of the number of neutron and proton bosons present and thus κ for the intruder configuration should differ from that of the normal configuration. Spectra of Ru (6h) and Ba (6p) isotopes served as an approximate guide for adjusting parameters ε and κ for 4p – 2h states in Te isotopes, since, as the IBM does not distinguish between particles and holes, these should show a certain similarity with the unknown intruder configuration in Te. The values of $\varepsilon(3\pi) = 0.50$, 0.54, 0.58 and 0.58 MeV and $\kappa(3\pi) = -0.16$, -0.18, -0.20 and -0.20 MeV for ^{118,120,122,124}Te, respectively, were used. Wave functions obtained for both configurations (normal and intruder) were then mixed using Hamiltonian [DUV81]

$$H = H_s + \alpha(s_\pi^\dagger s_\pi^\dagger + s_\pi s_\pi)^{(o)} + \beta(d_\pi^\dagger d_\pi^\dagger + \tilde{d}_\pi \tilde{d}_\pi)^{(o)}$$

where H_S is the usual IBM2 Hamiltonian. H was diagonalized in the basis provided by the lowest eigenstates of the $N_\pi = 1$ and $N_\pi = 3$ configurations.

In the present case, four states of each spin were taken into account. The procedure is dependent upon three additional parameters, α, β and Δ, which represents an energy needed to excite the intruder configuration and must be added to the eigenvalues of the N_π = 3 configuration. The parameter Δ can be expressed in terms of the two-proton separation energies [HEY85]

$$\Delta = S_{2p}(Z,N) - \{ S_{2p}(Z+2,N) + [S_{2p}(Z+2,N) - S_{2p}(Z+4,N)] \}$$

and is $\sim$ 5 MeV in the Z = 50 region. In our calculations, Δ = 4.65 – 5.0 MeV for [118-124]Te. The interaction strength parameters α and β were kept the same, α = β = 0.19 MeV for all nuclei. The value is somewhat larger than the published values for Cd, (α = β = 0.08 MeV) but explains very well a more complicated mixing of intruder and normal states in Te isotopes. The results of the mixing calculation are illustrated in Fig. 3 where selected experimental data are added for comparison. The mixing in [118]Te can explain the position of 0_2^+ state and leads to inversion of the intruder level order (2_3^+ lies lower than 0_3^+). In [120]Te, the intruder component is almost equally ($\sim$50%) present in both 2_3^+ and 2_4^+ states and then the intruder configuration rises up in energy in heavier isotopes. This picture of mixing, which was not observed in other nuclei, could explain, for example, the difficulty in the search for rotational bands, which would be very distorted by mixing and rather high in energy.

The mixed wave functions were used to calculate B(E2), B(M1) and E0 transition strengths. The transition operator for E2 transition is of the form

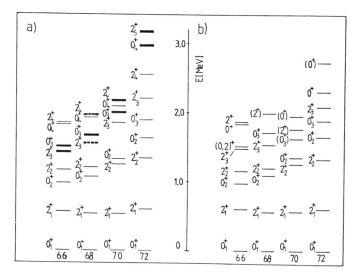

Fig. 3.   Results of the IBM2-mixing calculations (a) in comparison with experiment (b). Thick lines show states with large intruder configuration components.

$$T(E2) = e_1(Q_{\pi 1} + Q_{\nu 1}) + e_3(Q_{\pi 3} + Q_{\nu 3})$$

$$Q_\delta = (s_\delta^+ \tilde{d}_\delta + d_\delta^+ s_\delta)^{(2)} + \kappa_\delta (d_\delta^+ \tilde{d}_\delta)^{(2)}, \quad \delta \equiv \pi \; \nu$$

is the usual quadrupole operator and 1,3 refers to normal and intruder configuration, respectively. Similarly,

$$T(M1) = g_1 (d_\pi^+ \tilde{d}_\pi)_1^{(1)} + g_3 (d_\pi^+ \tilde{d}_\pi)_{(3)}^{(1)}$$

where $g_i = \sqrt{30\pi/4} \; (g_\pi - g_\nu)_i$ is an effective g-factor for i-configuration
(i=1,3), and

$$T(E0) = a\hat{N}_\pi + b(d_\pi^+ \tilde{d}_\pi)^{(0)}$$

where $\hat{N}_\pi = (s_\pi^+ s_\pi + d_\pi^+ \tilde{d}_\pi)$ counts the number of proton bosons. The term $a\hat{N}_\pi$
differs in a mixing calculation for each state and moreover, a and b are
different in the $N_\pi = 1$ and $N_\pi = 3$ subspaces. Thus, interpretation of $T(E0)$
in mixing calculations depends on a number of additional constants. No data
exist on parameters (such as isotopic or isomeric shifts) on Te isotopes
which would yield these constants for either mixed or unmixed states. Con-
sequently a simplified expression $\rho_{if}(E0) = A_1 (d_\rho^+ d_\rho)^{(o)} + A_3 (d_\rho^+ d_\rho)^{(o)}$
(subscript $\rho \equiv \pi$ or ν, subscripts 1,3 as before) was used with different
combinations A_1, A_3 for investigation of qualitative trends. The E2/M1
mixing ratios for $2_3^+ \to 2_1^+$ and $2_2^+ \to 2_1^+$ transitions were calculated using

$$\delta = 0.00832 \times E_\gamma \text{ (MeV)} \times <f\|\,T(E2)\,\|\,i> \left[\text{efm}^2\right]/<f\|\,T(M1)\,\|\,i> \left[\mu_N\right]$$

where E_γ is the photon energy of the transition involved.

Results for B(E2) and B(M1) probability ratios, as well as the mixing
ratio δ, involving 0_3^+ and 2_3^+ levels in ^{118,120}Te show considerably closer
agreement with experiment as compared with calculations without mixing. For
the sake of simplicity, $e_1 = e_3 = 12.4$ efm^2 and $g_1 = g_3$ ($g_\pi = +1.1\ \mu_N$, g_ν(N)
varied with N within -0.10 to $-0.15\ \mu_N$) was taken in all calculations. The
lack of data in the energy region of interest does not allow such comparison
for ^{122,124}Te. Experimental information is also rather limited for ^{114,116}Te
and does not provide a solid basis for mixing calculation despite rather
reasonable results without mixing. Examination of the two lighter Te isotopes would be very important for understanding of variations in the intruder
configuration either side of the middle of the neutron 50-82 shell.

The present situation regarding possible mixing in Te isotopes high-
lights several experimental questions, especially the uncertainty about the
1517 and 1845 keV levels in ^{118}Te. More complete data on all levels above
2 MeV is needed in ^{120,122}Te. In ^{120}Te a search for all 0^+ states, and in
the other nuclei, for E0 transitions between 0_3^+ and 0_2^+ states, by electron
conversion measurement would be valuable. The ^{120}Te (n,n'γ) reaction should
be another useful source of information on excited 0^+ states in ^{120}Te.
However even the present stage of the investigation shows a possibility of
understanding the problem of intruder configurations in even Te isotopes.

References

[CHO82] P. Chowdhury, W.F. Piel, Jr., and D.B. Fossan, Phys. Rev. C25 813 (1982).

[DUV81] P. Duval and B.R. Barrett, Phys. Lett. 100B 223 (1981).

[FIE78] H.W. Fielding, R.E. Anderson, P.D. Kunz, D.A. Lind, C.D. Zafiratos and W.P. Alford, Nucl. Phys. A304 520 (1978).

[GRE85] V.R. Green, J. Rikovska, T.L. Shaw, N.J. Stone, I.S. Grant, P.M. Walker and K.S. Krane, Annual Report, Daresbury Laboratory, 1984/85.

[HEY82] K. Heyde, P. Van Isacker, M. Waroquier, G. Wenes and M. Sambataro, Phys. Rev. C25 3160 (1982).

[HEY83] K. Heyde, P. Van Isacker, M. Waroquier, J.L. Wood and R.A. Meyer, Phys. Reports, 102 291 (1983).

[HEY85] K. Heyde, P. Van Isacker, R.F. Casten and J.L. Wood, Phys. Lett. 155B 303 (1985).

[MEY77] R.A. Meyer and L. Peker, Z. Phys. A283 379 (1977).

[MHE84] A. Mheemeed et al., Nucl. Phys. A412 113 (1984).

[ROB83] S.J. Robinson, W.D. Hamilton and D.M. Snelling, J. Phys. G: Nucl. Phys. 9 961 (1983).

[SHA85a] T.L. Shaw, V.R. Green, N.J. Stone, J. Rikovska, P.M. Walker, S. Collins, S.A. Hamada, W.D. Hamilton and I.S. Grant, Phys. Lett. 153B 221 (1985).

[SHA85b] T.L. Shaw, private communication, (1985).

[WEN81] G. Wenes, P. Van Isacker, M. Waroquier, K. Heyde and J. Van Maldeghem, Phys. Rev. C23 2291 (1981).

RECEIVED May 12, 1986

Spectroscopy of Light Br and Rb Isotopes
Onset of Large Quadrupole Deformation

K. P. Lieb[1,2], L. Lühmann[1], and B. Wörmann[1]

[1]Universität Göttingen, D-3400 Göttingen, Federal Republic of Germany
[2]State University of New York, Stony Brook, NY 11794

1. Introduction   Among the nuclei off the line of stability, the N $\cong$ Z = 34-40 region offers some very interesting and unique features: large quadrupole deformations, triaxiality, shape coexistence and transitions, and reductions of pairing correlations leading to rigid body rotation |BEN84, LIS84a, LIE84, HAM85|. We have performed over the last years at the University of Göttingen a systematic study of level energies and lifetimes of rotational bands in this mass region. In particular, we have investigated the effects which one or several aligned $g_{9/2}$ proton(s) or neutron(s) have on the collective parameters; we present here recent results in the odd-A proton nuclei $^{73-75}$Br and ^{77,79}Rb |PAN82, LUH85, LUH86, WOR85|.

2. Experiments   These nuclei have been studied via the heavy ion fusion evaporation reactions with ^{12}C, ^{16}O, ^{19}F, ^{36}Ar and ^{40}Ca beams provided by the tandem accelerators at Cologne, Daresbury and Brookhaven and the VICKSI cyclotron at Berlin. Besides the conventional γ-spectroscopic techniques (γγ-coincidences, excitation functions, angular distributions) we also employed neutron and charged particle gating and Compton suppression in order to enhance the channels and γ-ray lines of interest and to clean the spectra from Coulomb excitation, inelastic scattering and ß-decay processes. Lifetimes in the 10^{-13}- 10^{-7} sec range have been measured with the Doppler shift attenuation, recoil distance and electronic timing methods, as described in detail in |PAN82, LUH85|. In two of the reactions studied, we have been able to measure the spin dependence of the side feeding time τ_f namely for the reaction ^{62}Ni(^{16}O,p2n)^{75}Br at 60 MeV beam energy (LUH85| and the reaction ^{40}Ca(^{40}Ca,3p)^{77}Rb at 122 MeV beam energy |LUH86|. A summary of the experiments is given in Table 1.

Table 1: Reactions and techniques used to study the Br and Rb

Nucleus	Reaction	E (MeV)	I_{max} (h)	Technique[a]
^{73}Br	^{40}Ca(^{36}Ar,3p) ^{40}Ca(^{40}Ca,α3p)	105 155, 170	33/2	γγ, pγ, RD, AD, PT
^{75}Br	^{62}Ni(^{16}O,p2n) ^{66}Zn(^{12}C,p2n)	45-65 35-55	33/2	γγ, nγ, RD, AD, DSA
^{77}Rb	^{40}Ca(^{40}Ca,3p)	120-170	37/2	γγ, pγ, RD, AD, DSA
^{79}Rb	^{63}Cu(^{19}F,p2n) ^{70}Ge(^{12}C,p2n)	45-65 35-45	29/2	γγ, nγ, RD, AD, DSA, PT

a) RD=recoil distance method, DSA=Doppler shift attenuation method, AD=angular distributions, PT=pulsed beam electronic timing

0097-6156/86/0324-0233$06.00/0
© 1986 American Chemical Society

3. Quadrupole deformation parameters   Before discussing the individual nuclei let us look at the deformation parameters β_2 deduced from the
measured transitional quadrupole moments $|Q_t| = (16\pi/5\ B(E2,I \to I-2))^{1/2}$ /
$< IK20\,|I-2\ K >$: As a starting point we assumed axial symmetry ($\gamma = 0$) and
a fixed K value suggested by the Nilsson scheme, thus neglecting K mixing
due to triaxiality and rotation. In Table 2, these $|\beta_2|$ values are compa-
red with the results of Strutinsky Bogolyubov calculations with a defor-
med Saxon Woods potential performed by Nazarewicz et al. $|NAZ85|$.

Table 2: Experimental and predicted deformation parameters β_2

Nucleus	Experiment Spin range	K a)	$\|\beta_2\|$ a)	Ref.	Therory d) b) Configuration	β_2
^{72}Se	2^+	0	0.20(2)	$\|$HAM74$\|$		
	6^+-10^+	0	0.31(2)	$\|$LIE77$\|$		
^{73}Br	$9/2^+13/2^+$	1/2	0.40(2)	$\|$WOR85$\|$		
^{74}Se	2^+-10^+ c)	0	0.28(1)	$\|$BAR74$\|$		+0.31/-0.25
^{75}Br	$9/2^+,13/2^+$	1/2	0.35(2)	$\|$LUH85$\|$	$|$431 $3/2^+$	+0.35
	$\leq 17/2^-$	3/2	0.36(2)		$312\ 3/2^-\|$	+0.33
	$21/2^+,25/2^-$	1/2	0.29(2)			
	$19/2^-$-$27/2^-$	3/2	0.29(2)			
^{76}Kr	2^+-8^+	0	0.35(2)	$\|$KEM84$\|$ $\|$WOR84$\|$		+0.35/-0.31
^{77}Rb	$9/2^+$-$25/2^+$	3/2	0.40(2)	$\|$LIS84b$\|$	$|$431 $3/2^+\|$	+0.35
	$9/2^-$-$21/2^-$	3/2	0.45(2)	$\|$LUH86$\|$	$312\ 3/2^-\|$	+0.38
^{78}Kr	2^+-8^+	0	0.31(1)	$\|$HEL79$\|$ $\|$WIN85$\|$		+0.30/-0.25
^{79}Rb	$9/2^+$-$17/2^+$	3/2	0.36(2)	$\|$PAN82$\|$	$|$431 $3/2^+\|$	+0.33
	$9/2^-$-$21/2^-$	3/2	0.41(3)		$312\ 3/2^-\|$	+0.36
	$21/2^+,25/2^+$	3/2	0.24(2)			

a) Assumed fixed K and $\gamma = 0°$; the deformation parameters are the weighted
   average values deduced from the B(E2) of stretched E2 transitions in
   the spin range given
b) In even-even nuclei, deformations of prolate and oblate minima $|NAZ85|$
c) $B(E2,6^+ \to 4^+)$ not included
d) ^{74}Se, ^{75}Br from $|LUH85|$; all others from $|NAZ85|$.

The following conclusions can be drawn from this comparison:
   (1) The nuclei show consistently large deformations of $\beta_2 = 0.35 -$
0.45. It thus can be inferred that the predicted prolate to oblate shape
transition in this mass region $|BEN84, NAZ85|$ occurs by a change of γ
rather than by large variations of β_2.
   (2) The overall agreement between the experimental and calculated
β_2 values is good. Extrapolations to the conjectured large deformations
of the $N = Z = 34$-40 nuclei in their ground states therefore seem to be
well justified $|BEN84, MOL81|$. This agreement also implies that all odd-A
nuclei discussed here are prolate.- Furthermore, it appears that the positive and negative yrast bands have very similar β_2 values. In their
band head wavefunctions the $|431\ 3/2^+|$ and $|312\ 3/2^-|$ components dominate,
their energies being nearly degenerate.

(3) In all cases the additional proton induces a substantial increase of the core deformation by some 15 - 30 %. This "polarization" effect is stronger for the Se-Br isotopes than for Kr-Rb. The most dramatic polarization occurs between ^{72}Se and ^{73}Br as illustrated in Fig. 1b: due to the prolate-oblate shape coexistence at low spin, the deformation in ^{72}Se starts at only $\beta_2 = 0.20(2)$ and than increases to $\cong 0.31$ in the region where the ground band has reached a well deformed (prolate) shape |HAM77, LIE77| The recently measured lifetimes of the 9/2$^+$ and 13/2$^+$ levels in ^{73}Br determine $\beta_2 = 0.40(2)$, nearly double that of the ^{72}Se ground state |WOR85|.

(4) The dependence of the transitional moment $|Q_t|$ of the $g_{9/2}$ bands in ^{75}Br and ^{77}Rb is plotted in Fig. 2 and 3 as function of the rotational frequency $\hbar\omega$. While $|Q_t|$ is essentially constant in ^{77}Rb, it drops in ^{79}Rb and ^{73,75}Br above spin 17/2$^+$. Several explanations have been proposed to explain this reduction of Q_t: in the Interacting Boson (Fermion) model |IAC79 |, the finite boson number leads to reductions of the B(E2), as for example suggested for ^{79}Rb |PAN82|; alternatively, changes of β_2 and γ associated with the alignment of $g_{9/2}$ nucleons have been considered. Calculations in ^{75}Br suggest that Q_t decreases at $\hbar\omega = 0.45$ MeV as γ changes sign due to $g_{9/2}$ proton alignment |LUH85|.

4. Shape stabilisation in ^{73}Br So far the $g_{9/2}$ band up to spin 33/2 $\hbar$ (signature $\alpha = 1/2$) has been identified |WOR85|. The $g_{9/2}$ proton not only dramatically increases the deformation of the core as mentioned before, but also stabilizes the collective shape. Fig. 1a illustrates the $I_x(\omega)$ plot for the yrast bands in ^{72}Se, ^{74}Kr and ^{73}Br |HAM74, LIE77, ROT 84|. Prolate-oblate shape coexistence below 0.4 MeV and $(g_{9/2})^2$ alignment at 0.55 MeV produce irregularities in the ground bands of the even-even cores. The $g_{9/2}$ band in ^{73}Br in contrast features a nealy constant moment of inertia $\mathscr{J}/\hbar^2 = 21$ MeV^{-1} close to the rigid body value. The disappearance of the first backbend in ^{73}Br proofs the $g_{9/2}$ (2qp) proton character of the s-band in ^{74}Kr. A similar interpretation based on the blocking of the s-band by the $g_{9/2}$ proton also applies to ^{76,78}Kr |PAN82, LUH86|.

5. Rigid rotation in ^{77}Rb Among the four odd-A nuclei discussed here, ^{77}Rb offers a surprisingly simple rotational structure |LIS84b, LUH86|. As shown in Fig. 3, both the quadrupole and inertial moments of the yrast bands with signature $\alpha = 1/2$ vary little over a large frequency range. For the $g_{9/2}$ band, we find $\mathscr{F}^{(1)}/\hbar^2 = \mathscr{F}^{(2)}/\hbar^2$ at $\hbar\omega = 0.4$-0.7 MeV, indicating rigid rotation. If we assume equal triaxially shaped neutron and proton distributions, we can deduce the deformation parameters $\beta_2 = 0.37(2)$ and $\gamma = -25(7)^0$ from the measured values $|Q_t| = 3.13(12)$ b and $\mathscr{F}/\hbar^2 = 21.1(4)$ MeV^{-1}. This γ-value is supported by triaxial rotor plus quasiparticle calculations |TOK76|. The average transitional quadrupole moment of the negative parity band, $|Q_t| = 3.52(12)$ b, is in excellent agreement with that of the 3/2$^-$ ground state, $Q = 3.48(16)$ b, determined by laser spectroscopy |THI81|.- The suppression of pairing correlations as evidenced by rigid rotation can be qualitatively explained by the single particle structure at $\beta_2 = 0.4$. The N = Z = 38 energy gap of 2 MeV is considerably larger than the pairing gap parameters $\Delta_n \cong 1.4$ MeV and $\Delta_p \cong 1.1$ MeV (at $\beta_2 = 0$) resp. $\Delta_p \cong 0.6$ MeV at $\beta_2 = 0.4$ |HEY84, NAZ85|. Proton pairing correlations seem to be suppresed because the |431 3/2$^+$ | Nilsson orbit is blocked and the |422 5/2$^+$| orbit is too far away energetically. At $\beta_2 = 0.4$, the density of single particle levels near the

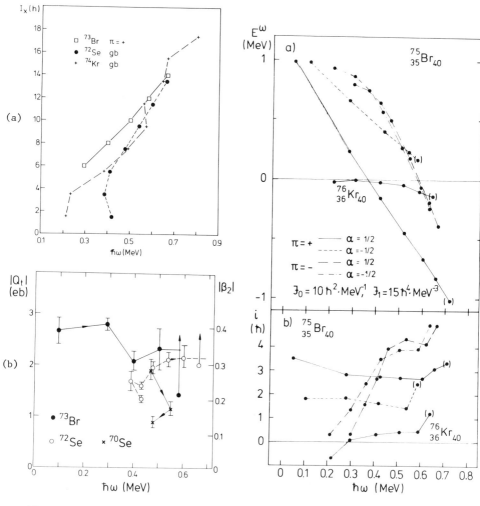

Fig. 1: Spin projection I_x (a) and transitional quadrupole moment $|Q_t|$ (b) of the $g_{9/2}$ band in ^{73}Br |WOR85| and the ground bands in ^{72}Se |LIE77| and ^{74}Kr |ROT84|.

Fig. 2: (a) Cranked shell model analysis of yrast bands in ^{75}Br. (b) Aligned single particle angular momentum $i_x(\omega)$ of these bands. Reproduced with permission from LUH85. Copyright 1985 American Physical Society.

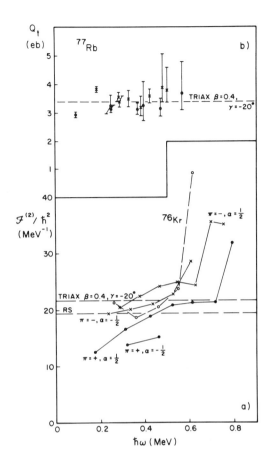

Fig. 3: Dynamical moment of inertia $(^2)/h^2$(a) and transitional quadrupole moment Q_t (b) of the yrast bands in ^{77}Rb |LUH86|.

the Fermi energy is reduced by a factor of 2 - 3 relative to β_2 = 0. We finally like to point out that rigid rotation sets in at rather low spin.

6. The transitional nucleus ^{75}Br   In this nucleus we have extended the yrast bands up to spin 33/2 and have determined 17 lifetimes |LUH85|. A cranked shell model analysis of the bands is presented in Fig. 2a. The $g_{9/2}$ band shows a large signature splitting with nearly constant aligned single-particle angular momenta $i_x(\alpha=+1/2)$ = 3.0 h and $i_x(\alpha=-1/2)$ = 1.7 h. The calculated total energy surface shows a well pronounced minimum at β_2 = 0.35, $\gamma \cong$ -15^0. The negative parity band has very little signature splitting ($\gamma \cong 0^0$), here i_x increases gradually and reaches 4.0 h at $h\omega$ = 0.5 MeV. This indicates strong mixing between the $p_{3/2}$ (1qp) and the spin aligned $(g_{9/2})^2 p_{3/2}$ (3qp) s-band, accompanied by a drop of B(E2) values (see Fig. 2b). A detailed discussion of the γ-driving forces and coexistence phenomena in ^{75}Br has been given in |LUH85|.

7. Conclusions   We have established that the light Br and Rb isotopes presented here have very large quadrupole deformations of $\beta_2 \approx 0.4$ and
moments of inertia close to the rigid body values. The odd proton in the
$|431\ 3/2^+|$ Nilsson orbit polarizes and stabilizes the γ-soft, shape coexistent Se and Kr cores into definite prolate triaxial shapes. This effect
sets in at rather low spin and seems to be intimately connected with the
suppression of pairing correlations near the $N = Z = 38$ gap developing
at $\beta_2 = 0.4$. We thus face a cumulative suppression of both proton and
neutron pairing correlations in the same oscillator shell, a fairly unique
feature in the periodic table.

Acknowledgments   The experiments reported here have been performed
in collaboration with J. Heese, F. Raether (Göttingen), D. Alber, H. Grawe, B. Spellmeyer (Berlin), C. J. Lister, B. J. Varley (Manchester), H.
G. Price (Daresbury), J. Eberth (Köln) and J. W. Olness (Brookhaven). It
is a pleasure to acknowledge the excellent spirit of team work and dedica-
tion. Discussions with R. Bengtsson, K. H. Bhatt, F. Iachello, W. Nazarewicz and T. Otsuka have been stimulating. One of the authors (KPL) acknowledges the hospitality of the State University of New York at Stony Brook
where this article has been prepared. This work has been also funded by
the German BMFT.

References

|BAR74| J. Barrette, et al., Nucl. Phys. A235 154 (1974)
|BEN84| R. Bengtsson, et al., Phys. Scr. 29 402 (1984)
|HAM74| J. H. Hamilton, et al., Phys. Rev. Lett. 32 239 (1974)
|HAM85| J. H. Hamilton, P. G. Hansen and E. F. Zganjar, Rep. Progr.
           Phys. 48 631 (1985)
|HEL79| H. P. Hellmeister, et al., Nucl. Phys. A332 241 (1979)
|HEY84| K. Heyde, J. Moreau and M. Waroquier, Phys. Rev. C29 679 (1984)
|IAC79| F. Iachello and O. Scholten, Phys. Rev. Lett. 43 679 (1979)
|KEM84| P. Kemnitz, et al., Nucl. Phys. A425 493 (1984)
|LIE77| K. P. Lieb and J. J. Kolata, Phys. Rev. C15 939 (1977)
|LIE84| K. P. Lieb, XIX Winter School on Physics, Zakopane/Poland (1984)
|LIS84a| C. J. Lister, et al., "Nuclera Physics with Heavy Ions", P. Braun
           Munzinger, ed. (Harwood Acad. Publ. Chur, 1984) p. 257
|LIS84b| C. J. Lister, et al., "Atomic Masses and Fundamental Constants"
           O. Klepper, ed. (THD Schriftenreihe, Darmstadt, 1984) p.300
|LUH85| L. Lühmann, et al., Phys. Rev. C31 828 (1985)
|LUH86| L. Lühmann, et al., Europhys. Lett. 1 623 (1986)
|NAZ85| W. Nazarewicz, et al., Nucl. Phys. A435 397 (1985)
|PAN82| J. Panqueva, et al., Nucl. Phys. A389 424 (1982)
|PIE76| R. B. Piercey, et al., Phys. Rev. Lett. 37 496 (1976)
|ROT84| J. Roth, et al., J. Phys. G10 L25 (1984)
|THI81| C. Thibault, et al., Phys. Rev. C23 2720 (1981)
|TOK76| H. Toki and A. Faessler, Phys. Lett. 63B 121 (1976)
|WIN85| G. Winter, et al., J. Phys. G11 277 (1985)
|WOR84| B. Wörmann, et al., Nucl. Phys. A431 170 (1984)
|WOR85| B. Wörmann, et al., Z. Phys. A321 171 (1985

RECEIVED July 16, 1986

Intruder States in Highly Neutron-Deficient Pt Nuclei
Evidence from Lifetime Measurements?

U. Garg[1], M. W. Drigert[1], A. Chaudhury[1], E. G. Funk[1], J. W. Mihelich[1], D. C. Radford[2,4], H. Helppi[2,5], R. Holzman[2], R. V. F. Janssens[2], T. L. Khoo[2], A. M. Van den Berg[2,6], and J. L. Wood[3]

[1]Physics Department, University of Notre Dame, Notre Dame, IN 46556
[2]Physics Division, Argonne National Laboratory, Argonne, IL 60439
[3]School of Physics, Georgia Institute of Technology, Atlanta, GA 30332

The recoil distance technique has been employed following the ^{154}Sm(^{34}S,4n)^{184}Pt reaction to measure the lifetimes of states in the ground state band (GSB) of the highly neutron deficient nucleus ^{184}Pt. Lifetimes have been extracted for up to the 18^+ state in the GSB. The B(E2) values exhibit a marked increase between spins 2 and 10, suggesting a strong mixing between the GSB and a less collective band at low spin and, thus, providing indirect evidence for an "intruder" $(6h-2p)0^+$ ground state. This is further confirmation of the shape coexistence occuring at low energy in this region.

The structure of very neutron-deficient Pt isotopes is most unusual and needs to be better clarified for a number of reasons: (1) There is an emerging picture of an extensive occurence of shape coexistence at low energy in the region near Z=82 and N=104 (see, for example,[HEY83,DUP84] and the references therein). This region needs to be carefully mapped away from closed shells to determine the systematics of bandhead energies and mixing matrix elements for the coexisting bands. The present view [WOO81]

[4]Current address: Chalk River Nuclear Laboratories, Chalk River, Ontario K0J 1J0, Canada
[5]Permanent address: Lappeenranta University of Technology, Finland
[6]Current address: Rijks Universiteit Utrecht, Utrecht, the Netherlands

0097-6156/86/0324-0239$06.00/0
© 1986 American Chemical Society

The lifetimes obtained from a preliminary analysis are summarized in Table 1 which also includes the B(E2) values extracted from the lifetime data. Fig. 2 shows a plot of B(E2)'s (in Weisskopf Units) vs. spin of the depopulating state.

TABLE I

E_γ (keV)	$I_i \rightarrow I_f$	τ^* (ps)	B(E2)† (W.U.)	B(E2)§ Theory (W.U.)
163	$2^+ \rightarrow 0^+$	582±50	112±10	85.3
272	$4^+ \rightarrow 2^+$	39± 3	200±17	96.5
363	$6^+ \rightarrow 4^+$	9± 2	220±49	107.8
433	$8^+ \rightarrow 6^+$	3.2±0.5	261±41	115.8
476	$10^+ \rightarrow 8^+$	1.7±0.4	309±73	120.7
497	$12^+ \rightarrow 10^+$	2.3±0.4	184±33	122.3
522	$14^+ \rightarrow 12^+$	2.0±0.4	166±34	104.6
555	$16^+ \rightarrow 14^+$	1.6±0.5	153±47	106.2
586	$18^+ \rightarrow 16^+$	1.3±0.8	144±90	107.8

*Errors are conservative estimates.

†Internal conversion has been taken into account in calculating B(E2) values.

§From (KUM80)

It is clear from Fig. 2 that there is a significant increase in the B(E2) values between spins 2^+ and 10^+ (by a factor of >2.5). This provides an unequivocal confirmation of the behavior expected because of the mixing between the 0^+ states and between the 2^+ states in the coexisting bands and, thus, provides strong evidence for the presence of shape coexistence in ^{184}Pt. A separate and intriguing aspect of our measurements is the decline in B(E2) values beyond spin 10^+. This might be indicative of the onset of triaxiality. It coincides with the beginning of backbending in the GSB and is, possibly, the result of a large induced triaxiality due to the addition of a pair of $i_{13/2}$ quasineutrons. Such behavior has recently been observed in the deformed rare-earth nuclei [FEW85]. Also, a possible emergence of triaxiality has been reported for ^{186}Hg [JAN 83]. Finally, our B(E2) values in ^{184}Pt can be compared (Table 1) with B(E2) values calculated [KUM80] using a microscopic approach of variation with number-conserving projection from a pairing-plus-quadrupole Hamiltonian.

of the even-mass Os, Pt, Hg and Pb isotopes is very puzzling in that the more deformed band is an excited band in Pb, Hg and Os and appears to be the ground band in Pt(for A<186). (2) There are a number of detailed calculations of the even-mass Pt isotopes using the Interacting Boson Approximation (IBA). Specifically, the heavier Pt isotopes have been considered [CAS78] as good examples of the O(6) limiting case of the IBA1. However, descriptions of the neutron-deficient Pt isotopes using IBA1 [BIJ80] and the IBA2 [BIJ80, CHI85] do not consider the possibility of coexisting bands. The neighboring even-Hg isotopes are well described [BAR83, BAR84] by an IBA band-mixing picture. The existence of co-existing bands in the light even-Pt isotopes is suggested by boson expansion theory calculations [WEE80], fitted potential energy surfaces [HES81], Hartfree-Fock plus BCS calculations [SAU81], and calculations [MAY77] using the shell correction method of Strutinsky.

It has been argued [WOO81] that the experimentally observed features in the light Pt isotopes support coexisting bands at low energy albeit marked by band mixing at low spin. The pattern proposed [WOO81] would place the more deformed band lower in energy than the less deformed one. A simple prediction of this picture is that the B(E2) values in the yrast bands should undergo a rapid increase with increasing spin. This increase would reflect the appearance of the full collective strength of the more deformed band, which is masked at low spin due to mixing with the less deformed band. In this paper we report measurements of the lifetimes in the yrast band of ^{184}Pt using the recoil distance technique [ALE79].

The high spin states in ^{184}Pt were populated via the ^{154}Sm(^{34}S,4n) ^{184}Pt reaction following bombardment, with 160 MeV ^{34}S ions from the Argonne Tandem-LINAC system, of a stretched, thin (.5 mg cm^{-2}) foil of enriched ^{154}Sm mounted in a recoil distance (plunger) apparatus. The target was made by evaporation onto a gold foil of ~1.3 mg cm^{-2} thickness. Another gold foil (10.2 mg cm^{-2}) was used as the stopper. Gamma rays were detected at 0° with a Ge detector in coincidence with an 8-element large NaI(Tℓ) sum spectrometer (SS) which surrounded the plunger. Population of high spin states was insured by requiring that at least 2 elements of the SS fire for an event to be acceptable and appropriate cuts were made in the sum energy to maximize the yields for the 4n reaction channel. The use of SS essentially eliminates from the final spectrum all the γ rays attributable to radioactivity and Coulomb excitation.

Data were taken at several target-to-stopper distances ranging from
4.0μ to 10 mm giving an effective range of lifetimes between .5 ps and
several ns. The ratios of intensities of the "unshifted" and "shifted"
components of the γ rays previously known [BES76] to belong to ^{184}Pt
were extracted and fitted for lifetimes using a computer code which incorporates the effects of cascade feeding and takes into account corrections
due to target thickness, detector solid angle, detector efficiency, etc.
Fig. 1 presents typical Ge detector spectra for a number of target-stopper
distances showing "unshifted" and "shifted" peaks for several γ rays in
^{184}Pt; the relevant level scheme is also shown.

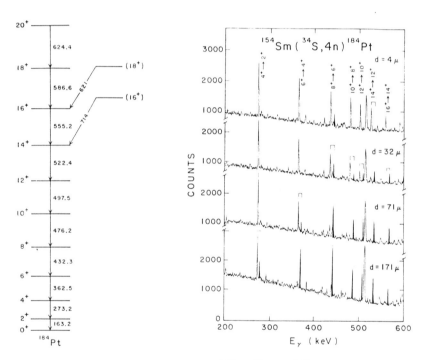

Figure 1. Portions of γ-ray spectra from ^{154}Sm(^{34}S,4n) ^{184}Pt at target-to-
stopper distances as indicated. The "shifted" components are shaded and
the γ rays belonging to ^{184}Pt GSB are marked. Also shown is the level scheme
for ^{184}Pt from [BES76,LAR85].

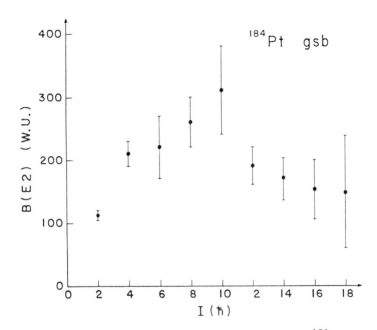

Figure 2. B(E2) values (in Weisskopf Units) extracted for ^{184}Pt GSB from the present measurements. The errors shown represent conservative upper-limit estimates.

The present results provide a strong confirmation of shape co-existence in ^{184}Pt. This paves the way for a theoretical treatment of the very neutron-deficient even-Pt isotopes using the IBA band-mixing picture, such as has been used [BAR84, BAR83] in the neigbhoring Hg isotopes, and contrary to the work in [BIJ80, CHI85]. For example, with the intruder $(6h-2p)0^+$ states available for extra π-bosons, it would be possible to re-produce in a straight forward manner the large deformations observed in these Pt isotopes. According to the systematics [HAG78, HAG79, WOO81] of yrast bands in the neigbhoring Pt isotopes, a similar behavior of B(E2)'s in the yrast band should be seen in ^{182}Pt, ^{180}Pt,...; a study of these isotopes would further test this picture of shape coexistence. These measurements are currently being planned.

Acknowledgments

This work was supported in part by the National Science Foundation (Grant No. PHY82-00426) and the Department of Energy (contract nos. W-31-109-Eng-38 and DE-AS05-80ER10599).

References

[ALE79]   T.K. Alexander and J.S. Foster, Adv. Nucl. Phys. $\underline{10}$ 197 (1979).

[BAR83]   A.F. Barfield et al., Z Phys. $\underline{A311}$, 205 (1983).

[BAR84]   A.F. Barfield and B.R. Barrett, Phys. Lett. $\underline{149B}$, 277 (1984).

[BES76]   S. Beshai et al., Z. Phys. $\underline{A277}$, 351 (1976).

[BIJ80]   R. Bijker et al., Nucl. Phys. $\underline{A344}$, 207 (1980).

[CAS78]   R.F. Casten and J.A. Cizewski, Nucl. Phys. $\underline{A309}$, 477 (1978).

[CHI85]   H.C. Chiang et al., Nucl. Phys. $\underline{A344}$, 54 (1985).

[DUP84]   P. Van Duppen et al., Phys. Rev. Lett. $\underline{52}$, 1974 (1984).

[FEW85]   M.P. Fewell et al., Phys. Rev. C $\underline{31}$, 1057 (1985).

[HAG78]   E. Hagberg et al., Phys. Lett $\underline{78B}$, 44 (1978).

[HAG79]   E. Hagbert et al., Nucl. Phys. $\underline{A318}$, 29 (1979).

[HES81]   P.O. Hess et al., J. Phys. $\underline{G7}$, 737 (1981).

[HEY83]   K. Heyde et al., Phys. Rep., $\underline{102}$, 291 (1983).

[JAN83]   R.V.F. Janssens et al., Phys. Lett. $\underline{131B}$, 35 (1983).

[KUM80]   A. Kumar and M.R. Gunye, Pramana $\underline{15}$, 435 (1980).

[LAR85]   A. Larabee, private communication.

[MAY77]   F.R. May et al., Phys. Lett $\underline{68B}$, 113 (1977).

[SAU81]   J. Saurage-letessier et al., Nucl. Phys. $\underline{A370}$, 231 (1981).

[WEE80]   K.J. Weeks and T. Tamura, Phys. Rev. C $\underline{22}$, 1323 (1980).

[WOO81]   J.L. Wood in Proc. 4th International Conf. on Nuclei Far from Stability, Helsingor, June 1981, CERN Report 81-09, p. 612.

RECEIVED May 8, 1986

Shape Coexistence in ^{185}Au

E. F. Zganjar [1], C. D. Papanicolopoulos [2], J. L. Wood [2], R. A. Braga [3], R. W. Fink [3], A. J. Larabee [4],
M. Carpenter [4], D. Love [4], C. R. Bingham [4], L. L. Riedinger [4], and J. C. Waddington [5]

[1]Department of Physics and Astronomy, Louisiana State University, Baton Rouge, LA 70808
[2]School of Physics, Georgia Institute of Technology, Atlanta, GA 30332
[3]School of Chemistry, Georgia Institute of Technology, Atlanta, GA 30332
[4]Department of Physics and Astronomy, University of Tennessee, Knoxville, TN 37996
[5]Department of Physics, McMaster University, Hamilton, Ontario L85 4K1, Canada

The β-decay of ^{185}Hg $\to$ ^{185}Au has been studied following
on-line mass separation at UNISOR/HHIRF. Transitions with
strong E0 components were observed feeding the $h_{9/2}$ and
$h_{11/2}$ bands and are interpreted as resulting from the
coupling of the $h_{9/2}$ and $h_{11/2}$ single-proton
configurations in the ^{184}Pt and ^{186}Hg cores, respectively.

The first indication of shape coexistence near Z=82, N=104 came from
the observation [BON72] of a large increase in the mean-square charge radius
in going from ^{187}Hg to ^{185}Hg. This result was obtained [BON76] by optical
pumping of on-line separated Hg isotopes at the ISOLDE facility. Later
measurements [KUH77, BON79] included many isomeric states as well as the
ground states of ^{184,183,181}Hg. Interestingly, while the 55 sec 1/2$^-$ ground
state of ^{185}Hg shows a large increase in the mean-square charge radius, the
28 sec 13/2$^+$ isomer does not.

Early in-beam γ-spectroscopy on ^{184}Hg and ^{186}Hg showed that their yrast
bands become deformed above the 2$^+$ state [PRO73, RUD76]. Studies of the
^{188}Tℓ $\to$ ^{188}Hg, ^{186}Tℓ $\to$ ^{186}Hg and ^{184}Tℓ $\to$ ^{184}Hg decays at UNISOR and ISOCELE
clearly revealed two coexisting bands which approach each other with
decreasing mass [HAM75, BON76, COL76, COL77, BER77, COL84]. Recent in-beam
work [MA 84] on ^{182}Hg indicates that the deformed band continues to drop in
energy relative to the ground state band, although the rate of approach
appears to have greatly diminished. These data [HAM85] are shown in fig. 1,

0097-6156/86/0324-0245$06.00/0
© 1986 American Chemical Society

which is presented here as the classic example of a coexistence of levels
built on different shapes. Shape coexistence appears to be a general feature
of nuclei in the neutron deficient region near Z=82.   Concurrently, but

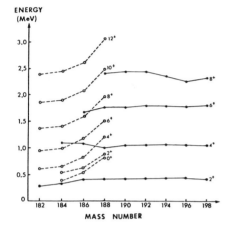

Fig. 1. The energies of the
spherical and deformed states in
even-even $^{182-198}$Hg. Deformed
states for A > 188 have not been
observed [MA 84, HAM85].

independent of these developments, coexisting bands in the odd-proton $T\ell$ and
Au isotopes were established by radioactive decay and in-beam reaction
studies [HEY83].

For the even-even nuclei in the region, one has the possibility that a
pair of particles or holes can couple to quite different cores. This has
been shown [DUP84] to occur even in proton magic Pb isotopes, where
excited 0_2^+ deformed states are observed in $^{192-198}$Pb with a continuous drop
in energy such that the 0_2^+ level become the first excited state in ^{192}Pb and
^{194}Pb. Most theoretical calculations indicate [HAM85] that the near-
spherical shape is oblate and the well-deformed shape is prolate, although
this is not firmly established experimentally. For odd-A nuclei in the
region, it then becomes possible to form multiple shape coexisting
configurations. For example, in the odd-mass Au isotopes the odd proton
particle can couple to the two configurations with different deformation in
the Pt cores, and the odd-proton hole to the two structures with different
deformation in the Hg cores.

The energy level spacings of the 0^+-2^+-4^+ members of the shape
coexisting structures observed in the even-even Pt isotopes do not provide
as clear an indication of which band is the more spherical as do the
corresponding levels in the even-even Hg isotopes. Evidence from

systematics indicates [WO081] that for neutron deficient isotopes of Pt with A ≤ 186, a deformed configuration is present in the ground state.

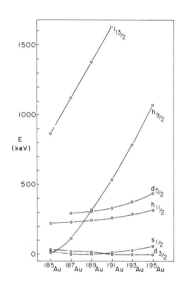

Additionally, a staggering of rotational band parameters has been found [HAG78] for $^{176-182}$Pt, where the odd-mass Pt isotopes appear more deformed than the even-mass ones.

A consistent picture of the details presented above, based on the idea that nuclear deformation is due to the residual force between valence neutrons and valence protons, has been proposed [WO082, HEY83]. In this picture, the excitation of nucleons into shell-model intruder states leads to the coexistence of states with different deformations. Shell-model intruder states for the odd-mass Au isotopes, for example, are presented in fig. 2. Note that the $h_{9/2}$ and $i_{13/2}$ intruders drop rapidly as one goes more neutron deficient [ZGA80]. In ^{185}Au the $5/2^-$ member of the $h_{9/2}$ intruder band actually becomes the ground state

Fig. 2. Shell model states in $^{185-195}$ Au.

[EKS76]. The corresponding data for the odd-mass Tℓ isotopes also show the $h_{9/2}$ and $i_{13/2}$ intruders, but in that case the $h_{9/2}$ reaches a minimum near ^{189}Tℓ and then rises again as the system becomes more neutron deficient [HEY83, HAM85]. That the intruder energy should increase as one moves away from mid-shell (N ≈ 104 in this case) is consistent with the above picture [WO082, HEY83]. Also shown in fig. 2 are the proton hole states. Note particularly the unique parity $h_{11/2}$ state. The coexistence observed in the odd-mass Au isotopes can be understood as arising from the coupling of the proton particle ($h_{9/2}$) and the proton hole ($h_{11/2}$) to the different core shapes in the even-even Pt and Hg isotopes, respectively. This is shown schematically for the Pt ⊕ $\pi h_{9/2}^{+1}$ coupling in fig. 3. These ideas were stimulated by the study [ZGA81] of the low-energy structure of ^{187}Au at UNISOR following the decay of ^{187g,m}Hg. In that case the $h_{9/2}$ particle couples to the ^{186}Pt core to form two bands with states of the same spin connected in many cases by transitions of E0+M1+... multipolarity [ZGA81].

These two bands can be interpreted (see fig. 3) as arising from the couplings ^{186}Pt $[0_1^+ \ \pi(2p\text{-}6h)] \oplus \pi h_{9/2}^{+1}$ and ^{186}Pt $[0_2^+ \ \pi(4h)] \oplus \pi h_{9/2}^{+1}$. When the odd proton is in the $h_{9/2}$ intruder orbital, the formation of the (2p-6h)

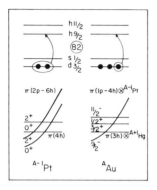

Fig. 3. A schematic view of intruder states near Z=82. The normal proton configurations for Au and Pt are $\pi(3h)$ and $\pi(4h)$ respectively. The proton intruder configurations are $\pi(1p\text{-}4h)$ and $\pi(2p\text{-}6h)$ respectively, where p=particle and h=hole.

configuration in the ^{186}Pt core is blocked due to the Pauli principle and the core configuration for ^{187}Au will then be correspondingly affected. The data indicate that it is the ground state of ^{186}Pt which is blocked and, consequently, that the ground state of ^{186}Pt consists largely of the (2p-6h) configuration resulting from the promotion of a pair of protons into the intruding $h_{9/2}$ orbital [WOO81, ZGA81].

We initiated a careful and detailed study of the ^{185g,m}Hg decay at UNISOR for the following reasons: To explore further the coexistence picture described above; to resolve the disagreement between our earlier data on this decay and that of the Orsay group [BOU82]; and to provide complementary information for an in-beam study [LAR85] of ^{185}Au. The decay of ^{185m}Hg (28s, 13/2$^+$) and ^{185g}Hg (55s, 1/2$^-$) to excited states in ^{185}Au were studied by γ-ray and conversion-electron singles, time-sequence spectroscopy and coincidence measurements. The ^{185}Hg sources were produced by the ^{176}Hf(^{16}O,7n) reaction using 140 MeV beams of ^{16}O^{+7} and were mass separated with the UNISOR on-line isotope separator. As previously observed in the study on ^{187}Au [ZGA81], this coexistence of particle and hole states, for the odd proton as well as the ^{184}Pt core, leads to a large level density and, consequently, to a very high spectral line density. It is only with data of great statistical quality coupled with an extremely careful analysis procedure (e.g. running coincidence gates, [ZGA81]) that one can obtain reliable and consistent results. For example, an important feature of the two $h_{9/2}$ bands are the E0+M1+... transitions. Preliminary analysis of our results on ^{185}Au supports our earlier [ZGA81] interpretation of such transitions as E0+M1+E2 in ^{187}Au and contradicts the interpretation [BOU82] that these transitions in ^{185}Au have anomalous M1 components. The Orsay group [BOU82] were forced to conclude that anomalous M1 components were involved since they obtained conversion coefficients larger than M1 or E2 for a number of transitions between levels of different spin. Our coincidence data were of sufficient statistical quality that we could quantitatively distinguish unresolved doublets. In essentially all cases, we were able to account for the anomalous internal conversion [BOU82] by locating another transition of nearly the same energy which connected states of the same spin and could thus contain an E0 component.

As in the ^{187}Au study, transitions in ^{185}Au with strong E0 components were observed to feed the $h_{9/2}$ and $h_{11/2}$ bands. One example for each is

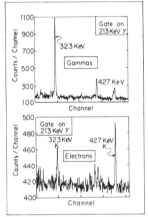

Fig. 4. Portions of the electron and γ-spectra gated on the 213 keV γ which is in cascade with both the 323 and 427 keV transitions. With E2 multipolarity for the 323, the 427 $\alpha(K)$ is determined to be 0.33 ± 0.03. The $\alpha(K)$ for M1 is 0.25.

presented in figs. 4 and 5, respectively. These transitions are therefore interpreted as decays out of the $h'_{9/2}$ and $h'_{11/2}$ bands to the $h_{9/2}$ and $h_{11/2}$ bands respectively. The $h'_{9/2}$ and $h'_{11/2}$ bands, of course, result from the coupling of the $h_{9/2}$ and $h_{11/2}$ single proton configurations to the excited 0^+ configurations in the ^{184}Pt and ^{186}Hg cores, respectively. These results are shown in fig. 6.

In summary, ^{185}Au lies in a most interesting region of shape coexistence where many different shapes are observed at low energies. Recent in-beam work [LAR85] on ^{185}Au suggests that these various shapes are made up of both oblate and prolate structures which can be understood as arising from the competition between close lying K=1/2 (oblate) and K=11/2 (prolate) states for $h_{11/2}$; and K=1/2 (prolate) and K=9/2 (oblate) for $h_{9/2}$. While the results of these

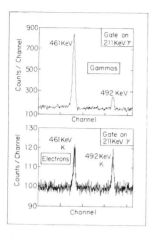

Fig. 5. Portions of the electron and γ-spectra gated on the 211 keV γ which is in cascade with both the 461 and 492 keV transitions. With E2 multipolarity for the 461, the 492 $\alpha(K)$ is determined to be 0.14 ± 0.02. The $\alpha(K)$ for M1 is 0.072.

studies are consistent with the explanation [WOO82, HEY83] of shape coexistence based on the promotion of a pair of particles into an intruder orbital, additional experimental and theoretical work is needed to fully determine which mechanism gives rise to shape coexistence in the Z=82 region.

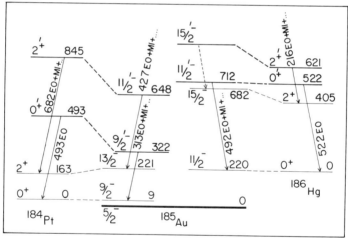

Fig. 6. Portions of the $h_{9/2}$, $h_{11/2}$, $h'_{9/2}$ and $h'_{11/2}$ bands in ^{185}Au compared to the 0^+, 2^+ and $0^{+'}$, $2^{+'}$ levels in ^{184}Pt and ^{186}Hg. The corresponding core-particle couplings are noted with dashed lines. Only transitions with E0 components are shown.

Acknowledgments

Work supported in part by the U.S. Department of Energy Contracts and Grants DE-FG05-84ER40159 at Louisiana State University, DE-AS05-80ER10599 and DE-AS05-76ER03346 at the Georgia Institute of Technology, DE-AS05-76ER04936 at The University of Tennessee, and DE-AC05-76OR00033 for UNISOR at the Oak Ridge Associated Universities.

References

[BER77] R. Beraud, et al., Nucl. Phys. A284, 221 [1977].

[BON72] J. Bonn, et al., Phys. Lett. 38B, 308 (1972).

[BON76] J. Bonn, et al., Z. Phys. A276, 203 (1976).

[BOU76] C. Bourgeois, et al., J. de Phys. 37, 49 (1976)

[BOU82] C. Bourgeois et al., Nucl. Phys. A386, 308 (1982).

[COL76] J. D. Cole, et al., Phys. Rev. Lett. 37, 1185 (1976).

[COL77] J. D. Cole, et al., Phys. Rev. C16, 2010 (1977).

[COL84] J. D. Cole, et al., Phys Rev. C30, 1267 (1984).

[DAB79] P. Dabkiewicz, et al., Phys. Lett. 82B, 199 (1979).

[DUP84] P. van Duppen et al., Phys. Rev. Lett. 52, 1974 (1984).

[EKS76] C. Ekstrom et al., Phys. Lett. 60B, 146 (1976).

[HAG78] E. Hagberg, et al., Phys. Lett. 78B, 44 (1978).

[HAM75] J. H. Hamilton, et al., Phys. Rev. Lett. 35, 562 (1975).

[HAM85] J. H. Hamilton et al., Rep. Prog. Phys. 48, 631 (1985).

[HEY83] K. Heyde et al., Phys. Rep. 102, 291 (1983).

[KUH77] T. Kühl et. al., Phys. Rev. Lett. 39, 180 (1977).

[LAR85] A. J. Larabee et al., preprint, Oct. 1985. (Sub. Phys. Lett.).

[MA 84] W. C. Ma, et al., Phys. Lett. 139B, 276 (1984).

[PRO73] D. Proetel, et al., Phys. Rev. Lett. 31 896 (1973).

[PRO74] D. Proetel, et al., Phys. Lett. 48B, 102 (1974).

[RUD73] N. Rud et al., Phys. Rev. Lett. 31, 1421 (1973).

[WOO81] J. L. Wood, Proc. 4th Int. Conf. on Nuclei Far From
         Stability, Helsingør, June, 1981, CERN 81-09, p. 612.

[WOO82] J. L. Wood, in Lasers in Nuclear Physics, C. E. Bemis et
         al., editors, Harwood Academic Pub., New York, 1982, p. 481.

[ZGA80] E. F. Zganjar, in Future Directions in Studies of Nuclei
         Far From Stability, J. H. Hamilton et al. ed., North-
         Holland, Amsterdam, 1980, p. 49.

[ZGA81] E. F. Zganjar et al., Proc. 4th Int. Conf. on Nuclei Far
         From Stability, Helsingør, June, 1981, CERN 81-09, p. 631.

RECEIVED May 15, 1986

38

Shape Coexistence in Neutron-Deficient Pb Isotopes

J. Penninga[1], W. H. A. Hesselink[1], A. Stolk[1], H. Verheul[1], J. K. Ho[2], J. van Klinken[2],
H. J. Riezenbos[2], M. J. A. de Voigt[2], and A. Zemel[2]

[1]Natuurkundig Laboratorium, Vrije Universiteit Amsterdam, the Netherlands
[2]Kernfysisch Versneller Instituut, Groningen, the Netherlands

In search for rotational bands on low lying intruder states in ^{194,196}Pb the
^{188}Hg$(\alpha,xn)^{194,196}$Pb and the ^{188}Os$(^{12}C,4n)^{196}$Pb reactions have been studied.
On the basis of $e^{-}-\gamma$ and $\gamma-\gamma$ coincidence data levels with spin up to $J^{\pi}=6^{+}$
and $J^{\pi}=14^{-}$ have tentatively been assigned to bands on the $J^{\pi}=0^{+}$ and $J^{\pi}=11^{-}$
excited states in ^{196}Pb. Furthermore, the g-factor of the $J^{\pi}=11^{-}$ isomer in
^{196}Pb has been measured. The experimental value g = 0.96(8) indicates that this
state has a proton 2p-2h configuration.

In many nuclei near single closed shells low lying intruder states have
been observed. These intruder states are known to be due to particle-hole
excitations across the closed shell. The additional particle-hole degree of
freedom causes dramatic changes in the properties of the nuclear states. It
is for example a characteristic feature of the intruder states that they act
as band heads for rotational bands.
    Recently Heyde et al. have reviewed the experimental evidence for shape
coexistence in odd A nuclei and the theoretical approaches which are made to
describe the experimental data [HEY83]. Also in several even mass nuclei
there is evidence for shape coexistence. A nice example are the rotational
bands on $J^{\pi}=0^{+}$ intruder states in the even mass Sn isotopes [BR079].
    Van Duppen et al. have recently reported on the observation of low
lying $J^{\pi}=0^{+}$ states in neutron deficient Pb isotopes [DUP84]. The latter
states are similar to those in the Sn nuclei and their observation is thus
not unexpected. The systematic trend of the observed $J^{\pi}=0^{+}$ states through the
various Pb isotopes and Sn isotopes is the same. However, the excitation
energy of the intruder states in Pb is much lower than in the Sn isotopes.
This is due to the larger number of valence neutrons causing a stronger residual interaction between the proton particles and holes and the neutrons. In
this respect it is of interest to investigate the protoperties of the rotational bands which will likely occur on top of $J^{\pi}=0^{+}$ intruder states in Pb.
In $^{112-118}$Sn rotational bands were excited up to $J^{\pi}=12^{+}$ states.
    The experimental study on rotational bands on $J^{\pi}=0^{+}$ intruder states in
the Pb isotopes is however more difficult than in Sn. In the first place the
rotational bands in the neutron midshell nuclei of Sn could be studied by
means of the $(\alpha,2n)$ reaction, whereas in Pb these nuclei are far away from
the stability line. In the second place neutron $(i_{13/2})^{2}$ states with angular
momentum $J^{\pi}=10^{+},12^{+}$ occur at relatively low excitation energy. This implies
that the predicted excitation energy for band members with spin J>10 is more
than 1 MeV above the Y-rast line. These states will therefore not likely be
excited in fusion-evaporation reactions. One might however have some hope to
populate the lower part of the bands with reactions induced by light ions.
    In search for rotational bands on $J^{\pi}=0^{+}$ states in ^{194,196}Pb we have
measured coincidences between γ-rays following the ^{198}Hg(α,xn) reactions and
conversion electrons corresponding to $E0_{0^{+}\to 0^{+}_1}$ transitions in these nuclei.
The α-particle energy was 75 MeV and $2^{+}\to 0^{+}_1$ 92 MeV, respectively. The metallic ^{198}Hg (98 % enriched) target was produced by evaporating mercury
oxalate onto a C-backing. The averaged target thickness was about 0.7 mg/cm^2.
The γ-ray spectra (cf. Fig. 1) were taken with two (^{196}Pb) and four (^{194}Pb)

0097-6156/86/0324-0252$06.00/0
© 1986 American Chemical Society

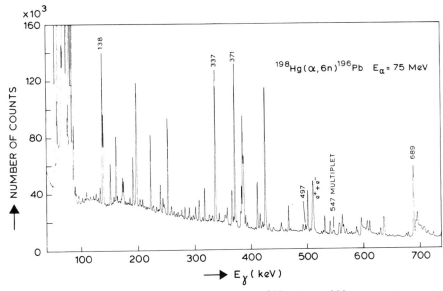

Fig. 1 Spectrum of γ-rays following the ^{198}Hg(α,6n)^{196}Pb reaction.

Ge detectors each provided with a cylindrical BGO + NaI Compton suppression shield. The conversion electrons were measured with two spectrometers each consisting of a mini orange filter and a Si(Li) detector. The transmission of the filter was optimized for detection of electrons with energies in the range 0.9 - 1.3 MeV. For ^{196}Pb the intensity of the E0 $0_2^+ \rightarrow 0_1^+$ transition in the electron spectrum was about the same as for the L-electrons of the $2_1^+ \rightarrow 0_1^+$ transition. This indicates that the intensity ratio $I_{0_2^+ \rightarrow 0_1^+}/I_{2_1^+ \rightarrow 0_1^+} \simeq 10^{-3}$. Thus, the expected electron-gamma coincidence intensity is about a factor 10^2 smaller than the γ-γ coincidence intensity for the strongest γ-ray transitions, taking into account a $\varepsilon = 0.03$ for the electron spectrometers. For ^{194}Pb this factor is even $3 \cdot 10^2$.

Fig. 2 shows the results for ^{196}Pb (the data analysis for ^{194}Pb is in progress). In addition to the 288, 307 and 754 keV transitions which have also been observed by Van Duppen et al. in the β-decay of Bi isotopes we have found evidence for a 413 keV and a 563 keV transition. These γ-rays are clearly visible in the coincidence spectrum between delayed γ-rays and delayed conversion electrons. This does not exclude that they are also present in the prompt-prompt spectrum. If the intensity in the latter spectrum would approximately be the same as in the delayed spectrum these γ-rays would have been escaped from our observation due to the higher background in the prompt spectrum. We have tentatively assigned the 413 keV and the 565 keV transitions to the decay of a $J^\pi=4^+$ and $J^\pi=6^+$ levels of the rotational band built on the $J^\pi=0^+$ intruder state on the basis of the coincidence intensity and the transition energies (cf. Fig. 3). The regular increasing transition energies between the members of the proposed band agree with those of the ground state band in ^{192}Pt. A similar observation has been made for the Sn isotopes.

It will be very difficult to obtain further evidence for the rotational band on the $J^\pi=0^+$ state in ^{196}Pb. A γ-γ coincidence measurement to establish the coincidence relations between the proposed intraband transitions seems

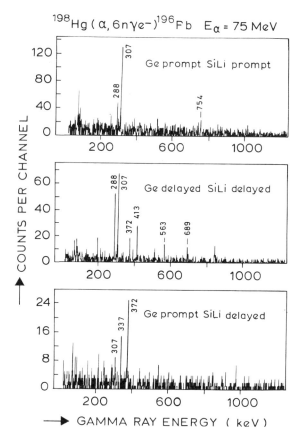

Fig. 2   Spectra of γ-rays coincident with the $E0_{0_2^+ \to 0_1^+}$ transition in ^{196}Pb.

hardly to be feasible due to the extremely low intensity of the γ-rays involved. The use of another reaction to excite the levels of the band is not
promising as well for reasons which have been pointed out already before.
    In a previous study on neutron excitations in ^{194,196}Pb we have identi-
fied two new isomers, which have most likely a proton 2p-2h character too
[NES83], [RUY85]. To obtain more evidence about the nature of the $J^{\pi}=11^-$
isomer in ^{196}Pb we have measured the g-factor of this state using the spin
precession method. The ^{198}Hg target was the same as the one used in the electron-gamma coincidence measurements. The experiment was performed with 75 MeV
α-particles using the ^{198}Hg(α,6n)^{196}Pb reaction. The applied magnetic field
was 1.5 T. To reduce the background in the spectra, which is crucial in this
experiment since the intensity of the weak 497 keV γ-ray is only 5 % of the
intensity of the $2^+ \to 0^+$ transition, we have not stopped the beam at the target
position but instead we have transported the bended beam to the Faraday cup.
This implies that we were not able to do a measurement with two field direc-

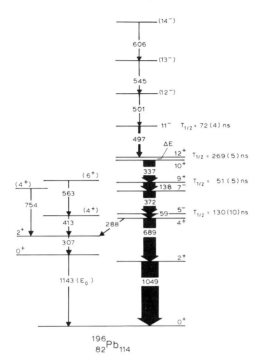

Fig. 3 Partial level scheme of ^{196}Pb.

$^{196}_{82}$Pb$_{114}$

tions otherwise we should have rotated the beamline 16° during the experiment.
    The γ-ray spectra were taken with two Ge detectors both positioned at
135° with respect to the beam direction at the target position. One of the Ge
detectors was provided with a BGO Compton suppression shield.
    In an independent run we have measured the angular distribution for the
497 keV-γ-ray using the same reaction and target. Fig. 4 shows the R(t) function deduced from the experimental data. Although the experimental errors are
large due to the relatively low number of events in the peak compared to the
number of background events, a least squares fit through the data only yields
one solution i.e. g = 0.96(8), A_2 = ‥0.14(6) and T_{1_2} = 74(10) ns. The values
obtained for A_2 and T_{1_2} are in agreement with the results of the angular
distribution measurements (A_2 = -0.09(6) and lifetime measurements (T_{1_2} = 72(5)
ns). The measured g-factor g = 0.96(8) indicates that the J^π = 11$^-$ isomer
has indeed a proton 2p-2h nature. The theoretical value for a $(\pi h_{9/2} \pi i_{13/2})_{11^-}$
configuration taking g_ℓ = 1 and g_s = 0.7g free is g = 1.02
    In explaining the low excitation energy of this proton 2p-2h intruder
state one should bear in mind that the excitation energy of particular proton
2p-2h configurations in this mass region is lowered remarkebly in an oblate
deformed potential [HEY83]. From the adjacent odd T_ℓ odd Bi isotopes one
can deduce that the configuration $[9/2^-[514]^2 1/2[440]^{-2}]_{0^+}$ has the lowest proton
excitation energy. In case of a broken proton pair one obtains a J^π = 11$^-$ state
with low excitation energy if the proton particles are in a 9/2$^-$[514] and
13/2$^+$[606] Nilsson orbit. Note that the g-factor in this case is equal for a
sferical and a deformed state since the Nilsson configuration only contains

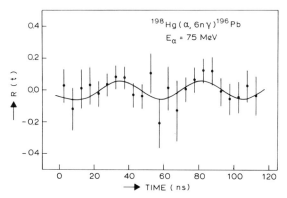

Fig. 4   R(t) function of the 497 keV transition in ^{196}Pb.

the $h_{9/2}$ and $i_{13/2}$ shells. Using the empirical relation between the energies of 2p-2h states in single closed shell nuclei and 1p-1h configuration in adjacent odd mass nuclei derived in ref. [DUP84] one finds the following approximation for the energy of the $J^{\pi}=11^-$ intruder states in the even mass Pb isotopes.

$$E_{ex}(J^{\pi} = 11^-,\ Z = 82, N) = E_{ex}(J^{\pi} = 13/2^+;\ Z = 81, N) +$$

$$E_{ex}(J^{\pi} = \tfrac{1}{2}^+,\ Z = 83, N) + [Sp(Z = 84, N) - Sp(Z = 83, N)]$$

This yields $E_{ex} \simeq 3.6$ MeV for the $J^{\pi} = 11^-$ state in ^{196}Pb which is in reasonable agreement with the experimental energy $E \simeq 3.2$ MeV in view of this crude approximation neglecting the residual interaction between the proton particles.

Assuming equal deformation for the $J^{\pi}=0^+$ intruder states and the newly found isomers, one may expect that these isomers also act as bandheads for a rotation band. We have studied these bands by measuring coincidences between γ-rays following the ^{198}Hg$(\alpha,8n)^{194}$Pb and the ^{188}Os$(^{12}$C$,4n)^{196}$Pb reactions. In the latter experiment we have identified a sequence of γ-rays with energies of 501, 545 and 606 keV which all have a prompt-delayed coincidence relation with the 497 keV transition. Moreover these γ-rays are coincident with each other. We have tentatively assigned these transitions to a band on top of the $J^{\pi}=11^-$ isomer. The moment of inertia derived from the intraband transitions assuming that the angular momentum for the subsequent band members differs by $\Delta J=1$, is in good agreement with the moment of inertia deduced for the band on the $J^{\pi}=0^+$ intruder state and that of the gsb in ^{192}Pt. In a preliminary analysis of the $\gamma-\gamma$ coincidence data obtained for ^{194}Pb we have found indications for a similar regular increasing sequence of γ-ray transitions in coincidence with the 351 keV transition, which depopulates the $J^{\pi}=10^-$ isomer in this nucleus. This supports the assignments made for ^{196}Pb. However, more experiments are needed to establish the spins proposed for this band.

Acknowledgments

The authors are indebted to K. Heyde and M. Huyse for stimulating discussions.

References

[BRO79] J. Bron, W.H.A. Hesselink, J.J.A. Zalmstra, M.J. Uitzinger, H.
         H. Verheul, K. Heyde, M. Waroquier, P. van Isacker and H. Vinx,
         Nucl. Phys. A318 335 (1979).
[DUP84] P. van Duppen, E. Coenen, K. Deneffe, M. Huyse, K. Heyde and
         P. van Isacker, Phys. Rev. Lett. 52 1974 (1984).
[HEY83] K. Heyde, P. van Isacker, M. Waroquier, J.L. Wood and A.A. Meyer,
         Phys. Rep. 102
[NES83] P. van Nes, Z. Sujkowski, W.H.A. Hesselink, J. van Ruyven, H. Verheul
         and M.J.A. de Voigt, Phys. Rev. C27 1342 (1983).
[RUY85] J.J. van Ruyven, J. Penninga, W.H.A. Hesselink, P. van Nes,
         K. Allaart, E.J. Hengeveld, H. Verheul, M.J.A. de Voigt,
         Z. Sujkowski and J. Blomqvist, to be published in Nuclear Physics.

RECEIVED May 28, 1986

39

Intruder States in the $Z = 82$ Region by the β^+/EC and α Decay of Neutron-Deficient Bi, Po, and At Nuclei

M. Huyse, E. Coenen, K. Deneffe, P. Van Duppen, and J. L. Wood[1]

Leuven Isotope Separator On Line (LISOL), Instituut voor Kern- en Stralingsfysika, B-3030 Leuven, Belgium

The α-decay study of mass-separated nuclei in the region around $Z=82$ provides a strong spectroscopic tool to investigate intruder states. Here this method is applied to odd mass At nuclei and to odd-odd Bi nuclei.

1. Introduction

The experimental evidence for shell-model intruder states in the $Z=82$ region has been significantly increased by recent studies at LISOL [DUP84, DUP85, COE85]. We identified and studied low-lying excited 0^+ states in the singly-closed-shell nuclei $^{190-200}$Pb. This was done by using γ-ray and conversion-electron spectroscopy on mass-separated sources of $^{192-200}$Bi [DUP84] and by using α-ray spectroscopy on mass-separated ^{194,196}Po nuclei [DUP85a]. These 0^+ states are interpreted as proton-pair excitations across the $Z=82$ closed shell: this interpretation is supported by the observation in ^{192,194,196}Pb of the beginning of a rotational-like band built on top of it [DUP84, DUP85b]. The systematics of the excitation energies of the 0^+ intruder states in $^{190-200}$Pb are given in fig. 1.

We also studied the β^+/EC and α decay of the odd-mass Bi nuclei (COE85): allowed α decay is observed between the $^{189-195}$Bi $\pi h_{9/2}$ ground states and the $^{185-191}$Tl $\pi h_{9/2}$ intruder states and between the $^{191-197}$Bi $\pi s_{1/2}$ intruder states and the $^{187-193}$Tl $\pi s_{1/2}$ ground states. The observation of forbidden α branches provides excitation energies for the intruder states in ^{189,191}Tl and ^{189,191,193,195}Bi and confirms the intruder-state excitation energies in ^{185,187}Tl. A summary of our results is given in fig. 1. The strong resemblance of the excitation-energy systematics of the intruder

[1]Permanent address: School of Physics, Georgia Institute of Technology, Atlanta, GA 30332

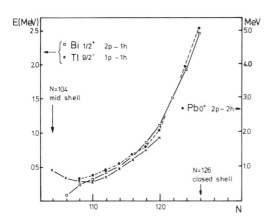

Figure 1. The systematics
of excitation energies for
the 0^+ intruder states in
the even Pb nuclei and of
the intruder states in the
odd-mass Tl and Bi nuclei.
For more details, see
[COE85, DUP85a].

states in odd-mass Bi and Tl and even-even Pb nuclei is discussed in [HEY85].
It is shown in [COE85] that absolute α-hindrance factors provide a strong
spectroscopic fingerprint for the identification of intruder states in the
Z=82 region. In this contribution we report on the identification of intruder states in the odd-mass At nuclei and odd-odd Tl nuclei using the same
method.

2. The odd-mass At nuclei

Odd-mass At nuclei ranging from mass 203 down to mass 197 were produced
in the bombardment of a 7μm thick natural Re target with a beam of ^{20}Ne (240
MeV, 200 enA). Production rates of the different At nuclei were optimized by
varying the ^{20}Ne beam energy using degrading foils. Multiscaled α-singles
spectra were taken on each mass with a Si surface-barrier detector (450 mm^2-
500μm, resolution 19 keV for the 5.486 MeV α line in ^{241}Am).
Fig. 2 shows the sum of a 7 x 1s cycle on mass 197. Most of the α lines can
be identified by using the compilation of Schmorak [SCH80]. But two lines
remain unidentified: the line at 6.707 MeV and the line at 6.457 MeV. By
checking the α spectra taken at neighbouring masses it is clear that they
belong to the decay chain with mass 197.
     It was possible to deduce the half-life of the 6.707 MeV line :
3.7 ± 2.5s. As its energy is more than 300 keV higher than the well-known α
lines of the ^{197}Po decay it is most likely that this α line belongs to the

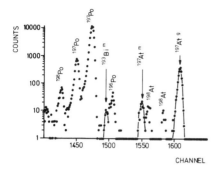

Figure 2. The α spectrum obtained
for A = 197.

decay of an isomer of ¹⁹⁷At (the half-life of the ¹⁹⁷ᵍAt decay is
0.35 ± 0.4s). The natural interpretation for this isomer is a 1/2⁺ intruder
state, analogous to the Bi isotopes. Strong support for this assignment is
the fact that the energy of the second unknown line matches the energy of the
α decay of the ¹⁹³ᵐBi intruder state. By comparing the intensity of the
¹⁹³ᵐBi α line with the ¹⁹⁷ᵐAt line and correcting for the different half-
lives with a mother-daughter relation it is possible to deduce the α-
branching ratio of ¹⁹³ᵐBi. The obtained value, 90 ± 20%, is consistent with
that obtained previously (50<α$_{br}$<100) [COE85]. Figure 3 gives the
resulting decay scheme of ¹⁹⁷ᵍ,ᵐAt. Assuming for both decays a 100% α-

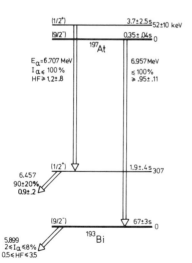

Figure 3. Decay scheme of ¹⁹⁷ᵍ·ᵐAt.

branching ratio, hindrance factors of 1.2 ± 0.8 and 0.95 ± 0.11 are obtained
for [197m]At and [197g]At, respectively. The fact that the hindrance factors
are near to unity indicates strongly that the initial and final state linked
by the α decay have the same spin and parity. The production of the 1/2[+]
isomeric state of [197]At is, compared to the 9/2[-] ground state, less than 1%
in the heavy ion reaction used: this is typical. The obtained value is in
agreement with the production ratio of the 1/2[+] isomer and 9/2[-] groundstate
in the Bi nuclei [COE85]. This low feeding is probably the reason why we do
not observe this intruder state in the heavier At nuclei.

3. The α decay of [190,192,194] Bi

A systematic decay study of the [190-200]Bi isotopes was carried out by
means of the heavy ion reactions [nat]Re(16mg/cm^2)[[16]O(<180 MeV), xn] and
[181]Ta(8mg/cm^2)[[20]Ne(<230 MeV),xn]. Multiscaled α-decay spectra for the
masses 190, 192 and 194 were taken with the same α detector as used in the At
experiment, together with α-γ coincidences. A Ge detector (resolution 2 keV,
relative efficiency 22% at 1332.5 keV) was used to detect γ rays. In coincidence mode the lower limit on the γ detector was set at ~ 40 keV.

Our β^+/EC decay studies [DUP85c] indicate for most of the neutron-deficient odd-odd Bi nuclei the existence of three β-decaying states: (10[-],11[-])
(6[+],7[+]) and (2,3): the low spin state is only weakly fed. So far only one α
branch was reported in the literature for the decay of the [190-194]Bi isotopes
[SCH80]: it was interpreted as the decay of the high spin (10[-]) state of Bi
to the groundstate (7[+]) of Tl. We observe four α branches. The two most
intense ones are nearly degenerate (and were previously observed as one line)
and they have a hindrance factor near unity. This conflicts with the assignment as a 10[-]-7[+] transition. Such a transition is of a $\pi h_{9/2}$-$\pi s_{1/2}$ character
and from the odd Bi decay we know that its hindrance factor lies around 600
[COE85]. As an example of the used arguments on which the α-decay schemes of
[190,192,194]Bi (see fig.4) are constructed, we discuss [190]Bi: the intense

6.453 MeV α line does not show any coincidences but its intensity explains
the observation of the 2.9s E3 transition of 374 keV linking a(10[-]) intruder
state in [186]Tl with the (7[+]) groundstate [KRE81]. The weak 6.819 MeV α line
can be placed as the cross over of the $10^- \overset{\alpha}{\to} 10^- \overset{374}{\to} 7^+$ cascade. The intense

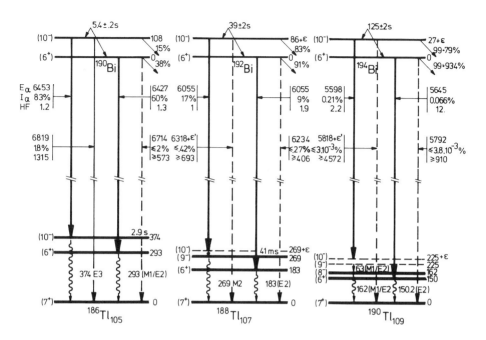

Figure 4.   The α-decay schemes of ^{190,192,194}Bi; the assigned spins are only
tentative but are linked to each other by the multipolarity of the γ transitions and by α branches with hindrance factors (HF) around unity. The
dashed γ transitions are observed but uncertainty exists because of doublet
structure and the missed γ transitions (indicated by ε).

6.427 MeV α line is coincident with a 293 keV γ ray. The M1/E2 multipolarity
of the 293 keV γ line is established by comparing the intensities of the x
rays and 293 keV γ line in the coincident γ-ray spectrum. The weak 6.714 MeV
α line can be placed as the cross over of the $6^+ \overset{\alpha}{\to} 6^+ \overset{293}{\to} 7^+$ cascade. The
low hindrance factors of the two intense α lines indicate that the α decay
links across the Z=82 shell closure initial and final states with the same
character: the most probable candidates are the $(\pi h_{9/2} \otimes \nu i_{13/2})10^-$ state
for the 6.453 MeV line and the $(\pi h_{9/2} \otimes \nu p_{3/2})6^+$ state for the 6.427 MeV α
line. The two weak α lines are forbidden α branches to the
$(\pi s_{1/2} \otimes \nu i_{13/2})7^+$ groundstate of ^{186}Tl. For the ^{192,194}Bi decay the
situation is very similar except that the lowest member of the intruder based

(πh9/2 $\otimes$ νi$_{13/2}$) multiplet in ^{188,190}Tl is not anymore 10$^-$ but 9$^-$ and 8$^-$, respectively [KRE81]. The 63 keV line in ^{190}Tl, coincident with the 5.598 MeV α line, is then interpreted as a transition inside the multiplet.

4. Conclusion

By studying the α decay of mass-separated nuclei in the region around Z=82 we have extended the knowledge of shell-model intruder states. The allowed α-decay branches of the odd-odd Bi nuclei connect across the Z=82 shell closure initial and final states of the same single-particle character.
The observation of allowed α decay of an isomer in ^{197}At is the first strong spectroscopic evidence of intruder states in the At nuclei [HEY83]. The low excitation energy (only 52 keV) of the 1/2$^+$ intruder in ^{197}At suggests that for ^{195}At the intruder state may become the groundstate (see fig. 1).

Acknowledgments

We would like to thank B. Brijs and J. Gentens for their technical assistance and the Interuniversitair Instituut voor Kernwetenschappen and the Nationaal Fonds voor Wetenschappelijk Onderzoek for financial support. We are indebted to K. Heyde, W.B. Walters and M.R. Schmorak for interesting discussions. One of us (JLW) was supported in part by US Department of Energy contract DE-AS05-80ER-10599.

References

[COE85] E. Coenen, K. Deneffe, M. Huyse and J.L. Wood, Phys. Rev. Lett. 54, 1783 (1985)

[DUP84] P. Van Duppen, E. Coenen, K. Deneffe, M. Huyse, K. Heyde and P. Van P. Van Isacker, Phys. Rev. Lett. 52, 1974 (1984).

[DUP85a] P. Van Duppen, E. Coenen, K. Deneffe, M. Huyse and J.L. Wood, Phys. Lett. 154B, 354 (1985)

[DUP85b] P. Van Duppen, E. Coenen, K. Deneffe, M. Huyse and J.L. Wood XX Winter School on Physics, Selected Topics on Nuclear Structure, Zakopane, Poland (1985)

[DUP85c] P. Van Duppen, Ph. D. Thesis, K.U. Leuven (1985), unpublished

[HEY83] K. Heyde, P. Van Isacker, M. Waroquier, J.L. Wood and R.A. Meyer, Phys. Repts. 102, 291 (1983)

[HEY85] K. Heyde, P. Van Isacker, R.F. Casten and J.L. Wood, proceedings this conference

[KRE81] A.J. Kreiner, C. Baktash and G. Garcia Bermudez, Phys. Rev. Lett. 47, 1709 (1981)

[SCH80] M.R. Schmorak, Nucl. Data Sheets 31, 283 (1980)

RECEIVED May 16, 1986

Section II: Experimental Exploration of Current Issues
Part 2: Octupole Modes in Nuclei
Chapters 40-43

40

Microscopic, Semiclassical, and Cluster Treatments of Low-Lying Reflection Asymmetric States in the Light Actinides

R. R. Chasman

Chemistry Division, Argonne National Laboratory, Argonne, IL 60439

The 1^- states in the even-even nuclides $220 < A < 230$ are the lowest lying non-rotational states found in even-even nuclides. These states, and the equivalent states in odd mass nuclides, are discussed in terms of octupole correlations, from both a two-body interaction and an octupole deformed one-body potential point of view. Some discussion of a cluster model treatment of these states is given.

The low-lying 1^- states in the even-even Ra and Th isotopes have been known [STE54] for many years and clearly signal the presence of strong octupole correlation effects in this mass region. There are a few nuclides in this region in which the 1^- state is below 300 keV, and none in which the excitation energy is less than 200 keV. In the case of strong octupole deformation, we expect to see a ground state rotational band sequence of 0^+, 1^-, and 2^+. There are no nuclides known in which the 1^- state is below the 2^+ state. There are several cases in which a 1^- state has been found slightly above or slightly below the 4^+ level. These data argue against permanent octupole deformation in even-even nuclides. On the other hand, there is no evidence [KUR81] for a two phonon 0^+ octupole state at twice the energy of the 1^- state in the cases when the 1^- energy is less than 300 keV. This means that the octupole correlation effects are much stronger than vibrational; i. e. we are in a transitional region for these even-even nuclides. Early calculations [MOL72] using the shell correction method of Strutinsky [STR67] did not find any nuclides in this mass region with a reflection asymmetric ground state shape.

My original interest [CHA79] in octupole correlation effects came from trying to understand the low-lying 0^+ excited state of ^{234}U. In ^{234}U, one calculates an excitation energy of the first 0^+ excited state of ~1350 keV with a conventional pairing force model. Experimentally, this state is found at 810 keV. Using a calculational approach that places pairing interactions and octupole interactions on an equal footing, the calculated excitation energy of the 0^+ excited state in ^{234}U is lowered to ~900 keV, in reasonably good agreement with the experimental value of 810 kev. Piepenbring [PIE83] has shown that the inclusion of octupole correlation effects generally gives a fairly good description of the 0^+ excited states in the mass region $224 < A < 228$, with the exception of ^{228}Ra.

In 1980, we extended our approach to include the calculation of states in odd mass nuclides [CHA80]. We included the quadrupole-quadrupole particle-hole interaction as well as the pairing force and the octupole-octupole interaction. In an odd mass system, the signature of octupole deformation is a parity doublet. This doublet consists of a pair of states having the same spins but opposite parities, almost degenerate in energy, with a large E3 matrix element connecting the two levels. The strong E3 transitions associated with octupole deformation do not compete with collective E2 transitions or with dipole transitions and would not generally be seen. In some cases, there are collective E1 transitions [AHM84],[AHM85] associated

0097-6156/86/0324-0266$06.00/0
© 1986 American Chemical Society

with the strong octupole correlations. In our study, we found several cases where the calculation predicts parity doublets. The most notable case is a 5/2+ ground state doublet in ^{229}Pa, with a splitting predicted to be less than 1 keV. The experimental study of the structure of ^{229}Pa by Ahmad et al. showed [AHM82] that the ground state of ^{229}Pa is indeed a 5/2+ parity doublet, with a splitting considerably smaller than 1 keV (~200 eV). This work demonstrates the existence of ground state octupole deformation in nuclei.

In 1981, Moller and Nix [MOL81] noted that the ground state binding energies of the nuclides near A=224 are increased by almost 1.5 MeV, when the octupole degree of freedom is introduced into the parameterization of nuclear shapes. Their results were obtained using the shell correction method [STR67]. The single particle energy level spectrum used in their calculations was generated from a folded Yukawa potential [BOL72]. This finding of Moller and Nix has been extended in the work of Leander et al. [LEA82], who have made an extensive survey of the light actinide region. Using the folded Yukawa potential, with the Strutinsky method, they have found many nuclides in this region with a reflection asymmetric ground state shape. These calculations studied octupole deformation in lighter nuclides than had been considered earlier. This approach has been extended to odd mass nuclides [LEA84],[RAG83].

In recent calculations that take the 2^6-pole degree of freedom into account, we have found [CHA85] that the binding energy increases associated with 2^6-pole deformation is on the order of 1 MeV in the light actinides. This increase in binding energy is greatest for nuclides with neutron numbers of 134 and 136. The inclusion of 2^6-pole effects reduces the energy gains due to reflection asymmetric shape deformation in the light actinides to ~0.6 MeV. With the inclusion of these 2^6-pole effects, there is a convergence in the magnitude of octupole correlation effects predicted with the two-body methods [CHA80] and the Strutinsky method [LEA82].

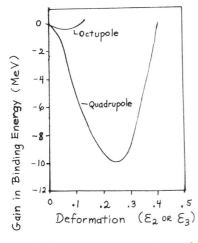

Fig. 1) A comparison of binding energy increases arising from Octupole and Quadrupole Deformation.

It is instructive to compare the largest binding energy arising from octupole correlation effects with the binding energy increase due to quadrupole interactions in the mid-actinide nuclides. In Fig. 1, we sketch such a comparison. The increase shown for octupole deformation is shown relative to the binding energy associated with reflection symmetric deformation; the increase shown for quadrupole deformation is relative to a spherical shape. These calculations are carried out using the Strutinsky procedure. Because this minimum associated with reflection asymmetric shapes is so shallow, it is worthwhile to deal with octupole correlation effects using a microscopic, two-body interaction treatment of the octupole-octupole residual interaction [CHA80]. The pairing force matrix elements, $G_{i,j}$, come from a density dependent delta interaction. This set of matrix elements [CHA77] explains many features of the actinides at low and

high spin. The quadrupole strength was adjusted to give the observed onset of
quadrupole deformation in the light actinides. The octupole strength is
adjusted to obtain the known energies of the low-lying 1^- states in the Th
isotopes.
    To solve the Hamiltonian, we exploit [CHA79],[CHA80] the fact that the
interaction is cylindrically symmetric and separable. We denote the deformed
orbitals with a given value of T_z and Ω as an Ω group. Orbitals with both
positive and negative parity are included in the Ω group; and the number of
configurations rises rapidly with the number of doubly degenerate orbitals in
the Ω group. Parity and particle number are not conserved within any one of
the Ω groups; however, we project states of good proton number, neutron
number, and parity to construct the wave function. The wave function
amplitudes are obtained by minimizing the energy of the fully projected wave
function. We can handle up to five doubly degenerate levels in a (Ω,T_z)
group. In the even particle number group, this amounts to 252 configurations
with 142 independent amplitudes. There are 210 independent amplitudes in an
odd particle number Ω group. By varying the octupole, quadrupole or pairing
strengths, we generate many different wave functions with different collective
properties. Our final solution is a linear combination of such wave
functions, taking their non-orthogonality fully into account. The structure
of these solutions is sufficiently rich to describe states that are spherical,
vibrational or deformed in these three degrees of freedom.
    From a microscopic point of view, octupole deformation in even-even
nuclides is signalled by 0^+ and 0^- bands that have the same properties. This
implies a ground state rotational band with alternating even and odd parity
states, i.e. 0^+, 1^-, 2^+, ... Such ground state bands have not been found in
any nuclides to date. In Fig. 2, we show the results of a microscopic calculation of octupole correlation energies in the 0^+ and the 1^- states of ^{226}Ra
as a function of the octupole interaction strength. At the bottom of this
figure, the calculated excitation energy of the 1^- state is shown. An arrow
is used to denote the experimentally known excitation energy. We note the
large difference of ~4 MeV in the octupole correlation energy for the two
states at this octupole interaction strength. For V>0.060, we see the onset
of octupole deformation. We do not show the single particle and pairing
contributions to the energy in Fig. 2; which accounts for most of the
remaining difference in the energy of the two states.
    In contrast to the situation in even-even nuclides, we see octupole
deformation in odd mass nuclides. The signal of octupole deformation in odd
mass nuclides is the parity doublet discussed above. The calculations [CHA80]
that we have made for odd mass actinides show the onset of octupole
deformation for several values of Ω in both odd proton and odd neutron
nuclides. The results of the calculations for protons are shown in Fig. 3.
There are predictions of parity doublets in several instances. The most
notable is the 5/2± ground state doublet in ^{229}Pa, with a splitting predicted
to be less than 1 keV; and shown to be ~0.2 kev by Ahmad et al. [AHM82]. The
calculation also gives a 3/2± doublet, with a 23 keV splitting, as the ground
state of ^{227}Ac, in good agreement with the known levels. It is noteworthy
that there is a large splitting between the $1/2^+$ and $1/2^-$ bands in ^{227}Ac
[SHE83] as suggested by these calculations. Recently, using our many-body
wave functions, we have calculated the decoupling constants of these two 1/2
bands and obtained values of -1.8 and 5.0 in reasonable agreement with the
experimental values of -2.02 and 4.55 obtained by Sheline [SHE83]. This
provides a nice example of the coexistence of octupole deformed and octupole
undeformed states in the same nucleus. As yet there has been no study of the
nuclide ^{227}Pa. In the microscopic calculations, we find that the presence of

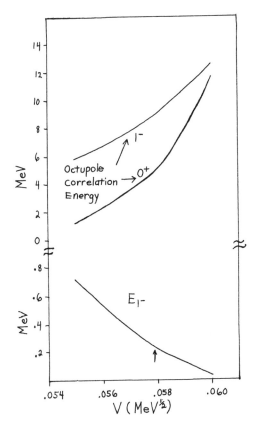

Fig. 2) Octupole correlation energy for 0^+ and 1^- states in ^{226}Ra as a function of octupole interaction strength.

an odd paticle in an appropriate orbital can polarize the nuclear core strongly. The octupole correlation effects are stronger in these states than they are in the ground states of even-even nuclides. In the shell correction calculations such effects are not so apparent.

A different approach to the description of the low-lying 1^- states has been proposed by Daley, Iachello and coworkers. In their approach, which is an extension of the Interacting Boson Model, two new bosons the σ and π bosons are introduced to explain low-lying 1^- states in the light actinides [IAC82],[DAL83],[DAL84]. The goals of this approach are to account for the fast E1 transitions and the alpha decay rates of the nuclides in this mass region. The σ and π bosons are used to describe the motion of an alpha particle cluster relative to a nuclear core. In the versions of this model that have been considered, the number of σ and π bosons is restricted to two. The σ and π bosons generate the group U(4). In a group U(N), the number of elements is N^2 and in O(N) it is $N(N-1)/2$. In boson models, the basic idea is to write a Hamiltonian in terms of the Casimir invariants (constants of the motion) associated with the groups of interest. A simple example of such a Hamiltonian is given by utilizing the groups O(3) and O(2). O(3) is the rotation group with elements L_x, L_y and L_z and has the Casimir invariant L^2. O(2) contains the single element L_z with a constant of motion L_z. A general Hamiltonian with dynamical symmetries using the group chain O(3) $\supset$ O(2) is $H=aL^2+bL_z$ and one can immediately give an eigenvalue spectrum for this Hamiltonian. A most helpful introduction to these concepts is given in [IAC79]. In going from the group U(4) to the rotation group O(3), there are two possible routes, and accordingly two sets of Casimir invariants that can be used to construct Hamiltonians with dynamical symmetries. The two routes are

$$U(4) \supset U(3) \supset O(3) \qquad I$$

and

$$U(4) \supset O(4) \supset O(3) \qquad II$$

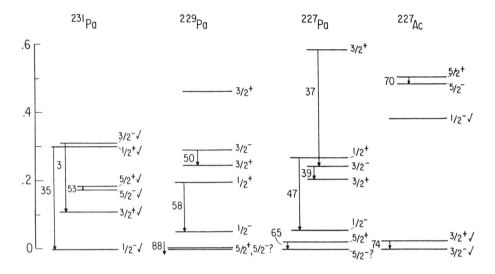

Fig. 3)   Calculated bandhead energies of odd proton nuclides. The numbers
          next to the arrows are proportional to the square of the $\langle r^3 Y(3,0) \rangle$
          matrix element.

It is the group O(4), containing the collective E1 operator of this boson
model ($\sigma^+\pi + \pi^+\sigma$), that is needed to give enhanced E1 transitions. In the
cluster model calculations that have been carried out, however, it is the
Casimir invariants of the chain involving U(3) that are used to construct a
Hamiltonian. This gives rise to large variations in the parameters of the
model in fitting the properties of the different nuclides of this mass region
[DAL83],[DAL84]. Although the present cluster model calculations do not
provide a good description of the light actinides, a cluster model that
utilizes the O(4) Casimir invariants and allows a larger number of σ and π
bosons might be more successful. At this time, the relation between a cluster
model description and an octupole correlation description of the light
actinides is not clear. In the octupole model, there are also quadrupole,
hexadecapole and 2^6 pole deformations that give rise to a shape that might be
thought of as having a cluster like bulge on one end. Such a cluster,
however, is one with the neck filled in; and one that is larger than an alpha
particle. Of course, there is no need for a cluster model to be restricted to
alpha clusters.
          In terms of the octupole-octupole interaction, the properties of the
low lying 1⁻ state in the even even nuclides and the properties of parity
doublets in odd mass nuclide are fairly well understood. Although we
understand the E1 transitions qualitatively, a quantitative treatment of the
E1 rates remains an open problem.
          I thank I. Ahmad, A. Friedman, B. Wilkins, and G. Leander for
stimulating discussions on many aspects of this work.

Acknowledgments

This work is supported by the U. S. Department of Energy, Chemistry Division, under contract W-31-109-ENG-38.

References

[AHM82] I. Ahmad et al., Phys. Rev. Lett. 49 1758 (1982).
[AHM84] I. Ahmad et al., Phys. Rev. Lett. 52 503 (1984).
[AHM85] I. Ahmad, Proceedings of this Symposium.
[BOL72] M. Bolsterli, E.O. Fiset, J.R. Nix and J.L. Norton, Phys. Rev. C5 1050 (1972).
[CHA77] R.R. Chasman, I. Ahmad, A.M. Friedman and J.R. Erskine, Rev. Mod. Phys. 49 833 (1977).
[CHA79] R.R. Chasman, Phys. Rev. Lett. 42 630 (1979).
[CHA80] R.R. Chasman, Phys. Rev. Lett. 96B 7 (1980).
[CHA85] R.R. Chasman, (submitted for publication).
[DAL83] H. Daley and F. Iachello, Phys. Lett. 131B 281 (1983).
[DAL84] H. Daley and M. Gai, Phys. Lett. 149B 13 (1984).
[IAC79] F. Iachello Lecture Notes in Physics, Vol. 119 140 (1979) (Proceedings of Gull Lake Workshop edited by G.F. Bertsch and D. Kurath, Springer-Verlag, Berlin, Heidelberg, New York).
[IAC82] F. Iachello and A. Jackson, Phys. Lett. 108B 151 (1982).
[KUR81] W. Kurcewicz et al., Nucl. Phys. A356 15 (1981); A383 1 (1982).
[LEA82] G. Leander et al., Nucl Phys. A388 452 (1982).
[LEA84] G.A. Leander and R.K. Sheline, Nucl. Phys. A413 375 (1984).
[MOL72] P. Moller, S.G. Nilsson and R.K. Sheline, Phys. Lett. 40B 329 (1972).
[MOL81] P. Moller and J.R. Nix, Nucl. Phys. A361 117 (1981).
[PIE83] R. Piepenbring, Phys. Rev. C27 2968 (1983).
[RAG83] I. Ragnarsson, Phys. Lett. 130B 353 (1983).
[SHE83] R.K. Sheline and G. Leander, Phys. Rev. Lett. 51 359 (1983).
[STE54] F. Stephens, F. Asaro and I. Perlman, Phys. Rev. 96 1568 (1954); 100 1543 (1955).
[STR67] V.M. Strutinsky, Nucl. Phys. A95 420 (1967).

RECEIVED July 14, 1986

41

Fast Electric Dipole Transitions in Ra-Ac Nuclei

Irshad Ahmad

Chemistry Division, Argonne National Laboratory, Argonne, IL 60439

Lifetimes of levels in ^{225}Ra, ^{225}Ac, and ^{227}Ac have been measured by delayed coincidence techniques and these have been used to determine the E1 gamma-ray transition probabilities. The reduced E1 transition probabilities in ^{225}Ra and ^{225}Ac are about two orders of magnitude larger than the values in mid-actinide nuclei. On the other hand, the E1 rate in ^{227}Ac is similar to those measured in heavier actinides. Previous studies suggest the presence of octupole deformation in all the three nuclei. The present investigation indicates that fast E1 transitions occur for nuclei with octupole deformation. However, the studies also show that there is no one-to-one correspondence between E1 rate and octupole deformation.

1. Introduction

Recent theoretical calculations [Cha80, Lea82, Lea84a] and measurements [Ahm82, She83, Ahm84] show that strong octupole correlation and/or octupole deformation effects play an important role in the description of nuclides in the mass 220-230 region. A signature of octupole deformation in an odd-mass deformed nucleus is the occurrence of a parity doublet which consists of two almost degenerate levels with the same spin but opposite parities. It has been observed [She83, Ahm84] that the presence of strong octupole correlations in the nuclear ground state modifies many single particle properties considerably. Nuclear properties which are affected are M1 transition rates, decoupling parameters, Coriolis matrix elements, and E1 transition rates.

In heavy elements the E1 transition probabilities are typically 1.0 x 10^{-6} Weisskopf units for $\Delta K=1$ transitions and 1.0 x 10^{-4} w.u. for $\Delta K=0$ transitions. We define fast E1 transitions as transitions with rates of >1.0 x 10^{-3} w.u. Recently we have measured [Ish85] level lifetimes in ^{225}Ra and ^{225}Ac which indicte the presence of fast E1 transitions in these nuclei. In the present article we present the measurements of these fast E1 transitions and discuss recent theoretical calculations.

2. Source preparation

Thin sources of the radioactive materials on thin backing were prepared for the alpha-electron and electron-electron delayed coincidence measurements. Sources of ^{229}Th and ^{231}Pa were prepared in the Argonne electromagnetic isotope separator by depositing the material on 40-μg/cm^2 carbon foils. The ^{225}Ra samples were obtained by depositing freshly purified ^{225}Ra on 100 μg/cm^2 polypropylene films. The 14.8-d ^{225}Ra was obtained by separating it from approximately one mg of ^{229}Th (7300 y). The Th sample had almost equal amount of alpha activity from ^{229}Th and the shorter lived ^{228}Th (1.91 y). The purified sample was allowed to decay for 2 months so that the shorter lived ^{224}Ra could decay out. After this decay period the sample was repurified to remove any Th left over from the frst separation. This material

0097-6156/86/0324-0272$06.00/0

was placed on a 100 μg/cm^2 polypropylene film and dried. Gamma-ray analysis
showed that the sample contained less than 1% ^{224}Ra activity.

3. Experimental results

The level lifetimes described here were measured by delayed coincidence
techniques. Since the electron energies involved in the measurement were
quite low (up to 10 keV) and high resolution was needed to isolate the
individual levels, the source and the detectors were placed in a vacuum
chamber. Two bare pilot B detectors (1 mm and 2 mm thick), each mounted on an
RC8575 photomultiplier tube, were used to detect the electrons. For the α-e$^-$
coincidence setup, one of the pilot B detectors was replaced by a high-
resolution Si detector. Fast timing signals, derived with constant fraction
discriminators, were used to start and stop a time-to-amplitude converter
(TAC). The TAC linear output and the two energy signals were connected to
ADC's which, in turn, were interfaced to a PDP 11/23 computer. The strobe
signal was provided by the triple (two energy signals and the TAC output)
coincidence output. The three parameter events were collected on a tape in
the event-by-event mode and were later sorted out with appropriate gates. The
TAC spectra were calibrated with a time calibrator unit and the half-lives
were computed with a least-squares analysis program.

3.1. Half-life of the 40.0-keV state in ^{225}Ac

A freshly prepared ^{225}Ra source on a 100 μg/cm^2 polypropylene backing was
used for the measurement of the level lifetime in ^{225}Ac. The decay scheme of
^{225}Ra is shown in Fig. 1. The excited state at 40.0 keV receives most of the

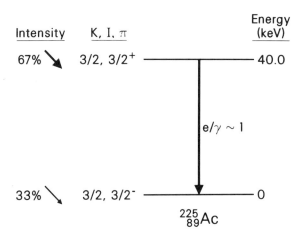

Fig. 1. Decay scheme of ^{225}Ra.

β^- population and it deexcites by a 40.0 keV El transition which has a total conversion coefficient of ~1. Since the pilot B detector is sensitive to α-particles (6 Mev α-particles produce signals of the same size as 500 keV electrons), a 12 mg/cm^2 Al foil was used to prevent the α-particles from reaching the start detector. This foil also absorbed electrons upto 60 keV energy. No absorber, except the source backing, was used on the stop side; the threshold on this detector was set at 10 keV. The three parameter events were collected over a three-day period producing several spectra. The TAC spectrum measured during the first half hour of the experiment is displayed in Fig. 2. A least-squares fit to the data gave a half-life of 0.72 ± 0.03 ns.

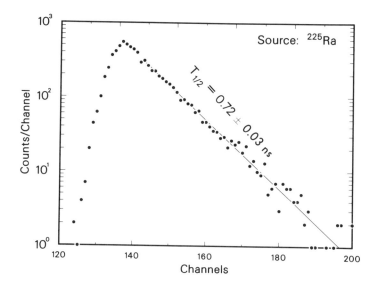

Fig. 2.   Time spectrum showing the decay of the 40.0-keV level in ^{225}Ac. Start signals included 80-300 keV electrons and stop signals were 20-40 keV electrons.

3.2.   Half-life of the 31.6-keV state in ^{225}Ra

The half-lives of levels in ^{225}Ra were measured by α-e$^-$ delayed coincidence technique. The α-particles were detected with a 6-mm diameter Au-Si surface barrier detector and the elctrons were detected with the 1 mm thick pilot B detector. α-particle singles [Bar70, He185] and coincidence [He185] studies show that only < 0.2% alpha intensity occurs at the 31.6-keV level. Therefore the start signals were obtained from the α_{236} peak (the 236 keV level partly decays via the 31.6-keV level). The time spectrum obtained in coincidence with α_{236} peak and 15-25 keV electrons contained three main components. The longest component in the spectrum was not present when the electron gate was set above 35 keV. This clearly indicates that the longest lifetime belongs to the 25.4 or 31.6 keV level. The half-life of the 25.4-keV level was obtained by measuring the TAC spectrum in coincidence with α_{25} and 15-25 keV electrons and was found to be 0.88 ± 0.04 ns. Therefore, the longest component in the spectrum belongs to the 31.6 keV level. A least-squares analysis gave a half-life of 2.1 ± 0.2 ns.

3.3 Half-life of the 27.4 keV state in ^{227}AC

We have performed coincidence measurements to establish that the 38 ns half-life previously measured [Led78] belongs to the 27.4 keV level. The relevant portion of the ^{231}Pa decay scheme is shown in Fig.3. The major alpha transitions populate the 29.9 and 46.4 keV levels; very little intensity occurs at the 27.4 keV level. The 29.9 keV level decays to the ground state by a fast highly converted M1 transition and the 46.4 keV level decays via the 27.4 keV level. Both the 29.9 and 18.9 (46.4-27.4) keV transitions generates Ac L X-rays.

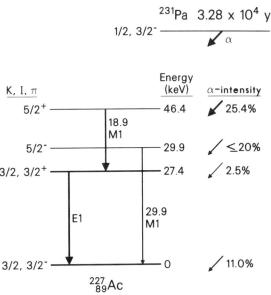

Fig. 3. Partial decay scheme of ^{231}Pa.

An α-γ coincidence experiment was performed using a cooled Si(Li) detector for the detection of photons and a Si detector for the detection of α-particles. Three parameter events were collected on tape and one dimensional spectra were later generated in coincidence with various gates. The spectra showed that the α_{46} and α_{30} are in prompt coincidence with L X-rays and the delay occurs at the 27.4 keV level. The analysis of the time spectrum between the α_{46} group and the 27.4 keV photopeak gave a half-life of 38.3 ± 0.3 ns, in agreement with previous measurements.

Discussion

From the measured level half-life, $T_{1/2}$, we have derived the gamma transition probabilities, T_{ex}, of the E1 γ-rays. These are given by the equation

$$T_{ex} = \frac{0.693 \cdot f}{T_{1/2} \cdot c^2} \qquad (1)$$

where f is the fraction of the total decay from the particular level which occurs by the given transition, and c^2 is the square of the appropriate Clebsch-Gordon coefficient between the two levels. In cases, where a single transition deexcites the level, $f = (1 + \alpha_T)^{-1}$, α_T being the total conversion coefficient. The above transition rate was divided by the Weisskopf estimate (w.u.) to obtain an energy-independent quantity. We have plotted this quantity for all known $\Delta K=0$ E1 transitions in heavy elements against the mass number in Fig. 4. The values for ^{225}Ac, ^{227}Ac, and ^{225}Ra are from the present work; other data are taken from literature [Led78, Asa60]. In Fig. 4 there is a definite enhancement in the E1 rate for the ^{225}Ac, ^{225}Ra, and ^{229}Pa nuclei.

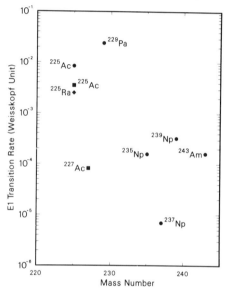

Fig. 4. Known transition rates of the $\Delta K = 0$ E1 transitions in odd-mass heavy nuclei.

Theoretical calculations [Cha77] indicate that the E1 matrix elements do not change much with the decreasing β_2 deformation between a given pair of states. One, therefore, expects that the E1 rate should not change with mass number. Thus the large enhancement in E1 rates for nuclei with A<230 suggests that these nuclei have some different nuclear structure properties. It has been established in previous studies that these three nuclei possess strong octupole correlations in their ground states. This indicates that enhancements in E1 rate are associated with octupole deformation. However, in ^{227}Ac, which also has strong ground state octupole deformation related properties, there is no such enhancement. Thus the present data on E1 rates in odd-mass nuclei show that there is no one-to-one correspondence between octupole deformation and the enhanced E1 rates.

So far, El transition rates have not been calculated with wavefunctions including octupole deformation. However, semiquantitative calculations performed by Chasman [Ahm84] and Leander [Lea84b] indicate that octupole deformation causes large enhancemets and large fluctuations in El rates.

Acknowledgments

The author wishes to thank his colloborators T. Ishii, J. E. Gindler, A. M. Friedman, R. R. Chasman, and S. B. Kaufman.

This research was supported by the U. S. Department of Energy under Contract W-31-109-Eng-38.

References

[Ahm82]   I. Ahmad, J. E. Gindler, R. R. Betts, R. R. Chasman, and A. M. Friedman, Phys. Rev. Lett. 49 (1982) 1758.

[Ahm84]   I. Ahmad, R. R. Chasman, J. E. Gindler, and A. M. Friedman, Phys. Rev. Lett. 52 (1984) 503.

[Asa60]   F. Asaro, F. S. Stephens, J. M. Hollander, and I. Perlman, Phys. Rev. 117 (1960) 492.

[Bar70]   S. A. Baranov, V. M. Shatinskii, V. M. Kulakov, and Yu. F. Rodionov, So. J. Nucl. Phys. 11 (1970) 515 [Yad. Fiz. 11 (1970) 925].

[Cha77]   R. R. Chasman, I. Ahmad, A. M. Friedman, and J. R. Erskine, Rev. Mod. Phys. 49 (1977) 833.

[Cha80]   R. R. Chasman, Phys. Lett. 96B (1980) 7.

[Hel85]   R. G. Helmer, M. A. Lee, C. W. Reich, and I. Ahmad, to be published.

[Ish85]   T. Ishii, I. Ahmad, J. E. Gindler, A. M. Friedman, R. R. Chasman, S. B. Kaufman, Nucl. Phys. A444 (1985) 237.

[Lea82]   G. A. Leander, R. K. Sheline, P. Moller, P. Olanders, I. Ragnarsson, and J. Sierk, Nucl. Phys. A388 (1982) 452.

[Lea84a] G. A. Leander and R. K. Sheline, Nucl. Phys. A413 (1984) 375.

[Lea84b] G. A. Leander, Proceedings of Fifth International Symposium on Capture Gamma-Ray Spectroscopy and Related Topics, Nashville, Tennessee, Sept. 1984.

[Led78]   C. M. Lederer and V. M. Shirley, Table of Isotopes (Wiley, New York, 1978).

[She83]   R. K. Sheline and G. A. Leander, Phys. Rev. Lett. 51 (1983) 359.

RECEIVED July 22, 1986

42

Dipole Collectivity in Nuclei and the Cluster Model

Moshe Gai

A. W. Wright Nuclear Structure Laboratory, Yale University, New Haven, CT 06511

Measurements on enhanced El deexcitation transitions in light and heavy nuclei are discussed. In
light nuclei these appear only from core-excited
cluster collective states that have large alpha particle widths (θ_α^2) as well as large B(E2) deexcitation
widths. They suggest the presence of an $\alpha + {}^{14}C$
cluster dipole degree of freedom as suggested by
Iachello and Jackson. In heavy nuclei the enhanced
B(E1) are found in the light Ra-Th isotopes where small
alpha particle hindrance factors for low-lying
$J^\pi = 1^-$ states and large ground state reduced alpha
particle widths are also found. In addition, the B(E1)
show a smooth dependence on J and on isotopic number
N. A phenomenon reminiscent of backbending is observed
for El deexcitations in ${}^{218}Ra$. The data suggest the
presence of a new low-lying collective dipole mode in
both heavy and light nuclei. In heavy nuclei this may
reflect either cluster states or stable octupole.
Analysis of the data on Ra isotopes in terms of the
Vibron cluster model yields a resonable fit with only
one variable parameter.

The nucleus ${}^{18}O$ is crucial for understanding of the T=1
effective nucleon-nucleon interaction as well as for the study of
coexistence of simple configurations of the nuclear many-body
system, such as the $\hbar\omega$ shell model configurations [1] and core
excited collective cluster configurations [2,3,4,5]. We have
carried out a complete spectroscopic study of ${}^{18}O$ [6] with special attention to measuring weak decay branches that, however,
correspond to large decay matrix elements. The bound states of
${}^{18}O$ were populated by alpha transfer reactions and their deexcitation was studied in a coincidence experiment using the
${}^{14}C({}^7Li,t\gamma){}^{18}O$ reaction. Quasi-bound resonant states above the
$\alpha + {}^{14}C$ threshold at 6.2 MeV and below 8.5 MeV of excitation, were
studied via the ${}^{14}C(\alpha,\gamma){}^{18}O$ radiative capture reaction. The
gamma ray transitions among the 16 lowest states of ${}^{18}O$ were
examined with the results shown in Fig. 1. A surprising result
is that only very few of the El deexcitations in ${}^{18}O$ appear to be
enhanced, and the enhanced ones link only states, which have
large alpha particle widths and which are understood theoretically to be of cluster structure. These states also exhibit
enhanced E2 deexcitations to lower collective states. Moreover,

0097-6156/86/0324-0278$06.00/0
© 1986 American Chemical Society

the enhanced Els appear to be very selective. For example, the collective 3⁻₂state at 8.29 MeV in ¹⁸O with a large alpha particle width (θ_α^2=20%) shows El deexcitation only to the collective third 2⁺ state at 5.26 MeV. The energetically favored El decays to the two-neutron 2⁺ states at 1.98 and 3.92 MeV were not observed, requiring smaller reduced matrix elements by at least a factor of 100. The core-excited 1⁻ state at 4.45 MeV exhibits enhanced El deexcitation to the collective 4p-2h 0⁺ state [2,3,4] but not to the ground state of ¹⁸O. The regular behavior of the enhanced El deexcitation in ¹⁸O, the correlation of enhanced Els and E2s and the large cluster widths suggest that these enhanced Els in ¹⁸O reflect clustering as recently proposed by Iachello and Jackson [7] in heavy nuclei. The enhanced Els correspond to large fractions of the molecular El sum rule [8] as expected for cluster states and the data on ¹⁸O provide the best example, thus far, of a low-lying cluster dipole band in nuclei.

The nuclei ²¹⁸Ra,²²⁰Ra: The importance of alpha clustering in heavy nuclei [7] was originally suggested to explain the presence of low-lying 1⁻ states having small alpha particle hindrance factors (F_α) [9]. We have thus undertaken [10] a detailed study of ²¹⁸Ra utilizing the ²⁰⁸Pb(¹³C,3n)²¹⁸Ra reaction at 69 MeV. In order to search for the low-lying non-yrast 1⁻,3⁻ states, we have carried out gamma-ray spectroscopy at the Coulomb barrier (59.0 and 59.5 MeV), where the incoming angular momentum is low and the non-yrast states are directly populated. Several changes and additions to a previously published [11] level spectrum for ²¹⁸Ra are included in Fig. 2, where we show our level spectrum for ²¹⁸Ra and several other Ra isotopes, including new data [12] on ²²⁰Ra.

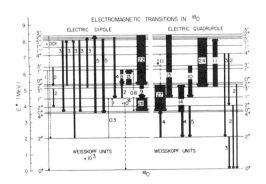
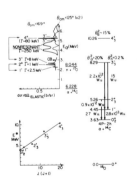

Fig. 1: El and E2 deexcitation matrix elements for ¹⁸O, and the suggested α + ¹⁴C (4p-2h) band in ¹⁸O of J$^\pi$= 0⁺₂, 1⁻, 2⁺₃, 3⁻₃, (4⁺₃).

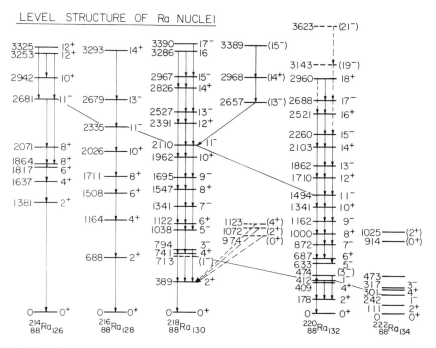

Fig. 2:   The level structure of the Ra isotopes. Note the low-lying 1⁻ and high-lying 11⁻ states, including a side band based on the 11⁻ state in ²¹⁸Ra.

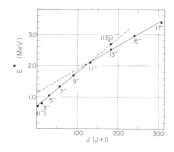
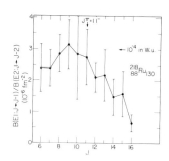

Fig. 3:   The moment of intertia and B(E1)/B(E2) ratios for negative parity states in ²¹⁸Ra. Note the changes above the 11⁻ state.

As we shall discuss later, the Vibron cluster model of Iachello-Jackson successfully reproduces the alternating parity sequence and low-lying (1^-), 3^- states in ^{218}Ra. In Fig. 2 we also note the systematic occurrence of 11^- states in the light Ra isotopes. In ^{214,216}Ra, these 11^- states are known to be two-quasiparticle states. In ^{218}Ra, a second band appears above the 11^- state. In addition, the dependence of excitation energy on $J(J+1)$ for the negative parity states shown in Fig. 3, appears to reflect a distinct change in the moment of inertia above the 11^- state. Such phenomena, arising from two-quasiparticles, are well recognized in quadrupole bands as backbending. In addition, in Fig. 3, we observe a systematic decrease in the B(E1)/B(E2) deexcitation ratios above the 11^- state. These data suggest that two-quasiparticle states are deexcited by E1 transitions which are not enhanced and thus to a reduction in the B(E1)/B(E2) ratios. Similar phenomena are well recognized for collective E2 transitions in the region of backbending.

In Fig. 4 we show the systematic behavior of B(E1)/B(E2) ratios in the Ra isotopes. While the B(E2) varies among these nuclei, it clearly cannot account for these large differences of the ratios. The data suggest a maximum B(E1) enhancement in ^{218}Ra. In Fig. 5 we show a similar plot for the Th isotopes. A measurement of the B(E1)s in ^{218}Ra [13] via the ^{13}C(^{208}Pb,3n) ^{218}Ra reaction at 5.3 MeV/u, with RDM Doppler shift techniques, is now in progress as a Yale-GSI-Munich collaboration. Preliminary results indicate enhanced B(E1) $\approx 6 \times 10^{-3}$ W.u. in ^{218}Ra corresponding to ~8% of the E1 molecular sum rule [8]. In ^{218}Ra an enhanced ground state alpha particle decay width is already known and the systematics of the Ra isotopes suggest a decrease in the alpha hindrance factors for the 1^- and 3^- states--as predicted by the cluster model.

It is well known that molecular states must display non-vanishing octupole as well as dipole transition moments. Indeed, it has been suggested that the yrast levels of nuclei in this region can be interpreted using models [14,15] which involve the assumption of static octupole shapes [16,17].

Calculations of the level spectra of Ra isotopes using the Vibron model [7] were performed [18]. In these calculations only one parameter was varied between the different isotopes of radium and the results are shown in Fig. 6. The good fits obtained suggest that the cluster model can reproduce the available data of heavy nuclei.

To conclude, the regular and smooth behavior of enhanced E1 deexcitations in light and heavy nuclei suggest that these arise from a collective degree of freedom, not studied previously. In light nuclei these E1s clearly arise from cluster states and in heavy nuclei both cluster and stable octupole shape models give adequate descriptions of the data.

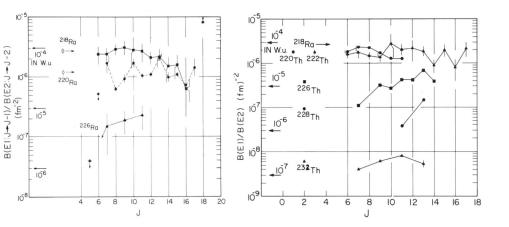

Fig. 4:   B(El)/B(E2) ratios for the   Fig. 5:   B(El)/B(E2) ratios for the Th
          Ra Isotopes.   isotopes.

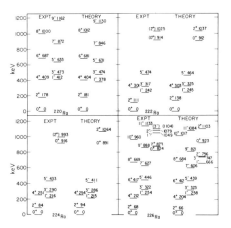

Fig. 6:   Cluster model prediction for Ra isotopes. In this calculation, only
          one parameter is varied, as discussed in text.

ACKNOWLEDGMENTS

I would like to acknowledge discussions with D.A. Bromley, F. Iachello and J.F. Shriner, and the cooperation of M. Ruscev, J.F. Ennis and P.D. Cottle, who worked on ^{18}O, ^{218}Ra and ^{220}Ra respectively, and of H.J. Daley, who worked on the Vibron model code. I also wish to thank R. Bonito and S. Sicignano for preparation of the camera-ready manuscript. This work was supported in part by U.S. Department of Energy Contract # DE-AC02-76ER03074.

REFERENCES

[1] R.D. Lawson, **Theory of the Nuclear Shell Model**, Clarendon Press, Oxford (1980), p.30.

[2] P. Federman and I. Talmi, Phys. Lett. **19** (1965) 490.

[3] G.E. Brown and A.M. Green, Nucl. Phys. **85** (1966) 87.

[4] A. Arima, H. Horiuchi and T. Sebe, Phys. Lett. **24B** (1967) 129.

[5] B. Buck and A.A. Pilt, Nucl. Phys. **A295** (1978) 1.

[6] M. Gai, M. Ruscev, A.C. Hayes, J.F. Ennis, R. Keddy, E.C. Schloemer, S.M. Sterbenz, and D.A. Bromley, Phys. Rev. Lett. **50** (1983) 239.

[7] F. Iachello and A.D. Jackson; Phys. Lett. **108B** (1982) 151.

[8] Y. Alhassid, M. Gai, and G.F. Bertsch; Phys. Rev. Lett. **49** (1982) 1482.

[9] C.M. Lederer and V.S. Shirley, **Table of Isotopes** (Wiley, New York, 1978).

[10] M. Gai, J.F. Ennis, M. Ruscev, E.C. Schloemer, B. Shivakumar, S.M. Sterbenz, N. Tsoupas and D.A. Bromley Phys. Rev. Lett. **51** (1983) 646.

[11] J. Fernandez-Niello, H. Puchta, F. Riess and W. Trautmann, Nucl. Phys. **A391** (1982) 221.

[12] P.D. Cottle, J.F. Shriner, Jr., F. Dellagiacoma, J.F. Ennis, M. Gai, D.A. Bromley, J.W. Olness, E.K. Warburton, L. Hildingsson, M.A. Quader, and D.B. Fossan; Phys. Rev. **C30** (1984) 1768.

[13] J.F. Ennis, M. Gai, D.A. Bromley, F. Azgui, H. Emling, E. Grosse, G. Seiler-Clark, H.J. Wollersheim, C.M. Mittag, F. Riess, Bull, Amer. Physical Soc. (1984).

[14] K. Neergard and P. Vogel, Nucl. Phys. **A149** (1970) 209.

[15] R.R. Chasman, Phys. Lett. **96B** (1980) 7.

[16] G.A. Leander, R.K. Sheline, P. Moller, P. Olanders, I. Ragnarsson and A.J. Sierk; Nucl. Phys. **A388** (1982) 452.

[17] W. Nazarewicz, P. Olander, I. Ragnarsson, J. Dudek, G.A. Leander, P. Moller and E. Ruchowska; Nucl. Phys. **A429** (1984) 269.

[18] J.F. Shriner, P.D. Cottle, J.F. Ennis, M. Gai, D.A. Bromley, J.W. Olness, E.K. Warburton, L. Hildingsson, M.A. Quader and D.B. Fossan; Phys. Rev. **C**, December issue 1985, in press.

RECEIVED August 27, 1986

43

Decay of ^{145}Cs to Levels of ^{145}Ba

J. D. Robertson[1], W. B. Walters[1], E. F. Zganjar[2], R. L. Gill[3], H. Mach[3], A. Piotrowski[3], H. Dejbakhsh[4], and R. F. Petry[5]

[1]Department of Chemistry, University of Maryland, College Park, MD 20742
[2]Department of Physics, Louisiana State University, Baton Rouge, LA 70805
[3]Physics Department, Brookhaven National Laboratory, Upton, NY 11973
[4]Cyclotron Laboratory, Texas A&M University, College Station, TX 77643
[5]Department of Physics, University of Oklahoma, Norman, OK 73019

An investigation of the β^- decay of the 0.59s ^{145}Cs was made at the TRISTAN on-line mass separation facility. The level scheme for ^{145}Ba has been constructed. The proposed spin and parity assignments are based upon transition multipolarities and $\gamma\gamma$ angular correlation measurements.

In 1950, Rainwater proposed that the discrepancy between the measured and calculated values of the quadrupole moments of some odd-A nuclei could be resolved by allowing those nuclei to take on a spheroidal shape. It is now well-known that the properties of many nuclei can be described on a deformed quadrupole basis. Over the past three years, much attention has been focused on the question of whether or not some nuclei, in addition to breaking rotational symmetry, also break reflection symmetry in the intrinsic frame. The nuclei in question are found in the Ra-Th region with A=220 -228 and a review of the extensive literature on this region can be found in references [LEA82], [IAC82], [GAI83], [NAZ84], and [LEA84].

In a recent paper, Leander et. al. introduce the idea that nuclei with N=88-90 in the immediate vicinity of ^{145}Ba might exhibit the same type of "octupole deformed" character that is observed in the Z=88-90 region.[LEA85] Three reasons are given as to why this might be a new region which can also be described by a basis which breaks reflection symmetry in the intrinsic frame. First of all, the systematics of the $J^\pi=1^-$ and 3^- levels in the neutron rich e-e Ba isotopes suggest a minimum in the splitting of the $K^\pi=0^+$ and 0^- bands near ^{146}Ba. The 1^- level in ^{146}Ba at 738 keV, in units of the first 2^+ level, is as low as the 1^- levels in the Ra and Th isotopes. The presence of these low-lying negative parity states in the structure of the light actinides was one of the first indications that the Ra-Th region was octupole deformed.[MOL72] Secondly, the deformed-shell-model calculations of Nazarewicz et. al. indicate that Z=56, as well as Z or N=88-90, is an optimal particle number for octupole deformation.[NAZ84] The Strutinsky type calculations predict an octupole deformed equilibrium shape for ^{146}Ba and a potential which is very soft towards octupole deformations for ^{144}Ba. The results are independent of the three single-particle potentials employed in the calculations. Finally, the large differences observed between the experimental and theoretical masses calculated by Moller and Nix in the Ra-Th region are also found in the region near ^{145}Ba.[LEA82] The difference is greatest for ^{145}Ba and, as in the light actinides, can be accounted for by the extra binding energy obtained by including octupole correlations in the mass calculations.

The ground state (g.s.) spin of ^{145}Ba has been determined to be 5/2 by laser spectroscopy at ISOLDE[MUE83]. The same group reports a g.s. magnetic moment of -0.272n.m. and a spectroscopic quadrupole moment (Qs) of 1.15b.

0097-6156/86/0324-0284$06.00/0

From the measured g.s. spin and Qs value, Leander calculates that ^{145}Ba may
be octupole deformed with an equilibrium value of $\beta_3 \approx 0.08$ ($\beta_2 = 0.172$ and
$\beta_4 = 0.069$).[LEA85] These results, however, are not conclusive. From a
simplified view, the g.s. spin of 5/2 for ^{145}Ba can be explained by
allowing the odd neutron to occupy the 5/2[521] Nilsson orbital with
$\beta_2 \approx 0.12$. In addition, the measured magnetic moment for the g.s. is consistent with both $\beta_3 = 0$ and $\beta_3 \neq 0$ in Leander's calculations.[LEA85]
      Another indication of octupole deformation in the structure of ^{145}Ba
would be the presence of a pair (or more) of strongly perturbed opposite
parity rotational bands connected by collective E1 transitions. Even more
specifically, these bands would have a small energy splitting and the
decoupling parameters for the two bands would be equal in magnitude but
opposite in sign.[RAG83] This type of collective structure is observed in
both ^{225}Ra and ^{227}Ac (fig. 1). In the study of the β^- decay of ^{145}Cs by
Rappaport et. al., only the 435.7 keV gamma was identified as an E1
transition.[RAP82] In their study, however, only 31 gamma rays were
assigned to the decay of ^{145}Cs and they determined the level structure of
^{145}Ba up to 785keV. A later study by Dejbakhsh found 77 new gamma rays and
defined levels up to 2.8MeV.[DEJ85] Unfortunately, the latter investigation was not able to determine the multipolarities of any of the new transitions.

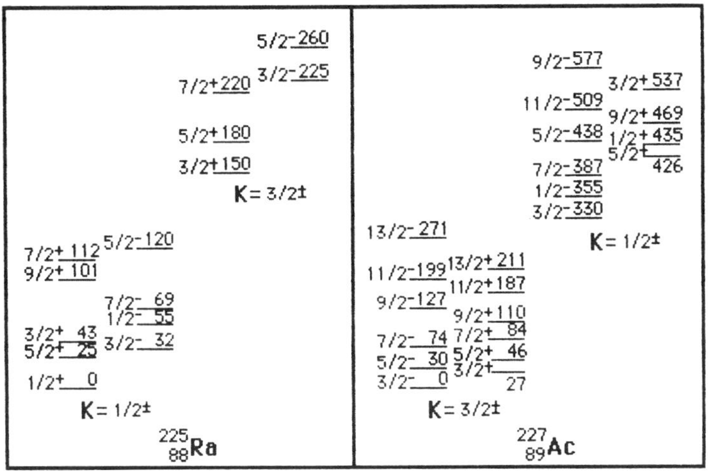

Figure 1. Parity doublets in ^{225}Ra and ^{227}Ac [15,16]

EXPERIMENTAL PROCEDURE
      The study of the β^- decay of 0.59s ^{145}Cs was conducted at the TRISTAN
mass separator on-line to the high flux reactor at BNL. A detailed
description of the TRISTAN facility can be found in reference [GIL81]. The
radioactive samples were produced by fissioning a uranium target integrated
in a positive surface ionization source.[PI084] In this experiment, the
usual Re surface ionizer was replaced with a Ta ionizer. The lower work
function of Ta ensured no independent production of Ba from the ion source
at a low power operation ($\approx 1200^0$C).

Two Ge(Li) detectors and two HpGe detectors were used to collect singles gamma spectra and 10^8 three parameter $\gamma\gamma t$ coincidence events. The resolution of the germanium detectors ranged from 1.7 to 2.2 keV FWHM at 1332 keV. In addition, a LEPS detector with 0.55 keV FWHM at 122 keV was used in the low-energy γ-ray measurements. Table 1 summarizes the results obtained in the angular correlation measurements. The electron spectra and the γ-e coincidence data were acquired using both a Si(Li) and HpGe detector. The results of the conversion electron measurements are given in table 2.

TABLE 1 - EXPERIMENTAL ANGULAR CORRELATION COEFFICIENTS

Transition	A22	A44	χ^2
323→112→0	0.27±0.04	≡ 0	23.0
368→199→0	0.13±0.05	0.21±0.05	1.4
317→435→0	-0.21±0.09	0.06±0.10	0.4
155→455→0	0.22±0.08	0.23±0.10	3.6

TABLE 2 - EXPERIMENTAL INTERNAL CONVERSION COEFFICIENTS

$E\gamma(keV)$	$\alpha_{EXP} \times 10^2$	$\alpha_{THEOR} \times 10^2$ *		
		α_{M1}	α_{E2}	α_{E1}
112.5-K	107.2±12.0	58.0	84.0	13.9
112.5-L	12.8±1.2	8.4	34.0	1.9
175.4-K	16.9±2.2	17.2	20.4	4.2
175.4-M	0.74±0.11	0.52	1.1	0.11
198.9-K	18.1±2.6	12.3	13.6	3.0
198.9-L	3.1±0.6	2.0	3.2	0.40
207.1-K	0→6.4	11.0	12.0	2.7
241.0-K	8.8±1.2	7.3	7.4	1.8
435.7-K	0.48±0.06	1.5	1.1	0.38
454.7-K	3.6±0.5	1.4	1.0	0.34

*Calculated from NDT 21, 225-230 (1978).

EXPERIMENTAL RESULTS
    The low-energy portion of the level scheme deduced for ^{145}Ba is shown in figure 2. A saturated sample of the A=145 isobars was collected and counted to determine the absolute intensity of the 175.4 keV γ-ray (19.8 ± 2.4/100 decays). This value, along with the γ and c-e intensities, was used to calculate the β^- feeding to the g.s. and excited levels in ^{145}Ba. The large uncertainty in the β^- feeding to the 198.9 keV level is a direct result of the large uncertainty in the relative intensity of the 198.9 keV γ-ray. ^{145}Cs has a 12% delayed neutron branch [RIS79] and the major transition in ^{144}Ba is a 199 keV -ray. As a result, the intensity of the 198.9 keV transition in ^{145}Ba could only be determined from the coincidence data. Our value of 54.7 for I_{199} is lower than the 68.5 value previously reported by Rappaport et. al. [RAP82].
    As noted above, the g.s. spin of ^{145}Ba has been measured to be 5/2. The negative parity assignment for the g.s. is based upon the systematics of the N=89 isotones and the Z=56 isotopes. The g.s. J^π of both ^{149}Nd and ^{151}Sm is 5/2$^-$. The g.s. of J^π of the odd-A Ba isotopes is: ^{139}Ba=7/2$^-$, ^{141}Ba=3/2$^-$, and ^{143}Ba=5/2$^-$.
    From the internal conversion coefficients given in table 2, the following gammas were determined to be M1/E2 transitions in agreement with

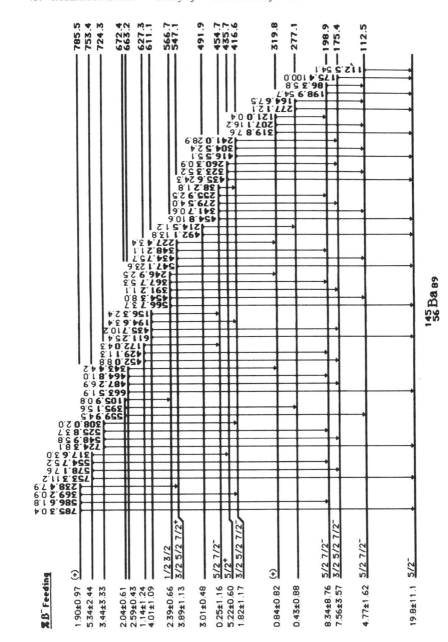

Figure 2. Low-Energy portion of ^{145}Ba level scheme.

the earlier work: 112.4, 175.4, 198.9, 241.0, and 454.7. Rappaport also reports that the 86.3 and 238.4 ϑ-rays are M1/E2 transitions. The non-zero A44 values for both the 368-199-0 cascade and the 155-455-0 cascade rules out the possibility that J^π of the 199 keV level or the 455 keV level is 3/2$^-$. In addition, the A22/A44 values for the 368-199-0 cascade limit the spin of the 567 level to either 1/2 or 3/2 once J^π of the 199 level is known not to be 3/2$^-$. The large A22 value of 0.24 for the 323-112-0 cascade, coupled with the fact that the 435 level is 5/2$^+$, rules out the possibility that the 112 level has J^π=3/2$^-$; the largest A22 can be for a 5/2$^+$-3/2$^-$-5/2$^-$ cascade is 0.08. The only transition that was determined to be E1 by the measured conversion coefficients was the 435.7 keV γ-ray. Setting an upper limit on the area expected for the 547.1 K-conversion peak in the electron spectrum indicates that the 547 is also an E1 transition. Likewise, an upper limit of 0.064±0.008 for α_K for the 207.1 keV transition can also be established by setting the mixing of the 175.4 keV transition to zero. This low α_K value suggests that the 207.1 γ-ray is an E1 transition. A spin of 3/2$^-$ for the 435 level is ruled out by the large negative A22 value for the 317-435-0 cascade. In addition, a log(ft) of approximately 6 for the 435 keV level makes the β^- decay to this level an allowed transition. Because the g.s. spin of ^{145}Cs is 3/2$^+$, this rules out the possibility that the spin of the 435 level is 7/2$^+$ and subsequently limits the spin of the 435 level to 5/2$^+$.

CONCLUSION

Clearly the rotational band structure observed in the light odd-A actinides is not seen in the structure of ^{145}Ba. Like the N-89 isotones ^{149}Nd[PIN77] and ^{151}Sm[CO076], the negative parity states in ^{145}Ba can not be grouped into any obvious rotational bands and seem to be very hard to explain in the framework of the Nilsson model. The levels at 319.8, 547.1, and 785.5 keV might be members of a strongly perturbed positive-parity rotational band, but both the 319 and 547 appear to be 5/2 levels. Aparently the 435.7 keV level can not be a member of the positive parity band including the 547 and 785 because neither of these positive parity levels decay to it.

The absence of any clearly defined negative-parity rotational bands in ^{145}Ba suggests that it is not strongly quadrupole deformed. The clear rotational structure observed in ^{153}Gd and ^{155}Dy completely dissapears as the N=89 isotones cross the Z=64 shell. This conclusion is also supported by the β_2 values calculated from the lifetimes of the first 2$^+$ states in ^{144}Ba and ^{146}Ba. The ratio β_2/β_{2sp} for ^{144}Ba and ^{146}Ba is 6.8 and 7.5 respectively. This is compared to the ratio of 12.2 and 10.7 observed in the quadrupole deformed nuclei ^{154}Gd and ^{156}Dy. Thus, although ^{145}Ba does not have the parity doublets expected for an octupole deformed nucleus, it is possible that (because β_2 is so small) its structure is determined by higher orders of deformation; i.e. the structure is very dependent on β_3, β_4, et.. The great similarity between the structure of ^{145}Ba, ^{147}Ce, and ^{149}Nd added to our inability to understand these N=89 (and N=87) nuclei in the framework Nilsson model points to the need for extended theoretical and experimental work in this transitional region.

ACKNOWLEDGMENTS

We would like to thank Dr. Leander for his helpful discussions. This work has been supported by the U.S. Department of Energy.

REFERENCES

[LEA82] G.A. Leander, R.K. Sheline, P. Moller, P. Olanders, I. Ragnarsson, and A.J. Sierk, Nucl. Phys. A388 452 (1982).
[IAC82] F. Iachello and A.D. Jackson, Phys. Lett. 108B 151 (1982).

[GAI83] M. Gai, J.F. Ennis, M. Ruscev, E.C. Schloemer, B. Shivakumar, S.M. Sterbenz, N. Tsoupas, and D.A. Bromley, Phys. Rev. Lett. 51 646 (1983).
[NAZ84] W. Nazarewicz, P. Olanders, I. Ragnarsson, J. Dudek, G.A. Leander, P. Moller, and E. Ruchowska, Nucl. Phys. A429 269 (1984).
[LEA84] G.A. Leander and R.K. Sheline, Nucl. Phys. A413 375 (1984).
[LEA85] G.A. Leander, W. Nazarewicz, P. Olanders, I. Ragnarsson, and J. Dudek, Phys. Lett. 152B 284 (1985).
[MOL72] P. Moller, S.G. Nilsson, and R.K. Sheline, Phys. Lett. 40B 329 (1972).
[MUE83] A.C. Mueller, F. Buchinger, W. Klempt, E.W. Otten, R. Neugart, C. Ekstrom, and J. Heinemeier, Nucl. Phys. A403 234 (1983).
[RAG83] I. Ragnarsson, Phys. Lett. 130B 353 (1983).
[RAP82] M.S. Rappaport, G. Engler, A. Gayer, and I. Yoresh, Z. Physik. A305 359 (1982).
[DEJ85] H. Dejbakhsh, private communication.
[GIL84] R.L. Gill, M.L. Stelts, R.E. Chrien, V. Manzella, H. Liou, and S. Shostak, Nucl. Instr. and Meth. 186 243 (1981).
[PIO84] A. Piotrowski, R.L. Gill, and D.C. McDonald, Nucl. Instr. and Meth. 22 1 (1984).
[RIS79] C. Ristori, J. Crancon, K.D. Wunsh, G. Jung, R. Decker, and K.L.-Kratz, Z. Physik. A290 311 (1979).
[PIN77] J.A. Pinston, R. Roussille, G. Sadler, W. Tenten, J.P. Bocquet, B. Pfeiffer, and D.D. Warner, Z. Physik. A282 303 (1977).
[COO76] W.B. Cook, M.W. Johns, G. Louhoiden, and J.C. Waddington, Nucl. Phys. A259 461 (1976).

RECEIVED August 1, 1986

Section II: Experimental Exploration of Current Issues
Part 3: High-Spin-State Investigations Using Heavy Ions
Chapters 44–52

44

Spectroscopic Consequences of Shape Changes at High Angular Momenta

Ingemar Ragnarsson and Tord Bengtsson

Department of Mathematical Physics, Lund Institute of Technology, P.O. Box 118, S-22100 Lund, Sweden

We discuss so called terminating states which occur because of the finiteness of rotational bands and which are associated with a gradual shape change until the rotation takes place around a symmetry axis. Predictions on terminating configurations in nuclei with 10-12 valence particles outside the ^{146}Gd core are compared with recent experimental data. Observed and calculated spectra show a remarkable similarity, at least for ^{156}Er and ^{158}Er which nuclei are discussed in detail here.

It is by now well-established that substantial shape changes occur at high spin for nuclei with a few particles outside the $^{146}_{64}Gd_{82}$ core. Such shape changes appear to be associated with so called band terminations as pointed out a few years ago [BEN83] and recently some experimental evidence for the occurence of terminating bands have been presented [BAK85, RAG85, STE85]. The states which terminate a band and which one could hope to observe in experiment appear to be very favoured energetically, i.e. relative to some average trend they are exceptionally low-lying. Indeed, a straightforward way to localise band terminations seems to be to analyse the energy systematics at high spin and compare with the general trends predicted by calculations.

With many enough particles (and holes) building an aligned state, it becomes possible to form also collective structures within the configuration. This leads to a terminating band where it is possible to follow the gradual alignment of the spin vectors until full alignment is achieved in a state where all spin vectors are quantised along a symmetry axis. The alignment process is accompanied by a gradual shape change, typically from prolate at lower spins over triaxial to oblate shape at the termination (fig. 1). The deformation changes give an energy gain relative to rotation at fixed deformation. Many terminating bands are predicted at spins I=40-60 for nuclei with 10-12 nucleons outside the ^{146}Gd core. Note also that this alignment process is since long known in ^{20}Ne [BOH75, RAG81].

Let us now consider ^{158}Er as an illustrative example of the competition between particle-hole states and collective bands. This nucleus is collective at low spins with a deformation corresponding to $\varepsilon=0.20-0.25$.

0097-6156/86/0324-0292$06.00/0
© 1986 American Chemical Society

TERMINATING BAND

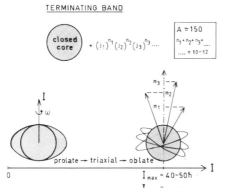

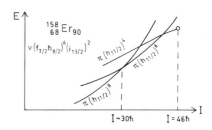

Fig.1. Schematic illustration of a band terminating configuration and its evolution with spin.

Reproduced with permission from [RAG85]. Copyright 1985 North-Holland Physics Publishing Company.

Fig.2. Schematic illustration of calculated (unpaired) bands in the yrast region of ^{158}Er.

Reproduced with permission from [RAG85]. Copyright 1985 North-Holland Physics Publishing Company.

With pairing neglected, the ground stateconfiguration would be described as $\pi(d_{5/2}g_{7/2})^{-4}(h_{11/2})^8 \nu(f_{7/2}h_{9/2})^6(i_{13/2})^2$. With increasing spin (and alignment) the deformation decreases and a band with the proton configuration $\pi(d_{5/2})^{-2}(h_{11/2})^6$ comes lower in energy [BEN85a]. For even higher spins and at a deformation close to spherical, it becomes favourable to close the Z=64 core leading to the $\pi(h_{11/2})^4$ proton configuration while the neutron configuration remains unchanged, $\nu(f_{7/2}h_{9/2})^6(i_{13/2})^2$. In this configuration, the maximum spin is $I_{max} = I_{max}^p + I_{max}^n = 16+30=46$. This band structure of ^{158}Er is schematically illustrated in fig. 2.

The full calculational result [BEN83] for ^{158}Er is compared with experiment in fig. 3. The calculations are based on the Nilsson-Strutinsky cranking method where we follow individual configurations as functions of spin [BEN85]. The energy of each state is minimised with respect to deformation, ε, γ, and ε_4. The calculated bands denoted by 1, 2 and 3 are the $\pi(h_{11/2})^8$, $\pi(h_{11/2})^6$ and $\pi(h_{11/2})^4$ configurations of fig. 2. Compared with the most recent experiment [SIM84, TJO85, DIA85], there is indeed a remarkable agreement. The calculated crossing between bands 1 and 2 at I≈26 (outside fig. 3) can be identified with an observed backbend at I=24 while the more irregular structure of the observed lowest I=40 and 42 states and those assigned I=44 and 46 is naturally identified with the calculated band 3 with a band termination at I=46. The higher-lying states assigned 40^+ and

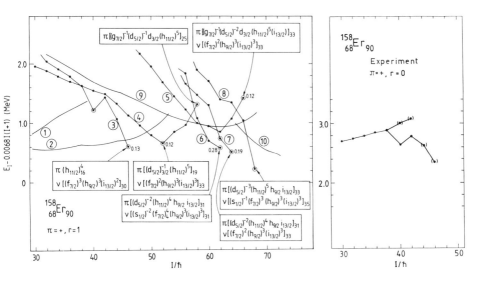

Fig.3. Calculated [BEN83] positive parity even spin configurations in the yrast region of ^{158}Er compared with recent experimental data [SIM84, TJO85, DIA85] in the I≥30 region. For states in paranthesis, the spin assignment is uncertain. The energies are given relative to a rotational term, $(\hbar^2/2J)I(I+1)$ with the moment of inertia taken as the rigid body value at a small deformation, $\varepsilon = 0.2$. Note that the calculated bands are strongly down-sloping when approaching their termination (encircled states), i.e. the last spin units are obtained very cheap energetically. The same feature is seen in the observed band extending up to a suggested 46$^+$ state.

Reproduced with permission from [RAG85]. Copyright 1985 North-Holland Physics Publishing Company.

42$^+$ should then be identified with the calculated more collective band 2. This is supported by the fact that the feeding from these latter states is observed to be faster than that from the lowest I≥40 states. The disagreement in relative energy between bands 2 and 3 can be traced back to the (uncertain) magnitude of the Z=64 single-particle gap.

The favoured energy of the 40$^+$ state in band 3 is not present in the simplified picture of fig. 2. It can be understood as indicated in fig. 4. Within the configuration which terminates at I=46 it is also possible to form an aligned I=40 state if the spin vector of one of the negative parity valence neutrons is put in opposite direction, i.e. moving one neutron from the $h_{9/2\ 5/2}$ to the $f_{7/2-7/2}$ orbital. This particle-hole state is then also associated with a terminating band. In the present computer calculations, it is only possible to identify the lowest state in a configuration leading to a calculated "yrast line" as shown by the dashed line in the figure.

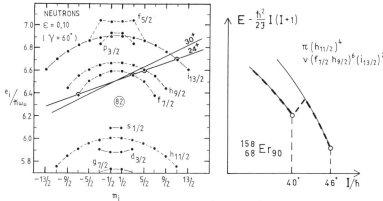

Fig.4. Illustration of aligned 24^+ and 30^+ states in the $(f_{7/2}h_{9/2})^6$ $(i_{13/2})^2$ N=90 configuration with all spin vectors of the valence particles aligned in the 30^+ state while one $f_{7/2}$ neutron is "anti-aligned" in the 24^+ state. Combined with the $\pi(h_{11/2})^4$ Z=68 16^+ state it becomes possible to form aligned 46^+ and 40^+ states in ^{158}Er. Thus, not considering band-mixing, an "yrast line" according to the thick dashed line results.

Reproduced with permission from [RAG85]. Copyright 1985 North-Holland Physics Publishing Company.

A negative parity band is formed in ^{158}Er if one more neutron is lifted to the $i_{13/2}$ shell. Combined with the same proton configurations as the positive parity bands, similar band-crossings are obtained with a fully aligned terminating state at 49^- and a low-lying 43^- state ressembling the positive parity I=40 state. These findings are consistent with present experimental data suggesting that above the collective 41^- state, some kind of branching occurs.

In ^{156}Er, a sequence of states which is unusually strongly populated has been observed [STE85] in the spin region I = 30-42. The 12 spin units are obtained at an energy cost which is approximately 1.5 MeV lower than expected for rigid rotation (fig. 5). Furthermore, as the next spin state at I=43 is observed relatively much higher in energy than the 42^+ state, the observed energy systematics strongly suggests that the 42^+ state is indeed a band termination. The observed structure is in general agreement with theoretical predictions [RAG84] as seen in fig. 5, showing a very low-lying fully aligned 42^+ state with the configuration $\pi(h_{11/2})^4$ $\nu(f_{7/2})^2(h_{9/2})^2(i_{13/2})^2$. Furthermore, the observed 36^+ and 34^+ states come relatively low in energy, a fact which is also seen in the calculations and

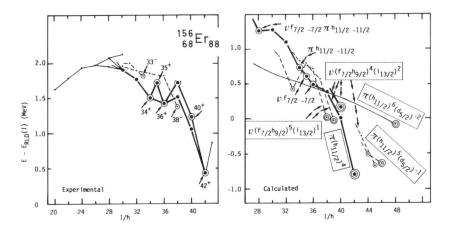

<u>Fig.5</u>. Observed [STE85] and calculated [RAG84] states in the yrast region of ^{156}Er with the configuration terminating at 42$^+$ specially marked. In a similar way as for ^{158}Er in fig. 3, an energy expression corresponding to rigid rotation, E_{RLD}, has been subtracted. The calculated aligned states and the corresponding observed states are encircled. Two circles are used for the calculated fully aligned states. For states with one anti-aligned spin vector the corresponding orbital is given.

Reproduced with permission from [RAG85]. Copyright 1985 North-Holland Physics Publishing Company.

explained in a similar way as the low-lying 40$^+$ state in ^{158}Er. Thus, compared to the 42$^+$ state, 36$^+$ and 34$^+$ states are formed from rearrangements of one particle, $\nu(f_{7/2\ 5/2} \rightarrow f_{7/2-7/2})$ and $\pi(h_{11/2\ 5/2} \rightarrow h_{11/2-11/2})$ respectively. Similarly, a 35$^+$ state is formed from the rearrangement $\nu(h_{9/2\ 7/2} \rightarrow f_{7/2-7/2})$, a fact which is consistent with an observed rather strong branch through a relatively low-lying 35$^+$ state. A second non-negligable branching observed in experiment involves a 40$^+$ state and a 38$^+$ state. A natural candidate for such a 40$^+$ state is the $\pi(h_{11/2})^4$ $\nu(f_{7/2})^3 h_{9/2}(i_{13/2})^2$ configuration and then the 38$^+$ state can be understood as part of the band terminating in this 40$^+$ state. This however requires that the "branched" 40$^+\rightarrow$38$^+$ and 38$^+\rightarrow$36$^+$ gamma rays come in reversed order compared to that given in ref. [STE85]. Indeed, the ordering of these two transitions cannot be determined [STE85a] from the present experiment and one can also note that the low-lying 38$^+$ state which results from the ordering given in ref. [STE85] seems very difficult to explain theoretically.

The I=34, 35, 36, 40 and 42 states essentially exhaust the possibilities to form low-lying positive parity states with I$\geq$30 within the ten valence particles of ^{156}Er. The most low-lying fully aligned negative parity states are calculated for I$^\pi$ = 38$^-$, 39$^-$ corresponding to the rearrangements $\nu(i_{13/2} \rightarrow f_{7/2})$ and $\nu(i_{13/2} \rightarrow h_{9/2})$ relative to the 42$^+$ state. This seems consistent with the fact that the highest spin observed for negative parity is at I=38 and that this 38$^-$ state is very favoured energetically (fig. 5). One more negative parity state, 33$^-$, is observed relatively low in energy, and indeed, the calculations give a low-lying aligned 33$^-$ state which compared to the 42$^+$ state is formed by the rearrangement $\nu(i_{13/2\ 11/2} \rightarrow f_{7/2-7/2})$.

In summary we have discussed the shape transition associated with the evolution of finite rotational bands to their points of termination. Many such terminating bands are predicted in configurations with 10-12 valence particles (and holes) relative to a ^{146}Gd core. According to the calculations, the very low energy of the terminating states is a prominent feature in these configurations. For the nuclei discussed here, ^{158}Er and ^{156}Er, it is found that not only the predicted trends are confirmed by recent experiments but also many of the detailed features in the calculations appear to be present in the observed high-spin spectra.

References

[BAK85]   C. Baktash et al., Phys. Rev. Lett. 54 (1985)978.

[BEN83]   T. Bengtsson and I. Ragnarsson, Phys. Scripta T5 (1983)165.

[BEN85]   T. Bengtsson and I. Ragnarsson, Nucl. Phys. A436 (1985)14.

[BEN85a] T. Bengtsson and I. Ragnarsson, Phys. Lett. B, to appear.

[BOH75]   A. Bohr and B.R. Mottelson, Nuclear Structure, vol. II (Benjamin, New York, 1975)

[DIA85]   R.M. Diamond, these proceedings.

[RAG81]   I. Ragnarsson, S. Åberg and R.K. Sheline, Phys. Scripta 24(1981)215

[RAG84]   I. Ragnarson and T. Bengtsson, Proc. 1984 INS-RIKEN Int. Symp. on Heavy Ion Physics, Mt. Fuji, Aug. 1984, J. Phys. Soc. Jpn 54 (1985) Suppl. II, p. 495.

[RAG85]   I. Ragnarsson et al., Phys. Rev. Lett. 54 (1985)982.

[SIM84]   J. Simpson et al., Phys. Rev. Lett. 53 (1984)648;

[STE85]   F.S. Stephens et al., Phys, Rev. Lett. 54 (1985)2584.

[STE85a] F.S. Stephens, priv. comm., June 1985.

[TJO85]   P. Tjøm et al., subm. to Phys.Rev. Lett.

[RAG85]   I. Ragnarsson and T. Bengtsson, Nucl. Phys. A447 (1985) 253c-255c.

RECEIVED July 21, 1986

45

Evolution of Nuclear Shapes at High Spins

Noah R. Johnson

Oak Ridge National Laboratory, Oak Ridge, TN 37831

Outstanding progress has been made during the past ten years on an understanding of the properties of nuclei excited into states of high angular momentum. Much of the experimental progress has resulted from γ-γ coincidence measurements utilizing complex detector arrays. Many of the properties of the yrast and near-yrast bands in nuclei revealed in these measurements have become reasonably well explained by current theory. Both cranked shell model (CSM) and cranked Hartree-Fock Bogoliubov (CHFB) calculations have enjoyed considerable success in accounting for many aspects of high spin behavior. However, for a detailed understanding of the structure of these high spin states and for a stringent test of these models, it is necessary to resort to measurements of their static and dynamic electromagnetic multipole moments. During the past few years we at Oak Ridge have concentrated on studies of the latter quantity, the dynamic electric quadrupole (E2) moments which are a direct reflection of the collective aspects of the nuclear wave functions. For this, we have carried out Doppler-shift lifetime measurements utilizing primarily the recoil-distance technique.

The nuclei with neutron number N ≈ 90 possess many interesting properties. These nuclei have very shallow minima in their potential energy surfaces, and thus, are very susceptible to deformation driving influences. It is the evolution of nuclear shapes as a function of spin or rotational frequency for these nuclei that has commanded much of our interest in the lifetime measurements to be discussed here. There is growing evidence that many deformed nuclei which have prolate shapes in their ground states conform to triaxial or oblate shapes at higher spins. Since the E2 matrix elements along the yrast line are sensitive indicators of deformation changes, measurements of lifetimes of these states to provide the matrix elements has become the major avenue for tracing the evolving shape of a nucleus at high spin. Of the several nuclei we have studied with N ≈ 90, those to be discussed here are ^{160,161}Yb [FEW82], [JOH82], [FEW82], [FEW85] and ^{158}Er [OSH84a], [OSH84b]. In addition, we will discuss briefly the preliminary, but interesting and surprising results from our recent investigation of the N = 98 nucleus, ^{172}W [RAO85].

0097–6156/86/0324–0298$06.00/0
© 1986 American Chemical Society

II. Experimental Aspects and Data Analyses

Both ^{160}Yb and ^{161}Yb were produced by the reactions ^{116}Cd(^{48}Ti,xn) and its inverse ^{48}Ti(^{116}Cd,xn), at a center-of-mass energy of 145 MeV in each case. The recoil-distance device used in the measurements is discussed in [JOH81]. It was designed to fit inside the annular opening of a 25-cm x 25-cm NaI crystal which acts as a total-energy filter. In this way it was possible to gate on given regions of the total γ-ray energy spectrum and get enhancement of the desired reaction channel. Spectra were obtained for a total of 17 different flight times.

Lifetimes were extracted from the resulting decay curves by the computer program, LIFETIME [WEL85] which includes all of the usual corrections [STU76] to the data. Knowing the general features of the level scheme, one usually models a two-step cascade side feeding to each level to account for population from undefined transitions. The program then solves the Bateman equations while adjusting the lifetimes and initial populations of the levels to obtain the best fits to the decay curves. Both the shifted and unshifted γ-ray intensities are used in the fitting procedure. Uncertainties in the lifetimes were determined by the method of the subroutine MINOS, described in [JAM75].

Excited ^{158}Er nuclei for these studies were produced via the reaction ^{128}Te(^{34}S,4n) at a bombarding energy of 155 MeV. In these experiments the large NaI detector was removed in order to place an array of five Ge detectors at 90° with respect to the beam direction and at close geometry (6 cm) to the target. This was done in order to test if a high-efficiency coincidence measurement could offer simplifications in the analysis and interpretation of the experimental data, especially with respect to simplifying the analyses problems associated with side feeding. Direct side feeding makes no contribution to a γ-ray transition if the spectrum is gated by band members higher than the transition of interest.

Measurements were taken on ^{158}Er at a total of 14 target-stopper distances. A portion of the total-projected spectra taken at four of the target-stopper separations is shown in Fig. 1. These spectra are of excellent statistical quality and reveal a favorable peak-to-background ratio in the 0° detector as a result of using the Compton suppression shield.

Lifetimes for the yrast sequence of ^{158}Er were determined from four different sets of coincidence data: 1) that from gating on the first

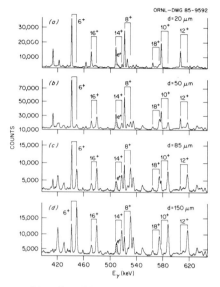

Fig. 1. Illustrative "total-projected" coincidence spectra of ^{158}Er covering the 400-650 keV region for four of the fourteen distances measured.

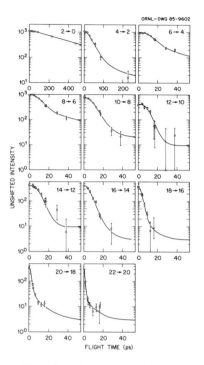

Fig. 2. Decay curves for members of the yrast sequence in ^{158}Er. The points represent the experimental data with the appropriate corrections. The solid curves are the fitted time distributions determined by the program LIFETIME.

transition above the one of interest; 2) that from gating on the second transition above the one of interest; 3) that from the sum of all gates below the transition of interest; and 4) that from the total-projected coincidences. Although side feeding to the state of interest is not eliminated in the last two types of data, they have an advantage in being of excellent statistical quality. In fact, we found that the program LIFETIME handled their more complex side feeding conditions quite well, based on comparisons of lifetimes from all four sets of data. In Fig. 2 are shown experimental data for each state and the program fits to these data.

For the ^{172}W studies, the ^{124}Sn (^{52}Cr, 4n) reaction was utilized at a beam energy of E_{Lab} = 230 MeV. The experimental arrangement was similar to that for ^{158}Er measurements, except here we used six large volume Ge detectors at 90° for coincidence gating. Data were collected for 18 recoil flight distances ranging from 16 μm to 7mm. To this point, the preliminary lifetime analyses are available only for the total projected coincidence spectra and the sums of gates below the transition of interest.

III. Discussion

Experimental transition quadrupole moments, Q_t, were obtained from the reduced electric quadrupole transition probabilities, B(E2), according to the expression

$$B(E2:I{\rightarrow}I{-}2) = \frac{5}{16\pi} <I\ 2\ 0\ 0\ |\ I{-}2\ 0>^2\ Q_t^2,$$

where the term in brackets is a Clebsch-Gordon coefficient. In Fig. 3, these Q_t values are plotted as a function of the rotational frequency for some yrast and near-yrast states in ^{160}Yb and ^{161}Yb. Figure 4 shows a plot of Q_t values for the yrast sequence of ^{158}Er. The latter data show clear evidence of centrifugal stretching in the ground band as expected for an N = 90 nucleus. The data for the ground band of ^{160}Yb are, unfortunately, too limited to indicate whether this effect is present there.

A most interesting feature of the data in Fig. 3 is that the quasiparticle bands in both ^{160}Yb and ^{161}Yb show an overall trend of loss of collectivity with increasing ω. For ^{160}Yb there are three different two-quasiparticle bands, with each showing Q_t values behaving somewhat similarly as a function of the rotational frequency. Likewise, ^{161}Yb shows one- and three-quasiparticle bands with rather similar behavior. As seen in Fig. 4, the ^{158}Er Q_t values also show a dropoff in the s band. In fact, the Q_t values for the 8^-, 9^- and s bands

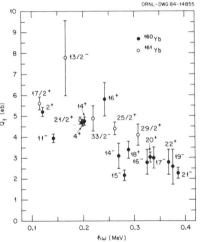

Fig. 3. Transition quadrupole moments of some yrast and near-yrast states of ^{160,161}Yb vs rotational frequency.

of ^{158}Er show a reducing trend when plotted as a function of rotational frequency, although it is less pronounced than is the data of Fig. 3. The common aligned quasiparticle in all of these bands is the lowest energy $i_{13/2}$
orbital with parity and signature $(\pi,\alpha) = (+,+1/2)$. This quasiparticle is
coupled to the $(-,+1/2)$ and $(-,-1/2)$ $h_{9/2}$ orbitals to form the 6^- and 9^- side
bands in ^{160}Yb and the 8^- and 9^- bands in ^{158}Er. The similarities in the
behavior of Q_t with ω for these bands suggest the possibility that the $i_{13/2}$
$(+,+1/2)$ quasineutron has a dominant influence on the core.

Addressing the ideas discussed above, the group at Lund has reported
[BEN83] results of self-consistent cranked Hartree-Fock-Bogoliubov (HFB)
calculations of the shape of ^{160}Yb as a function of angular momentum. They
find that the $i_{13/2}$ alignment produces a quadrupole deformation ε_2 which is
near the ground-state value and which remains approximately constant with
spin. However, the triaxiality parameter γ shows a steady increase, reaching
about $10°$ by $I = 18^+$, the highest spin they reported. (In the Lund convention, $\gamma = 0°$ corresponds to collec-
tive rotations of a prolate ellipsoid
and $\gamma = 60°$ corresponds to noncollec-
tive rotations of an oblate ellipsoid.)

Another theoretical approach
has also been taken [LEA83] [FRA83]
and it involves a more phenomenological examination of the effects of the
γ degree of freedom on the energy of
the quasiparticles and the core. The
cranked shell model (CSM) is used to
calculate the quasiparticle energies
which are added to the energy of the
rotating core as a function of γ. In
the N~90 transition region, the $i_{13/2}$
quasiparticles drive the equilibrium
value of γ to around $5°$. So, at
least qualitatively, both the self-
consistent results for ^{160}Yb and the
CSM results (calculated for ^{160}Yb,
but will be very similar for ^{158}Er)

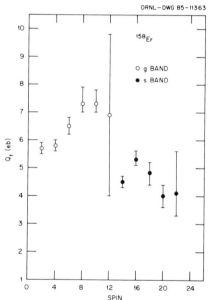

Fig. 1. Transition quadrupole
moments of yrast states in ^{158}Er.

are seen to be in agreement with our experimentally observed loss of collectivity at high rotational frequencies in the quasiparticle bands. However,
we point out that should all of the loss of collectivity in these two nuclei
be attributed to triaxiality, it is necessary to invoke values of γ in excess
of 20° at spin 18$^+$.

An interesting aspect of the CSM approach above is that the deformation
driving tendency is strongly affected by the location of the Fermi surface of
a nucleus. If it lies low in the shell, then the high-j quasiparticle has
its lowest energy for $\gamma > 0$. When the Fermi surface lies near mid shell, γ
tends to move to the negative sector, giving rise to triaxiality conforming
to collective rotations of an oblate ellipsoid. Admittedly, a nucleus whose
neutron Fermi surface lies near the middle of the $i_{13/2}$ shell should be a
well-deformed rotor with a pronounced minimum in its potential energy surface
and, therefore, it is expected to be relatively resistant to triaxial deformation driving influences.

In measuring the lifetimes of high-spin states in ^{172}W (N = 98, which
is near the middle of the $i_{13/2}$
shell), we did not anticipate a
significant difference in the collectivity at spins 18-20 in the yrast
sequence from that found low in the
ground band. At this point we
[RAO85] have only partially completed
the data anlayses, but to our surprise, the Q_t values of ^{172}W shown
in Fig. 5 display a very similar
behavior to the much softer nuclei
near N = 90. This result raises some
very interesting questions relating
to which degrees of freedom are
changing to produce this effect. Obviously, additional lifetime measurements of nuclei near N = 96-100 are
important; but the search for better
theories to account for such phenomena is equally demanded.

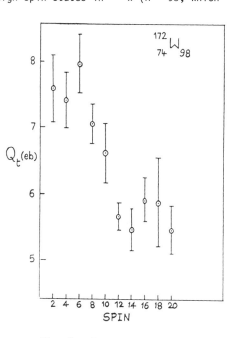

Fig. 5. Transition quadrupole
moments of yrast states in ^{172}W.

Acknowledgments

The following collaborators are to be thanked for their roles in carrying out the research discussed here: M.P. Fewell, F.K. McGowan, J.H. Hattula, I.Y. Lee, C. Baktash, Y. Schutz, J.W. Johnson, J.C. Wells, L.L. Riedinger, M.W. Guidry, S.C. Pancholi, M. Oshima, R.V. Ribas, M.N. Rao, K. Erb, J.W. McConnell and A. Larabee.

References

[BEN83] R. Bengtsson, Y-D. Chen, J-S. Zhang and S. Aberg, Nucl. Phys. A405, 221 (1983).

[FEW82] M.P. Fewell, J. Hattula, D.R. Haenni, N.R. Johnson, J.W. Johnson, I.Y. Lee, F.K. McGowan, L.L. Riedinger, J.C. Wells, and S.C. Pancholi, Bull. Am. Phys. Soc. 27, 522 (1982).

[FEW82] M.P. Fewell, J.S. Hattula, N.R. Johnson, I.Y. Lee, F.K. McGowan, H. Ower, S.C. Pancholi, L.L. Riedinger, and J.C. Wells, in Proc. of the Conf. on High Angular Momentum Properties of Nuclei, Oak Ridge, TN, Nov. 1982, ed. by N.R. Johnson, Nuclear Science Research Conf. Series, Vol. 4 (Harwood Academic, New York, 1983), p. 69.

[FEW85] M.P. Fewell, N.R. Johnson, F.K. McGowan, J.S. Hattula, I.Y. Lee, C. Baktash, Y. Schutz, J.C. Wells, L.L. Riedinger, M.W. Guidry, and S.C. Pancholi, Phys. Rev. C, 31, 1057 (1985).

[FRA83] S. Frauendorf and F.R. May, Phys. Lett. 125B, 245 (1983).

[JAM75] F. James and M. Roos, Comput. Phys. Commun. 10, 343 (1975).

[JOH81] N.R. Johnson, J.W. Johnson, I.Y. Lee, J.E. Weidley, D.R. Haenni, and J.R. Tarrant, ORNL Phys. Div. Prog. Report No. ORNL-5787, 1981, p. 147.

[JOH82] N.R. Johnson, in Proc. of the 1982 Inst. for Nuclear Study Intnl. Symposium on Dynamics of Nuclear Collective Motion, ed. by K. Ogawa and K. Tanahe (Inst. for Nuclear Study, Univ. of Tokyo, 1982), p. 144.

[LEA83] G.A. Leander, S. Frauendorf and F.R. May, Proc. Conf. on High Angular Momentum Properties of Nuclei, Vol. 4, Nuclear Science Research Conf. Series, ed. by N.R. Johnson (Harwood Academic Publishers, New York, 1983) p. 281.

[OSH84a] M. Oshima, N.R. Johnson, F.K. McGowan, I.Y. Lee, C. Baktash, R.V. Ribas, Y. Schutz, and J.C. Wells, Bull. Am. Phys. Soc. 29, 1043 (1984).

[OSH84b] M. Oshima, N.R. Johnson, F.K. McGowan, I.Y. Lee, C. Baktash, R.V. Ribas, Y. Schutz, and J.C. Wells, ORNL Phys. Div. Prog. Report, ORNL-6120 (1984) p., 77.

[RAO85] M.N. Rao, N.R. Johnson, C. Baktash, F.K. McGowan, I.Y. Lee, K. Erb, J.W. McConnell, M. Oshima, J.C. Wells, A. Larabee and L.L. Riedinger, unpub.

[STU76] R.J. Sturm and M.W. Guidry, Nucl. Instrum. Methods 138, 345 (1976).

[WEL85] J.C. Wells, M.P. Fewell and N.R. Johnson, ORNL/TM-9105 (1985).

RECEIVED July 15, 1986

Signature Splitting in ^{135}Pr

T. M. Semkow[1], D. G. Sarantites[1], K. Honkanen[1], V. Abenante[1], C. Baktash[2], Noah R. Johnson[2], I. Y. Lee[2], M. Oshima[2], Y. Schutz[2], C. Y. Chen[3], O. Dietzsch[3], J. X. Saladin[3], A. J. Larabee[4], L. L. Riedinger[4], Y. S. Chen[5], and H. C. Griffin[6]

[1]Washington University, St. Louis, MO 63130
[2]Oak Ridge National Laboratory, Oak Ridge, TN 37831
[3]University of Pittsburgh, Pittsburgh, PA 15260
[4]University of Tennessee, Knoxville, TN 37996-1200
[5]Joint Institute for Heavy Ion Research, Oak Ridge, TN 37831
[6]University of Michigan, Ann Arbor, MI 48109

In-beam spectroscopic study of ^{135}Pr was made using 91 MeV ^{120}Sn(^{19}F,4n) reaction. A strong negative parity proton band based on the $h_{11/2}^-$ 3/2[541] configuration with $(\pi,\alpha)=(-,-1/2)$ was observed. $(-,+1/2)$ unfavored band is probably observed. Also two positive parity proton bands are observed and are based on the $g_{7/2}^+$ 3/2[422] configuration with $(\pi,\alpha)=(+,\pm1/2)$. In the case of $\pi(+)$ bands the backbending is caused by the alignment of two $h_{11/2}^-$ 3/2[541] protons. For the $(-,-1/2)$ band the backbending is caused by the alignment of two $h_{11/2}^-$ protons: 3/2[541] and 1/2[550]. This is considerably different than in ^{134}Ce core, where the $h_{11/2}$ neutron-holes are aligning.

Nuclei in the vicinity of La, Ce, Pr are predicted to be soft in γ-deformation [CHE83]. An investigation was undertaken to study the signature splitting between rotational bands in ^{135}Pr, which is related to γ-deformation. ^{135}Pr was produced by the ^{120}Sn(^{19}F,4n) reaction at 91 MeV using the tandem accelerator at the Brookhaven National Laboratory. Coincidence data were recorded between 4 Compton-suppressed Ge detectors at $\sim 40°$ relative to the beam and 2 unsuppressed Ge detectors at $\sim 85°$. In addition an array of 11 NaI detectors around the target were used as a multiplicity selector.

The ^{135}Pr decay scheme is shown in Fig. 1 and it was constructed from $\gamma\gamma$-

0097-6156/86/0324-0305$06.00/0
© 1986 American Chemical Society

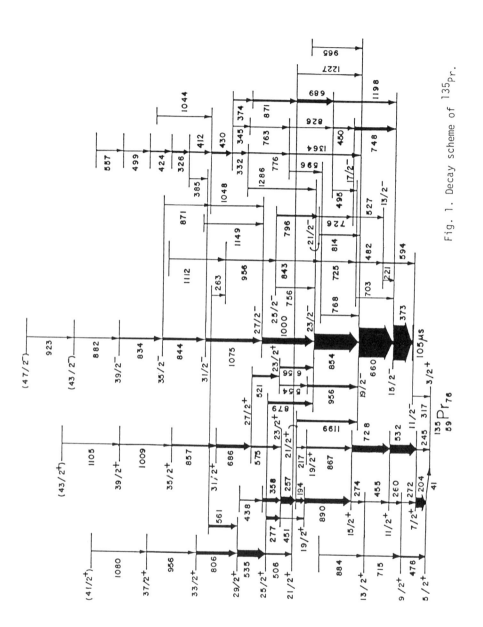

Fig. 1. Decay scheme of ^{135}Pr.

coincidence, intensity, and angular correlation information. Low-spin studies of ^{135}Pr from in-beam and radioactive decay work [KOR85, WIS75, CON73, EKS72] were valuable in assigning spins and parities of high-spin states. Most of the intensity goes to the proton band which is based on $h_{11/2}^-$ $3/2[541]$ configuration, with signature $\alpha=-1/2$. There is also some evidence for the $\alpha=+1/2$ unfavored band. The path through the 843, 725, and 482 keV cascade was chosen because the 482 keV transition was known before [KOR85], over more intense path through 796 and 726 keV cascade. The two bands are strongly decoupled because of the K=3/2 band head. However, by adding 4 neutrons the 13/2$^-$ level moves below 15/2$^-$ level in ^{139}Pr [PII80] due to the increase of γ from 21° to 34° (slightly oblate, see Fig. 2). The $\alpha=\pm1/2$ positive parity bands are based on $\pi g_{7/2}^+$ $3/2[422]$ configuration. There are several M1 connecting transitions observed. The M1 branching fractions from favored band are: 0.05 for the 274 keV, 0.12 for the 260 keV, and 0.91 for 204 keV transitions. On the other hand transitions from the unfavored to favored band are so weak, that they can be seen only from the favored band below. This is in agreement with B(M1)/B(E2) systematics for connected rotational bands [HAG82]. Above the backbending most of the intensity goes to the $(\pi,\alpha)=(-,-1/2)$ band similarly as it is observed in ^{133}Pr [HIL85].

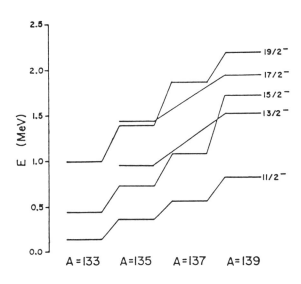

Fig. 2. Systematics of $\Pi(-)$ levels in odd-A Pr isotopes. References: A=133 [HIL85], A=135 [KOR85], A=137 [KLE75], A=139 [PII80].

Changes of γ-deformation in rotational nuclei are caused by the balance between driving forces resulting from the quasi-particle properties before backbending, the behavior of aligning quasi-particles at the backbend, the γ-softness of the core, and the collective rotation [LEA82, FRA83, CHE83, BEN84]. These changes in γ-deformation are thus observed as changes or even inversions [BEN84] of signature splitting. The core is expected to be driven by rotation towards $\gamma \sim -30°$ (in the Lund convention [AND76, LEA82]). Quasi-particles occupying high-j shell cause the driving force towards $\gamma > 0°$ when the Fermi level is at the beginning of the shell, towards $\gamma \leq 0°$ with Fermi level in the middle of the shell, and towards $\gamma < 0°$ with Fermi level close to the top of the shell [LEA82].

^{135}Pr is in the transitional region, where both the $h_{11/2}$ protons and $h_{11/2}$ neutron-holes can align. For example in the lighter Ce isotopes, $^{128-132}$Ce, the $h_{11/2}$ protons align causing $\gamma > 0°$ shift, while in the heavier ^{134,136,138}Ce the $h_{11/2}$ neutron holes were found to align causing $\gamma < 0°$ shift towards the oblate shapes. This has been concluded from both the spectroscopic [MUL78, MUL84], as well as from the g-factor [ZEM82] measurements.

Fig. 3 shows the aligned angular momentum for the three bands in ^{135}Pr. For the $\pi(+)$ bands the alignment is ~ 9.2 h at $\hbar\omega_c \cong 0.32$ MeV. Preliminary cranking-shell-model calculations show that this is consistent with the alignment of two $h_{11/2}^-$ 3/2[541] protons at $\hbar\omega_{AB} = 0.3$ MeV. The calculations were done with $\varepsilon_2 = 0.25$, $\varepsilon_4 = 0.02$, $\Delta_p = \Delta_n = 1.2$ MeV taken from systematics, and the crossing was found at $\gamma \cong 0°$. The $\pi(-)$ band crosses at $\hbar\omega_c \cong 0.46$ MeV with alignment of ~ 10 h. Using blocking arguments the calculation gives $\hbar\omega_{BC} \cong 0.46$ MeV with the alignment of two $h_{11/2}^-$ protons: 1/2[550] and 3/2[541]. At the higher frequencies both the $\pi(+)$ and $\pi(-)$ bands show an upbend. This may be due to the BC-alignment of protons in $\pi(+)$ bands and the AB-alignment of neutrons in $\pi(-)$ bands, with the increase of the frequency being related to higher interaction strengths. The alignment of $h_{11/2}$ protons in ^{135}Pr is surprising in comparison with ^{134}Ce and further calculations at the different γ's are in progress to see if another picture would be consistent with the Pr data. ^{133}Pr [HIL85] behaves similarly to ^{135}Pr, except the ω_c's are slightly lower and the $(-,-1/2)$ band shows upbend rather than the backbend.

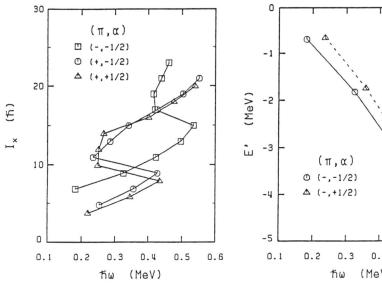

Fig. 3. Aligned angular momentum vs.
rotational frequency.

Fig. 4. Energy in the rotating frame
vs. rotational frequency.

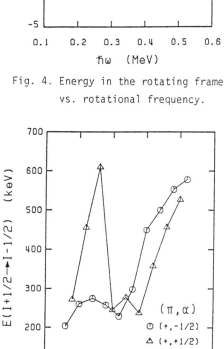

Fig. 5. Energy in the rotating frame
vs. rotational frequency.

Fig. 6. Energy between levels I+1/2
and I-1/2 vs. middle spin.

Experimental energies in the rotating frame, E', are plotted on Figs. 4 and 5 for the $\pi(-)$ and $\pi(+)$ bands, respectively. The signature splitting is defined as the difference between E' for opposite signatures. The signature splitting decreases with increasing ω for the $\pi(-)$ bands. The signature splitting for the $\pi(+)$ bands can be illustrated better in Fig. 6, where the energy difference between $\Delta I = 1$ levels is plotted vs. middle spin for both signatures [RIE83]. Large signature splitting before the backbending inverts after the backbending. This behavior is different than in ^{133}Pr, where the signature splitting is smaller and does not invert.

Acknowledgments

The authors wish to acknowledge L. A. Adler and O. El-Ghazzawy for making the computer codes available. This work is supported by the U. S. Department of Energy, Division of Nuclear Physics.

References

[AND76] G. Andersson et al., Nucl. Phys. A268 205 (1976)

[BEN84] R. Bengtsson et al., Nucl. Phys. A415 189 (1984)

[CHE83] Y. S. Chen et al., Phys. Rev. C 28 2437 (1983)

[CON73] T. W. Conlon, Nucl. Phys. A213 445 (1973)

[EKS72] C. Ekström et al., Nucl. Phys. A196 178 (1972)

[FRA83] S. Frauendorf and F. R. May, Phys. Letters 125B 245 (1983)

[HAG82] G. B. Hagemann et al., Phys Rev. C 25 3224 (1982)

[HIL85] L. Hildingsson et al., private communication (1985)

[KLE75] H. Klewe-Nebenius et al., Nucl. Phys. A240 137 (1975)

[KOR85] M. Kortelahti et al., Z. Phys. A 321 417 (1985)

[LEA82] G. A. Leander et al., Proceedings of the Conference on High Angular Momentum Properties of Nuclei, Oak Ridge (1982), p. 281

[MUL78] M. Müller-Veggian et al., Nucl. Phys. A304 1 (1978)

[MUL84] M. Müller-Veggian et al., Nucl. Phys. A417 189 (1984)

[PII80] M. Piiparinen et al., Nucl. Phys. A342 53 (1980)

[RIE83] L. L. Riedinger, Phys. Scripta 15 36 (1983)

[WIS75] K. Wisshak et al., Nucl. Phys. A247 59 (1975)

[ZEM82] A. Zemel et al., Nucl. Phys. A383 165 (1982)

RECEIVED July 17, 1986

Massive Transfer Reactions and the Structure of Transitional $A \simeq$ 100 Nuclei

D. R. Haenni, H. Dejbakhsh, and R. P. Schmitt

Cyclotron Institute, Texas A&M University, College Station, TX 77834

Band crossings and other features of the mass 100 transitional nuclei have been studied with massive transfer and fusion reaction based γ-ray spectroscopy. Experimental blocking argument results and calculations agree that $\nu h_{11/2}^2$ alignment is responsible for the band crossing in ^{102}Ru. The soft core of ^{103}Rh is influenced by the configuration of the odd proton; however, the mechanism for this is not clear.

High-spin features of the transitional nuclei in the mass 100 region (Z<50, N>50) are interesting but not well studied. There is an abrupt shape change between 58 and 60 neutrons. These nuclei are also transitional to three different closed orbital configurations (Z=40, Z=50, and N=50). Backbending is known for ^{104,106}Pd [GRA76] but only schematic comparisons with cranked shell model (CSM) predictions have been made [STA84]. The present discussion will center around ^{102}Ru and ^{103}Rh which have N=58 and are midway between the 40 and 50 proton orbital closures. Most of the nuclei in this mass region are difficult to study since they lie on the neutron rich side of the valley of stability. The investigation of high-spin phenomenon via in-beam γ-ray spectroscopy with the usual (HI,xnγ) fusion reactions is in general hindered by the lack of suitable targets.

Part of the spectroscopy program at TAMU is the development of the so called massive transfer (MT) or break-up fusion reactions for discrete-line structure studies. Basically these heavy-ion reactions involve the transfer of a projectile fragment to the target with the remaining energetic fragment being emitted in the forward direction. With MT reactions one tends to observe a better population of high-spin states when compared to the corresponding fusion reaction. Detection of the emitted light fragment can also provide a sort of exit channel filter. These features can be useful in the study of nuclei which are more neutron rich than normally accessible by

(HI, xnγ) reactions. Further details concerning γ-ray spectroscopy with
MT reactions can be found elsewhere [HAE82].

Experimentally particle-γ and particle-γ-γ coincidences are measured.
With beams heavier than Li a γ-ray multiplicity filter is needed to
select high-multiplicity MT events from other reactions which result in
energetic particles but low multiplicities. From a single MT spectroscopy
measurement it is possible to simultaneously obtain much of the usual

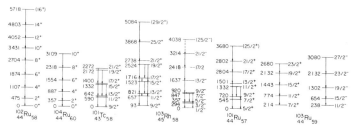

Fig. 1.   Partial level schemes for ^{102}Ru and neighboring nuclei.

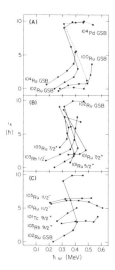

Fig. 2.   Particle alignment vs
rotational frequency for bands
^{102}Ru and neighboring nuclei.

experimental γ-ray data (singles, excitation function, cross bombardment, angular
distribution, and coincidence) for several
adjacent exit channels.

Partial level schemes for ^{102}Ru and
its neighbors are given in Fig. 1. Each
band contains new levels. These results
were obtained at the TAMU Cyclotron
Institute with ^{7}Li and ^{11}B MT reactions
induced on a metallic ^{100}Mo target at 49
and 77 MeV, respectively and conventional
(^{7}Li,xnγ) spectroscopy at 45 MeV. For
^{101}Ru other bands were also populated but
they could not be extended to higher spins
than previously reported [KLA82]. The
^{104}Ru level scheme agrees with recent
Coulomb excitation work [STA84].

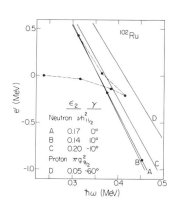

Fig. 3. Experimental routhian for ^{102}Ru compared with CSM.

The yrast cascade in ^{102}Ru shows a backbend with a crossing frequency of 0.37$\hbar\omega$ and an alignment gain of 9.5$\hbar$. The usual alignment plots [BEN79] in Fig. 2a compare ^{102}Ru with adjacent even-even nuclei. Except for ^{104}Pd the reference band for ^{102}Ru is used to generate these plots. Both ^{100}Ru [VOI76] and ^{104}Ru show upbends with the former at a higher frequency than ^{102}Ru. The experimental Routhian [BEN79] for ^{102}Ru is compared with CSM predictions (triaxial version) [FRA83] in Fig. 3 assuming either the $\nu h_{11/2}$ or $\pi g_{9/2}$ alignment. From these calculations the backbend in ^{102}Ru should arise from $\nu h_{11/2}$ alignment with a small positive ε_2 deformation and perhaps a slightly positive γ deformation.

This conclusion should be tested through blocking arguments based on the single quasiparticle bands in the adjacent odd-A nuclei. New data has been obtained through this work for all the neighbors of ^{102}Ru. In Fig. 2b bands based on $\nu d_{5/2}$, $\nu g_{7/2}$, and $\pi p_{1/2}$ exhibit crossings at about the same frequency as ^{102}Ru and thus are not the cause of the observed alignment. Bands based on the $\nu h_{11/2}$ orbital, Fig. 2c, block the ^{102}Ru backbend. Interpretation of the results for the $\pi g_{9/2}$ bands is not as straight forward. To first order a plot of the favored states for ^{101}Tc and ^{103}Rh indicate that this orbital also blocks the backbend.

In other transitional odd-A nuclei, however, band crossings are known to change the signature splitting (staggering between the favored and unfavored states) and the B(M1)/B(E2) ratios. This occurs when the odd nucleon and the aligning pair arise from the unique parity orbits in their respective shells. In both ^{81}Kr [FUN83] and ^{159}Tm [LAR84] for example the signature splitting is reduced above the band crossing and the B(M1)/B(E2) ratios are enhanced. Frauendorf [FRA84] has explained these features with the assumption that the unpaired nucleons produce configuration dependent γ deformations of the soft transitional core. For the mass 100 region an odd

$\pi g_{9/2}$ would generate a negative γ deformation and large signature splitting. Aligning a $\nu h_{11/2}$ pair with it would push γ to near 0° and remove the signature splitting. A $\Delta I=1$ band feeds the $19/2^+$ and $21/2^+$ levels of the $\pi g_{9/2}$ band in ^{103}Rh and may represent such a $\pi g_{9/2}\nu_{11/2}^2$ band. This band

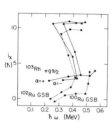

Fig. 4. Particle alignment vs rotational frequency for the $\pi g_{9/2}$ band in ^{103}Rh assuming a change in signature splitting.

has levels at 3396, 3630, 3937, 4320, 4705, 5195, 5662, and 6205 keV with spins $23/2^+$ to $37/2^+$, respectively. An alignment plot for this band crossing with the $\pi g_{9/2}$ band is shown in Fig. 4. A backbend occurs at a frequency consistent with ^{102}Ru. The $\pi g_{9/2}$ band is observed above the band crossing region. The start of an up-bend is found above 0.5$\hbar\omega$. The present level scheme for ^{101}Tc does not extend high enough to show a similiar crossing band.

The complexity of our angular distribution spectra do not permit reliable extraction of mixing ratios for the $I\to I-1$ transitions in the $\pi g_{9/2}$ band. Assuming that these are pure M1, a 10 fold increase in the experimental B(M1)/B(E2) is observed between the one and three quasiparticle bands. Donau and Frauendorf [DON83] have suggested that since the rotation axis is not the same as the alignment axis in a high K band, particle alignment can enhance M1 transition rates. This occurs when the g factor (g_i-g_R) has an opposite sign for the initial and the aligning quasiparticles. They gave a semi classical expression for the B(M1)/B(E2) ratios. With a strong coupled $g_{9/2}$ proton (K=9/2) this expression gives a factor of 3 increase in the B(M1)/B(E2) ratio over the same range of spin.

The signature splitting observed in the $g_{9/2}$ band does not necessarily imply configuration dependent γ deformations as proposed by Frauendorf. A similiar splitting in ^{105,107}Ag was reproduced [POP79] with a quasiparticle-plus-rotor (SR) model including a VMI treatment of the soft core. The level scheme for ^{103}Rh shows many of the same features as the Ag isotopes but is more complete. To explore this further the levels in ^{103}Rh have been calculated using the SR, IBFM-1 [JOL85], and triaxial rotor (AR) [MAY 75] models. Parameters for the SR model were chosen in a manner similiar to the

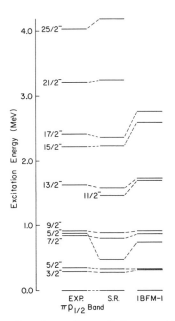
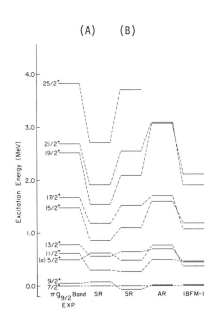

Fig. 5. Comparison of the negative parity $\pi p_{1/2}$ band in ^{103}Rh with model calculations.

Fig. 6. Comparison of the positive parity $\pi g_{9/2}$ band in ^{103}Rh with model calculations.

Ag isotopes [POP79] but using an average ^{102}Ru - ^{104}Pd core. The IBFM-1 parameters are from a recent calculation [JOL85] for odd-A Rh isotopes assuming that Rh is a hole in the corresponding Pd core. Parameters for the AR model were again based on an average ^{102}Ru - ^{104}Pd core.

Results of these calculations are compared to ^{103}Rh in Figs. 5, 6. For the negative parity levels the SR and IBFM-1 models produce a reasonable agreement with the data. The fits to the $g_{9/2}$ band with the same parameter sets [(A) for the SR model] are not as good and appear to indicate a softer core than observed. The AR model predicts signature splitting for the $g_{9/2}$ band as do the other models but its assumption of a rigid rotor core is not appropriate for a transitional nucleus. The SR model fit to the $g_{9/2}$ band can be improved (B) if the deformation ε_2 is decreased (0.2-0.15) and the VMI parameter C is increased (0.006-0.02 MeV^3). These changes are in the direction of a stiffer more spherical core.

With core parameters derived from neighboring even-even nuclei both the SR and IBFM-1 models generate a similiar pattern of results. The $p_{1/2}$ band is more or less reproduced while the $g_{9/2}$ band is compressed. This provides evidence that the soft core in ^{103}Rh is influenced by the odd proton. It is more apparent for the $g_{9/2}$ band than the $p_{1/2}$ band. The cause for this is not clear. It may arise from γ deformation or perhaps results from n-p interactions. This interaction has already been shown [FED79] to be the deformation driving force in this mass region with the $\pi g_{9/2}$ and $\nu g_{7/2}$ orbitals playing key roles. Bands based on the $\pi g_{9/2}$ orbital would likely show effects from this interaction. More data, particularly transition rates for the $g_{9/2}$ band and better models for such transitional odd-A nuclei are needed to draw more definite conclusions.

References

[BEN79] R.Bengtsson and S. Frauendorf, Nucl. Phys. A327 139 (1979).

[DON83] F. Dönau and S. Frauendorf, High Angular Momentum Properties of Nuclei edited by N. R. Johnson (Harwood Academic, New York) 281 (1983).

[FED79] P. Federman and S. Pittel, Phys. Rev. C 20 820 (1979).

[FRA83] S. Frauendorf and F. R. May, Phys. Lett. 125B 245 (1983).

[FRA84] S. Frauendorf, International Symposium on In-Beam Nuclear Spectroscopy, Debrecen, Hungary (1984).

[FUN83] L. Funke et al., Phys. Lett. 120B 301 (1983).

[GRA76] J. A. Grau et al., Phys. Rev. C 14 2297 (1976).

[HAE82] D. R. Haenni et al., Phys. Rev. C 25 1699 (1982).

[JOL85] J. Jolie et al., Nucl. Phys. A438 15 (1985).

[KLA82] W. Klamra et al., Nucl. Phys. A376 463 (1982).

[LAR84] A. J. Larabee et al., Phys. Rev. C 29 1934 (1984).

[MAY75] J. Mayer-ter-Vehn, Nucl. Phys. A249 111 (1975). ibid. 141.

[POP79] Rakesh Popli et al., Phys. Rev. C 20 1350 (1979).

[STA84] J. Stachel et al., Nucl. Phys. A419 589 (1984).

[VOI76] M. Λ. J. de Voight et al., Nucl. Phys. Λ270 141 (1976).

RECEIVED July 14, 1986

High-Spin Structure of [163]Lu

K. Honkanen[1], H. C. Griffin[2], D. G. Sarantites[1], V. Abenante[1], L. A. Adler[1], C. Baktash[3], Y. S. Chen[6], O. Dietzsch[4], M. L. Halbert[3], D. C. Hensley[3], Noah R. Johnson[3], A. J. Larabee[5], I. Y. Lee[3], L. L. Riedinger[5], J. X. Saladin[4], T. M. Semkow[1], and Y. Schutz[3]

[1]Washington University, St. Louis, MO 63130
[2]University of Michigan, Ann Arbor, MI 48109
[3]Oak Ridge National Laboratory, Oak Ridge, TN 37831
[4]University of Pittsburgh, Pittsburgh, PA 15260
[5]University of Tennessee, Knoxville, TN 37996-1200
[6]Joint Institute for Heavy Ion Research, Oak Ridge, TN 37831

The $\pi 9/2^-[514]$ ground band structure of [163]Lu shows a large signature splitting with inversion above the backbend suggesting a shape change associated with the triaxiality degree of freedom. The $\pi 1/2^-[541]$ band shows no backbending up to $\hbar\omega = 0.4$ MeV indicating a more deformed structure.

Recently, the cranked shell model has been very successful in interpreting the spectra of the deformed nuclei at high spins. In this picture, the nucleus is assumed to have an intrinsic nonspherical shape and to rotate adiabatically around a space fixed axis.[BEN79] Traditionally, most experimental and theoretical efforts were focused on the asphericity parameter ε_2 and its influence on the rotational spectra. In contrast, only very recently attention has been directed to the triaxiality parameter γ, which plays an important role on the evolution of shapes as a function of spin. More thorough investigations are needed to provide a detailed understanding of the behavior of nuclei at high spins as a function of γ.

It has been pointed out by several authors that the population of rotationally aligning high-j orbitals may provide a sensitive probe of the triaxiality parameter γ, and could be used to trace the evolution of γ with angular momentum [BEN83], [FRA83], [LEA83]. These observations are based on .the fact that a particle (hole) in a high-j orbital polarizes the core toward

0097-6156/86/0324-0317$06.00/0
© 1986 American Chemical Society

oblate (prolate) shape; whereas, a quasi-particle in a half-filled high-j
orbital shell generates the intermediate situation leading to a triaxial
shape. These expectations where confirmed in a cranked shell-model calculation by Leander [LEA83], who concluded that in a γ-soft nucleus the presence
of a high-j quasi-particle with favored signature will stabilize the shape of
the nucleus at a γ value that depends initially on the position of the Fermi
level in the shell. For non γ-soft nuclei that have a potential-energy minimum at some other γ value the quasi-particle exerts a driving force toward
positive γ values when the shell is less than half full or toward negative γ
values when it is more than half full. In contrast the unfavored signature
states are much less sensitive to variations of γ and λ, the position of the
Fermi level in the shell. This results in signature splittings that become a
sensitive measure of the variation of γ-deformation with spin and λ.
Inversion of signature splitting above the backbend in $^{157}_{67}Ho_{90}$ [HAG82] and in
$^{159}_{69}Tm_{90}$ [RIE83] in the $\pi7/2^-[523]$ band was explained by the positive driving
influence of the two $i_{13/2}$ neutrons above the backbend which results in a
change from negative to slightly positive γ above the backbend.

In order to explore the possibility of such effects in the γ plane as a
function of the proton number we have studied the high spin structure of
$^{163}_{71}Lu_{92}$ which has two protons and two neutrons more than ^{159}Tm. In this case
the $\pi9/2^-[514]$ and possibly the $\pi7/2^+[404]$ orbitals would be predominantly populated. The former of these may be a new candidate for such a signature inversion.

The high spin states in ^{163}Lu were populated via the $^{122}Sn(^{45}Sc,4n)$ reaction at 192 MeV with the tandem accelerator at the Holifield Heavy-Ion
Facility. A stack of four (250 $\mu g/cm^2$ each) ^{122}Sn self-supporting foils were
used as a target. A short run with a thicker (1.2 mg/cm^2) target gave similar energy resolution in the Ge γ-ray energy spectra.

The γ-ray spectroscopic information was obtained using an array of five
Ge detectors with pentagonal NaI anti-Compton shields located at 63° to the
beam and three additional Ge detectors at 24°. Two-fold or higher coincident
events from these detectors were used to trigger the 72 NaI detectors of the
Spin Spectrometer (SS) at ORNL. [JAA83] An average Compton suppression factor of 3.5 for the ^{60}Co spectrum was obtained. The Ge detectors were placed
at 20.8 cm from the target.

The event tapes were processed in order to (1) linearize and match the gains of the NaI detectors, (2) separate the neutron from the γ pulses in the SS, and (3) construct the total pulse height, H, and the coincidence fold k. The processed events were then sorted by placing 2-dimensional gates in (H,k) space in order to separate the cascades associated with the 4n from the 5n channel. This procedure reduces the background from the competing channels by a factor of 2 or more. The gated events were sorted into an Eγ-Eγ matrix, which was corrected for Ge detector efficiency. A two-dimensional background subtraction procedure was developed to subtract the Compton-Compton and peak-Compton backgrounds from each gate on the observed peaks. Angular correlation information was obtained from a matrix with the 24^0 and 63^0 detectors on separate axes.

In Fig. 1 are shown three band structures believed to be associated with ¹⁶³Lu. Energy systematics in the heavier odd A Lu isotopes show that the π9/2⁻[514] band would be the yrast band, although the π1/2⁺[411] decreases rapidly in energy with decreasing A relative to the π7/2⁺[404] state and in the ¹⁶⁵Lu and ¹⁶³Lu may become the ground state. In ¹⁶⁵Lu [JON84] the π9/2⁻[514] band is the yrast band and in ¹⁶³Lu it is also the strongest band populated in the present work. The decay scheme shown in Fig. 1 is preliminary and further analysis is in progress to connect the different structures based on observed coincident γ-rays. The transitions from all these structures have yields associated with (H,k) distributions characteristic of the (⁴⁵Sc,4n) reaction and thus were assigned to ¹⁶³Lu.

There is a striking similarity between some of the bands in ¹⁶³Lu and ¹⁶⁵Lu. In ¹⁶⁵Lu the 9/2⁻[514] band was seen up to I = 43/2⁻ and it has a significant signature splitting [JON84]. Based on systematics from the odd-A Lu isotopes, we have assigned this yrast band as the π9/2⁻[514]. The back-bending in this region is due to breaking and alignment of an $i_{13/2}$ neutron pair, which places the backbending at about 12 units of rotational angular momentum. A second band was assigned as π1/2⁻[541] based on systematics and in analogy with ¹⁶⁵Lu. A third band structure shown in Fig. 1 is most likely a part of the 7/2⁺[404] band.

The favored and unfavored bands (α = $-\frac{1}{2}$ and $\frac{1}{2}$) gain 7.2 and 8.2 units of angular momentum above the backbend, respectively (Fig. 2a). This is con-

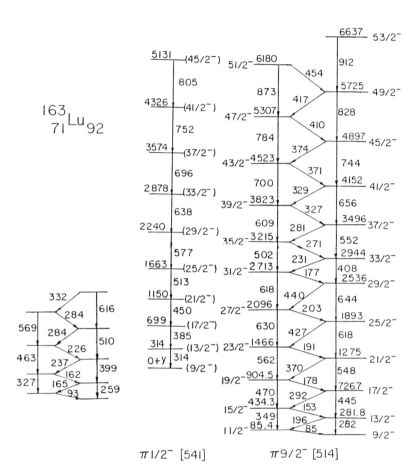

Fig. 1   Tentative decay scheme for the three band structures assigned to
          163Lu. The spin 9/2- for the lowest observed level in the π9/2-[514]
          band was assumed based on systematics.

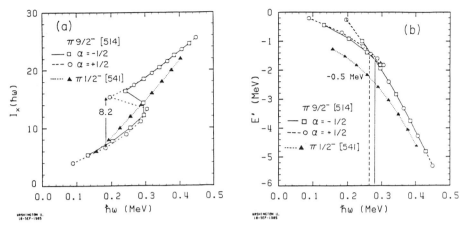

Fig. 2. Panels a and b show the aligned angular momentum and the energy in the rotating frame (E' = E_{Level} - $\hbar\omega I_x$) for the bands indicated.

sistent with the decoupling and alignment of an $i_{13/2}$ neutron pair. The crossing frequencies for the $\alpha=-\frac{1}{2}$ and $\alpha=\frac{1}{2}$ bands are 0.263 and 0.280 MeV, respectively, as is seen from Fig. 2b. Below the backbend the $\pi 9/2^-[514]$ band shows a large negative signature splitting which increases with ω from ~ 55 to 120 keV (Fig. 2b), but above the backbend a signature inversion occurs with a splitting of 10 keV which remains constant with increasing ω. This is seen in detail in Fig. 3.

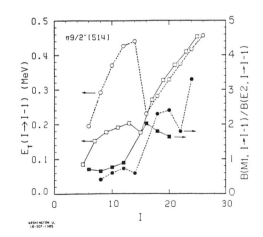

Fig. 3 Signature splitting indicated by the differences in Eγ (I→I-1) values for transitions starting from the $\alpha=-\frac{1}{2}$ levels (open squares) and the $\alpha=\frac{1}{2}$ levels (open circles). The corresponding full points give the B(M1,I→I-1)/B(E2, I→I-2) ratios obtained from branching ratios assuming $\delta=0$.

NUCLEI OFF THE LINE OF STABILITY

A cranked shell-model calculation for ^{163}Lu gave a $\sim$ 80 keV negative signature spitting for the quasi-protons for a γ value of $-10°$ at $\hbar\omega$ = 0.24 MeV. A signature inversion is predicted for the quasi-proton configuration at $\gamma \approx 0°$. Above the backbend the two $i_{13/2}$ quasi-neutrons are driving toward positive γ values. When the routhians for these 3-quasiparticles are added and the effect of the core is included, a γ value for this con-figuration can be obtained. Qualitatively, it appears that a significant shape change from negative to slightly positive γ values is present in this case. Thus, the signature inversion in the $\pi9/2^-[514]$ band is quite similar to that observed in ^{155}Ho on ^{159}Tm where the $7/2^-[523]$ quasi-proton con-figuration was pushed toward slightly positive γ values by the positive driving influence of the two strongly aligning $i_{13/2}$ quasi-neutrons.

The gradual gain in alignment of the $\pi1/2^-[541]$ band with increasing ω (Fig. 2a) suggests that either the aligning quasiparticles exhibit a strong interaction, or that a larger ϵ_2 value compared to 0.2 for the $\pi9/2^-[514]$ band and possibly a different γ value are needed in order to explain its behavior.

Acknowledgments

Work supported by the U. S. Department of Energy, Division of Nuclear Physics.

References

[BEN79]   R. Bengtsson and S. Frauendorf, Nucl. Phys. A 314, 27 (1979); A314, 139 (1979).

[BEN83]   R. Bengtsson, et al. Bormio, 1982 p. 144; and in "High Angular Momentum Properties of Nuclei", Ed. N. R. Johnson (Harwood, New York, 1983) p. 161.

[FRA83]   S. Frauendorf and F. R. May, Phys. Lett. 125B, 245 (1983).

[LEA83]   G. A. Leander, et al. in "High Angular Momentum Properties of Nuclei", Ed. N. R. Johnson (Harwood, New York, 1983), p. 281.

[HAG82]   G. B. Hagemann, et al. Phys. Rev. 25C, 3324 (1982).

[JAA83]   M. Jääskeläinen et al., Nucl. Instr. 204, 385 (1983).

[JON84]   S. Jonsson et al. Nucl. Phys. A422, 397 (1984).

RECEIVED April 11, 1986

Shape Competition and Alignment Processes in Light Au and Pt Nuclei

L. L. Riedinger [1], A. J. Larabee [1], and J.-Y. Zhang [2]

[1]University of Tennessee, Knoxville, TN 37996-1200
[2]Joint Institute for Heavy Ion Research, Oak Ridge, TN 37831

Calculations are presented and data are reviewed on the properties of the high-j states in the light Au nuclei. Both prolate and oblate structures are observed in this region. It is found that the collective model describes well the band- head and the high-spin properties of the $h_{9/2}$ and $i_{13/2}$ proton states, without resort to an "intruder state" phenomenology.

There has been much discussion at this and other conferences about the existence of nuclear intruder states, i.e. levels outside the model space for a given nucleus. There is ample evidence [HEY83] for the existence of intruder states at or near closed shells, which demonstrates the importance of the varying particle-hole composition of these states. As one proceeds away from the closed shell by adding or subtracting nucleons, at some point there occurs normal deformed nuclei adequately described by the Nilsson model, in which case the shell-model particle-hole structure of states becomes irrelevant. One important question is for how many nucleons beyond a closed shell does the intruder description of certain states give way to the collective description. We pursue here this question for the Z = 79 isotopes of gold and describe experiments and calculations performed on both the bandhead energies of high-j states (occasionally called intruders) and the bands built on these states. We find that the collective picture adequately describes these levels in a transition region of competing nuclear shapes. While oblate structures occur in the heavier gold isotopes, the lighter ones are dominated by prolate bands.

Our group has performed measurements on a number of Pt and Au nuclei. High-spin states in ^{185,186}Au [LAR85] and in ^{185}Pt [WAD85] have been studied at the McMaster University Tandem Accelerator with ^{19}F induced reactions. An array of 5 Ge and 6 NaI counters was used to collect gamma-gamma coincidence data. Angular distribution measurements were also performed. Bands in ^{183,184}Pt were studied [CAR85] with a ^{34}S induced reaction at the Holifield Heavy Ion Research Facility. The Spin Spectrometer was used with 9 Ge counters, six of which were Compton suppressed with NaI annuli.

The systematics of the prolate and oblate states observed in the light

0097-6156/86/0324-0323$06.00/0
© 1986 American Chemical Society

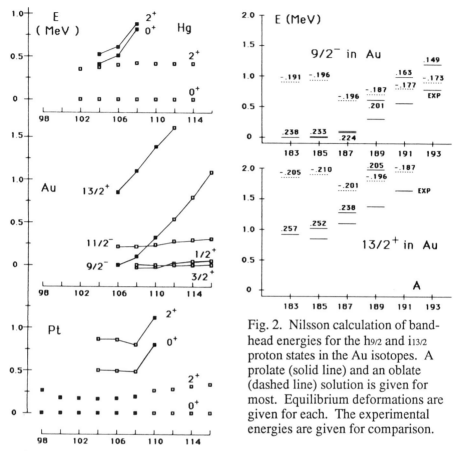

Fig. 2. Nilsson calculation of band-head energies for the $h_{9/2}$ and $i_{13/2}$ proton states in the Au isotopes. A prolate (solid line) and an oblate (dashed line) solution is given for most. Equilibrium deformations are given for each. The experimental energies are given for comparison.

Fig. 1. Experimental systematics of some prolate (closed squares) and oblate (open squares) states in Hg (Z=80), Au (79), and Pt (78) nuclei.

Hg, Au, and Pt nuclei are shown in Fig. 1. The first 2^+ state of the Hg isotopes is remarkably constant in energy and has been described as a slightly deformed oblate state. As mapped in detail in UNISOR measurements [HAM75], there exist rapidly falling prolate states in $^{184-188}$Hg. The heavy Pt isotopes are evidently similar in ground properties to the Hg nuclei, but the isotopes below N = 110 may be prolate in shape, judging by the structure of the $i_{13/2}$ band in adjacent odd-N Pt isotopes. As summarized by Wood [WOO81], low-lying 0^+ states in $^{182-186}$Pt may be the oblate states similar in structure to the Hg ground configuration. Thus, Z = 79 around N = 106 appears to be the point of crossing for the oblate and prolate minima. As seen in Fig. 1, the $\pi h_{9/2}$ and $\pi i_{13/2}$ states fall rapidly in the light Au nuclei, which suggests the intruder description, i.e. a different behavior of the normal 3-hole states ($h_{11/2}$, $d_{3/2}$, $s_{1/2}$) compared to the 1-particle 4-hole levels ($h_{9/2}$, $i_{13/2}$). The structure of the bands built on these states suggests the prolate or oblate nature described in Fig. 1.

A Strutinsky-type Nilsson calculation of the bandhead energies for the Au isotopes is shown in Fig. 2. The parameters used in this Nilsson calculation were those suggested by the Lund group [BEN85], except for $\mu = 0.52$ for the N = 5 and N = 6 proton shells. This value better describes levels in this region and became necessary in order to explain in ^{185}Au the newly found 1/2[530] band, based partially on the $f_{7/2}$ shell state [LAR85]. As seen in Fig. 2, there are both prolate and oblate minima calculated for each of the quasiparticle states. While the oblate $h_{9/2}$ and $i_{13/2}$ states are rather constant in energy as a function of N, the prolate minimum falls rapidly, becoming lower at ^{189}Au for the former and at ^{187}Au for the latter. These calculations match rather well the observed trend (see Fig. 1), and suggest that it is changing deformation which is responsible for the "intruding" high-j states, as opposed to the differing particle-hole character. This collective explanation requires a deformation of -0.173 for the lowest $h_{9/2}$ state in ^{193}Au, but unfortunately the band structure built on this state is unknown and thus it is difficult to surmise if this deformation is reasonable. Calculations with a Woods-Saxon potential give results very similar to those shown in Fig. 2 [NAZ85].

It is clear from the calculations of Fig. 2 that the band structures seen in the light Au nuclei should be dominated by prolate shapes, as is observed experimentally in ^{185}Au [LAR85]. Rotation-aligned bands are built on $h_{9/2}$, $i_{13/2}$, and $f_{7/2}$ Nilsson states, indicative of K = 1/2 bands and prolate shapes. However, the stucture built on the $h_{11/2}$ bandhead is more complex, as shown in the partial level scheme of Fig. 3. The 220-keV bandhead is known [BER83] to be a 26 ns isomer, explained as due to an oblate $h_{11/2}$ to prolate $h_{9/2}$ transition. The sequence of E2 transitions to the right in Fig. 3 is similar to that seen in the heavier odd-A Au nuclei [GON79], and is explained as the weak coupling of the $h_{11/2}$ hole state to the slightly oblate Hg core. However, the sequence of levels beginning at 1210 keV is different from anything seen in the heavier isotopes. The strongly-coupled nature of this new band suggests a prolate $h_{11/2}$ structure, which is expected to lie close in energy to the oblate band according to the calculations. The complicated structure shown in Fig. 3 is therefore quite compatible with the predicted near degeneracy of prolate and oblate bands in ^{185}Au.

Four bands are observed to high spins in ^{185}Au and are shown in the

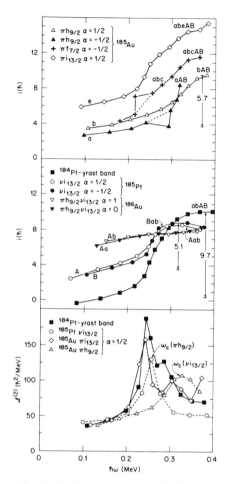

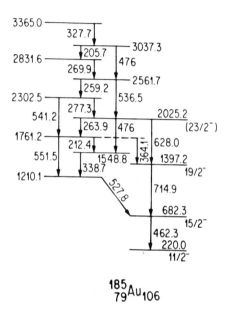

Fig. 3. A partial level scheme for ^{185}Au emphasizing the structure built upon the $h_{11/2}$ state [LAR85]. The 220-keV state has an isomeric decay of 26 ns to the $h_{9/2}$ level.

Fig. 4. In the top two panels, the experimental alignment (i) as a function of rotational frequency for bands in ^{185,186}Au [LAR85], ^{184}Pt [CAR85], and ^{185}Pt [WAD85]. Reference parameters of $\mathscr{I}_o$= 22 $\hbar^2$/MeV and $\mathscr{I}_1$ = 110 $\hbar^4$/MeV^3 are used. In the bottom panel, the second moment of inertia is plotted versus frequency for selected bands.

alignment vs. rotational frequency graph of Fig. 4. The α = -1/2 negative parity bands cross and interact around I = 23/2, which explains the perturbations in those two bands in Fig. 4. Alignment gains of different magnitudes are observed in each of these bands, suggesting differing alignment processes taking place.

By comparison, the yrast band of the core nucleus, ^{184}Pt, is shown in the second panel of Fig. 4, as are the two signatures of the 9/2[624] i13/2 band in ^{185}Pt. It was previously thought by some (e.g. [BES76]) that the crossing in ^{184}Pt resulted from the alignment of i13/2 neutrons, as is the case throughout the N=90 to 106 deformed region. On the other hand, our earlier measurements on ^{185}Au [KAH78] suggested that the crossing in ^{184}Pt had to result from πh9/2 alignment. It is now apparent from the data shown in Fig. 4 that both of these quasiparticle alignments take place at nearly the same frequency. The third panel in Fig. 4 contains a plot of the second moment of inertia versus frequency, which more sensitively shows crossings for these bands. The yrast band of ^{184}Pt and the πi13/2 band of ^{185}Au have two peaks in this second-moment plot. The γi13/2 band of ^{185}Pt shows only the first peak, while the πh9/2 band of ^{185}Au displays the second. These data thus indicate the existence of a πh9/2 crossing at ℏω = 0.25 MeV with a Δi = 5.1 ℏ, and a γi13/2 crossing at ℏω = 0.31 MeV with a Δi = 5.7 ℏ. The yrast band of ^{184}Pt is affected by both crossings, which explains the large alignment gain (9.7 ℏ) compared to the neighboring odd-A nuclei. Our measurements on ^{186}Au [LAR85] indicate a πh9/2γi13/2 band which shows no crossing, since both alignment processes are blocked (see Fig. 4).

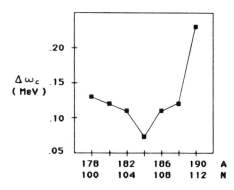

Fig. 5. Cranked Shell Model frequency for the proton h9/2 crossing minus that for the neutron i13/2 crossing plotted as a function of neutron number for the Pt isotopes.

Cranked Shell Model calculations have been performed to learn if these two close-lying band crossings can be explained with reasonable shape parameters. The difference in the calculated crossing frequencies is shown in Fig. 5 as a function of neutron number for the Pt isotopes. Values of ε_2 and ε_4 were obtained from a potential-energy-surface calculation (with χ and μ values as given above) and then used in the CSM calculation. The minimum in the curve occurs for ^{184}Pt, in agreement with the data. No attempt has yet been made to exactly reproduce these experimental crossing frequencies, since other parameters such as pairing gaps are poorly known in this region. Nevertheless, the calculations do show that it is reasonable for the $\pi h_{9/2}$ and γ $i_{13/2}$ crossings, rather widely split in other nuclei, to be very close in frequency in ^{184}Pt. In conclusion, our data on high-spin states in ^{185,186}Au and in ^{184,185}Pt are mostly indicative of band structures and alignment processes in prolate nuclei. The oblate $h_{11/2}$ band is observed, but is crossed at intermediate spin by what is probably a prolate $h_{11/2}$ structure. Collective model (Nilsson) calculations of bandhead properties of the Au isotopes agree well with the observed systematics of the "intruding" $h_{9/2}$ and $i_{13/2}$ proton states, calling into question the particle-hole interpretation of these states.

Acknowledgments

The authors wish to thank Witek Nazarewicz and Jim Waddington for many helpful discussions. Research at the University of Tennessee is supported by the U.S. Department of Energy under contract No. DE-AS05-76ERO- 4936. The Joint Institute for Heavy Ion Research has as member institutions the Univ. of Tennessee, Vanderbilt Univ., and the Oak Ridge National Laboratory; it is supported by the members and by the U.S. DOE.

References
[BEN85] T. Bengtsson and I. Ragnarsson, Nucl. Phys. A436 14 (1985).
[BER83] V. Berg, Z. Hu, J. Oms, and C. Ekström, Nucl. Phys. A410 445
         (1983).
[BES76] S. Beshai, K. Fransson, S.A. Hjorth, A. Johnson, T. Lindblad, and J.
         Szarkier, Z. Physik A277 351 (1976).
[CAR85] M.P. Carpenter, C.R. Bingham, L.H. Courtney, S. Juutinen, A.J.
         Larabee, Z.M. Liu, L.L. Riedinger, C. Baktash, M.L. Halbert,
         N.R.Johnson, I.Y.Lee, Y. Schutz, A. Johnson, J. Nyberg, K.
         Honkanen, D.G. Sarantites, Bull. Am. Phys. Soc. 30 762 (1985).
[GON79] Y. Gono, R.M. Lieder, M. Muller-Veggian, A. Neskakis, C.
         Mayer-Böricke, Nucl. Phys. A327 269 (1979).
[HAM75] J.H. Hamilton et al., Phys. Rev. Lett. 35 562 (1975).
[HEY83] K. Heyde, P. Van Isacker, M. Waroquier, J.L. Wood, and R.A.
         Meyer, Phys. Rpts. 102 291 (1983).
[KAH78] A.C. Kahler, L.L. Riedinger, N.R. Johnson, R.L. Robinson, E.F.
         Zganjar, A. Visvanathan, D.R. Zolnowski, M.B. Hughes, and T.T.
         Sugihara, Phys. Lett. 72B 443 (1978).
[LAR85] A.J. Larabee, M.P. Carpenter, L.L. Riedinger, L.H. Courtney, J.C.
         Waddington, V.P. Janzen, W. Nazarewicz, J.Y. Zhang, R. Bengtsson,
         and G.A. Leander, submitted to Phys. Lett.
[NAZ85] W. Nazarewicz et al., to be published.
[WAD85] J.C. Waddington, S. Monaro et al., to be published.
[WOO81] J.L. Wood, 4th International Conference on Nuclei Far From
         Stability, Helsingør, 1981, pg 612.

RECEIVED July 31, 1986

Rotational Bands in Deformed Odd-Odd Nuclei

William C. McHarris, Wen-Tsae Chou, Jane Kupstas-Guido, and Wade Olivier

National Superconducting Cyclotron Laboratory and Departments of Chemistry and of Physics and Astronomy, Michigan State University, East Lansing, MI 48824

In-beam γ-ray studies of deformed odd-odd nuclei in the neutron-deficient Re region produce simpler spectra than expected. Almost all of the deexcitation feeds through a few rotational bands based on high-Ω proton and/or neutron states (K itself is not necessarily large). A similar phenomenon is found with high-spin bands in the near-closed-shell Sb nuclei. This can be explained on the basis of highly-aligned single-particle states being most efficient at sharing large amounts of angular momentum with the collective core modes: Once locked into such states, the nucleus tends to deexcite through related, highly-aligned states, thus picking out a select subset of possible states. In addition, such odd-odd nuclei tend to have multi-particle states at lower energies than other nuclei, often with the particles forming a sort of oblate girdle about the prolate core -- these can be thought of as oblate-|| pseudo-rotational bands. Very-heavy-ion beams are necessary to bring in enough angular momentum to excite much in the way of high-spin collective modes in these odd-odd nuclei.

Until recently in-beam γ-ray spectroscopy of odd-odd nuclei was thought to be unrewardingly cumbersome. Spectra were unduly messy, and the resulting level schemes were mostly lists of numbers, for it was difficult to assign quantum numbers, much less meaningful configurations to the states. During the last several years, however, a number of experimental groups have discovered that, under certain conditions, in-beam γ-ray studies of odd-odd nuclei produce far simpler spectra than expected, with the resulting level schemes yielding a considerable amount of worthwhile information about details of nuclear structure. The necessary conditions appear to be deformed nuclei and/or high-spin states.

In Fig. 1 we show the level scheme produced by the ^{181}Ta(α,3nγ)^{182}Re reaction [SLA84]. This reaction brings in only a moderate amount of angular momentum; yet most of the deexcitation proceeds through only two high-spin rotational bands, the remainder through two additional low-spin but high-Ω bands. All four bands show considerable distortion -- in particular, a compressed A-term and a non-negligible positive B-term when their energies are fitted to the standard equation,

$$E_J = E_0 + AJ(J + 1) + BJ^2(J + 1)^2.$$

This type of distortion is characteristic of Coriolis coupling, where the matrix elements have the form,

0097-6156/86/0324-0329$06.00/0

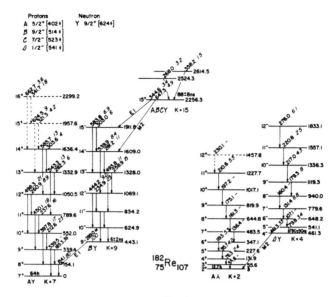

Fig. 1

$$-\hbar^2/2\mathscr{J}[J(J + 1) - \Omega(\Omega \pm 1)]^{1/2}[J(J + 1) - K(K \pm 1)]^{1/2}$$

Thus, a perturbation expansion results in a negative A-term (effectively increasing the moment of inertia, so that $\hbar^2/2\mathscr{J}$ appears smaller than normal) and a positive B-term. If we examine the single-particle states involved in these bands, we find that the neutron state for all of them is the $9/2^+[624]$ Nilsson state, which originates from the $i_{13/2}$ spherical state and has very large Coriolis matrix elements; also, three of the four proton states originate from the $h_{11/2}$ or $h_{9/2}$ spherical states and have large Coriolis matrix elements. It is these states that decouple and align their spins with that of the rotating core to produce "backbending" [STE72]. In other words, an efficient mechanism for carrying large amounts of angular momentum is for it to be shared between $\vec{R}$ of the core and $\vec{j}$ of the single nucleons. In even-even nuclei this requires the uncoupling of a pair, but in odd-odd nuclei it merely requires the selective population of aligned single-particle states. A whimsical illustration of this effect is given in Fig. 2. [Such states can be designated "klokast" (cleverest) states, in the "yrast" spirit of misusing Swedish words.] Once such states are populated in the highly-excited nucleus, normal γ-ray selection rules tend to keep the deexcitation locked into related states, the result being that only a small subset of the possible states are seen. This mechanism means that a great deal of angular momentum is tied up in the single-particle modes. For example, in the ^{182}Re level scheme the largest value of $\vec{R}$ is only 10 (in the K = 2 band), not enough to produce any drastic shifts in the moments of inertia. It will require much heavier beams to bring in larger amounts of angular momentum before phenomena such as backbending can be studied in odd-odd systems.

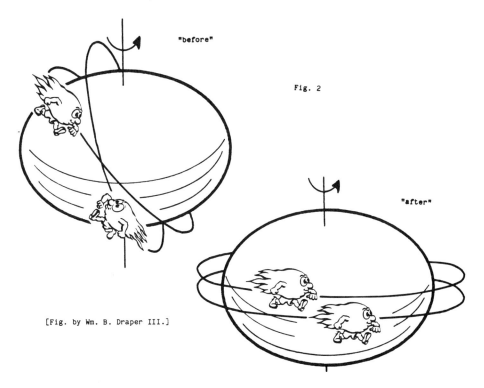

"before"

Fig. 2

"after"

[Fig. by Wm. B. Draper III.]

Odd-odd nuclei also have a head start on multi-particle states. An example of this can be seen in the 15$^+$ state at 2256.3 keV in ^{182}Re. A stylized illustration of this is shown in Fig. 3, where the four particles form a sort of oblate girdle about the prolate

Fig. 3

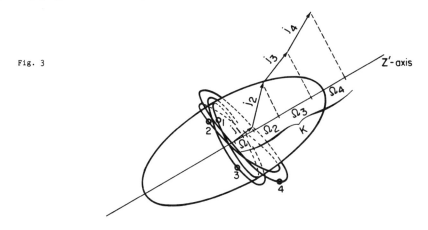

core. Classically, an oblate spheroid rotating about its symmetry axis (oblate-||) has the largest moment of inertia and consequently produces the lowest-lying rotational states, i.e., is the most efficient in handling high angular momentum. Quantum mechanically, such rotations are not defined, of course, but it should be possible [BOH75] to produce pseudo-rotational oblate-|| bands by having each successive "member" result from a different coupling or combination of single particles. The 15$^+$ state in ^{182}Re could be the beginning of such a band.

Because so much angular momentum is tied up in single-particle motion, heavier beams are needed to extend the studies. ^{182}Re is difficult to produce by heavier beams, but the lighter Re nuclei can easily be produced by beams such as ^{14}N on Er targets. Preliminary results [LIE85] for ^{180}Re are shown in Fig. 4, and an example of a γ-ray spectrum for ^{178}Re [OLI85] is shown in Fig. 5. Both are somewhat more complex than ^{182}Re; yet they indicate

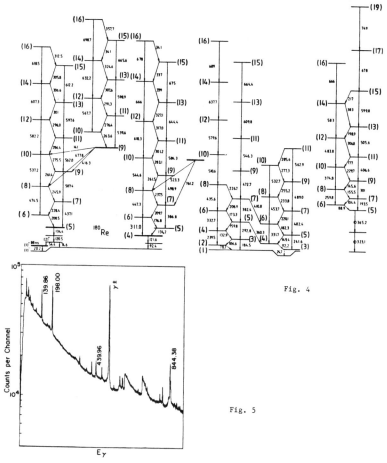

Fig. 4

Fig. 5

the same sort of behavior: Most of the deexcitation occurs through a select few bands, which are composed of highly-aligned odd-odd states.

Although odd-odd rotational bands are best understood in well-deformed nuclei, for quite some time they have also been known to exist, based on excited states, in nuclei very close to closed shells [SAM77]. Rather intriguing examples of this occur in many of the odd-odd Sb isotopes. The high-spin level scheme of ^{116}Sb, resulting from the ^{115}In(α,3nγ) reaction [BEN85], is shown in Fig. 6. It contains at least two rotational bands, the J = 7

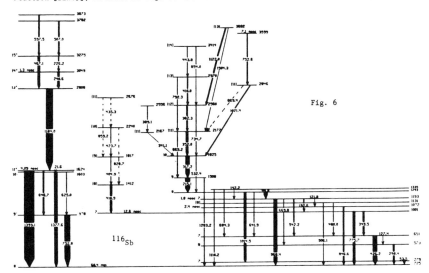

Fig. 6

^{116}Sb

bands based on states at 1001 and 1193 keV. The former can be characterized as a relatively straightforward strongly-coupled rotational band, with the proton in the $9/2^{+}$[404] orbital (originating from the $g_{9/2}$ spherical state) and the neutron in the $5/2^{+}$[402] orbital (from $d_{5/2}$). The proton state again is associated with large Coriolis matrix elements. The second band is a bit more complicated, but it can be explained in a fashion silimar to that by which bands were explained in ^{198}Tl [TOK77]. The proton can be taken basically as being in the $g_{9/2}$ orbital, which is deformation aligned (but split into the proper deformed states), while an $h_{11/2}$ neutron splits away from the core to become rotationally aligned. Calculations reproduce the observed spacings remarkably well, even down to the squashed appearance of the first several members of the band. That these rotational bands figure prominently in the deexcitation pattern is again consistent with our proposed mechanism of populating "klokast" states. Less can be said about the most prominent high-spin deexcitation pattern (at the left side of Fig. 6), for there are too many possible constructs. However, oblate-|| pseudo-rotation and rotationally-aligned particles could certainly play major roles. And the other odd-odd Sb show remarkably parallel behavior.

This paper is necessarily a very abbreviated survey, but we hope we have demonstrated by these few examples that Coriolis coupling and rotational alignment play an important part in high-spin structures of odd-odd nuclei. This can be true for nuclei near closed shells, as well as for nuclei in well-deformed regions. Selectivity caused by their operation actually simplifies rather than complicates spectra obtained in in-beam γ-ray spectroscopy. This, coupled with the abundance of relatively low-lying multi-particle states, makes odd-odd nuclei a fertile field for further study.

Acknowledgments

This work was supported in part by the U. S. National Science Foundation under Grant No. PHY-83-12245.

References

[BEN85] W. H. Bentley, R. A. Warner, Wm. C. McHarris, and W. H. Kelly, submitted to Phys. Rev. C (1985).

[BOH75] A. Bohr and B. R. Mottelson, Nuclear Structure (Benjamin, Reading, Mass., 1975), Vol. II, pp. 43ff, 72ff.

[LIE85] R. M. Lieder, XXIII International Winter Meeting on Nuclear Physics, Bormio, Italy, Proceedings, p. 276 (1985).

[OLI85] W. A. Olivier, W.-T. Chou, J. Kupstas-Guido, and Wm. C. McHarris, work in progress (1985).

[SAM77] L. E. Samuelson, W. H. Bentley, W. H. Kelly, R. A. Warner, F. M. Bernthal, and Wm. C. McHarris, Phys. Rev. C 15, 821 (1977).

[SLA84] M. F. Slaughter, R. A. Warner, T. L. Khoo, W. H. Kelly, and Wm. C. McHarris, Phys. Rev. C 29, 114 (1984).

[STE72] F. S. Stephens and R. Simon, Nucl. Phys. A183, 257 (1972).

[TOK77] H. Toki, H. L. Yadav, and A. Faessler, Phys. Lett. 71B, 1 (1977).

RECEIVED July 14, 1986

Heavy-Ion-Induced Transfer Reactions

A Spectroscopic Tool for High-Spin States

P. D. Bond

Brookhaven National Laboratory, Upton, NY 11973

One of the early hopes of heavy ion induced transfer reactions was that new states in nuclei would be preferentially populated. The fact that this hope diminished was due primarily to insufficient understanding of the reaction mechanism, but also to the generally poorer energy resolution obtained with heavy ions as compared to light ions. In this paper, I hope to demonstrate that with the proper kinematical conditions there is a remarkable selectivity which can be obtained with a proper choice of the reaction and that these reactions can be valuable spectroscopic tools. The data in this talk have been taken using beams from the Brookhaven National Laboratory double MP tandem facility with particles identified in the focal plane of a QDDD spectrometer.

Shown in Fig. 1 are spectra for three single neutron transfer reactions leading to known states in the same final nucleus ^{149}Sm. The (^{13}C, ^{12}C) in reaction in Fig. 1 populates final states much as the (d,p) reaction and

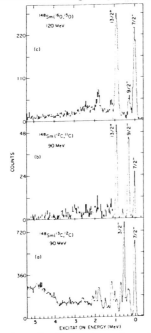

because of the worse energy resolution this heavy ion reaction is thus not very useful for spectroscopic purposes. The (^{12}C, ^{11}C) and (^{16}O, ^{15}O) reactions on the other hand show a very strong selectivity for high spin states with the latter reaction also showing a strong preference for $j=\ell+1/2$ final states. The reasons for this selectivity are discussed elsewhere [BON83], the purpose here is to use this selectivity for spectroscopic studies.

Shown in Fig. 2 is a schematic representation of the shell model states for the region near ^{146}Gd. There has been a great deal of interest in this mass region since it was proposed [OGA78] that a shell closure occurs for Z=64. As in the ^{208}Pb region high spin single particle states are available so that simple configurations are likely to contribute to the structure of high spin states. In particular, near ^{146}Gd the proton $h_{11/2}$ orbital and the neutron $i_{13/2}$ orbital should play an important role, however, there is little direct evidence about the states based on these orbitals. For the Nd isotopes considered here the protons have partially filled the $g_{7/2}-d_{5/2}$ levels and the neutrons are beginning to fill orbitals above N=82.

Fig. 1. Single neutron transfer reaction for ^{148}Sm → ^{149}Sm for three different projectiles.

0097-6156/86/0324-0335$06.00/0
© 1986 American Chemical Society

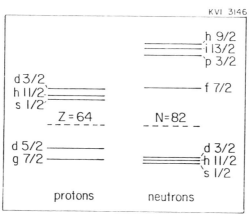

Fig. 2.   Schematic representation of shell model orbitals near ^{146}Gd.

We first focus on the previously unknown $i_{13/2}$ neutron strength in the N=84 nucleus ^{144}Nd. An earlier (d,p) study [RAM76] observed no $i_{13/2}$ strength except in the low lying 3$^-$ state. Gamma ray experiments following compound nucleus formation [BER76, GEE76, QUA82] found several negative parity states but all with odd spin (natural parity). The findings are consistent with the supposition that the negative parity states are formed by a 3$^-$ core excitation coupled to two $f_{7/2}$ neutrons producing spins of 3$^-$, 5$^-$, 7$^-$, 9$^-$. Single nucleon transfer would only weakly populate these states through multistep processes. If, on the other hand, these negative parity states were due primarily to $\nu f_{7/2} * \nu i_{13/2}$ configurations the ordering of the natural parity states would be the same but transfer to these states should be strong. In addition, the unnatural parity states (J$^\pi$ = 4$^-$,6$^-$,8$^-$, 10$^-$) of this multiplet should also be seen with the strongest state in the transfer spectrum being the unnatural parity state J$^\pi$=10$^-$. This state would not be easily seen in the xn experiments because it is not yrast.

The selectivity shown in Fig. 1 demonstrates the (^{16}O, ^{15}O) transfer strongly favors transfer to states of $f_{7/2}$ or $i_{13/2}$ character. In bombardment of a ^{143}Nd target, which has J$^\pi$=7/2$^-$, the states in ^{144}Nd which should be populated are therefore ($\nu f_{7/2}$)^2=0$^+$, 2$^+$,4$^+$, 6$^+$ and $\nu f_{7/2}$*$\nu i_{13/2}$ = 3$^-$,...,10$^-$. In Fig. 3 the spectrum of the ^{143}Nd(^{16}O, ^{15}O)^{144}Nd reaction [COL85] is shown with the location of known states indicated on the figure. For excitation energies up to about 1.5 MeV there is no ambiguity in assignments since other states are not present. However, above that excitation energy the resolution of roughly 100 keV is not sufficient to uniquely identify the states. In order to help determine the spin and excitation of the levels, gamma rays in coincidence with the particle peaks have been measured with an intrinsic Ge detector.

Shown in Fig. 4 are two gamma ray spectra [COL85] one for the peak labeled 9$^-$ in Fig. 3 and one for the particle peak at about 3.8 MeV, which was previously unknown. The upper gamma ray spectrum in Fig. 4 confirms that the peak at 2.902 MeV is indeed the 9$^-$ state seen in previous work [BER76, GEE76, QUA82]. The particle peak at 3.8 MeV shows a nearly identical gamma ray spectrum with the addition of one gamma ray of 900 keV, just the difference in energy between the 9$^-$ state and this state. Since this peak is so

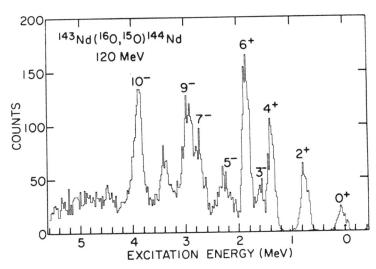

Fig. 3.   Spectrum of the ^{143}Nd(^{16}O, ^{15}O)^{144}Nd reaction.
Reproduced with permission from [BON83].   Copyright 1983 Gordon
& Breach Science Publishers.

strongly populated in particle transfer and decays primarily to the 9⁻ state
it is almost certainly the previously unobserved 10⁻ state. Gamma rays in
coincidence with the other labeled peaks in the spectrum confirm they are
indeed the states indicated on Fig. 3.

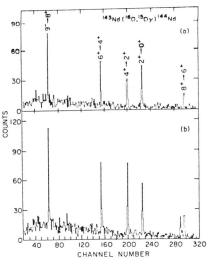

Fig. 4.   Gamma rays in coincidence with a) the state at 2.9 MeV and b) the
state at 3.8 MeV.
Reproduced with permission from [BON83].   Copyright 1983 Gordon
& Breach Science Publishers.

The level scheme of states in ^{144}Nd seen in this experiment is shown in Fig. 5. The structure of the negative parity states is as follows. The low lying 3$^-$ and 5$^-$ states appear to be rather complicated as their population does not follow the (2J+1) population expected for a simple multiplet. In contrast, the 7$^-$, 9$^-$ and 10$^-$ appear to have equivalent $f_{7/2} * i_{13/2}$ strength and thus have rather simple structure. On the other hand, no evidence is found for the 4$^-$ or 8$^-$ states and only weak evidence that the state at 3.3 MeV is a 6$^-$ state. Configuration mixing is a possible, though not a certain, reason for the absence of these states. In any case, there does not appear to be a simple $\nu f_{7/2} * \nu i_{13/2}$ multiplet.

For the positive parity states, the population of the 0$^+$ → 6$^+$ states follows (2J+1) which indicates that these states are consistent with being an $(f_{7/2})^2$ multiplet. As can be seen in Fig. 1, the spectrum of the (^{12}C, ^{11}C) reaction leading to ^{144}Nd enhances $h_{9/2}$ transfer. Comparison of the (^{16}O, ^{15}O) and (^{12}C, ^{11}C) spectra clearly shows that the 8$^+$ state at 2.709 MeV is primarily an $f_{7/2} * h_{9/2}$ configuration. Unfortunately, there are no extensive shell model calculations with which to compare these results.

We turn now to the proton levels in the N=82 nucleus ^{142}Nd. Similar spectra to those of Fig. 1 for proton transfer demonstrate that the (^{18}O, ^{17}N) reaction strongly enhances transfer to $d_{5/2}$ and $h_{11/2}$ proton orbitals. Since the ground state of ^{141}Pr is 5/2$^+$ the (^{18}O, ^{17}N) reaction on that nucleus strongly enhances final states in ^{142}Nd of ($\pi d_{5/2}$)^2=0$^+$, 2$^+$, 4$^+$ and $\pi d_{5/2} * \pi h_{11/2} = 3^- \ldots 8^-$. The single proton transfer spectrum [BON85] to ^{142}Nd is shown in Fig. 6. Gamma rays in coincidence with these particle peaks have been measured and lead to the level scheme shown in Fig. 5.

The gamma decay of the level at 3.24 MeV is the same as was seen in (α,xn) [GEE75]. The parity of this level, previously assigned a spin of 7 [GEE75], can be assigned as negative since positive parity states up to only spin 4 can be populated. This conclusion is consistent with the results of the (^{3}He,d) [JON71] stripping reaction which saw a strong L=5 peak in this region of excitation energy. The previously unassigned fully aligned 8$^-$ state from the $\pi d_{5/2} * \pi h_{11/2}$ configuration is identified by the strong population in the particle spectrum together with the observed gamma decay to the 7$^-$ state. The assignment of the 8$^-$ state is also consistent with an L=5 stripping peak seen in the (^{3}He,d) reaction [JON71]. The assignment of the 6$^-$ state is more uncertain. A gamma decay to the presumed 5$^-$ state is observed but another unobserved high energy transition must also depopulate this level. A spin 6 level at about the same excitation energy was seen in the (α,xn) experiment and decayed via a 1.245 MeV gamma ray to the 6$^+$ state but other proposed branches are not seen here. The 6$^-$ assignment is consistent with an L=5 peak seen in (^{3}He,d) [JON71] and further indications that the 3.44 MeV peak is a 6$^-$ peak comes from the expected 2J+1 population of the $d_{5/2} * h_{11/2}$ multiplet. The previously unassigned 5$^-$ state is tentatively assigned at 2.96 MeV but shows no clear cut coincident gamma rays which would be expected to be of rather high energy. The peak at 2.96 MeV does coincide with an L=5 peak seen in (^{3}He,d). Only very tentative evidence is found for the 4$^-$ state which is expected to be weakly populated. The 3$^-$ state, which is nearly degenerate with the first 4$^+$ state, involves several configurations.

The positive parity states are interesting because two 4$^+$ states are seen with about equal strength and the previously unobserved decay of the higher one at 2.436 MeV is almost exclusively via a transition to the lower 4$^+$ state. The positive parity of the higher 4$^+$ state is determined from the L=2 character of the (^{3}He,d) stripping reaction.

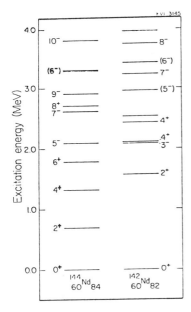

Fig. 5. Deduced levels schemes
for ^{144}Nd and ^{142}Nd from
the reaction studied here.

Reproduced with permission
from [BON83]. Copyright
1983 Gordon & Breach Science
Publishers.

Recent shell model calculations [KRU85] for the nucleus ^{142}Nd reproduce both the relative transfer strength to these 4^+ states and gamma ray branching of the upper state very well. In addition, the $d_{5/2}* h_{11/2}$ negative parity multiplet is predicted to be rather pure and the positions of the levels are reproduced to within roughly 100 keV. Thus the understanding of the levels based on these proton orbitals appears to be in rather good shape, in contrast to the neutron levels.

In summary, the selectivity of certain heavy ion reactions have been used to identify two proton and two neutron states of high spin (both yrast and non-yrast) in Nd nuclei. The first direct information about the configurations of some of these states has been obtained and the results suggest simple configurations for some but not all of them. At the same time certain members of the neutron $f_{7/2} * i_{13/2}$ multiplet are not seen and comprehensive shell model calculations would be very useful to determine the reason. Heavy ion induced transfer reactions, if chosen carefully, are valuable spectroscopic tools,

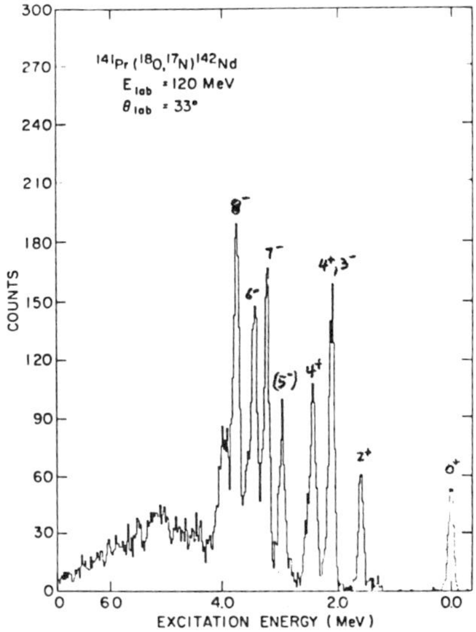

Fig. 6. Spectrum of the ^{141}Pr(^{18}O, ^{17}N)^{142}Nd reaction.

Reproduced with permission from [BON83]. Copyright 1983 Gordon & Breach Science Publishers.

and the unique possibilities available with heavy ions to transfer more massive clusters will assure their continued importance to the field.

Acknowledgments

This work was supported by the U. S. Department of Energy, Division of
Basic Energy Sciences under Contract No. DE-AC02-76CH00016.

References

[BER76]Ya Ya Berzin et al. Bull. Acad. Sci. USSR Phys. Ser. 40 80 (1976).
[BON83]P.D. Bond, Comments on Nuclear and Particle Physics 11 231 (1983).
[BON85]P.D. Bond, C.E. Thorn, J. Cizewski and M. Drigert, to be published.
[COL85]M.T. Collins, P.D. Bond, Ole Hansen, C.E. Thorn and J. Barrette, to be
     published.
[GEE75]L.E. de Geer et al., Nucl. Phys. A246 104 (1975).
[GEE76]L.E. de Geer et al., Nucl. Phys. A259 399 (1976).
[JON71]W.P. Jones et al., Phys. Rev. C 4 580 (1971).
[KRU85]H. Kruse, private communication.
[OGA78]M. Ogawa et al., Phys. Rev. Lett. 41 289 (1978).
[RAM76]S. Raman et al., Phys. Rev. C 14 1381 (1976).
[QUA82]M.A. Quader et al., Bull. Am. Phys. Soc. 27 519 (1982) and private
     communication.

RECEIVED July 31, 1986

Berkeley High Energy Resolution Array

Early Results

R. M. Diamond

Nuclear Science Division, Lawrence Berkeley Laboratory, University of California, Berkeley, CA 94720

As in all fields of spectroscopy, high resolution is of great importance in nuclear γ-ray studies. Also important are a good response function and good efficiency, so as to be able to obtain high-order coincidences when observing de-excitation cascades. The design of the Berkeley High Energy-Resolution Array is discussed and some first results are given.

Nuclear spectroscopic studies are often limited, as are most types of spectroscopy, by either insufficient resolution or insufficient statistics, or both. Four years ago Frank Stephens and I began planning a high-resolution, high-statistics γ-ray system in order both to push discrete-line work to higher spins and to help with continuum studies. Our three primary criteria for a high-resolution array were: high energy resolution, good response function, and good efficiency. Three secondary features we desired were: capability of a total-energy spectrometer, capability as a multiplicity filter, and a prompt initial timing signal.

Germanium detectors provide the highest resolution for γ-rays possible with reasonable efficiency, so our system is based on an array of such detectors. For cascade γ rays (in prompt coincidence) an improvement in the effective resolution is possible by using double- and higher-coincidence spectra rather than singles. For this reason, and to obtain good statistics, as many detectors as close to the target as is reasonably possible is desirable (see below).

Unfortunately, Ge detectors have poor response functions. For a 5 x 5 cm coaxial detector (nominal 20%), only about 3/4 of impinging 1.33 MeV γ rays do interact, and, of these, only 15-20% give useful full-energy peaks. In a double-coincidence measurement, then, only 2-4% of the events are good peak-peak values, and, in a triples measurement, much less than 1% are useful. The solution to this problem has been known for some time; put

0097-6156/86/0324-0341$06.00/0
© 1986 American Chemical Society

Compton-suppression shields around the Ge detectors. In the past this has meant using NaI scintillators, culminating in the array of six such shielded detectors, TESSA2, at Daresbury [TWl83]. But NaI shields are rather large, permitting only a small number of modules at a relatively large distance from the target. We decided to try to use a bismuth germanate (BGO) shield [DIA81], for, if successful, it offered a great advantage in compactness (it has a γ-ray absorption length 2-1/2 times shorter than NaI). But BGO also had two disadvantages: its light output is only about 15% that of NaI, and at that time (1981) no company had made BGO crystals large enough for this purpose. After more than a year of waiting, the latter problem was solved by making the shield out of six pieces that fit together to form a cylindrical unit, each with its own photomultiplier tube. By adding a NaI collimator or "cap" to the front of the shield, the angle of escape of the backscattered γ rays is narrowed, and the Compton edges, the worst remaining interference, are essentially removed. A peak/total ratio of 55% or greater for ^{60}Co above a 300 keV threshold is obtained; this means that triple-coincidence spectra, which have not been used before, become possible with a 16% yield, and even quadruple coincidences are usable with a 9% yield. Other designs using larger BGO shields can give P/T ratios as high as 70% for ^{60}Co, so tremendous improvement in the Ge response function is possible.

System efficiency involves the number of Ge detectors of a certain size at a certain distance from the target. With our shield design involving 13 x 13 cm tapered BGO cylinders, twenty-one 20% Ge detector units can be placed 15 cm from a target, giving a solid angle of 14% of 4π, and leaving the top and bottom around the target chamber clear (Fig. 2) to accommodate the photomultiplier tubes for the small, central BGO ball. It should be mentioned that γ-ray summing in the individual Ge detectors also limits their size and closeness to the target, but is 5% for γ-ray cascades of average multiplicity 20 in this arrangement. A final limitation on closeness is the magnitude of Doppler broadening for the detectors near 90° to the beam (recoil) direction. This must be considered for each nuclear system studied, as the effect can be quite large in (H.I.,xn) reactions where the recoils attain several percent the speed of light. It is, of course, negligible for transitions occurring after the recoiling nuclei have stopped in the target or backing material.

With these considerations in mind, we have designed the High-

Resolution Array to consist of 21 BGO-shielded Ge detectors arranged (in
three rings of seven detectors) around a small, central "ball" of 40 BGO
sectors. These sectors together form a sum spectrometer and multiplicity
filter around the target, as well as giving the angular pattern of the γ
rays emitted. A cut-away perspective drawing of the system is shown in Fig.
1 (without the PM tubes on the BGO shields or ball sectors). A view of the
actual system (but without the NaI caps and the BGO ball) is shown in Fig. 2
with six of the detectors pulled back into the left foreground to show the
small target chamber in the center. By the end of this year the 21 NaI caps
will be in place, and all that will remain to construct is the central BGO
ball whose design is already complete. With 21 of our detectors at 15 cm
from the target, an event rate of 10^5/second (the order of our usual rate in
a (^{40}Ar,xn) reaction with a 1 mg/cm^2 target and a 1-2 pna beam), and an
average γ-ray multiplicity of 20, we had estimated the double-, triple- and
quadruple-coincidence rates to be 11 K/s, 2.2 K/s and 280/s, respectively.
We have achieved these rates fairly routinely.

   I shall describe our first discrete-line results on ^{156}Er [STE85] and
^{158}Er. The ^{156}Er data were taken in two days of running at the LBL 88"
Cyclotron with a 170 MeV ^{40}Ar beam on a lead-backed ~1 mg/cm^2 ^{120}Sn target
using 9 Compton-suppressed Ge detectors, and consist of 1.2 x 10^8 double-

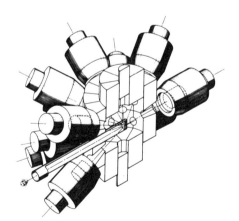

Fig. 1 Perspective view of ~1/2
the system.

Fig. 2. Photograph of 21-detector
array around small target chamber.
Six modules are pulled back into
left foreground.

coincidence events. From many gated spectra the level scheme shown in Fig. 3 was constructed, involving nearly 100 levels and ~130 transitions; approximately one-half are new and more than one-half of the transitions are multiple. There are a number of interesting features in this scheme, but time to describe only two.

The positive-parity yrast band behaves reasonably normally with a neutron $i_{13/2}$ backbend at spin 12ℏ and a second crossing probably the proton $h_{11/2}$, at spin 24ℏ. What is new is the nature of the band between the 9.6 MeV 30^+ and the 14.4 MeV 42^+ levels. The latter is most likely the maximally aligned configuration (relative to the doubly-magic ^{146}Gd core) predicted by theory, $\pi[(h_{11/2})^4]_{16}\nu[(h_{9/2})^2 (f_{7/2})^2 (i_{13/2})^2]_{26}$, a non-collective state of oblate shape. But decay of this state to the 30^+ level proceeds mainly by a cascade of six stretched

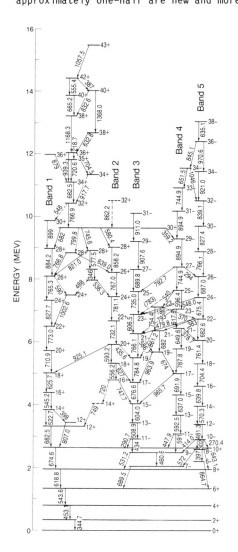

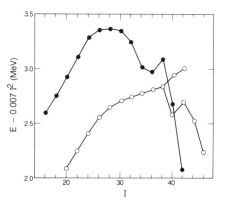

Fig. 3.   Level scheme for ^{156}Er.

Fig. 4.   Energy levels of ^{156}Er, -●- , and ^{158}Er, -○-, minus a rigid-rotor energy vs. spin, I.

E2 transitions, and suggests a band structure. An interpretation consistent with these features and with current theoretical calculations is that this is a triaxial band, with the parameter γ increasing with increasing spin, and terminating in the fully aligned and oblate 42^+ state. The energies of these levels relative to those of a rigid rotor, as shown in Fig. 4, support this idea, as their decrease in energy with spin compared to a rigid rotor or to the weakly prolately-deformed bands below $24\hbar$ is characteristic of that calculated for a band terminating in a fully-aligned state. In addition, the particularly low energy of the 42^+ state results in a relatively large population (10% of the $4^+ \rightarrow 2^+$ transition) rarely seen at high spin.

The other new feature observed is the large number of interband transitions seen at the backbends (probably BC) of the negative-parity bands around spin $22\hbar$. All four bands take part in this cross-feeding, indicating considerable mixing of the bands and suggesting a structural change. It may be related to the fact that four of the six valence neutrons have become aligned and so have to a large extent quenched the neutron pairing correlations. This could cause some change in shape, but really does not explain the extensive mixing of states involving both signatures. Yet above this region three of the bands continue in a somewhat less regular fashion, but still involving only stretched E2 transitions of roughly monotonically increasing energy. Clearly, much remains to be learned about the detailed structure of these bands.

In a study of ^{158}Er levels, of which only the positive-parity yrast band is completed, two interesting features, one new, again were found. The reaction used was 175 MeV ^{40}Ar from the 88" Cyclotron on several backed and unbacked ~1 mg/cm^2 ^{122}Sn targets. With the full 21-detector array we recorded ~2 x 10^8 triple coincidences (corresponding to 6 x 10^8 doubles) in a two-day run. A part of the spectrum (700-1400 keV) obtained with the unbacked target in coincidence with a gate on the 1058 keV, $38^+ \rightarrow 36^+$ transition is shown in Fig. 5a. The five higher transitions observed in [SIM84] at 827, 1031, 1203, 1210, and 1280 keV can be seen, and in addition there is a weak 971 keV line. Figure 5b shows the same region gated by the same transition but for a gold- backed target. The 1203 and 1210 keV lines are missing, and we believe this is because they are smeared out by Doppler shifts, that is, they have lifetimes (plus feeding times) shorter than the mean slowing time (0.5 picosecond). On the other hand, for the 827, 971,

1031, and 1280 keV lines to be sharp means that they have been emitted after an interval of a few picoseconds. A weak 1276 keV line is observed if the 1017 keV, $36^+ \rightarrow 34^+$ transition is used as a gate, and altogether the coincidence and angular correlation studies give the level scheme for the top part of the positive-parity yrast band shown in Fig. 6.

The most interesting result is the identification of fast and slow feeding components into the 38^+ level. Among the rare-earth nuclei having neutron numbers between 82 and 88, slow feeding times (>>1 ps) have been observed, presumably because of regions of non-collective behavior (oblate or spherical shapes) along the decay pathways. In contrast, the well deformed rare-earth nuclei (90 < N < 110) appear to have fast feeding times (<<1 ps), thought to be due to strongly collective rotational bands. It is therefore not so surprising that ^{158}Er, which lies on the boundary between the regions, has both fast and slow feeding components. But this is the first time resolved lines have been identified for both branches and shown how they feed into the yrast band.

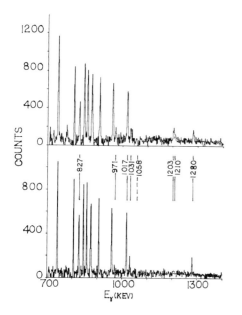

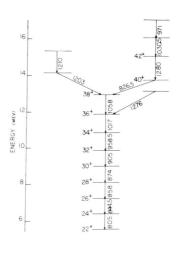

Fig. 5. Partial spectrum of transitions in ^{158}Er in coincidence with 1058 keV, $38^+ \rightarrow 36^+$ line for unbacked target, upper spectrum, and gold-backed one, lower spectrum.

Fig. 6. Partial level scheme for the top of the yrast band in ^{158}Er.

The other point of interest relates to the nature of the highest spin state observed, that decaying by the 971 keV transition. If the 971 and 1031 keV lines are stretched E2 transitions, as we suspect, this state is the 46^+ one that is the maximally aligned state predicted by theory and analogous to the 42^+ state in ^{156}Er. Figure 4 shows the level energy (minus a rigid-rotor energy) for ^{158}Er compared to the better established sequence in ^{156}Er, and there is a remarkable similarity for the slow branch, though displaced four units higher in spin.

If lifetimes can be measured, much more can be said about the nature of these states, and this is true in general for the states at high spin in other nuclei. But it should be remembered that the discrete lines involved are quite weak (in ^{158}Er, 1-5% of the $4^+ \rightarrow 2^+$ transition), and if we want to know what most of the states are like at these high spins we must develop our techniques to study continuum spectra, and in particular to determine the effects of excitation above the yrast line. The new Compton-suppressed Ge arrays are surely going to help us do that, as well as give us a new dimension in discrete-line spectroscopy.

Acknowledgments

A large number of people contributed to the successful completion of the 21-modular array; space limitations allow only the naming of R. Belshe, A. Dancosse, F. Gin, D. Landis, and M.K. Lee. The two experiments described were performed by P.O. Tjøm, F.S. Stephens, J.C. Bacelar, E.M. Beck, M.A. Deleplanque, J.E. Draper, and A.O. Macchiavelli.

This work was supported by the Director, Office of Energy Research, Division of Nuclear Physics of the Office of High Energy and Nuclear Physics of the U.S. Department of Energy under Contract DE-AC03-76SF00098.

References

[BEN83] T. Bengtsson and I. Ragnarsson, Phys. Scripta T5, 165 (1983).
[DIA81] R.M. Diamond and F.S. Stephens, The High Resolution Ball (proposal) (unpublished 1981).
[SIM84] J. Simpson et al., Phys. Rev. Lett. 53, 648 (1984).
[STE85] F.S. Stephens et al., Phys. Rev. Lett. 54, 2584 (1985).
[TWI83] P.J. Twin et al., Nucl. Phys. A409, 343e (1983).

RECEIVED September 2, 1986

Section II: Experimental Exploration of Current Issues
Part 4: Nuclear Moments

Chapters 53-60

53

On-Line Nuclear Orientation

N. J. Stone

Clarendon Laboratory, Parks Road, Oxford OX2 3PU, United Kingdom

A brief survey is given of the technique and its limiting
parameters. Recent results on moments of proton rich
iodine isotopes show shape coexistence and the presence of
the $g_{9/2}$ [404] intruder orbital for $A \geq 118$.

As a technique for measuring nuclear moments and investigating nuclear
level structure, low temperature nuclear orientation has long been kept from
areas of current activity in low energy nuclear physics by the half life
limitation caused by needs of sample preparation and cooling to below 1 K.
The development of $^3He/^4He$ dilution refrigerators capable of continuously
maintaining temperatures of order 10 mK in the presence of heat flux of order
1 μW opened up the possibility of using the method on-line to sources of
active isotopes. The magnetic hyperfine interaction experienced by nuclei of
all elements when present as dilute impurities in ferromagnetic metals is a
quasi-universal method of producing appreciable polarization at such temperatures. More recently single crystals of non-cubic metals have come to the
fore as suitable sources for electric quadrupole alignment, although the
temperature required is often lower than 10 mK. A final necessary innovation
has been the development of implantation techniques to introduce nuclei of
many elements, whether chemically compatible or not, into host lattices such
that a major proportion of the implants come to rest in essentially
undamaged substitutional sites and experience large hyperfine interactions.
Such samples are thin, with little self-absorption for alpha- and beta-
particles.
The combination of these three developments has led to the on-line
nuclear orientation method (OLNO). A further advance of the technique has
been the combination with NMR, which uses resonant rf perturbation of the
nuclear polarization, yielding moment values to a few parts in 10^4.
The wide range of elements in which magnetic polarization and electric
alignment of radioactive nuclei have been observed is shown in figure 1,
which gives also the number of isotopes in which NMR/ON has been detected.
As with conventional nuclear orientation, on-line work can yield the
strength of the hyperfine interaction, deduced from the temperature dependence of the radiation anisotropy. For the majority of elements in Fig. 1
host metals may be chosen in which the magnetic field, or electric field
gradient, is known, allowing extraction of the nuclear moment. Typically
extracted moments are accurate to 5-10%, possibly subject to various
systematic errors. By contrast, NMR/ON gives precise unambiguous results.
Simultaneously the different degrees of anisotropy shown by different transitions in the same decay contain information on the spins of the nuclear
levels and the multipolarities and mixing ratios of transitions linking them.
This separation can be seen in the familiar expression for the angular
distribution from an axially oriented ensemble

$$W(\theta,T) = 1 + \Sigma B_\lambda (T) A_\lambda P_\lambda (\cos\theta)$$

where $B_\lambda (T)$ are the temperature dependent orientation parameters of the

0097-6156/86/0324-0350$06.00/0
© 1986 American Chemical Society

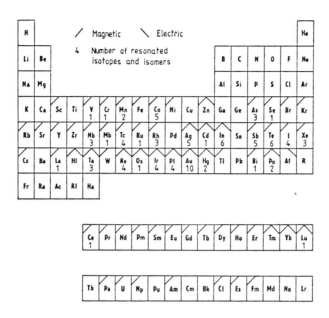

Fig. 1.   Elements studied by low temperature nuclear orientation
(magnetic and electric) and by NMR/ON.

parent state and A_λ are directional distribution coefficients describing the
decay sequence [KRA71].

A schematic OLNO experiment is shown in Fig. 2. The primary accelerator beam strikes a target incorporated in the ion source of an isotope
separator, where heavy ion beams can produce a wide range of isotopes predominantly on the proton rich side of the N/Z stability line. Alternatively
a spallation source or a neutron beam can be used with advantage to reach
additional neutron rich isotopes. The activities desorb from the target, are
ionised, and enter the separator. Ions of a chosen mass are focussed and
pass through a slit in the focal plane. Leaving the separator, having been
accelerated through a potential of 50 - 100 kV, they enter a cold (4 K) beam
tube and are implanted into a target foil, usually of iron, attached by a
copper cold finger to the mixing chamber of a dilution refrigerator and maintained close to 10 mK. The target foil is magnetised to provide an axis of
polarization for the implanted nuclei which are oriented in the magnetic
hyperfine field.

What are the critical parameters of such a system? Clearly we are
concerned with the attainable temperature, the effectiveness of the low temperature implantation, the required particle flux and the condition that the
nuclei, initially random, become polarized before they decay. Each of these
is briefly considered.

The current lowest fully on-line temperature is 10 mK, but, for somewhat longer half-lives, closing off the refrigerator after implantation can
give results to ∿ 7 mK in about 30 m (see Fig. 4). The Bonn group, who were

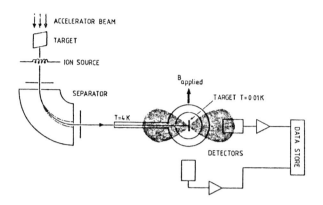

Fig. 2. Schematic on-line nuclear orientation experiment.

the first to operate on-line to an isotope separator, have shown for many
systems that implantation below 4K is at least as effective in producing highly
substitutional implants as room temperature implantation [HER78]. NMR/ON in
fully OLNO implanted systems has been observed in Leuven [VAN85a].
        A crucial parameter in implanted source preparation is the total
implant dose. For on-line work the flux of implant ions is very low as we
are far from any stable beam. Even a flux of 10^6 ions s^{-1} gives only 10^{11}
ions/day, well below the limiting concentration of order 0.1 atomic percent
if implantation is at 50 KeV or above. A valuable feature of OLNO is that a
sequence of isotopes may be simultaneously studied as the primary implant
decays towards the stability line. This frequently makes accessible isotopes
of elements which do not readily form ion beams.
        Nuclear orientation is a singles counting technique in which the
measured effects are typically of order 10 percent (but may be much larger).
Individual measurements accurate to 1 percent will generally give useful
results. This requires perhaps 3×10^4 counts to be recorded, if some allow-
ance is made for background subtraction. It is simple to mount four large
Ge(Li) detectors, in pairs, at $\theta = 0$ and $\theta = \pi/2$ relative to the polarisation
axis and about 7 cm. from the source. The combined photopeak efficiency is
then approximately 3 percent at each angle, so the desired recorded count
requires 10^6 disintegrations feeding the observed transition. An implantation rate of 10^4 nuclei per second, that is, assuming a 1 percent separator
efficiency, a reaction yield of 10^6 nuclei per second, will give the required
counts in 100 seconds for a transition fed in 100 percent of decays. It is
clear not only that measurements can be made by OLNO when implantation rates
fall as low as 100 nuclei per second for strongly fed transitions, but also
that at higher rates weak transitions can be studied with adequate accuracy
to yield useful information.
        NMR/ON requires that resonant destruction of the anisotropy be
detected. The fact that a sequence of counts as a function of changing frequency is necessary means that this precision method will be restricted to
nuclei with implantation rates greater than 10^3 per second, although all
transitions in the decay of the resonated isotope can be combined to give the
total resonance signal.

On implantation into the cold foil the nuclei are initially 'hot' i.e.
unpolarised. They approach thermal equilibrium with the foil lattice temperature through the Korringa relaxation mechanism via the conduction electrons.
A relationship of the form $T_1 T(h\nu)^2 = C$ where C is a constant for a specific
host and $h\nu$ is the Zeeman splitting has been established to give estimates of
T_1 to about a factor of two. Typical values at 10 mK lie in the range
1s-1000s.
   The half life $T_{\frac{1}{2}}$ must be $\geq T_1$. The leading orientation parameter B_2
is given approximately for low degrees of polarisation by $B_2 = 12(h\nu/kT)^2/5$.
Useful NO measurements require typically $B_2(min) = 0.2$, giving a maximum useful temperature of $T(max) = Ih\nu/k$, for which
$T_1 = C/\left[(h\nu)^2 T(max)\right] = Ck/I(h\nu)^3 = 2 \times 10^4 C/I(h\nu)^3$ where C is in MHz^2 sK and
$h\nu$ in MHz. For iron host $C = 5 \times 10^4$. The average value of B_2, taking the
simplifying assumption of a nuclear spin temperature T_N approaching the
lattice temperature T_L as $(1/T)_N(t) = 1/T_L (1-e^{-t/T_1})$ is
$\bar{B}_2 = B_2(T_L)\left[2X^2/(1 + 3X + 2X^2)\right]$ where $X = T_{\frac{1}{2}}/T_1 \ell n 2$ and $B_2(T_L)$ is the full
equilibrium limit. For cases with higher degrees of polarisation a better
approximation is $B_2 \alpha 1/T$ and we have $\bar{B}_2 = B_2(T_L)\left[X(1+X)\right]$. Results for
$\bar{B}_2/B_2(T_L)$ are shown in figure 2 as a function of X. These relations allow
estimates to be made of the mean polarisation achieved for any combination of
host C values, hyperfine splitting and half life. The limiting case is found
by putting T(max) equal to the lowest attainable temperature. This gives
$h\nu(min)$, hence $T_1(min)$ and $T_{\frac{1}{2}}(min)$. For different elements the hyperfine
field varies from <10 to >100 T, and with varying nuclear spins and moments
each case must be considered separately.

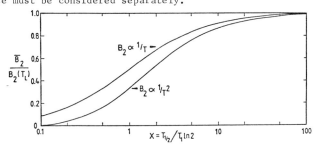

Fig. 3. Mean value of B_2 as a function of $T_{\frac{1}{2}}/T_1$ ratio.
Upper curve for large anisotropies (>10%), lower for small.

   The results of these considerations limit current methods of OLNO.
Representative estimates of lower limit half-lives are given in Table 1,
assuming iron host and taking I=g=1. These should be taken as a guide only,
variations with I and g being indicated. It is seen that, for many elements,
limiting lifetimes are less than 1 minute. In effect this means a range of
15-25 isotopes per element above Z $\sim$ 30.
   Comparison of OLNO with other methods of moment and level structure
study has been given briefly in $\left[STO85\right]$. Most alternative moment measurements
require considerably higher count rates. When working further from stability
decay schemes are often poorly established and, as Q values for decay increase, they become more complex. The occurrence of isomers is relatively
frequent, which has the advantage of populating states of a wide range of
spin values and the drawback of producing complicated parentage for transitions lower in the daughter decay scheme. The over all result is that nuclear

TABLE 1

Listing for each element of the temperature at which B_2 = 0.2 in iron host,
the estimated T_1 at that temperature and hence the minimum half-life
accessible. In Table, I = g = 1 is assumed, see footnotes.

Element	B_{hf} (T)	T_{max} (mK) +	T_1 at T_{max} = $T_{\frac{1}{2}}$ (min) seconds*	Element	B_{hf} (T)	T_{max} (mK) +	T_1 at T_{max} = $T_{\frac{1}{2}}$ (min) seconds*
21 Sc	-13.2	4.8	1.0 x 10^3	54 Xe	160	59	.6
22 Ti	-12.2	4.5	1.3 x 10^3	55 Cs	27.6	10.1	109
23 V	-8.7	3.2	3.5 10^3	56 Ba	8.5	3.1	3700
24 Cr	-6.6	2.5	7.8 x 10^3	57 La	-46	16.8	236
25 Mn	-22.8	8.4	194	58 Ce	41	15.0	33
26 Fe	-33.9	12.4	59	62 Sm	314	115	.074
27 Co	-28.7	10.5	97	63 Eu	148	54	.7
28 Ni	-23.4	8.5	179	65 Tb	380	139	.042
29 Cu	-21.8	8.0	222	66 Dy	610	223	.01
30 Zn	-19.1	7.0	330	68 Er	768	280	.005
31 Ga	-9.4	3.4	2.8 x 10^3	69 Tm	625	229	.009
32 Ge	6.0	2.2	1.1 x 10^3	70 Yb	125	46	1.2
33 As	34.4	12.6	57	71 Lu	61	22	10.1
34 Se	69	25.2	7	72 Hf	-63	23	9.2
35 Br	84	30.7	4	73 Ta	-64	23	8.8
36 Kr	66	24	8	74 W	-69	25	7.0
37 Rb	5.4	1.9	1.4 x 10^3	75 Re	-75	27	5.5
38 Sr	-10	3.7	2.3 x 10^3	76 Os	-109	40	1.8
39 Y	-22.6	8.3	200	77 Ir	-147	54	0.72
40 Zr	-27.4	10.0	112	78 Pt	-128	47	1.1
41 Nb	-26.6	9.7	122	79 Au	-115	42	1.5
42 Mo	-25.6	9.4	137	80 Hg	-84	31	3.9
43 Tc	-31.7	11.6	72	81 Tl	-18.5	6.8	363
44 Ru	-49	17.9	20	82 Pb	26.2	9.7	128
45 Rh	-55.7	20.4	13	83 Bi	119.1	44	1.4
46 Pd	-54.7	20.0	14	84 Po	238	87	0.17
47 Ag	-44.7	16.3	26	85 At	254	93	0.14
48 Cd	-39.2	14.3	38	86 Rn	170	62	0.47
49 In	-28.7	10.5	97	88 Ra	12.7	4.7	1100
50 Sn	8.5	3.1	3.8 x 10^3	90 Th	31	11	77
51 Sb	23.4	8.5	180	92 U	56	20.5	13
52 Te	68.1	24.9	7	94 Pu	113	41	1.6
53 I	114.6	41.9	1.5	96 Cm	<10	<3.6	>2300

* this column scales as $(g^3I)^{-1}$.
+ this column scales as (gI).

orientation alone will often not be adequate to elucidate decay sequences,
spins of levels, and multipole mixing ratios, although it gives information
on all these parameters. Frequently it will be necessary to perform associated experiments using other spectroscopic techniques such as gamma-gamma or
beta-gamma coincidence, angular correlation, electron conversion and gamma
ray linear polarization. These can be carried out simultaneously with OLNO,

indeed gamma-gamma coincidence and angular correlation (for decay sequence and support on mixing ratios) and linear polarization (for magnetic/electric character and again multipole mixing ratios) can be done at the refrigerator using the same decay events with additional electronics. Electron conversion is particularly useful in picking up EO transitions.

Several examples of the extraction of decay information from OLNO data are available, eg. for gamma decay in [VAN85b], and for alpha decay in [WOU85]. As a singles counting method OLNO is efficient at obtaining maximum information from weaker sources.

It is clear that the current methods of OLNO have extensive potential application. Looking further ahead, in-beam orientation, either by hfs pumping or by reflection from polarised metal surfaces may be the route to opening up shorter lived isotopes. Exploratory work on beam reflection methods is going on in Leuven [VAN85a]. At present it is not clear whether pumping can be made quantitative or that reflection gives effects above a few percent.

Recent results on Iodine Isotopes

In recent months results given in Table 2 have been obtained for magnetic dipole moments of Sb, I and Xe proton rich isotopes at the DOLIS-COLD facility at Daresbury. Examples of the data are given in Fig. 4. The ^{117}Sb result should be compared with the established value of 3.54 nm. All other results are new, or have been briefly reported [SHA75].

Table 2

Isotope	Half Life	I^π	X(15mK)	Magnetic Moment (nm)
^{116}Sb	16m	3^+	10	± 2.6(3)
^{117}Sb	2.8h	$5/2^+$	280	± 3.5(2)
^{117}I	2.3m	$(5/2^+)$	70	± 3.0(3)
^{118}Im	8.5m	(7^-)	60	± 4.2(2)
^{118}Ig	14.3m	2^-	270	± 1.9(3)
^{119}I	19.1m	$(5/2^+)$	550	± 2.9(1)
^{120}Im	58m	(7^-)	400	± 4.18(18)
^{120}Ig	1.35h	2^-	640	± 1.22(14)
^{121}I	2.12h	$(5/2^+)$	2200	± 2.3(1)
^{122}I	3.6m	1^+	75	± 0.93(9)
^{119}Xe	5.8m	$(5/2^+)$	16	± 0.68(4)
^{121}Xe	39m	$(5/2^+)$	100	± 0.70(4)

Discussion

Sb   The configurations of low-lying 3^+ states in light Sb isotopes have been assigned [$\pi g_{7/2}$ $\nu s_{1/2}$] in ^{122}Sb and [$\pi d_{5/2}$, $\nu s_{1/2}$] in ^{120,118}Sb based on comparison of their measured g-factors with calculations based on neighbouring odd-A nuclei which yield g = 1.06(3) for the former configuration and g = 0.82(3) for the latter. The present imprecise measurement, g = ± 0.87(9) in ^{116}Sb needs to be improved using the NMR/ON technique, but

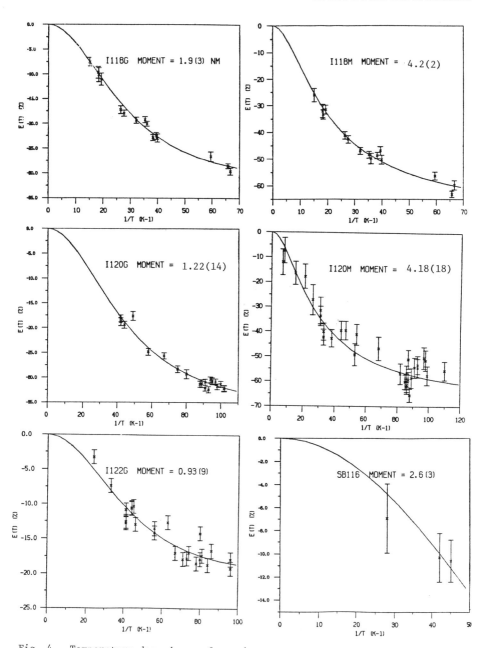

Fig. 4. Temperature dependence of γ-anisotropy for oriented I isotopes.

favours the $d_{5/2}$ proton state, found as ground state in all odd-A isotopes $^{111-121}$Sb.

__I.__   In $^{118}_{53}I_{65}$ and $^{120}_{53}I_{67}$ we have results on two isomers for each mass number. The high spin isomer (assigned 7$^-$ on the basis of levels fed in decay) has $\mu = \pm$ 4.2(2)nm in both cases. This is distinctive of the distorted $\pi(404)_{9/2}{}^+$ orbital which, coupled to $\nu(532)_{5/2}{}^-$ would give $I^\pi = 7^-$, $\mu = 4.8$nm. The occurrence of this orbital at low excitation is directly indicative of large equilibrium deformation associated with the soft mid-shell neutron number. The low spin isomer (2$^-$) shows a moment in ^{118}I, $\pm$ 1.9(3)nm, which is clearly larger than found for levels of the same I^π in ^{120}I $\pm$ 1.23(3)nm and ^{124}I $\pm$ 1.14(8)nm [DEJ83]. A possible explanation for this variation is that in ^{118}Ig we are close to the shell model configuration [$\pi g_{7/2}, \nu h_{11/2}$]2$^-$ ($\mu_{schmidt}$ = -2.5nm) with relatively small equilibrium deformation, suggesting that this isotope is a dramatic example of isomerisation with markedly different deformation in the two isomers. The same configuration [$\pi g_{7/2}, \nu h_{11/2}$]2$^-$ has been proposed for $^{122}_{51}$Sb$_{71}$ (μ_{meas} = -1.9 nm) and Callaghan et al have shown that detailed corrections to the Schmidt moment reduce the calculated value to -1.95(5)nm [CAL74].

The moment of ^{122}I(1$^+$) at $\pm$ 0.93(9)nm lies closer to the deformed analogue ^{124}Cs(1$^+$) [$\pi(420)\frac{1}{2}, \nu(411)\frac{1}{2}$] μ = 0.67 nm, than to the spherical ^{120}Sb(1$^+$) [$\pi d_{5/2}, \nu d_{3/2}$] μ = + 2.3(2)nm.

References

[CAL74] P.T. Callaghan et al. Nuclear Physics A221 1 (1974).
[DEJ83] J. de Jong and H. Postma, Hyp. Int. 15/16 69 (1983).
[HER78] P. Herzog et al. Nucl. Inst. and Meth. 155 421 (1978).
[KRA71] K. Krane, Los Alamos Report LA4677 (1971).
[STO85] N.J. Stone, Hyp. Int. 22 3 (1985).
[WOU85] J. Wouters et al. Hyp. Int. 22 527 (1985).
[VAN85a] D. Vandeplassche et al. Hyp. Int. 22 483 (1985).
[VAN85b] E. Van Walle et al. Hyp. Int. 22 507 (1985).

RECEIVED July 18, 1986

54

Spins and Moments Determined by On-Line Atomic-Beam Techniques

Curt Ekström

The Studsvik Science Research Laboratory, S-611 82 Nyköping, Sweden

Hyperfine structure measurements using on-line atomic-beam
techniques are of great importance in the systematic study
of spins and moments of nuclei far from beta-stability. We
will discuss the atomic-beam magnetic resonance (ABMR) method, and laser spectroscopy methods based on crossed-beam
geometry with a collimated thermal atomic-beam and collinear
geometry with a fast atomic-beam. Selected results from the
extensive measurements at the ISOLDE facility at CERN will
be presented.

Nuclear spins and moments of ground and isomeric states are basic properties probing the structure and shape of atomic nuclei. The systematic experimental study of these quantities along isotopic and isotonic chains thus
allows a mapping of the nuclear behaviour, to be compared with the predictions of different nuclear models.

The main experiments specifically aimed at revealing the nuclear structure by measuring nuclear spins and moments are those investigating the hyperfine structure (hfs) of free atoms using radio-frequency and optical spectroscopy techniques. Although a large amount of data has been obtained in off-
line experiments, it is only through the introduction of on-line techniques
at different ISOL-facilities that short-lived nuclides far from stability,
and thus long isotopic chains, have come within reach for study. The experiments performed at the ISOLDE facility, CERN, by the Gothenburg-Uppsala,
Orsay and Mainz groups constitute a main effort in this direction. Certainly,
these experiments profit from the excellent performance of the ISOLDE mass-
separator and from the wide range of elements available in high production
yields [RAV79, RAV84].

The hfs experiments at ISOLDE may be divided into techniques employing
atomic-beams and resonance-cells. Here, we will concentrate on the different
atomic-beam experiments performed by the groups mentioned above. The present
subject has been discussed in some detail in [EKS85], giving e.g. a full reference list on the atomic-beam works at ISOLDE.

In the optical spectroscopy experiments, data on the isotope shifts (IS)
may be obtained in addition to those on the hfs. The important nuclear information on the changes of mean square charge radii, deduced from the IS results, will be discussed by Kluge at this symposium [KLU85].

NOTE: The author of this chapter worked with members of the ISOLDE Collaboration, CERN,
CH-1211 Geneva 23, Switzerland.

0097-6156/86/0324-0358$06.00/0
© 1986 American Chemical Society

Atomic-Beam Experiments at ISOLDE

In this section we will briefly discuss the different atomic-beam methods used in hfs measurements at ISOLDE, and give information on the isotopic chains studied. The description, although classified according to the technique used, follow to a large extent the chronological order.

The atomic-beam magnetic resonance (ABMR) apparatus of the Gothenburg-Uppsala group, connected on-line to the ISOLDE mass separator, is given schematically in Fig. 1. A detailed description of the design and operation may be found in [EKS78, EKS81]. The ABMR method has been applied to several different elements at ISOLDE for measurements mainly of nuclear spins and magnetic dipole moments. The main results have been obtained in the alkali elements $_{37}$Rb, $_{55}$Cs and $_{87}$Fr, and in $_{79}$Au. Shorter sequences have been studied in $_{35}$Br, $_{49}$In, $_{53}$I, $_{63}$Eu, $_{69}$Tm and $_{81}$Tl. References to the ABMR works are given in [EKS85]. The ABMR project at ISOLDE is, in its present form, now essentially terminated. The possibilities for presicion hfs measurements to get information on the hyperfine anomaly, probing the distribution of nuclear magnetism, are under discussion.

Laser spectroscopy in crossed-beam geometry with a collimated thermal atomic-beam was introduced at ISOLDE by the Orsay group [THI81a, TOU81] (cf. Fig. 1). With this method, a considerable extension of the ABMR measurements in the alkali elements, mentioned above, was obtained. The nuclear spectroscopic quadrupole moments could be reached by studying the hfs of the $^2P_{3\,2}$ excited atomic state, as well as the IS in the isotopic sequences. A number of nuclear spins and magnetic moments was also added. The Orsay works on the alkali elements $_{37}$Rb, $_{55}$Cs and $_{87}$Fr at ISOLDE have been presented in [HUB78, LIB80, THI81b, THI81c, COC85]. The atomic-beam laser spectroscopy, in the Orsay version, is mainly restricted to the alkali elements, where, however, future double-resonance experiments, giving accurate hfs constants, will be of interest.

At present, the most powerful method for hfs and IS measurements is collinear fast-beam laser spectroscopy, developed and introduced at ISOLDE by the Mainz group [NEU81] (cf. Fig. 2). During the few years of operation, this experiments has produced a vast amount of data on nuclear spins, moments and mean square charge radii of long isotopic chains throughout the nuclear chart. The sensitivity of the method will probably be further increased by the introduction of non-optical detection techniques. The first measurements at ISOLDE were made in a long sequence of barium isotopes, $^{122-146}$Ba [MUE83], covering both sides of the N = 82 neutron-shell closure. Similarly, the chemical analogue radium has been studied in the mass range 208 - 232 [AHM83], i.e. on both sides of N = 126. A systematic investigation of the rare-earth nuclides, with special emphasis on those in the transitional region above N = 82, is in progress. At present, the measurements include $^{140-153}$Eu [AHM85], $^{142-160}$Gd, $^{146-164}$Dy, $^{150-170}$Er and $^{156-176}$Yb [BUC82, NEU85]. Recently, radon isotopes in the mass range 202 - 222 and indium isotopes between ^{111}In and ^{127}In were investigated [NEU85], the latter extending the sequence $^{107-111}$In studied by the same method at GSI [ULM85]. In addition to these systematic measurements, point efforts on nuclei of particular interest are made, like in the cases ^{182}Hg and ^{207}Tl [NEU85]. Furthermore, the optical lines connecting the 7s atomic ground state with the 8p excited states in francium were discovered by collinear laser spectroscopy, and the first hfs measurements on the heavy francium isotopes were made [NEU85].

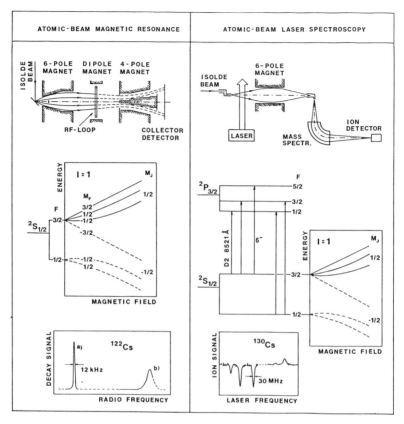

Fig. 1

Schematic view of the atomic-beam magnetic resonance and atomic-beam
laser spectroscopy set-ups at ISOLDE. In both experiments, the 60 keV
ion-beam from ISOLDE is converted to give a continuous flow of thermal
atoms through the machines. Resonances are observed in the ABMR experiments when rf-transitions are induced in the atomic ground-state hfs,
changing the atoms from a focusing ($M_J > 0$) to a defocusing ($M_J < 0$)
state. The $\Delta F = 1$ resonance transitions in ^{122}Cs ($I = 1$) are given at
the bottom left: a) (F = 3/2, M_F = 1/2) → (1/2, -1/2) and (3/2, -1/2) →
(1/2, 1/2), b) (3/2, 3/2) → (1/2, 1/2). The small line-widths, determined mainly by the transit time of the atoms through the rf-field,
should be noted. They give the dipole-constant in this particular case
with a precision of 5 ppm. In the atomic-beam laser experiments, the
resonances are recorded after state selection with a six-pole magnet;
negative signals are obtained when the population of a state with $M_J >$
0 in the strong-field limit is reduced by optical pumping with laser
light via an excited atomic state, and positive signals when the population is increased (cf. the laser-frequency scan in ^{130}Cs ($I = 1$)
at the bottom right).

Reproduced with permission from [EKS85]. Copyright 1985 J. C.
Baltzer AG Scientific Publishing Company.

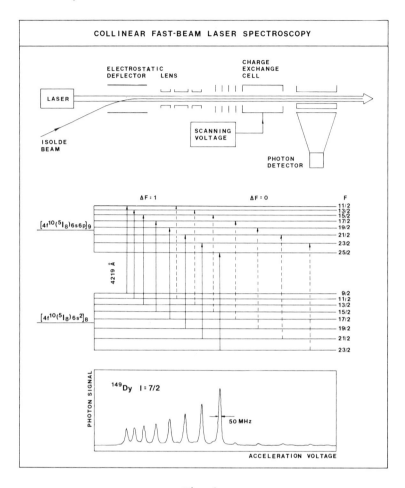

Fig. 2

Simplified drawing of the collinear fast-beam laser experiment at
ISOLDE. The ion-beam is deflected into the apparatus and passes,
superimposed on a laser beam, along the optical axis. The ions are
neutralized in a charge-exchange cell to give a beam of fast atoms.
These are excited by the laser light and the detection of the resonances are made by recording the emitted fluorescence photons in a
photomultiplier. The resonances are scanned by keeping the laser
frequency at a fixed value and varying the Doppler shift of the
absorption line by an additional voltage put on the retardation
and charge-exchange systems. The photon spectrum of ^{149}Dy (I = 7/2)
as a function of acceleration voltage is given at the bottom. The
$\Delta F = 1$ and $\Delta F = 0$ resonances observed in the 4212 Å transition are
indicated in the energy level diagram.

Discussion of Selected Results

The elements in which long isotopic chains have been subject to hfs and IS measurements at ISOLDE are indicated in Fig. 3. Here, we will choose two nuclear regions for short comments; the rare-earth region and the "heavy-radium" region.

As discussed in the preceeding section, an extensive investigation of the rare-earth region is being performed by collinear laser spectroscopy. It includes a two-dimensional mapping of the (N, Z)-plane, covering not only scattered isotopic sequences but isotopes of a range of elements.

The information on nuclear deformation from the IS-data indicates that the steep increase, observed between N = 88 and 90, is restricted to a few elements around gadolinium (Z = 64). The effect becomes less pronounced in the lighter as well as heavier elements. In barium (Z = 56) and ytterbium (Z = 70) the step vanishes completely and there is a smooth transition from spherical to strongly deformed shapes above N = 82. This behaviour indicates that the Z = 64 proton subshell closure, with its stbilizing effect for spherical shapes, plays an important role.

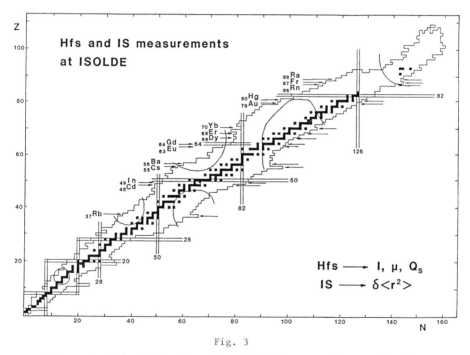

Fig. 3

Nuclear chart including the major shell closures and approximative isodeformation curves for ε = 0.2. Hfs and IS measurements have been performed at ISOLDE in long isotopic chains of the elements indicated by arrows. Shorter sequences of the elements $_{35}$Br, $_{53}$I, $_{69}$Tm and $_{81}$Tl, not indicated here, have also been studied at ISOLDE.

The picture above is supported by the results on nuclear spins and moments. Here, an interpretation within the particle-rotor model gives a rather complete mapping of the ground-state singel-particle structure in the N = 82 region. The model calculations reproduce well the transition from essentially shell-model states close to N = 82 to strongly coupled almost pure Nilsson states in the region of strong nuclear deformation [EKS85].

The nuclides in the "heavy-radium" region above N = 126 have a location on the nuclear chart similar to that of the rare-earths above N = 82. However, because of the presence of extremely short-lived nuclides just above N = 126, the systematic investigations of radon, francium and radium have been limited to nuclides above N = 131.

As mentioned above, the radon and radium sequences have been investigated by collinear fast-beam laser spectroscopy, whereas in francium all three atomic-beam methods, ABMR, atomic-beam laser spectroscopy and collinear laser spectroscopy, have contributed.

The "heavy-radium" region is characterized by an increasing quadrupole deformation with increasing neutron number. This is evidenced by the IS-data as well as by previous B(E2)-data. In addition, there is evidence for the occurence of the theoretically predicted [LEA84, NAZ84] octupole deformation in this region. The IS-data in radium is much better reproduced by including an octupole deformation of $\beta_3 \simeq 0.1$. Similar values are obtained from the magnetic moments in radium [AHM83, LEA84, RAG83] and in francium [COC85].

Acknowledgments

Fruitful discussions with members of the different hfs groups at ISOLDE, and in particular those with Dr. R. Neugart, are gratefully acknowledged.

References

[AHM83]   S.A. Ahmad et al., Phys. Lett. 133B(1983)47.
[AHM85]   S.A. Ahmad et al., Z. Phys. A321(1985)35.
[BUC82]   F. Buchinger et al., Nucl. Instr. and Meth. 202(1982)159.
[COC85]   A. Coc et al., submitted to Phys. Lett. B.
[EKS78]   C. Ekström et al., Nucl. Instr. and Meth. 148(1978)17.
[EKS81]   C. Ekström, Nucl. Instr. and Meth. 186(1981)261.
[EKS85]   C. Ekström, Hyp. Int. 22(1985)65.
[HUB78]   G. Huber et al., Phys. Rev. Lett. 41(1978)459.
[KLU85]   H.-J. Kluge, contribution to this symposium.
[LEA84]   G.A. Leander and R.K. Sheline, Nucl. Phys. A413(1984)375.
[LIB80]   S. Liberman et al., Phys. Rev. A22(1980)2732.
[MUE83]   A. Mueller et al., Nucl. Phys. A403(1983)234.
[NAZ84]   W. Nazarewicz et al., Nucl. Phys. A429(1984)269.
[NEU81]   R. Neugart, Nucl. Instr. and Meth. 186(1981)165.
[NEU85]   R. Neugart et al., to be published.
[RAG83]   I. Ragnarsson, Phys. Lett. 130B(1983)353.
[RAV79]   H.L. Ravn, Phys. Rep. 54(1979)201.
[RAV84]   H.L. Ravn, Proc. On-Line in 1985 and Beyond, Workshop on the
          ISOLDE Programme, Zinal, Switzerland (1984) p. D9.
[THI81a] C. Thibault et al., Nucl. Instr. and Meth. 186(1981)193.
[THI81b] C. Thibault et al., Phys. Rev. C23(1981)2720.
[THI81c] C. Thibault et al., Nucl. Phys. A367(1981)1.
[TOU81]   F. Touchard et al., Nucl. Instr. and Meth. 186(1981)329.
[ULM85]   G. Ulm et al., Z. Phys. A321(1985)395.

RECEIVED July 4, 1986

55

UNISOR Collinear Laser Facility
First Results

H. K. Carter[1], G. A. Leander[1], J. A. Bounds[2], and C. R. Bingham[2]

[1]UNISOR, Oak Ridge Associated Universities, Oak Ridge, TN 37831
[2]Physics Department, University of Tennessee, Knoxville, TN 37996-1200

The hyperfine structure and isotope shifts of $189m,191m,193m,193g$,Tl have been measured by means of collinear fast-beam/laser spectroscopy. Deformations for the 9/2⁻ isomers are determined to be larger than for the 1/2⁺ ground state and increase with decreasing neutron number. Despite different deformations, rotational properties are nearly identical in $185-199$Tl. Microscopic theory ascribes this to a systematic balance between changing deformation and neutron pairing.

Prior to about 1955 much of the nuclear information was obtained from application of atomic physics. The nuclear spin, nuclear magnetic and electric moments and changes in mean-squared charge radii are derived from measurement of the atomic hyperfine structure (hfs) and Isotope Shift (IS) and are obtained in a nuclear model independent way. With the development of the tunable dye laser and its use with the online isotope separator this field has been rejuvenated. The scheme of collinear laser/fast-beam spectroscopy [KAU76] promised to be useful for a wide variety of elements, thus UNISOR began in 1980 to develop this type of facility. The present paper describes some of the first results from the UNISOR laser facility.

The light thallium isotopes were chosen to be the first series of isotopes to be studied at the UNISOR laser facility. The behavior of the $h_{9/2}$ intruder states and the bands built on them has been well documented [HEY83]. This level and the band structure built on it, drops in energy with respect to the 1/2⁺ ground state as the neutron number N decreases until mass 189. At mass 187 and lower the state then rises. This suggests a rapidly changing deformation of this 9/2⁻ state. However, throughout this, the states built on the 9/2⁻ state maintain nearly constant energy level spacings. This suggests that the bands built on the 9/2⁻ state have relatively unchanging deformation as a function of neutron number. Thus thallium appeared to be a good candidate

0097-6156/86/0324-0364$06.00/0

for measurement of IS since we could then determine changes in quadrupole moments and rms charge radii.

Experimental Apparatus and Procedure

Figure 1 shows schematically the laser facility online to the UNISOR isotope separator. The collinear laser/fast-atom beams technique [KAU76] is used. Laser light of 535 nm excites the $6^2P_{3/2}$ metastable state to the $7^2S_{1/2}$ level. The resonant condition is detected by the 377 nm transition to the ground state. The resonant light is collected and channelled to the PMT by a fiber optic collector which is 10.8 cm long. With the cylindrical mirror the collector is estimated to accept 44% of the light emitted by an on-axis point source. Colored glass filters are used to reduce the laser light background. A collinear magnetic field is used to Zeeman shift the transition out of resonance until the atom reaches the detection region.

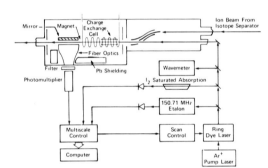

Fig. 1. Schematic of the laser system online to UNISOR isotope separator.

The data recorded as the laser frequency is scanned consists of the fluorscence signal from the PMT, a Doppler-free I_2 spectrum and frequency markers from the etalon. The etalon provides a calibration of the frequency scan. The Doppler-free I_2 spectra provides an absolute frequency reference used to correct for small laser frequency drifts, separator voltage drifts and to determine the absolute acceleration voltage of the separator for the Doppler shift corrections. We are thus able to record data over long periods of time, e.g. 3 hours, and maintain a reasonable resolution of 100 MHz. Some of the first online data recorded with this system is shown in Figure 2. The overall detection efficiency has been measured to be 1/1000, i.e. one detected photon per 1000 atoms, for the largest transition in the nuclear spin 1/2 isotopes.

Results

While data were obtained on isotopes from mass 189 to 193 resulting in moments and isotope shifts, the data will be discussed only in relation to the problem outlined above. Examination of Figure 2 reveals that both the characteristic three hfs transistions due to the 1/2 ground state and the six transitions due to the 9/2 isomeric state are observed in [193]Tl. In [189,191]Tl the 9/2 isomer has fallen below the 3/2 state and isomeric decay is insufficient to permit observation of the ground state.

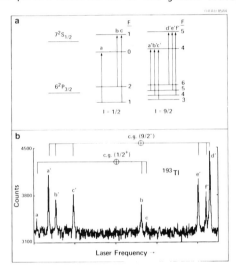

Fig. 2. Sample of hfs data recorded in present experiment of [193g,m]Tl. The upper portion shows the atomic hyperfine levels for the two spins.

The change in mean-squared-charge radius is obtained from the isotope shift using standard techniques [HEI74]. For thallium the normal mass shift is approximately 8 MHz between masses and the specific mass shift is smaller than the experimental error. The resulting field shift is proportional, to good approximation, to an electronic factor times $\delta<r^2>$. For the case of Tl the electronic factor is not directly calculable but should be virtually the same for all isotopes.

To deduce this proportionality factor for Tl we assumed zero deformation for [207]Tl (which is one proton from double magic [208]Pb) and [200]Tl. We then use the droplet model [MYE83] to extract deformations from the field shifts. Droplet model values of $\delta<r^2>$ with $\beta = 0$ are fit to the two points of zero deformation and predictions for different deformations are shown as solid lines in Figure 3.

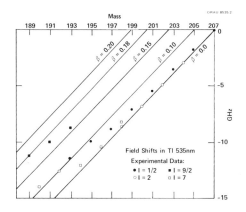

Fig. 3. Experimental field shifts in Tl. Solid lines represent dif- ferent droplet model deformations. Data for A ≤ 193 are from present experiment; remainder from previ- ous experiments. [HUL61].

The deformations of the isomers were also deduced from the spectrosco- pic quadrupole moment obtained from the hfs constant B. Since no quadrupole moments have been measured previously in Tl, hence no direct calibration could be made, we relied on the calculations of Lindgren and Rosen [LIN75]. Their relativistic calculations were used with our measured B's to give the values of Q_S listed in Table I.

TABLE I. Moments and deformations of light Tl isotopes

Tl	Q_S	Q_0	β_2	$\langle\beta_2^2\rangle^{1/2}$
	(eb)		from Q_0	from IS
193g	–	–	–	0.099(1)
193m	-2.20(2)	-4.02(3)	-0.144(1)	0.158(1)
191m	-2.27(3)	-4.16(5)	-0.150(2)	0.170(2)
189m	-2.29(4)	-4.19(7)	-0.153(3)	0.181(2)

A theoretical calculation was carried out to see if all the observed features of the Tl isotopes will emerge from the deformed shell model. First the equilibrium shape of the intrinsic mean field was determined by the

Strutinsky procedure for both the 1/2 and the 9/2 configurations of [185-199]Tl.
The energy as a function of quadrupole (ε_2) and hexadecapole (ε_4) deformation
was calculated. In Figure 4c the quadrupole deformamation coordinate ε_2 is
shown by the dashed line for the 9/2 configuration (for the 1/2 configuration
$\varepsilon_2 = -0.07$). Both the value and rate of change of deformation are seen to be
compatible with the experimental results from the present work. The main
reason for doubting the role of deformation in the past was based on the
notion that a simple correspondence exists between the deformation and the
rotational moment of inertia. The almost identical strong-coupled bands based
on the 9/2 level would then be incompatible with changing deformation. To
investigate this we calculated microscopic moments of inertia for the k = 9/2
bands at their respective equilibrium deformations determined above, using the
methods described in references [AND81] and [ARV83]. The results are plotted
in Figure 4b, where the 11/2 - 9/2 spacing is representative of the moment of
inertia. The significant result is that the rotational excitation energy
stays almost as constant in theory as in experiment despite the changing
deformation.

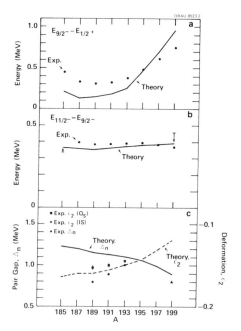

Fig. 4. Results of
theoretical calculations
for [185-199]Tl described
in the text in comparison
with experiment.

It is possible to understand this result on a microscopic basis. The microscopic mechanism for this lies in changing neutron pairing correlations, as can be verified by varying ε_2, Δ_p, Δ_n and N one at a time, holding the others fixed. The self-consistent neutron gap parameter, Δ_n, is plotted in Figure 4c and is seen to increase with decreasing A. Thus, when the number of valence neutron holes increases, both the deformation and the neutron pairing correlations increase with opposite and compensating effects on the moment of inertia. The arrows in Figure 4b show the shift that comes from the use of the self-consistent and N-dependent values of Δ_n, rather than a single intermediate value.

Conclusion

The experimental work here has shown that the deformation of the isomeric 9/2 states in the light thallium isotopes is increasing with decreasing neutron number while the 1/2 state remains relatively constant. This work shows the fallacy of assuming that a constant moment of inertia infers a constant deformation. Theoretically, this work has demonstrated an interesting aspect of the competition between quadrupole and pairing correlations in near-singly-closed-shell nuclei.

Acknowledgments

The authors wish to express appreciation to their colleagues: W.M. Fairbank, Jr., P. Juncar, R.L. Mlekodaj, and E.H. Spejewski. Support was furnished by the U.S. Department of Energy under contract numbers DE-AC05-76OR00033 and DE-AS05-76ERO-4936.

References

[AND81] C.G. Andersson et al. Nucl. Phys. A361 147 (1981).
[ARV83] P. Arve, Y.S. Chen and G.A. Leander, Phys. Scr. T5 157 (1983).
[HEI74] K. Heilig and A. Steudel, At. Data Nucl. Data Tables 14 613 (1974).
[HEY83] K. Heyde, P. Van Isacker, M. Waroquier, J.L. Wood, and R.A. Meyer, Phys. Repts. 102 291 (1983).
[HUL61] R.J. Hull and H.H. Stroke, J. Opt. Soc. Am. 51 1203 (1961). D. Goorvitch, S.P. Davis and H. Kleniman, Phys. Rev. 188 1897 (1969). R. Newgarut, priv. communication.
[KAU76] S.L. Kaufman, Opt. Commun. 17 309 (1976).
[LIN75] I. Lingren and A. Rosen, Case Studies in Atomic Collision Physics 4 93 (1975).
[MYE83] W.D. Myers and K.H. Schmidt, Nucl. Phys. A410 61 (1983).

RECEIVED July 18, 1986

Ground-State Studies at ISOLDE

H.-J. Kluge

ISOLDE, CERN, CH-1211 Geneva 23, Switzerland, and Institut für Physik, Universität Mainz, D-6500 Mainz, Federal Republic of Germany

A large number of nuclei far from the stability line have been studied in their ground and isomeric states at the ISOLDE on-line isotope separator at CERN. This report gives references to recent works, and describes future developments for a laser ion source, and further improvements of the techniques for determining nuclear masses, the spins, moments, and changes in nuclear charge radii. The odd-even staggering of the charge radii of the nuclei around Z=82 are briefly discussed.

1. INTRODUCTION

The ground state studies covered by this contribution concern the determination of the nuclear mass, the spin, the magnetic dipole and the electric quadrupole moments, and finally the change in charge radii. With the exception of masses the information is accessible by a measurement of the hyperfine structure (HFS) and isotope shift (IS) of optical transitions. These data are urgently needed to test nuclear models, to establish level schemes, or to extrapolate the properties of nuclear matter even further to the limits of nuclear stability against proton and neutron emission, i.e. to the drip lines. Also new phenomena might be discovered which are not expected from our knowledge of stable and long-lived nuclei.

Nuclear shape coexistence, which was extensively discussed during the symposium, might serve as an example to illustrate the importance of data from measurements of the mass, the HFS and IS. The most convincing signature of shape transitions or shape coexistence is the determination of the positions and depths of the minima in the energy-deformation curve. This information can indeed be obtained by a determination of the mass (depth), of the IS which yields the deviation from spherical shape ($\langle \beta^2 \rangle$) when measured relative to a spherical nucleus, and of the isomer shift which is a measure of the difference in deformation for the co-existing states ($\delta\langle \beta^2 \rangle$). Furthermore, the spectroscopic quadrupole moment indicates oblate or prolate shape and gives additional information on the location of the minima ($\langle \beta \rangle$). Finally, the determination of the nuclear spin and magnetic moment pins down the configuration and orbitals involved.

In fact, the first information on phase transitions and shape coexistence was obtained by optical spectroscopy: In 1949 Brix and Kopfermann [BRI49] found a

NOTE: The author of this chapter worked with members of the ISOLDE Collaboration, CERN, CH-1211 Geneva 23, Switzerland.

sudden jump of the IS's when going from N=88 to N=90 in the rare earth isotopes. This led to the discovery of the quadrupolar shape of nuclei. In 1972, our group found a strong discontinuity of the IS's of the neutron-deficient Hg isotopes [BON72] which was the first evidence for shape coexistence. Until recently, the sharpness of both shape transitions was theoretically not understood. Today we know that it is caused by the near neighbourhood of a magic (Z=82) or semimagic (Z=64) configuration which is now called reinforcement of magicity.

2. STATUS OF GROUND STATE STUDIES AT ISOLDE

Most data on ground state properties of nuclei far from stability have been obtained at the on-line mass separator ISOLDE at CERN, where the isotopes of over 60 elements are available with high yields (up to 10^{11} atoms per sec and mass number) and half lives down to 10^{-2}sec. [CER85]. Fig. 1 shows the chart of nuclei and indicates those regions where optical spectroscopy has been performed in long isotopic chains. Similar systematics of mass measurements have been restricted until now to the isotopes of the alkaline elements [AUD84].

Some of the results obtained at ISOLDE for ground and isomeric states are discussed in the contributions to this symposium by C. Ekstroem, H.T. Duong, and E. Roeckl. A description of the ISOLDE-2 (operating) and the ISOLDE-3 (under construction) facilities can be found in the contribution by B.W. Allardyce. Additional information and references are given in the proceedings of recent conferences on nuclei far from the stability line [OAK82, KAN84, LUE84, DAR84].

This contribution concentrates on a discussion of some on-line techniques for ground state studies which are now under development and will be applied at ISOLDE in the near future. The plans for a nuclear-orientation set up at ISOLDE are outlined in the contribution to this symposium by N. Stone.

3. MASS MEASUREMENTS

The Orsay group has demonstrated the power of on-line, direct mass measurements using a Mattauch-Herzog spectrometer [AUD84]. This program was terminated in 1983 because the masses of all accessible alkaline isotopes were measured. Several ISOLDE groups are now preparing the second generation of mass measurements in long isotopic chains.

USING THE ISOLDE-3 MASS SEPARATOR: The high resolution of ISOLDE 3 (design value = 3.10^4) enables direct mass measurements. Such measurements were proposed at the Workshop "On-line in 1984 and beyond" held at Zinal [ZIN84]. In these experiments, the transmission signals have to be split in order to obtain the necessary accuracy. However, the feasibility of this technique has already been demonstrated by the Chalk River group [SHA84].

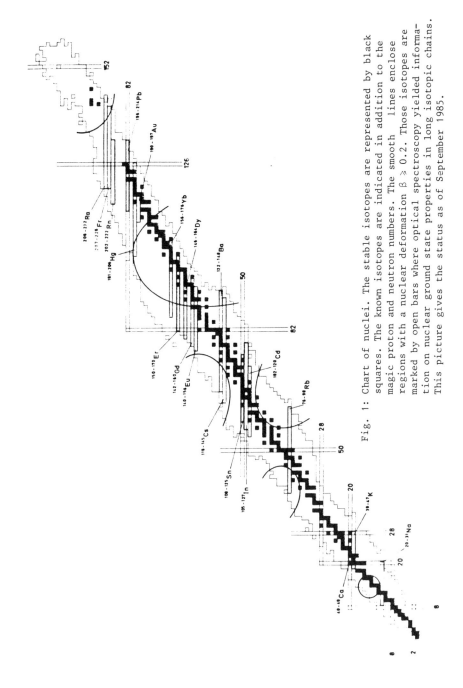

Fig. 1: Chart of nuclei. The stable isotopes are represented by black squares. The known isotopes are indicated in addition to the magic proton and neutron numbers. The smooth lines enclose regions with a nuclear deformation β ≥ 0.2. Those isotopes are marked by open bars where optical spectroscopy yielded information on nuclear ground state properties in long isotopic chains. This picture gives the status as of September 1985.

RF MASS SPECTROMETER: A very much improved version of Smith's RF spectrometer is under construction at Orsay [COC84]. First it will be used to measure the p-p̄ mass difference at the antiproton storage ring LEAR at CERN, with an accuracy of the order of $\Delta M/M = 10^{-9}$. Later on it is planned to install the apparatus at ISOLDE for mass measurements of radioactive isotopes with an accuracy better than 10^{-6}.

A high resolution of about 10^6 is achieved by passing the ions through electrostatic deflectors into 2 cyclotron orbits (r = 0.5m) in a homogenous magnet, and analyzing the outcoming beam by electrostatic deflectors. In the magnetic field the RF field changes the beam radius by 5 mm for ions in resonance, allowing them to pass through a slit (Fig. 2). Since the transmission of the ISOLDE beam is measured as a function of the applied RF frequency, and the expected transmission in resonance is about 10^{-4}, this technique can be applied to the whole variety of available ISOLDE beams.

PENNING TRAP: A different approach to mass measurements was developed at Mainz. Thermal ions can be confined in a static homogenous magnetic field and a superimposed electrostatic quadrupole field. A measurement of the cyclotron frequency enables the measurement of the mass of the stored ion. A resolving power of 10^6 to 10^8 and accuracies of 10^{-7} to 10^{-8} have been achieved with this technique for light masses [GRA80, VAN85].

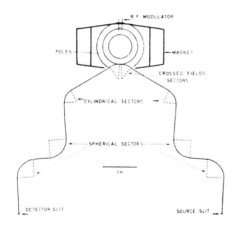

Fig. 2: RF mass spectrometer built at Orsay.

The experiment prepared for ISOLDE aims for a resolving power of 10^6 and an accuracy of better 10^{-6} which corresponds to an uncertainty less than 100 keV for an ion with mass number 100 [DAB84]. The set up is shown in Fig. 3. The ion trap for the measurement of the cyclotron frequency is placed in a super-conducting magnet. The resonance is measured by a time-of-flight technique [GRA80]. In order to fill the trap with about 10 ions from the outside world, the ion beam of ISOLDE is stopped on a filament mounted inside an auxilary ion trap. After heating the foil, the radioactive ions are surface ionized, stored, cooled and then ejected by a short pulse. The bunched ions can now be caught in the precision trap by retardation and by lowering for a short time the potential of its entrance electrode. These ions are then used to perform the resonance experiment, after which the process is repeated to refill the trap with a new sample. The time for each cycle is about 1 second. Clearly, the technique would greatly benefit from an ISOLDE beam bunched in the ion source (see below).

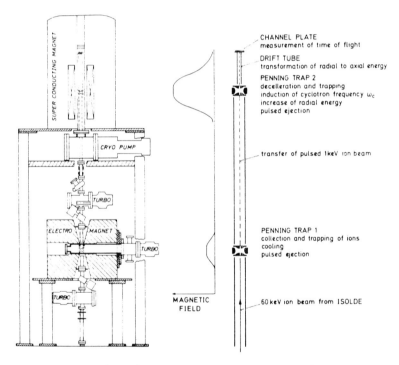

Fig. 3: Principle (right) and experimental set-up (left) for mass measurements in a Penning trap.

Both traps have been tested and perform almost as specified (ratio of incoming to ejected ion intensity about 10^{-4} for the bunching trap, resolution of the precision trap = 3.10^5 for A = 28). The transfer of the ions from the bunching trap to the precision trap is now under study.

4. LASER ION SOURCE

Resonance ionization spectroscopy (RIS) with pulsed tunable lasers offers new possibilities for constructing pulsed, highly selective ion sources with high efficiencies. Fig. 4 shows the principle of RIS and the planned set up. The efficiency of the ion source is determined mainly by the ratio of the repetition rate of the lasers to the frequency of collisions of the atoms with the walls of the ionization chamber [KLU85]. Hence, a pump laser with the highest available pulse rate has to be used. This can be realized by a copper vapor laser (v_{rep} = 6 kHz, length of laser pulse = 20 ns, pulse power = 300 kW). It was shown in the case of Gd that the three-step photoionization can be saturated via an autoionizing state using a laser beam of 1 cm diameter [PEU 85]. The efficiency of the laser ion source is expected to be up to 40% [KLU85]. Apart from high efficiency the method has an inherent high selectivity against the ionization of other elements. This is due to the low atomic level density reached by allowed E1 transitions (1 state/eV) and the comparably small Doppler width (10^{-5} eV).

The magnet of an on-line mass separator will generally assure the selectivity in respect to different isotopes. Hence, the isobaric selectivity of laser ion sources is of primary interest, although isotopic selectivity might also be achieved by RIS using the HFS and IS, which differ from isotope to isotope.

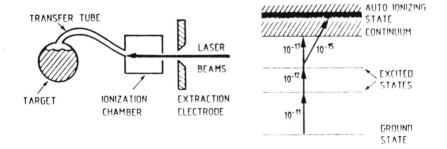

Fig. 4: Right: Principle of resonance ionization spectroscopy. Three tunable pulsed Dye lasers are used to stepwize excite and ionize the atom. The effective cross-sections are indicated for the different transitions in cm^2. Left: Layout of a laser ion source for on-line mass separators.

The time structure of the ion beam (every 150 µs a pulse with a length of 15 ns) enables the improvement of many techniques such as the filling of ion traps, the injection into storage rings or accelerators, or, for example, the suppressing of the background of collinear laser spectroscopy with detection of fluorescence light. In the latter case an improvement by two orders of magnitude will be obtained with a 10^{-4} duty cycle.

5. OPTICAL SPECTROSCOPY

Several groups at ISOLDE are planning further improvements of their techniques. For each element the most appropriate experimental scheme has to be found . Today, collinear laser spectroscopy is the most general high-resolution and sensitive method for optical spectroscopy on radioactive beams delivered by on-line mass separators. Its sensitivity ranges from 10^4 - 10^6 atoms/s depending on the strength and multiplicity of the optical transitions.

Until now, the main goal of the measurements was systematic ground state studies in long isotopic chains. In the future, however, more experiments will aim for the investigation of a specific nucleus. According to Murphy's law, these nuclei are generally far away from the valley of stability or are produced with low yield. Hence a special version of the technique or an improvement of the scheme are required. Table 1 gives a survey of these attempts and summarizes at the same time the on-going programme of ground state studies by optical spectroscopy at ISOLDE.

6. ODD-EVEN STAGGERING

The IS's of nuclei far from stability turned out to be the most informative data obtained by optical spectroscopy. This is because the nuclear charge radius depends on collective as well as on single-particle effects. The integral IS's ($\delta\langle r^2\rangle^{A,A'}$ with A' being a reference isotope) exhibit the gross behaviour of nuclear matter as a function of varying neutron number. These can be compared with predictions of macroscopic models like the Droplet Model [MEY83], which describes the overall trend quite well.

The differential IS's ($\delta\langle r^2\rangle^{A,A+1}$ or $\delta\langle r^2\rangle^{A,A+2}$) are measures of the radius change of the neighbouring nuclei and yield more clearly than the integral IS the effect of the addition of a neutron or a neutron pair. In this way, the increase or decrease of deformation is easily observable as a function of neutron number. A still more sensitive measure of the influence of an unpaired neutron on the charge radius is the odd-even staggering parameter γ introduced by H.H. Stroke [TOM64] and given by

$$\gamma^A = 2(\langle r^2\rangle^A - \langle r^2\rangle^{A-1})/(\langle r^2\rangle^{A+1} - \langle r^2\rangle^{A-1})$$

TABLE 1: On-line techniques for Optical Spectroscopy at ISOLDE

BASIC TECHNIQUE	INCREASE OF SENSITIVITY BY	APPLICATION	STATUS	REF.*
COLLINEAR SPECTROSCOPY IN FAST ATOMIC AND IONIC BEAMS				
- Observation of optical fluorescence	-	rare earth, Rn, In	data taking	(M203) (P77)
- Observation of optical fluorescence, with frequency doubled CW dye laser	-	D_2 line of Ra	first data taking Nov. 85	(IP-16)
- Observation of precession of optical polarization in a magnetic field	-	g_I of Ra isotopes	first data taking Dec. 85	(P76)
- Reionization by charge exchange after laser interaction	ion detection after charge exchange	Rn, Xe	tested	(M203)
- Resonant 3-photon ionization with pulsed laser (Cu-vapor pump laser)	ion detection after photo-ionization	rare earth	in preparation	(P78)
- Observation of fluorescence with pulsed ion beam	10^{-4} duty cycle of laser ion source	rare earth	in preparation	[KLU85]
- Two-step photo ionization with CW laser via Rydberg states	ion detection	In	tested	(P77)
- One-step photo ionization with CW laser via Rydberg states	ion detection	Fr, Cs	tested	(IP-17)
- Implantation of optically pumped beams in a crystal-observation of β asymmetry	asymmetry of β radiation	Li	first data taking Oct. 85	[BON84]
OPTICAL PUMPING IN RESONANCE CELLS				
- Polarization by spin exchange of optically pumped Rb observation of γ anisotropy	anisotropy of γ radiation	Rn	data taking	(P79)
RESONANCE IONIZATION SPECTROSCOPY IN THERMAL ATOMIC BEAMS				
- Resonant 3 photon ionization with pulsed laser, mass measurement of photo ions by time of flight	ion detection TOF	Au, Pt	first data taking Dec. 85	(IP-14)
- As above with pulsed thermal beam using laser	ion detection, TOF, duty cycle	Au	in preparation	[KLU84]

*) References in round parentheses give the CERN filing number of the proposal. Copies might be obtained by the ISOLDE Secretariat, EP Division, CERN.

γ =1 indicates that the effect of adding an odd neutron is just half of the effect
of adding a pair. Such a behaviour would be expected, e.g. by the Liquid Drop
Model. The other extreme case (γ = 0) is obtained when the charge distribution
of the nucleus with even N completely ignores the additional neutron. Generally,
γ < 1 is obtained throughout the chart of nuclei. In some rare cases, γ > 1 is
found. These exceptions can be associated with an irregularity of the IS of one
of the 3 isotopes entering the calculation of γ. Fig. 5 shows the staggering parameter in the region around N = 126. With the exception of ^{199m}Hg, γ < 1 is
found below N = 126 for all isotopes. In case of the I = 13/2 isomers of Hg, a
smooth decrease of γ is found to about γ = 0 at A = 185. Many explanations
have been tried to interpret the phenomenon: different shell filling in odd and
even isotopes [TOM64], blocking of zero-point quadrupole vibration in the odd
neutron isotope [REE71], different polarization of the core by the paired and
unpaired neutron [TAL84], and the influence of the type of coupling in the
particle-plus-rotor picture [STR79]. None of these interpretations gives a quantitative description of the staggering effect exept for Ca and Pb [TAL84].
Nevertheless, γ < 1 is an empirical fact found for almost all isotopes. However,
at N = 133 to 137, γ > 1 is observed. This anomaly cannot be attributed to an
irregularity of the IS of one individual isotope, but it is found for 7 nuclei of
Rn, Fr, and Ra. Above N = 137, the usual γ < 1 is observed again.

A similar anomaly was found in the masses of the Ra isotopes in this region
[LEA82], which stimulated the search for octupole deformation in these nuclei.
Again, the results of mass measurements and of optical spectroscopy might be
useful to fix the energy-deformation curve (now β_3 in addition). However, calculations are still missing. Perhaps the extended IBA model is the most appropriate way to calculate the effect of the additional or missing fermion on the
charge radii of the paired neighbouring nuclei. The regular behaviour of the
staggering parameter of the Hg isomers with their pure $i_{13/2}$ states eventually
offers the key for solving the puzzle.

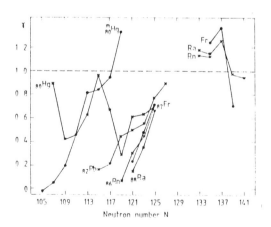

Fig. 5: Odd-even staggering
parameter of the iso-
topes around N = 126.
The data on Rn, Fr,
and Ra are obtained by
the Mainz and Orsay
groups and are pre-
liminary.

REFERENCES

[AUD84]   G. Audi et al., in [DAR84], p. 119 and references therein.

[BON72]   J. Bonn et al., Phys. Lett. 38B (1972) 308.

[BON84]   J. Bonn in [LEU84], p. 57.

[BRI49]   P.Brix et al., Z. Phys. 126 (1949) 344.

[CER85]   "ISOLDE USERS' GUIDE, ed. by H.-J. Kluge, CERN (1985);
          available from ISOLDE, CERN.

[COC84]   A. Coc et al. in [DAR84], p. 661.

[DAB84]   P. Dabkiewicz et al. in [DAR84], p. 684.

[DAR84]   Proceedings of the 7th Intern. Conf. on At. Masses and
          Fundamental Constants, Darmstadt 1984, ed. by O. Klepper,
          TH Darmstadt (1985); available from GSI/Darmstadt

[GRA80]   G. Graeff et al., Z. Phys. A 297 (1980) 35.

[KAN84]   Proceedings of the Intern. Workshop on Hyperfine Interactions,
          Kanpur 1984, Hyperf. Interactions 24, 25 and 26 (1985).

[KLU84]   H.-J. Kluge in [KAN84], p. 69

[KLU85]   H.-J. Kluge et al., Proceedings of the Accelerated
          Radioactive Beams Workshop, Vancouver 1985 (in print).

[LEA82]   G.A. Leander et al., Nucl. Phys. A388 (1982) 452.

[LUE84]   Proceedings of the Intern. Symposium on Nuclear Orientation
          and Nuclei far from Stability, Leuven 1984, ed. by
          B.I. Deutsch and L. Vanneste, Hyperf. Interactions 22 (1985).

[MYE83]   W.D. Meyers et al., Nucl. Phys. A 410 (1983) 61.

[OAK82]   Proceedings of the Conference on Lasers in Nuclear Physics,
          Oak Ridge 1982, ed. by C.E. Bemis, H.K. Carter, Harwood
          Academic Publishers, Chur-London-New York (1982).

[PEU85]   P. Peuser et al., Appl. Phys. B (in print 1985).

[SHA84]   K.S. Sharma et al. in [DAR84], p. 68.

[STR79]   H.H. Stroke et al., Phys. Lett. 82B (1979) 204.

[TAL84]   I. Talmi, Nucl. Phys. A423 (1984) 189.

[TOM64]   W.J. Tomlinson III et al., Nucl. Phys. 60 (1964) 614.

[REE71]   B.S. Reehal et al., Nucl. Phys. A161 (1971) 385.

[VAN85]   R.S. van Dyck et al., Int. Journal of Mass Spectrom.
          and Ion Processes, in print (1985).

[ZIN84]   Contributions to the Workshop on the ISOLDE programme "On-line
          in 1985 and Beyond", Zinal 1984; available from ISOLDE/CERN.

RECEIVED August 20, 1986

57

Laser Spectroscopy of Short-Lived Isotopes

H. T. Duong

Laboratoire Aimé Cotton, Centre Nationale de Recherche Scientifique, Bât. 505, 91405 Orsay, France

Nuclear properties (spins, moments, charge radii) revealed by the analysis of hyperfine structure and isotope shift of atomic levels have been obtained in decades of experiments. Since 1975 with the introduction of tunable dye laser, the rebirth of the methods, some already known since 1930, had led to many on line experiments on short lived isotopes not investigated before. I report here a sample of the experiments done by the Orsay, Mainz groups at CERN. Although experiments have been carried out by the Orsay group using the proton beam of the CERN Proton Synchrotron, most of the experiments have been done at Isolde, the on - line mass separator at CERN, whose radioactive beams are essential to the success of these experiments [RAV 84].

$20\text{-}31$Na

The method of particle detection of optical resonances, in alkali atoms, by means of magnetic deflection has been developed [DUO 74] and applied to 21.25Na at Orsay [HUB 75]. A target, placed inside a high temperature oven, is bombarbed with a proton beam. The thermalised reaction products are emitted from the oven and form an atomic beam, see Fig 1 a. The atoms interact with the light from a C.W. tunable single mode dye laser in the presence of a constant weak magnetic field H_c, which defines the quantization axis, parallel to the laser beam. At optical resonance with one of the hyperfine component of the D_1 or D_2 line, the laser light induces an optical pumping which changes the population distribution between the magnetic substates (F, m_F) of the ground state. The population changes are analysed by a six pole magnet which focuses those corresponding to $m_J = +1/2$ and defocuses those corresponding to $m_J = -1/2$. After the six pole magnet the focused atoms of the beam are ionized, mass separated and counted When the laser frequency is scanned the optical resonances appear as positive or negative peaks on a baseline ion signal. Between the interaction region with the laser light and the six pole magnet, see fig 1 b, a r.f magnetic field given by a r. f loop could be applied to the optically pumped atoms for magnetic resonance measurement of the hyperfine structure of the ground state [DUO 82] or the spin determination. Part of the optical resonances observed in the case of sodium [HUB 78] and the magnetic resonance curve in the ground state of 44K, [DUO 82] are shown in fig 1c and 1d respectively.

0097–6156/86/0324–0380$06.00/0

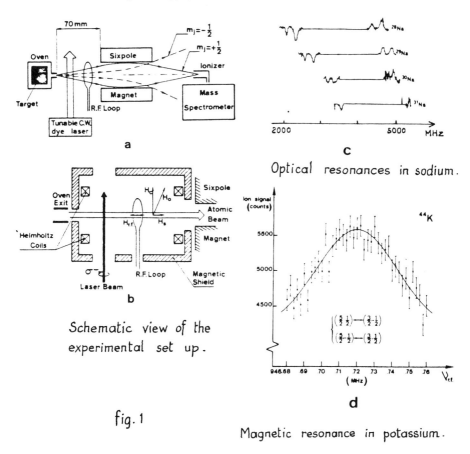

Optical resonances in sodium.

Schematic view of the experimental set up.

fig. 1

Magnetic resonance in potassium.

The only difference between the method described above and the ABMR method introduced by Rabi [RAB 38] is that the atoms are spin polarized by light interaction instead of magnetic deflection in the A magnet. Such a possibility of replacing deflecting inhomogeneous magnetic field by optical pumping had been pointed out several years ago [KAS 50] . The advantage is a much shorter apparatus, the overall length from the oven exit to the end of the six pole magnet is about 30 cm. The resulting gain in solid angle allowed the study of rare isotopes Optical resonances in sodium have been detected with 10^4 atoms coming out of the oven.

207 - 213, 220 - 228 Fr

When the apparatus is used at Isolde, where one deals with monoisotopic ion beam delivered by the mass separator, the ions are implanted then reemitted, as thermal atoms, out of the device shown in fig 2, [THI 81].

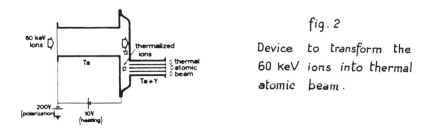

fig. 2

Device to transform the 60 keV ions into thermal atomic beam.

The rest of the apparatus is the same as when operated at the Proton Synchrotron. First tested on cesium [HUB 78], [THI 81] the apparatus was used to uncover the resonance lines of francium for which no optical transition had ever been observed. The CERN on line mass separator, Isolde, makes available a source of more than 10^9 atoms/sec of chemically and isotopically pure ^{213}Fr isotope. Such an amount is more than needed for a laser atomic beam spectroscopy. The first step is obviously to locate the resonance line at low resolution, using a broad band laser excitation. In a second step, once the line is located, a high resolution study is undertaken, [LIB 80] and [BEN 84]. The observed signal is displayed (fig 3a) at low resolution and (3 b) at high resolution.

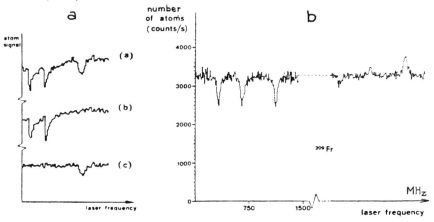

fig. 3 : Francium optical resonances.

And all the relevant data for francium are collected in [COC 85] .

Except for lithium, all the alkali atoms have been studied with the method developped in 1974. Data on hundreds of isotopes and isomers are obtained. Some of these data concerning the nuclear moments are theoretically analysed, see for example [CAM 80] , [MOL 80] ,[EKS 78] ,[EKS 79] and[RAG 79] .

When working on line with a mass separator, such as Isolde, the collinear laser spectroscopy is a method fully adapted. In a collaboration Orsay - Mainz, the second members of the principal series in francium have been located and studied at high resolution with this method. In table 1 the measured wavenumbers of the four lines are given.

^{212}**Fr**

D_1	$\sigma=$ 12 237.4093(20)cm^{-1}
D_2	$\sigma=$ 13 924.2984(20)cm^{-1}
D_1'	$\sigma=$ 23 113.0506(20)cm^{-1}
D_2'	$\sigma=$ 23 658.3918(20)cm^{-1}

Table 1

The schematic view of the Mainz apparatus for collinear laser spectroscopy, installed at Isolde is given in fig 4. The 60 keV ion beam is set collinear with the laser beam, then accelerated (or decelerated) and finally neutralized in charge exchange cell. By Doppler tuning the atomic absorption is set resonnant with the stabilized laser frequency, and the fluorescence emitted is detected.

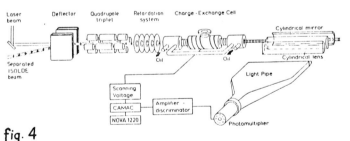

fig. 4

Schematic view of the Mainz collinear laser spectroscopy apparatus .

Because of the velocity bunching effect due to initial acceleration the ion beam is nearly monokinetic, and the neutralisation does not effect the velocity distribution The details of the method can be found in [KAUF 78], [NUE 78] By neutralisation in an alkali vapour, the atomic metastable states are preferentially populated since their energies match the ionisation potential of the corresponding alkali atom. Therefore this technic is ideally suited for laser spectroscopy of rare gas, and is recently successfully used to study the heaviest one, radon. Fig. 5 gives the recording of the fourteen hyperfine components in the J=2 → J=3 transition in 207 radon isotope.

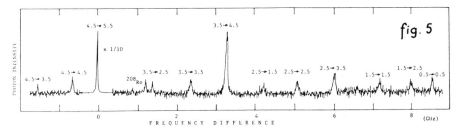

Fluorescence spectrum of ^{207}Rn in the transition $7s\ [3/2]_2 - 7p\ [5/2]_3$.

Analysis of the data obtained is under way. Preliminary results for isotope shift show the same inversion in odd-even staggering in the same neutron range as in francium and radium. This is a possible evidence for an octupole mode of deformation as suggested in [AHMAD 83]; see also [LEA 84] ,[NAZ 84] , and [SHE 83] .

Conclusion

In the past ten years a huge amount of data by laser spectroscopy of short lived isotopes have been collected and the field is still developping with new proposed technics. Instead of analysis of individual data, which is not desirable, trends may be found and thus raise up further theoretical interest. In this respect, the odd-even staggering in isotope shift, a feature well known long ago and very well documented, deserves certainly a close theoretical attention.

References

[AHM 83] S.A. AHMAD et al., Atomic Masses and fundamental constants 7 ed. by O. Kepper, Technische Hochschule Darmstadt Lehrdruckerei, 361 (1984)

[BEN 84] N. Bendali et al., C.R. Acad. Sc. Paris 299, 1157 (1984)

[COC 85] A. Coc et al., accepted for publication in Phys. Lett.

[CAM 80] X. Campi et M. Epherre, Phys Rev C 22, 2605 (1980)

[DUO 74] H.T. Duong and J.L. Vialle, Opt. Commun. 12, 71 (1974)

[DUO 82] H.T. Duong et al., J. Physique 43, 509 (1982)

[EKS 78] C. Ekstrom, S. Ingelman, G. Wanberg and M. Skarestad, Nucl. Phys. A 311, 269 (1978)

[EKS 79] C. Ekstrom, L. Robertson, G. Wanberg and J. Heinemeier, Phys. Scr 19, 516 (1979)   HUB 75   G. Huber et al., Phys. Rev. Lett. 34, 1209 (1975)

[HUB 78] G. Huber et al., Phys. Rev. Lett 41, 459 (1978)

[KAS 50] A. Kastler, J. Physique Radium, 11, 255 (1950)

[KAU 78] S.L. Kaufman, W. Klempt, G. Moruzzi, R. Neugart E.W. Otten and B. Schinzler, Phys. Rev. Lett 40, 642 (1978)

[LEA 84] G. A Leander and R.K. Sheline, Nucl. Phys. A413, 375 (1984)

[LIB 80] S. Liberman et al., Phys. Rev A22, 2732 (1980)

[MOL 80] P. Möller and J.R. Nix, Los Alamos Report LA - UR 80, 1996 (1980)

[MUE 83] A.C. Mueller, F. Buchinger, W. Klempt, E.W Otten, R. Neugart and B. Schinzler, Nucl. Phys. A403, 234 (1983)

[NAZ 84] W. Nazarewicz et al. Nucl. Phys A429, 269 (1984)

[RAB 38] I.I. Rabi, J.R. Zacharias, S. Milman and P. Kusch, Phys Rev 53, 318 (1938)

[RAG 79] I. Ragnarson, Symposium on future directions in studies of nuclei far from stability, Nashville Tennessee (1979)

[RAV 84] H.L. Ravn, Paper D9 in Appendix, A workshop on the Isolde programme, Zinal (1984)

[SHE 83] R.K. Sheline and G.A. Leander, Phys Rev Lett 51, 359 (1983)

[THI 81] C. Thibault et al., Nucl. Phys. A367, 1 (1981)

RECEIVED September 2, 1986

58

Magnetic Moments of Excited States in Nuclei Far from Stability

A. Wolf [1], Z. Berant [1], R. L. Gill [2], D. D. Warner [2], John C. Hill [3], F. K. Wohn [3], G. Menzen [4], and K. Sistemich [4]

[1] Nuclear Research Center Negev, Beer-Sheva 84190, Israel
[2] Brookhaven National Laboratory, Upton, NY 11973
[3] Ames Laboratory, Iowa State University, Ames, IA 50011
[4] Institut für Kernphysik, Kernforschungsanlage Jülich, D-5170 Jülich, Federal Republic of Germany

Magnetic moments of excited states in nuclei far from stability have been measured by gamma-gamma angular correlation at the output of the fission product separators TRISTAN and JOSEF. The results obtained until now will be reviewed. They provide important nuclear structure information about nuclei around closed shells, and transitional nuclei in the A=100 and 150 regions.

1. Introduction

A large variety of nuclei off the line of stability are produced by fission. These are nuclei on the neutron-rich side of the periodic table, most of which can not be reached by any other nuclear reaction. The most interesting nuclei produced in this way can be classified in two groups: a) nuclei around closed shells (i.e. the region of ^{96}Zr, ^{132}Sn, the N=82 isotones); b) nuclei around A=100 and A=150 where rapid transitions from spherical to deformed shapes are known to occur. We will refer to these nuclei as belonging to group (a) and group (b). The study of spectroscopic properties of nuclei in these groups provides valuable tests of different nuclear models such as the shell model, the geometrical collective model, IBA.

Magnetic moments are sensitive to the details of the nuclear wave function, and thus it is of considerable interest to measure g-factors of excited states in nuclei produced by fission. The first measurements of this type have been performed using a ^{252}Cf spontaneous fission source [CHE76, WOL76]. With the advent of high-intensity on-line isotope separator devices such as JOSEF [LAW76] and TRISTAN [GIL81], more magnetic moments of neutron-rich nuclei were measured. Until now, g-factors of excited states in ^{97}Zr [BER85a], ^{100}Zr [WOL80], ^{136}Xe, ^{138}Ba [BER85b], and ^{142,144}Ba [WOL83a], were measured using the perturbed angular corelation (PAC) method and the intense radioactive beams provided by JOSEF and TRISTAN. These measurements were carried out with conventional electromagnets, with which magnetic fields of up to about 3.0 Tesla could be obtained. Recently, superconducting magnets were installed at both TRISTAN and JOSEF, thus making possible g-factor measurements of states with half-lives down to about 0.1 nsec. The latest results are g-factors of 2_1^+ states of ^{102,104}Mo [MEN85] and ^{146,148}Ce [WOL85a].

In this talk we will briefly review magnetic moment results for excited states in nuclei around closed shells (group a), and their significance to the shell model. Then we will summarize the results for transitional nuclei, and discuss the systematics of g-factors of 2_1^+ states at the onset of deformation around A=100 and A=150.

2. Experimental techniques

A rather characteristic feature of nuclei off the line of stability is that many of them have relatively long-lived low-excited states, and others have prominent isomeric states. For example, even-even nuclei in the vicinity of shells belonging to group (a) above have 4_1^+ states with $T_{1/2} \sim 1-4$ nsec and

0097-6156/86/0324-0386$06.00/0
© 1986 American Chemical Society

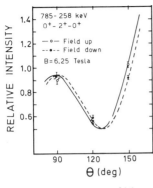

Fig.1 PAC results for ^{146}Ce

6^+_1 states with $T_{1/2}\sim100-200$ nsec. Even-even nuclei at the onset of deformation (group b) have 2^+_1 states with $T_{1/2}\sim0.1-4$ nsec. These half-lives are suitable for either time-differential or integral PAC measurements. This well-known technique [FRA66] involves the measurement of a $\gamma-\gamma$ correlation in an external magnetic field. The time-differential technique can be applied to states with half-lives larger than about 10 nsec. For shorter half-lives, the integral PAC method should be used, in which one measures either the angular shift $\Delta\theta$ of the correlation when an external field B is applied, or the double ratio:

$$R^2(\theta) = \frac{I(\theta,B)}{I(\theta,-B)} \Big/ \frac{I(-\theta,B)}{I(-\theta,-B)} \quad (1)$$

where $I(\theta,B)$ is the number of counts of the $\gamma-\gamma$ correlation at angle θ, with magnetic field up. Both $\Delta\theta$ and $R(\theta)$ are functions of the angular correlation coefficients A_{22}, A_{44} and of the product $gBT_{1/2}$, $T_{1/2}$ being the half-life of the state involved and g its g-factor. For short-lived states large values of the magnetic field B are needed. High magnetic fields (of the order of 5 Tesla) are necessary in two cases: a) for short-lived states ($T_{1/2}\sim0.1-0.5$ nsec); b) for cases where the beam intensity provided by the separator is weak and a large $R(\theta)$ is necessary to attain a reasonable error bar on the g-factor. In general, the maximum value of $R(\theta)$ is limited by the anisotropy of the angular correlation. For $0^+-2^+-0^+$ cascades, the maximum value is $R(150^0)\sim2.0$. This value takes into account the solid angle attenuation of the correlation in a typical experiment at TRISTAN.

Superconducting magnets providing fields of up to 6.25T were installed on output beam ports of the separators JOSEF and TRISTAN. The separation methods of these devices are different, but both finally produce radioactive sources of fission products on a moving tape device. The tape carries the activity to a position inside the superconducting magnet. The angular correlation system, consisting of three Ge detectors at JOSEF and four detectors at TRISTAN is centered around the magnet. The experimental systems were described in detail elsewhere [WOL83b, BER85b, MEN85]. As an example, we present in Figure 1 the PAC measurement for ^{146}Ce [WOL85a], obtained at TRISTAN with a field of 6.25T.

3. g-factors of excited states in nuclei around closed shells

3.1. g-factor of the $7/2^+$ 1264.4 keV level in ^{97}Zr.
Shell model calculations predict a quasi-shell closure at ^{96}Zr. Therefore, it is of interest to measure g-factors of states in ^{97}Zr and test whether they can be described by simple shell model configurations. The 1264.4 keV level has a half-life of 102 nsec, and its g-factor was measured by the time-differential PAC method at TRISTAN [BER85a]. The result, g=0.39(4), is consistent with the Schmidt value of 0.43, which assumes no core polarization and the free value for the neutron g_s factor, $g_s=g_s^{free}$. This indicates that the 1264.4 keV level is a very pure single-particle state, thus confirming the shell model prediction of a quasi-shell closure at ^{96}Zr.

3.2. g-factors of 4_1^+ and 6_1^+ states in N=82 isotones.
The 4_1^+ states of the N=82 isotones ^{136}Xe, ^{138}Ba have half-lives of 1.32

Table 1
g-factors of 4_1^+ and 6_1^+ states
in N=82 isotones

Nucleus	J^π	g(exp)	g(calc)
^{134}Te	6^+	0.846(25)	0.82
^{136}Xe	4^+	0.80(15)	0.84
^{138}Ba	4^+	0.80(14)	0.90
	6^+	·0.98(2)	1.01
^{140}Ce	4^+	1.11(4)	1.11

and 2.17ns, respectively. Their g-factors were measured at TRISTAN using the integral PAC method [BER85b]. The 6_1^+ state in ^{134}Te$_{82}$ is an isomeric state with $T_{1/2}$=163ns. The g-factor of this state was measured with a ^{252}Cf source, using the inherent alignment of prompt fission products and the time-differential technique [WOL76]. The results are presented in Table 1, together with g(4^+) for

^{140}Ce$_{82}$ and g(6^+) for ^{138}Ba [PEK79, IKE76].
A shell model calculation was carried out for the N=82 isotones from ^{133}Sb to ^{148}Dy using all $g_{7/2}$, $d_{5/2}$, $d_{3/2}$, $s_{1/2}$ and $h_{11/2}$ configurations outside the closed ^{132}Sn core. The resulting configuration mixed wave functions were used to calculate the g-factors in Table 1, and using two sets of elemental proton g-factors. The first set consists of the free values of g_ℓ and g_s. The calculated g-factors are significantly smaller than the experimental values[BER85b] This is due to the neglect of core polarization effects, as was discussed in detail for ^{134}Te [WOL76]. In order to account for these effects, we calculated a set of effective elemental proton g-factors from the ground state g-factors of ^{139}La and ^{141}Pr. The resulting values, g_ℓ'=1.12 and g_s'=4.12 were then again combined with the shell-model wave functions to calculate g(4_1^+) and g(6_1^+). The results, given in column 4 of Table 1, are in excellent agreement with the measured values. Thus, a single set of effective g-factors is sufficient to account for core polarization effects in the four N=82 isotones discussed here. Moreover, a close inspection of the contribution of various components of the wave-functions to the respective g-factors, reveals that in all cases a single component of the wave function contributes significantly (more than 75%) to the g-factor. More specifically, this component is $(g_{7/2})^n$ with n=2, 4,6 for ^{134}Te, ^{136}Xe, ^{138}Ba respectively.

4. g-factors of 2_1^+ states at the onset of deformation
The magnetic moments of collective states are mainly determined by the proton motion. The simple hydrodynamical model predicts g=Z/A for such states. Significant deviations from this relation were found experimentally. These deviations are due to: a) different pairing forces of protons and neutrons; this effect causes a reduction of about 15% with respect to Z/A, but does not affect the smooth behavior of g vs. A; b) changes in the number of protons that actually take part in the collective motion. This latter effect, which is particularly important in the vicinity of subshell closures, causes significant structure in the dependence of g vs. A. Such structure was observed for neutron-rich Nd and Sm isotopes [KOL84]. These nuclei are in the region A=150 where a transition from vibrational to rotational structure takes place around N=88. This onset of deformation causes the elimination of the Z=64 subshell closure, which means that around N=88 drastic changes in the number of active protons are expected to occur. Additional nuclei in this region are the neutron-rich Ba, Ce, Gd and Dy isotopes.
The A=100 region is similar to the A=150 region. The onset of deformation is observed here in Sr, Zr, Mo and Ru isotopes, when the number of neutrons increases beyond N=58. Also, the Z=40 subshell is eliminated for N≥60.
We mentioned in the Introduction that neutron-rich nuclei around A=100 and A=150 are produced by fission. The PAC method was used to measure

g-factors of 2_1^+ states in some of these nuclei. $g(2_1^+)$ for ^{100}Zr, ^{102,104}Mo were measured at JOSEF [WOL80, MEN85], and for ^{144,146}Ba, ^{146,148}Ce were measured at TRISTAN [WOL83a,WOL85a]. The experimental results are summarized in Table 2.

4.1. Systematics of $g(2_1^+)$ in the A=150 region.
In Figure 2 we plotted the experimental $g(2_1^+)$ values vs. A, for the neutron-rich Ba - Gd isotopes. The data was taken from Table 2, [KOL84] and [LED78]. We see that while for Ba and Gd the $g(2_1^+)$ values slowly decrease as a function of A, as expected for a normal Z/A dependence, the behavior for Ce, Nd and Sm is different. In fact, a pronounced increase of $g(2_1^+)$ vs. A is observed for these isotopes.

We now proceed to analyze the values of $g(2_1^+)$ in terms of IBA-2. The g-factor of a state which is purely symmetric in the proton and neutron degrees of freedom can be written [SAM84]:

Table 2
Experimental values of $g(2^+)$ for transitional nuclei, from JOSEF and TRISTAN

Nucleus	$T\frac{1}{2}(2_1^+)$ (nsec)	$g(2_1^+)$
^{100}Zr	0.71(3)	0.22(5)
^{102}Mo	0.114(13)	0.42(7)
^{104}Mo	0.91(3)	0.19(11)
^{144}Ba	0.70(3)	0.34(5)
^{146}Ba	0.85(6)	0.28(7)
^{146}Ce	0.25(3)	0.24(5)
^{148}Ce	1.01(6)	0.37(6)

$$g(2_1^+) = g_\pi N_\pi/N_t + g_\nu N_\nu/N_t, \qquad (2)$$

where g_π,g_ν are the proton boson, neutron boson g-factors, N_π,N_ν are the number of proton, neutron bosons, and $N_t=N_\pi+N_\nu$. When we use equation (2), the question of counting N_π,N_ν arises. The counting of N_ν is straightforward beyond N=82. The counting of N_π is somewhat complicated by the elimination of the Z=64 subshell for N≥90. Thus, N_π should be counted in the 50-64 subshell for N≤88, and in the major 50-82 shell for N≥90. According to this counting procedure, the Sm isotopes have N_π=1 for N≤88 and N_π=6 for N≥90, i.e. a drastic change ΔN_π =5 occurs as we add neutrons beyond N=88. For Ce, Nd this effect is smaller, i.e. ΔN_π=1,3 respectively, while for Ba ΔN_π=0 since Ba is below midshell for both 50-64 and 50-82 shells. From (2) it is obvious that for Ce, Nd and Sm, $g(2_1^+)$ will suddenly increase at N=90, due to the change ΔN_π in the number of active proton bosons. We have here a qualitative explanation for the structure observed in Figure 2, and also for the different behavior of Ba, where ΔN_π=0 and no drastic effect is expected.

The above counting procedure assumes a sudden dissipation of the Z=64

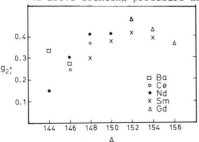

Figure 2. $g(2_1^+)$ systematics around A = 150.

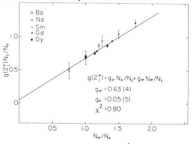

Figure 3. Plot of $g(2_1^+)N_t/N_\nu$ vs. N_π/N_ν. Reproduced with permission from [WOL85b]. Copyright 1985 North-Holland Physics Publishing Company.

Table 3
Values of N_π^{eff} in the
$A=150$ region

Nucleus	N_π^{eff}
^{146}Ce	1.5(6)
^{148}Ce	4.9(21)
^{144}Nd	0.2(1)
^{146}Nd	1.6(5)
^{148}Nd	5.0(15)
^{150}Nd	6.7(18)
^{148}Sm	1.5(3)
^{150}Sm	4.0(8)
^{152}Sm	6.8(13)
^{154}Sm	7.0(9)

subshell at N=90. Actually, it is reasonable to expect that this dissipation should be gradual. The experimental $g(2_1^+)$ values for Ce, Nd and Sm isotopes can be used to extract effective values of N_π (N_π^{eff}) across the region of interest (i.e. N=86-92). One can thus follow the dissipation process of the subshell. N_π^{eff} can be determined by using equation (2) and the experimental values of $g(2_1^+)$. In order to do this, we need the values of g_π, g_ν. We rewrite equation (2) :

$$g(2_1^+)N_t/N_\nu = (g_\nu + g_\pi N_\pi/N_\nu) \qquad (3)$$

If we assume g_π, g_ν to be constant in the region of A=150, then from (3) it follows that a linear relationship exists between $g(2_1^+)N_t/N_\nu$ and N_π/N_ν. To test this we use experimental values of $g(2^+)$ for $^{144,146}Ba$, ^{150}Nd, $^{152,154}Sm$, $^{154-160}Gd$, and $^{160-164}Dy$. For all these isotopes N_π is counted in the major 50-82 shell. The results are shown in Figure 3. We see that the linear relationship predicted by equation (3) is very well confirmed by the data, with $g_\pi = 0.63 \pm 0.04$, $g_\nu = 0.05 \pm 0.05$. These values of g_π, g_ν can now be used together with the experimental $g(2_1^+)$ from Table 2 and [KOL84] to obtain N_π^{eff} for the transitional Ce, Nd, and Sm isotopes. The results are given in Table 3 [WOL85b]. Although some of the error bars are large, we clearly observe an increase of N_π^{eff} across the region N=86-92, thus confirming the gradual dissipation of the Z=64 subshell. A similar result was obtained by Casten [CAS85a] from an analysis of energy level systematics.

Another interesting result of the above analysis is the relatively low value of g_π, much less than the expected $g_\pi = 1.0$. Detailed microscopic calculations are needed to explain this fact. Moreover, when we use $g_\pi = 0.63$, $g_\nu = 0.05$ to calculate the B(M1) strength of the recently observed isovector 1^+ states [BOH84], we obtain B(M1)$_{calc} = 0.9 - 1.3\mu^2 N$, for the Gd, Dy, Er, Yb isotopes. This value is in excellent agreement with the experimental results B(M1)exp=0.8-1.5$\mu^2 N$. However, when we use $g_\pi = 1.0$, $g_\nu = 0.0$, we obtain B(M1)$_{calc} = 2.7 - 3.8\mu^2 N$, i.e. much larger than the experimental values.

4.2. $g(2_1^+)$ results in the A=100 region.

The experimental data in the transitional region A=100 is scarce. A systematic study like in Figs. 2,3, is not possible here. The Z=40 subshell is eliminated for N≥60. Sr, Kr and Se are 1,2,3 bosons away from Z=40 and thus are analogous to Sm, Nd and Ce. Interesting isotopes are: $^{94-100}Sr$, $^{92-98}Kr$, etc. Unfortunately, no $g(2_1^+)$ data is available for these nuclei. The only existing results are for ^{100}Zr and $^{102,104}Mo$. (Table 2). The value of $g(2_1^+)$ for ^{100}Zr is quite small, and suggests, in analogy with the A=150 region, that a relatively small number of protons take part in the collective motion. It is therefore of interest to calculate N_π^{eff} for ^{100}Zr. To that purpose we need the values of g_π, g_ν. Because of insufficient data, a best fit procedure as in Figure 3 is not possible. However, if we assume $g_\nu = 0$, then g_π can be extracted from $g(2_1^+)$ of $^{102,104}Mo$, provided N_π, N_ν for this nucleus are known. We use $N_\pi = 3$, and $N_\nu = 5,6$ for $^{102,104}Mo$, obtained from IBA-2 calculations [SAM82], and calculate g_π using equation (2). The resulting value, $g_\pi = 1.00(23)$, is larger than g_π in the A=150 region. With $g_\pi = 1.0$ and $g_\nu = 0.0$, the experimental $g(2_1^+)$ of ^{100}Zr gives:

$$N_\pi^{eff} = 1.5 \pm 0.6$$

This value is much smaller than $N_\pi = 5$, which would be expected if the Z=40 subshell were eliminated. We conclude that the vibrational-rotational transition has not been completed in ^{100}Zr, and that the Z=40 subshell persists even for N=60. A similar result for ^{100}Zr was recently reported by Casten [CAS85a], from energy level systematics. Casten finds $N_\pi=3.3$ for ^{100}Zr, also less than 5, and thus in qualitative agreement with our conclusion.

5. Conclusions

We have shown that g-factors of excited states in nuclei far from stability can provide important nuclear structure information, such as: purity of wave functions, configuration-mixing, number of active protons, dissipation of shell closures, values of g_π, g_ν.

Further experiments should concentrate towards the double magic ^{132}Sn, nuclei around ^{96}Zr, and transitional even-even nuclei around A=100. Open questions are, for example: what are the values of $g(4_1^+)$ and $g(6_1^+)$ for ^{132}Sn and what is the structure of the respective states? Is the dissipation of the Z=40 subshell similar to that of the Z=64 subshell? Is g_π in the A=100 region really different from that in the A=150 region? Also, the transitional nuclei around A=130 and 190, recently discussed by Casten [CAS85b], should also be explored and will certainly provide a lot of interesting data.

References

[BER85a] Z. Berant et al., Phys. Lett. 156B, 159 (1985).
[BER85b] Z. Berant et al., Phys. Rev. C31, 570 (1985).
[BOH84]] D. Bohle et al., Phys. Lett. 148B, 260 (1984)
[CAS85a] R.F. Casten, Phys. Lett. 152B, 145 (1985)
[CAS85b] R.F. Casten, Phys. Rev. Lett. 54, 1991 (1985)
[CHE76] E. Cheifetz and A. Wolf, Cargese Conf., CERN 76-13 p.471 (1976)
[FRA66] H. Frauenfelder and R.M. Steffen, in : Alpha-, Beta- and Gamma-ray Spectroscopy, ed. K. Siegbahn (1966).
[GIL81] R.L. Gill et al., Nucl. Instr. Methods 186, 243 (1981)
[IKE76] H. Ikezoe et al., Hyp. Int. 2, 331 (1976).
[KOL84] N. Benczer-Koller et al., Ann. Israel Phys. Soc. 7, 133 (1984).
[LAW76] H. Lawin et al., Nucl. Instr. Methods 137, 103 (1976).
[LED78] C.M. Lederer and V.S. Shirley eds., Table of isotopes, 7th Ed.
[MEN85] G. Menzen et al., Z. Physik A 321, 593(1985).
[PEK79] L.K. Peker, Nucl. Data Sheets 28, 267 (1979).
[SAM82] M. Sambataro and G. Molnar, Nucl. Phys. A376, 201 (1982).
[SAM84] M. Sambataro et al., Nucl. Phys. A423, 333 (1984).
[WOL76] A. Wolf and E. Cheifetz, Phys. Rev. Lett. 36, 1072 (1976).
[WOL80] A. Wolf et al., Phys. Lett. 97B, 195 (1980).
[WOL83a] A. Wolf et al., Phys. Lett. 123B, 165 (1983)
[WOL83b] A. Wolf et al., Nucl. Instr. Methods 206, 397 (1983).
[WOL85a] A. Wolf et al., to be published
[WOL85b] A. Wolf, D.D. Warner and N. Benczer-Koller, Phys. Lett. 158B 7 (1985)

RECEIVED September 5, 1986

59

Spectroscopy and Measurement of Electromagnetic Moments in ^{198,200,210}Po

K. H. Maier

Hahn-Meitner-Institut Berlin, D-1000 Berlin 39, Glienicker Str. 100, Federal Republic of Germany

The quadrupole coupling constants for the ^{210}Po $I^{\pi}=8^+$, 11^-, 13^- isomers in Bi have been measured, and $Q(11^-) = 82(2)$ fm^2 and $Q(13^-) = 90(2)$ fm^2 normalized to $Q(^{210}$Po$8^+) = 57$ fm^2 are deduced. In beam γ-spectroscopy of ^{198,200}Po showed the ($\pi h_{9/2}$ 8^+) $\pi(h_{9/2}^2 i_{13/2}$ 11^-) and ($\nu i_{13/2}$ 12^+) isomers. The B(E2 $8^+\rightarrow 6^+$) and $Q(8^+)$ in ^{198}Po to ^{210}Po are discussed, a sudden drop is found for the B(E2) in ^{198}Po. The B(E3,$11^-\rightarrow 8^+$) rises very steeply in the light Po isotopes.

This contribution covers work done with the Nuclear Structure Group at Lawrence Livermore National Laboratory and at the Hahn-Meitner-Institut at Berlin. The former uses the triton beam of the Los Alamos tandem for γ-spectroscopy in beam to measure fairly basic properties of simple few particle states very close to ^{208}Pb. There are still many gaps in our knowledge, particularly on neutron particle and proton hole states, a few of these could be filled by this unique setup for γ-spectroscopy with neutronrich triton projectiles. Here I will cover recent measurements of the quadrupole moments for the 8^+, 11^- and 13^- isomers in ^{210}Po. At HMI we use the heavy ion beams from VICKSI for spectroscopy of nuclei further away from ^{208}Pb to observe the gradual transition from pronounced single particle structure to increasingly collective behaviour. As example spectroscopy of ^{198,200}Po is presented here. We can then look into the systematics of Po-isotopes from magic ^{210}Po with N=126 down to N=114 with a few holes in the $i_{13/2}$ neutron shell and therefore diminished stability of spherical shape.

A single crystal of ^{209}Bi metal was struck by a pulsed beam of 15 MeV tritons with about 1 ns pulsewidth. The bismuth crystal was heated to 478±5 K to avoid effects from radiation damage. Its c-axis pointed at 45° to the beam in the plane of the two detectors at 158° and 90°. Fig. 1 gives the relevant part of the ^{210}Po level scheme. For the high energy lines from the 11^- and

0097-6156/86/0324-0392$06.00/0

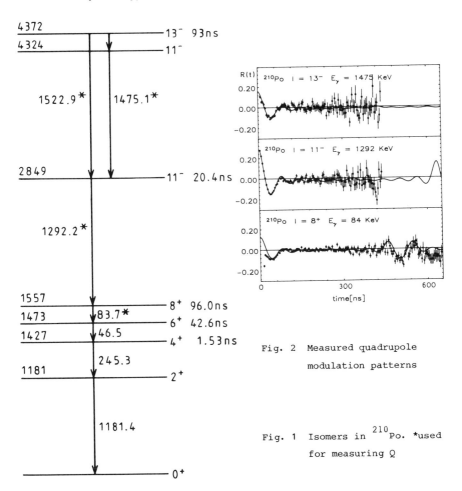

Fig. 2 Measured quadrupole modulation patterns

Fig. 1 Isomers in ^{210}Po. *used for measuring Q

13^- levels coaxial Ge-detectors with 6 mm lead absorbers were used, whereas for the 83.7 keV $8^+ \to 6^+$ transition, lying between the Bi- and Po- K_α and K_β X-rays, planar detectors were employed. Fig. 2 shows the quadrupole modulation patterns $1/2[N(158^\circ,t)/N(90^\circ,t)-1]$ and the results are presented in the table. The field gradient for Po in Bi eq(PoBi) = $11.7 \cdot 10^{17}$ V/cm^2 is normalized by calculating $Q(^{210}$Po $8^+)$ = 57 fm^2 from the measured B(E2, $8^+ \to 6^+$) assuming pure h9/2 configurations. As the structure of ^{210}Po is quite pure this is the best normalization available. Mahnke et al. [2] got

$14.5(15) \cdot 10^{17}$ V/cm^2 from the same procedure for ^{208}Po which is less safe on
the nuclear model assumption.

Table 1: Quadrupole moments in ^{210}Po. The coupling constants are for PoBi at
478(5) K. *Calculated as reference from B(E2,$8^+ \to 6^+$)

level	$e^2 Qq/h$		$|Q|$ present	$|Q|$ (ref. 3)
8^+	160.7(10)	MHz	57*	57*
11^-	229(4)	MHz	81(2) fm^2	82(19) fm^2
13^-	253(4)	MHz	90(2) fm^2	62(11) fm^2

In detail the time distribution of the 83.5 keV line at early times is
not reproduced by the fit, due to feeding from the 11^- isomer. However, the
distance to the next pronounced structure, at which all nuclear spins are in
phase again is practically independent of this. Feeding from 13^- into 11^- is
weak and spread out in time and therefore unimportant. Variations of the
fitted range and other parameters changed the results for the frequencies by
<1 %. Results of Dafni et al. [3] are also shown. Their experimental condi-
tions were less clean, in particular the combined effect of two isomers had
to be fitted.

Assuming pure configurations ($h_{9/2}^2$ 8^+), ($h_{9/2}^2$ $i_{13/2}$ 11^-) and
($h_{9/2}^2 \otimes {}^{208}$Pb $5^-,13^-$) we have $Q(11^-) = 3/4Q(8^+)+Q(i_{13/2})$ and
$Q(i_{13/2}) = -38$ fm^2 can be deduced, which is rather small. This might be
compared with -44 ± 20 fm^2 from a similar analysis of ^{211}At [4]. The ^{208}Pb 5^-
state is mainly ($\nu g_{9/2}^{-1} p_{1/2}$) with minor proton components of primarily
($\pi h_{9/2}^{-1} s_{1/2}$), which cannot be present in ^{210}Po 13^-. Therefore
$Q(13^-) = Q(8^+) + Q(\nu g_{9/2})$. From another experiment [5] with the triton beam
we could derive $Q(\nu g_{9/2}) = -29(2)$ fm^2 from the measured B(E2,^{210}Pb $8^+ \to 6^+$).
The summed moment of -86 fm^2 agrees beautifully with the direct measurement
-90 fm^2.

^{198,200}Po have been studied by ^{182,184}W(^{20}Ne,4n) reactions at beam
energies between 102 and 112 MeV. The decay of the 12^+ isomers (fig. 3)
proceeds through additional 11^- and 8^+ isomers as proven by time resolved γ-γ
coincidences with a pulsed beam. Lifetimes and g-factors have been measured.
In addition Mahnke et al. [6] measured the quadrupole interaction of ^{200}Po 8^+
in Bi. The g-factors determine the main configurations of the isomers as

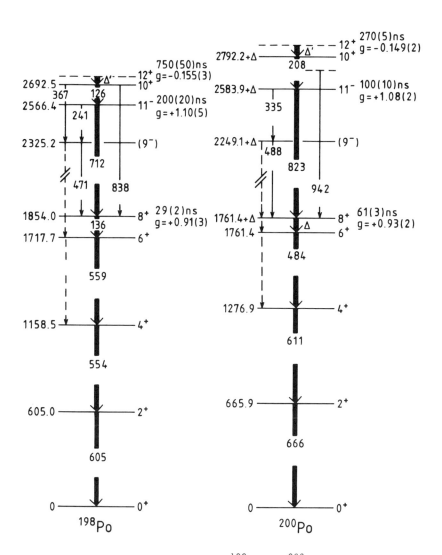

Fig. 3: Decay of the 12^+ isomers in ^{198}Po and ^{200}Po. The broken lines indicate weak side branches that are not given in detail here.

$(\nu i13/2\ 12^+)$, $(\pi h9/2\ i13/2\ 11^-)$ and $(\pi h9^2/2, 8^+)$ and show that they are quite
pure. The decay of the 12^+ isomers is peculiar as it proceeds $12^+\to 10^+\to 11^-$
while the $12^+\to 11^-$ transition is missing.

For the following the 11^- assignment is important. It is based on the
E1 character of the 126 rsp. 208 keV transitions that is evident from the
high γ-intensity implying low conversion. The angular distribution of these
lines then gives $|\Delta I| = 1$. Also all other angular distributions agree. The
lifetimes of the 12^+ levels agree with those of the $12^+\to 10^+$ transitions in
the isotonous Pb isotopes [7]. Therefore the unobserved $12^+\to 10^+$ transition in
Po is postulated here. In ^{198}Po also the decay from 2692.5 keV to the 9^- and
8^+ states rules out that this level is the 12^+ member of the $\nu i13/2^2$ configuration.

Assuming pure $\pi h9/2^2$ configurations of the 8^+ and 6^+ levels in Po
isotopes their E2 properties can be described by the effective charge as
single parameter. This is shown in fig. 4 based on the present calibration in
^{210}Po. B(E2)-values, assuming a transition energy below the L-electron
binding, follow the quadrupole moments niceley from ^{208}Po to ^{200}Po, the
slight deviation in ^{208}Po with a known transition [8] is not understood. But
then t gether with a jump of $E(8^+)$-$E(6^+)$ by ~100 keV the B(E2)-value drops in
^{198}Po by a factor 5. This might mean that the 6^+ level has become a neutron
or collective state.

The $11^-\to 8^+$ E3-transition proceeds from $i13/2\to h9/2$ and the spin
flip that is involved implies a hindrance by a factor 20. Regarding this
hindrance the measured B(E3) for ^{198}Po is very high (fig. 5). The already
large B(E3) in ^{200}Po is confirmed by the measurements of Weckström et al.
[9]. The steep rise of the B(E3) occurs when holes appear in the $\nu i13/2$
shell. This means we have a similar situation as around ^{220}Ra where octupole
instabilities are very important. In both cases the octupole pair of proton
orbitals $f7/2$ and $i13/2$ lies close in energy, while for the neutrons the
$g9/2$ and $j15/2$ orbitals are substituted in our case by $i13/2^{-1}$ and $f7/2^{-1}$ with an
energy difference of ~800 keV versus ~1500 keV. Question: Have we reached the
edge of a new region with strong octupole effects?

Fig. 6 compares the "neutron levels" of ^{198}Po with ^{196}Pb. Energies,
$B(E2, 12^+\to 10^+)$ and $g(12^+)$ agree extremely well [7, 10, 11]. On the other hand
we have seen a close resemblance of the 8^+ and 11^- proton states to ^{210}Po.
High spin neutron and roton states seem to coexist with little mutual influ-

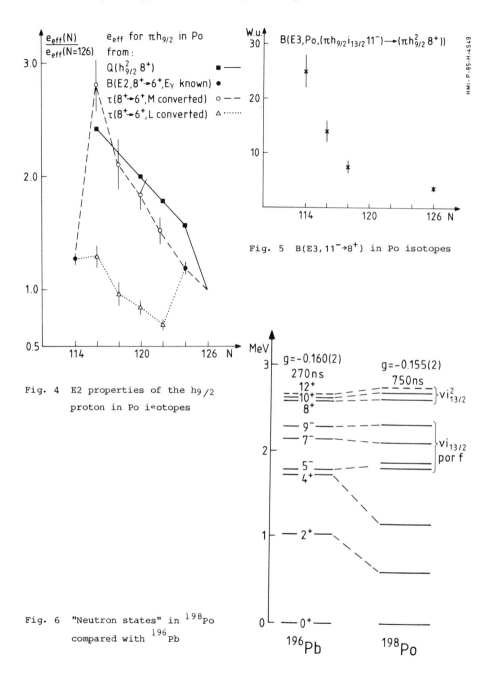

Fig. 5 $B(E3, 11^- \rightarrow 8^+)$ in Po isotopes

Fig. 4   E2 properties of the $h_{9/2}$
proton in Po isotopes

Fig. 6   "Neutron states" in ^{198}Po
compared with ^{196}Pb

ence. K. Heyde et al. [12] showed that for the relevant proton orbitals $h_{9/2}$ and $i_{13/2}$ a gain in total binding energy of ~4 MeV results for these levels at a small oblate deformation of $\varepsilon = -0.12$. For the $i_{13/2}$ neutrons the same should hold for a corresponding prolate deformation, since we deal with holes here. Opposite deformations could explain that neutron and proton states do not interfere. As a highly speculative summary, slight oblate and prolate deformations might coexist in this region with strong octupole components mixed in.

Acknowledgments

Collaborators in this work mainly on the Livermore part are: J.A. Becker, D.J. Decman, R. Estep, E. Henry, L.G. Mann, R.A. Meyer, G.L. Struble and W. Stöffl and mainly on the part at HMI A. Berger, H. Grawe, H. Kluge, A. Kuhnert, A. Maj, J. Recht, N. Roy.

References

[1] O. Häusser et al., Nucl. Phys. A 273 (1976) 253

[2] H.-E. Mahnke et al., Z.Phys. B 45 (19 82) 203

[3] E. Dafni et al., Nucl. Phys. A 394 (1983) 245

[4] H.-E. Mahnke et al., Phys. Let. 122B (1983) 27

[5] D.J. Decman et al., Phys. Rev. C28 (1983) 28

[6] H.-E. Mahnke et al., Wissenschaftlicher Ergebnisbericht 19 83, Bereich Kern- und Strahlenphysik, HMI (ISSN 0175-8349) p.78

[7] Ch. Stenzel et al., Z.Phys. A 321 (1985) in press

[8] O. Dragoun et al., Nucl.Phys. A391 (1982) 29

[9] T. Weckström et al., Z.Phys. A 321 (1985) 231

[10] Ch. Stenzel et al., Nucl.Phys. A 441 (1983) 248

[11] J.J. van Ruyven et al., preprint

[12] K. Heyde et al., Phys.Reports 102 (1983) Nr. 5, p.300

RECEIVED September 3, 1986

Electromagnetic Properties of Neutron-Deficient Pb Isotopes

H. Haas, Ch. Stenzel, H. Grawe, H. E. Mahnke, and K. H. Maier

Hahn-Meitner-Institut Berlin, D-1000 Berlin 39, Glienicker Str. 100, and Freie Universität Berlin, Federal Republic of Germany

The electromagnetic moments for a number of $(\nu i_{13/2})^n$ isomeric states in light Pb isotopes have been determined. The quadrupole moments show a pronounced shell filling effect with an increased E2 polarization charge for A ≤ 200. From the magnetic moments evidence for small wave function admixtures is obtained.

The neutron hole states in the lead isotopes present an ideal testing ground for a shell model description of semi-magic nuclei ranging from doubly closed ^{208}Pb to ^{190}Pb with 18 neutron holes. The primary purpose of the series of experiments summarized in this contribution was to study of electromagnetic properties of the high spin $i_{13/2}$ orbital over this large range.

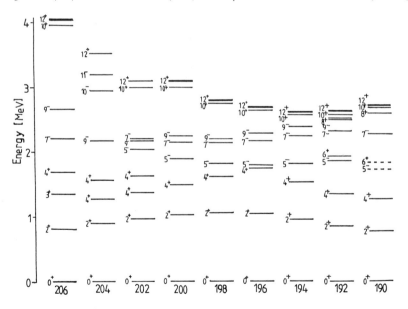

Fig. 1: Partial level schemes for even Pb isotopes

0097-6156/86/0324-0399$06.00/0

In Fig. 1 partial level schemes for the even-even Pb isotopes ^{206}Pb
to ^{190}Pb are shown. All of them have a number of long lived states in the
10 ns to 10 μs halflife range. Particularly the 12^+ states, formed by 2
neutron holes in the $i_{13/2}$ shell, constitute a unique series of high spin
isomers.

The measurement of nuclear moments of isomeric states allows to study
the M1 and E2 properties with great precision. Theoretical calculations of
core deformation have led to non-zero values of $\varepsilon \approx 0.05$ even for the semi-
magic light lead nuclei [AND78]. The measurement of nuclear quadrupole
moments can lead to a direct determination of even quite small deviations
from spherical symmetry of the core. Magnetic moments on the other hand are a
very sensitive measure of the purity of a shell model state. Even small
admixtures of core excited states can lead to significant changes. With this
motivation high precision nuclear moment measurements were performed with the
pulsed beam perturbed angular distribution (PAD) technique using various
nuclear reactions to excite the isomeric states.

Quadrupole Moments: The nuclear quadrupole interaction of several of
the 12^+ isomeric states was measured at ~220 K in solid Hg. The experiments
were either done with (α,xn) on thick isotopically enriched solid Hg targets
[MAH79] [ZYW81] or by recoil implantation following W(^{16}O,4n) [STE85]. In
all cases the nuclear quadrupole
coupling constant eQV_{zz}/h can be
extracted from the PAD modulation
pattern with high accuracy. An
example is shown Fig. 2, where
also the experimental advantage of
using single crystals for such
measurements can be clearly seen.
Small corrections due to the somewhat different target temperatures
employed in the various runs can
be made unambiguously, since the
temperature dependence of the
electric field gradient V_{zz} for
Pb in Hg has been experimentally
determined [MAH79].

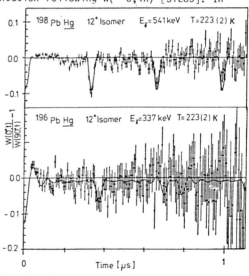

Fig.2: PAD quadrupole modulation functions

Unfortunately the absolute value of V_{zz} has not been separately measured up to now. One must therefore rely on the fact that for ^{206}Pb the 12^+ isomeric state and the 10^+ state originate from the same $(i_{13/2})^{-2}$ configuration. In this case the B(E2) value for the transition, experimentally known with reasonable accuracy is simply connected to the static quadrupole moment through vector coupling. One obtains $|e0(12^+)| = 10.38\sqrt{B(E2)}$. With this normalization the quadrupole moments for all the isomers are obtained as shown in Fig. 3.

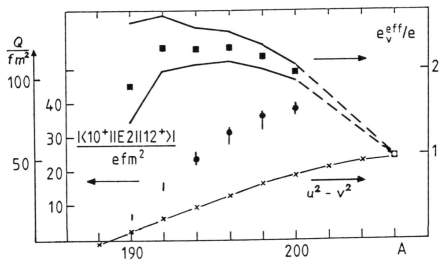

Fig. 3: Quadrupole moments (●), E2 matrix elements (|), quasiparticle amplitudes (— X —), and effective neutron charges (■). The limiting curve for e_{eff}/e indicates uncertainty due to analysis.

To check the consistency of the analysis, the $B(E2, 12^+ \rightarrow 10^+)$ values for the lighter isotopes are also plotted. They have been obtained from the halflives under the assumption that the $12^+ \rightarrow 10^+$ transition energy is between the K and L binding energy of Pb in all cases [ALB78]. The quantitative agreement is very strong evidence for the correctness of the analysis.

With this backing one can attempt to extend the treatment to the even lighter isotopes ^{192,190}Pb, where only the 12^+ lifetime could be determined, as a PAD measurement of the quadrupole moment would need excessive statistics. The time spectrum obtained with the conventional pulsed beam technique for ^{192}Pb (Fig. 4a) in parallel to the g-factor experiment to be

described below [STE83a] shows clear
evidence for an additional isomer.
This can be identified with the 10^+
state, since the calculated
$B(E2,10^+\to8^+)$ is very close to the
value expected for $(i_{13/2})^2$ configurations. For ^{190}Pb the measurement
of the 12^+ halflife was experimentally more demanding, due to the
weak production in ^{158}Gd(^{36}Ar,4n).
Using a recoil separation technique,
however, [STE85] very clean spectra
(Fig. 4b) could be obtained. The
B(E2) values extracted for
^{190,192}Pb perfectly extend the
trend observed in the heavier
isotopes (Fig. 3).

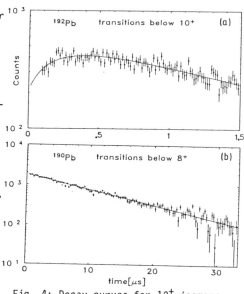

Fig. 4: Decay curves for 12^+ isomers

     The dominating tendency of the quadrupole moments toward a zero
crossing about ^{189}Pb may be directly understood as a consequence of a half
filled $i_{13/2}$ shell at this point. In the quasi-particle model the parameter
(u^2-v^2) is used a measure of the subshell filling. We have extended previous
theoretic calculations [PAU73] in the Tamm-Dancoff approximation using a
surface delta interaction to the lighter Pb isotopes. The values of (u^2-v^2)
obtained are shown in Fig. 3. They permit us to calculate the effective
charge e_{eff}/e associated with an $i_{13/2}$ neutron in the different
isotopes. It is this parameter, also graphed in Fig. 3, that allows to
extract information about contributions of the core to the E2 moments. It is
seen to increase sharply between A = 206 and 200 and then to level off at
$e_{eff}/e \simeq 2.5$. This value corresponds to a prolate core deformation of
$\beta = 0.04$ for the 12^+ states in $^{190-200}$Pb, in qualitative agreement with the
theoretical expectations.

     Magnetic Moments: The g-factors for the $(i_{13/2})^2$ states have been
measured by (α,xn) for $^{196-206}$Pb [STE83b], $(^{16}O,4n)$ for ^{194}Pb [STE85] and
$(^{40}Ar,4n)$ for ^{192}Pb [STE83a] with the PAD technique to high precision.
Typical results are displayed in Fig. 5. For the matrices employed, Hg and
Pb, Knight shift and diamagnetic shielding corrections could be applied.

From the data shown in Fig. 6 one can see that, in contrast to previous data [ROU77], these results exhibit a very clear trend: The (negative) magnetic moments remain constant between A = 206 and 200 and then show a continuous increase in magnitude toward the Schmidt value. Qualitatively such a trend has been partly predicted by calculations of the M1 core polarization [ROU77], the calculations, however, failing to yield quantitative agreement. For ^{206}Pb, on the other hand, a much more sophisticated theory [SPE77] was able to reproduce the g-factor correctly with an effective M1 operator. We

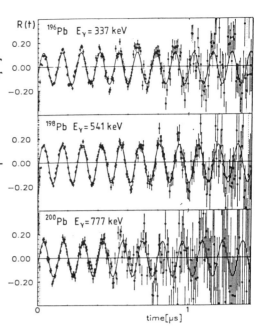

Fig. 5: Typical spinrotation curves

have scaled the core polarization calculations with this operator to get a description of the $i_{13/2}$ contribution to the g-factors. Such a treatment (solid line in Fig. 6) yields complete agreement for the lighter isotopes, while near to the closed shell small admixtures to the 13/2 wave function coming from 3$^-$ and 5$^-$ core excitations can account for the remaining discrepancies.

Fig. 6: g-factors
for $(i_{13/2})^n$
isomers.

Theory ^{207}Pb
[SPE77]
3$^-$,5$^-$ admixtures
subtracted
Effect of core
polarization

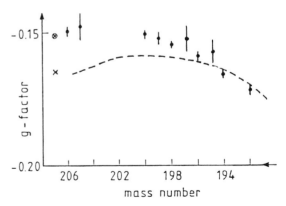

As a byproduct of the present experiments magnetic moments for several lower-lying isomeric states in the even-even isotopes were also obtained. In general a satisfactory explanation for the values can be reached by the use of effective g-factors for the low spin orbitals as determined experimentally [STE85]. A noteworthy exception are the 5^- states in ^{196,198}Pb where small positive values were found, while all neutron configurations predict small negative g-factors. A collective admixture, possibly connected with the proton intruder states recently discovered in this region [DUP84], could account for the discrepancy.

Furthermore, in the (α,xn) reactions with thick Hg targets many isomeric states in the odd Pb isotopes are also excited. The explanation is particularly straightforward for the $(29/2, 33/2)^+$ states in ^{195,197,199}Pb and the $33/2^+$ state in ^{205}Pb. The virtually pure $(i_{13/2})^n$ character of these isomers is supported by the fact that their g-factors, included in Fig. 6, follow the trend seen in the even isotopes very well.

Conclusions: The measured E2 moments for the 12^+ isomers clearly demonstrate the $(i_{13/2})^n$ quasiparticle nature of these states and can be quantitatively explained by the assumption of a small induced core deformation for the more neutron deficient Pb isotopes. 12^+ isomers in the halflife range of 10 µs and 1 µs are expected from extrapolations for ^{188}Pb and ^{186}Pb, respectively. Unfortunately initial experiments to identify these have not been successful. The magnetic moments of the $(i_{13/2})^n$ isomers are quantitatively explained in the theory of Speth et al. [SPE77] if core polarization is properly accounted for.

References

[ALB78] G. Albouy et al., Nucl.Phys. A303 521 (1978)

[AND78] C.G. Andersson et al., Nucl.Phys. A309 141 (1978)

[DUP84] P. van Duppen et al., Phys.Rev.Lett. 52 1974 (1984)

[MAH79] H.-E. Mahnke et al., Phys.Lett. 88B 48 (1979)

[PAU73] M. Pautrat et al., Nucl.Phys. A201 449 (1973)

[ROU77] C. Roulet et al., Nucl.Phys. A285 156 (1977)

[SPE77] J. Speth, E. Werner, W. Wild, Phys.Reports 33C 127 (1977)

[STE83a] Ch. Stenzel et al., Hyperfine Interactions 15/16 97 (1983)

[STE83b] Ch. Stenzel et al., Nucl.Phys. A411 248 (1983)

[STE85] Ch. Stenzel et al., Z.Phys. A322 83 (1985)

[ZYW81] S. Zywietz et al., Hyperfine Interactions 9 109 (1981)

RECEIVED July 15, 1986

Section III: New Facilities and Techniques
Part 1: Major Facilities
Chapters 61–66

Increasingly powerful tools are necessary if investigations of nuclei with lifetimes of tiny fractions of a second are to be pursued. Research today requires large and powerful accelerators and reactors, on-line mass separators, lasers, superconducting magnets and other exotic equipment, not to mention the sophisticated computers necessary to acquire and analyze the complicated data sets. Consequently, nuclear structure research has, to a great extent, become a collaborative effort of research teams, often crossing national boundaries and involving multinational financial support. The development of new, more powerful facilities is critical to future advances in nuclear structure research.

Chapters in this section describe the evolution of new major facilities at Chalk River and CERN (European Organization for Nuclear Research). Plans for an elaborate radioactive beams facility at TRIUMF (originally Tri Universities Meson Facility; now including four or five universities) and for a He-Jet/mass separator at LAMPF (Los Alamos Meson Physics Facility) are also presented. Ongoing development of major new instruments such as recoil mass spectrometers at Michigan State University and at Los Alamos National Laboratory promises to provide great quantities of new information in the next several years. Developments of techniques in fast radiochemistry and in the preparation of radioactive targets also promise to open new regions to investigation. This section, therefore, provides a vista for the future. The many developments about to be realized and those in various stages of design will define the regions for experimental development for the next decade.

61

ISOLDE Today and Tomorrow

B. W. Allardyce

ISOLDE, CERN, CH-1211 Geneva 23, Switzerland

    Isolde is known throughout the world as the on-line isotope separator at the 600 MeV synchrocyclotron at CERN, Geneva. Work has been going on there since 1967 on many topics including nuclear physics, solid state, atomic physics, etc. There have been great improvements made over the years but it became clear that the physics output could only be increased if there were to be an expansion of the facility. It was decided to build a second on-line separator to be used alternately with the existing Isolde, thus doubling the potential. The new separator is under construction, called Isolde 3, and is scheduled for first operation next year.

1.   Evolution of Isolde

    The Isolde separator started to work on-line to the CERN 600 MeV synchrocyclotron (SC) as long ago as 1967, when the accelerator was already 10 years old. In those days the SC could produce an extracted proton beam of up to about 50 nA which, combined with the early target/ion sources used by Isolde at that time, resulted in on-line ion beams of, for example, He isotopes with a maximum intensity of 10^7 ions/sec. However, this was adequate for the experiments planned then, and Isolde ran successfully for about 6 years with gradually improving techniques. Finally, about 12% of the SC's beam time was devoted to Isolde physics, or about 800 hours per year.

    In 1973/4 the long-awaited SC improvement programme was implemented, and in parallel with it the Isolde group upgraded their installation, from then on labelled Isolde 2. This upgrade converted what had been a fertile series of experiments into an experimental facility. On the accelerator side, the Improvement Programme consisted of fitting a new radiofrequency system, improving the beam optics from the ion source through to extraction, and increasing the extraction efficiency by an order of magnitude with a septum magnet. As a result the SC could deliver beam intensities of 2 µA to the Isolde target, an increase by a factor of more than 40.

0097-6156/86/0324-0406$06.00/0
© 1986 American Chemical Society

In the underground Isolde zone the changes were equally dramatic. The zone was completely reshuffled so as to gain space, and the separator magnet was moved close to the production target : previously the ion beams drifted 3m through a concrete shielding wall before reaching the magnet. The result was a doubling of the experimental area available for physics, which could now be fully exploited with four separate beam lines from the electrostatic switchyard after the magnet. Targets were another area of improvement with the introduction of disposable, self-contained target/ion source units and a remote target change mechanism, which permitted the full proton intensity to be used on a target of thickness of order 100 gm/cm^2. The first targets to be used were the highly successful alkali producing targets.

Demand for beam time at IS2 steadily increased after the upgrading, reaching 1900 hours of physics in 1980, or 35% of the SC's beam time. An additional experimental zone of 120 m^2 at ground level was opened in 1978, fed by a vertical extension of one of the beam lines in the underground zone. In 1982 an industrial robot was introduced as a more flexible way of exchanging target units. Meanwhile a great deal of work was done to improve the reliability of the facility and the lifetime of the targets, as well as to extend the range of elements available as beams. Encouraging tests were made with the 303 MeV/amu ^{3}He^{++} beam of the SC which showed that for certain elements (and especially for the isotopes on the neutron-deficient side of the yield curve), there were considerable gains to be made compared to using 600 MeV protons. Beams of 86 MeV/amu ^{12}C were also tried, with less encouraging results. The main limitation in both cases was the lack of primary beam intensity, which the SC Group promised to rectify, given sufficient development time.

By the early 1980's it became clear that the Isolde facility was saturated at the level of about 2000 hours of physics per year, and some thought had to be given to the future. Isolde was limited by the physical size of the experimental zones, by the primary proton intensity, and by the need to service the equipment in the highly radioactive target zone. There was also a limitation from the manpower side, since the staff were already fully occupied keeping the facility running, developing new techniques and producing the roughly 30 consumable target/ion source units needed each year.

Various scenarios for the future were investigated, including a move to SIN where very much higher proton intensities would have been available, but the proposal finally accepted in December 1983 was to build a second separator (IS3) at the SC and to run the accelerator primarily for Isolde for 4000 hours per year. A cut-away view of how the complete facility will look is shown in Fig. 1.

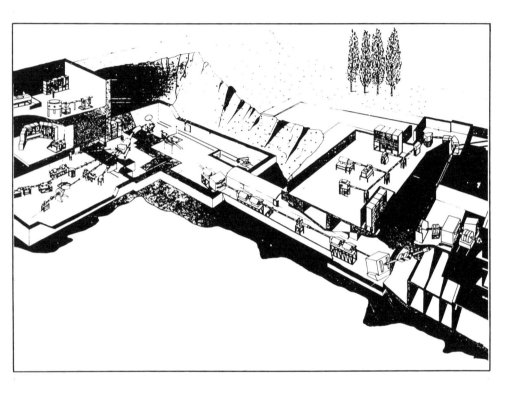

Fig. 1   The SC-Isolde facility

2.   Present status of IS2

Currently there are about 30 experiments in progress at Isolde, and the facility is used by almost 200 physicists. A detailed catalogue of the beams available at Isolde was presented last year at the Zinal workshop (Ra84), and a summary of which elements can be produced is shown in Fig. 2.

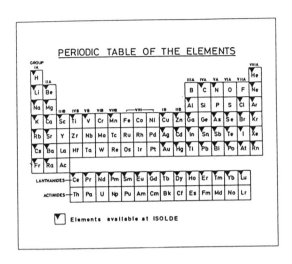

Fig. 2 Summary of the elements available at Isolde

Of the 66 elements which may be produced, roughly one half have a high yield from the target, which means $\sim10^{11}$ atoms/sec at the peak of the yield curve for a 1 μA proton beam, and are available with good chemical selectivity ; for these elements no further development work is necessary. The remainder could benefit from improvements in selectivity or in the release of the desired radioactivity from the target material into the ion source.

The most popular ion source at Isolde is the plasma source developed from FEBIAD sources found in many laboratories. The latest version of this source has recently been completed and will gradually become standard at Isolde ; it has direct resistive heating of the cathode surface. The target/ion source combination has to be tailored to the particular experiment being performed, in order to optimise the yield and/or the selectivity, and this results in there being several different versions of the target/ion source ; for example, different cases require different temperatures in the transfer tube between target and ion source. Sources other than the plasma source are also frequently used, notably the surface ionisation source, both positive and negative. Consequently the production of the target/ion source units is not a trivial task and requires a lot of manpower.

Target/ion source units are exchanged by means of the industrial robot which was purchased in 1982. This development has been very successful, especially in its effect on doses to the personnel. Another recent improvement to the facility has been the addition of a second beam line from the underground area to the experimental area at ground floor level.

3.   Progress on IS3

The ideal location for IS3 would have been in a new underground zone near to IS2, but the cost excluded this solution. The only way of keeping the cost to an acceptable level was to make use of existing buildings, and it was on this basis that the project was finally accepted (Ce83). As a result the target had to be situated inside the SC vault in spite of the inconvenience that this implied, and the separator had to be designed so that the experiments could be housed in the existing "Proton Hall" of the SC. This results in two separate Isolde installations which can receive beam alternately from the SC, with a switch-over time of an hour or so. It is planned to use the same target/ion source pots for the two separators, so as to maintain only one production line for the targets and thus keep costs to a minimum.

The new separator has the characteristics listed in the table, and the optical properties of the beam are shown in Fig. 3. Electrostatic lenses will be used throughout, with a wide aperture (14 cm) in order to accomodate the beam envelope from a slit source ; the intention is to be able to run with both the normal Isolde sources and a slit source developed from the Isocelle design. Unseparated ion beams are transported 8m through the SC shielding wall to a focus before the first 90° separator magnet, which has a highly uniform field. Auxiliary coils and an electrostatic multipole element will be used to correct aberrations in the optics and to permit the best per-formance of the separator to be realised. The focal plane chamber after the first magnet will be used to measure the beam properties, and to select the desired isotope. A second stage of separation then follows, via a similar 60° magnet, and the beam is finally distributed to the experiments in the Proton Hall.

Fig. 3 Calculations of the beam in the horizontal plane from the source at bottom right through to a focus after the second bending magnet.

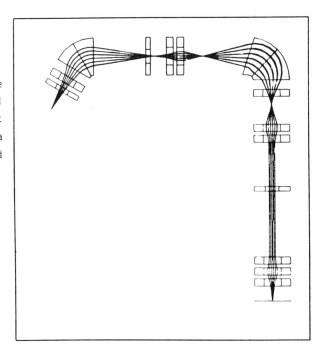

Table   -   Parameters of the Isolde 3 separator

- Two-stage separation with a combined resolving power of ~5000 at high transmission ; possibility to reach ~30,000 with slits.
- Maximum current to be transported to the first magnet : 3mA using a slit source.
- Acceleration voltage 60 kV maximum.
- Dispersion after the first magnet > 12 mm per %, perpendicular to the beam; the first focal plane to cover a mass range of ± 10%.
- Enhancement factor ~10^4 for the first stage, increasing by a further factor of 10^2 to 10^4 in the second stage and by tuning.

The power supplies for the target/ion source will be housed in a Faraday cage in the Proton Hall, joined to the target by a specially-designed high voltage tube of low capacity. All the supplies are commercial and are designed for very high stability and low ripple (1 in 10^5). Power is fed to the supplies from a commercial isolation transformer. The high voltage, 60 kV as at IS2, is generated in a home-made supply with equally demanding specifications. A great deal of attention is being paid to keeping the stored energy to a minimum and to introduce shielding and Faraday cages in order to minimise the effect of sparking on other experimental equipment, especially computers. This has been a considerable problem at IS2. One other important advantage which IS3 will have over IS2 is the ease of access to the high voltage supplies, since the Proton Hall is not a highly radioactive zone.

Control of the separator will be by a microprocessor based serial CAMAC system using standard CERN interfaces, and work is progressing rapidly in this domain. A control room, close to the SC control room, is being prepared, much of the CAMAC hardware is available already, and a start has been made on the software.

The separator is being constructed in a collaborative way by CERN and several participating laboratories. Most pieces are designed and some items have already arrived ; the magnets themselves are to be delivered in October, and all the civil engineering work has been completed. It is our intention to assemble the first part of the separator in the near future for tests in the Proton Hall prior to mounting it in its correct place during the Spring of 1986.

4.   Future possibilities

Although exotic possibilities such as the acceleration of radioactive ions or their storage in a storage ring have been discussed there are no firm plans to proceed in this way. However, two development areas are clear, and await only time and manpower.

The use of a ^{3}He^{++} beam of 303 MeV/amu has already been mentioned as an advantage in certain cases, especially for nuclei far from stability. Such a beam is also advantageous in that it produces less induced radioactivity in the SC vault. However, the ^{3}He^{++} beam has not so far been produced with sufficient intensity to make its frequent use an interesting proposition compared to 600 MeV protons. Improvements have been made to the r.f. system and to the SC's ion source which should now permit an intensity gain of a factor 3 or 4 to be achieved, but so far this has not been tested. If the ^{3}He^{++} intensity rises to the proton intensity (i.e. $> 10^{13}$ ions/sec), this beam could become the preferred Isolde production beam of the future.

Ion sources are also an area where it is planed to make improvements, once the new separator is working and manpower is liberated. New ideas for on-line sources include the laser source (which will probably not make new elements available to Isolde but will certainly improve the selectivity), and ECR sources (where the problem will be to adapt existing ECR technology to the high-temperature, high radiation environment of the Isolde separator).

References

[Ra84] H. Ravn in "Zinal 1984, Workshop on the Isolde Programme beyond 1985". Also as an Appendix in Proceedings of TRIUMF-ISOL Workshop, Mont Gabriel, June 1984, TRI-84-1.

[Ce83] CERN PSCC-83-27: The Isolde Project.

RECEIVED July 18, 1986

62

A New Heavy-Ion Facility at Chalk River

J. C. Hardy

Atomic Energy of Canada, Limited, Chalk River Nuclear Laboratories, Chalk River, Ontario K0J 1J0, Canada

The Tandem Accelerator Superconducting Cyclotron facility at Chalk River, which is nearing completion of phase 1, will be capable of accelerating all ions to at least 10 MeV/u. Together with the on-line isotope separator it will provide a powerful means of studying exotic nuclei.

During phase 1 of construction of the Tandem Accelerator SuperConducting Cyclotron (TASCC) facility at Chalk River, the existing tandem accelerator was reversed, a superconducting cyclotron built and some 60 metres of beam-transport line installed to connect the two and to deliver beams to an "interim target line". The result, which required an appreciable building extension, is shown in figure 1.

As of early September, 1985, the facility is being commissioned. Carbon, nitrogen and oxygen beams from the tandem have reached the interim target locations via the bypass line, and an iodine beam has been injected into the cyclotron. First attempts to accelerate iodine in the cyclotron are imminent. Some experimental equipment, including the on-line isotope separator [SCH81], is already set up and undergoing tests with tandem beams.

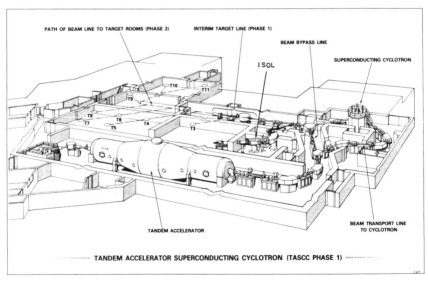

Fig. 1. Artist's impression of phase 1 of the TASCC facility. Shown in dashed lines is the beam delivery system to 11 target locations, planned for phase 2. This second phase, which is not yet funded, utilizes existing target rooms.

0097–6156/86/0324–0414$06.00/0
© 1986 American Chemical Society

When it reaches its full capability, TASCC will accelerate all ions between lithium and uranium to energies up to 50 MeV/u and 10 MeV/u, respectively. It will feed some major pieces of apparatus: the Q3D magnetic spectrometer, the isotope separator, a growing array of gas and solid-state detectors housed in a 1.5 m diameter scattering chamber, and the "8π" γ-ray spectrometer [AND 84]. All are currently operational except the 8π spectrometer, which is being built by a consortium of Canadian universities and AECL Chalk River, with completion scheduled for late 1986. It will comprise two subsystems: i) a spin spectrometer of 72 bismuth germanate (BGO) detectors, and ii) an array of 20 Compton-suppressed hyperpure (HP) Ge detectors.

At the completion of phase 2, these devices and others will be served by the beam lines shown dashed in figure 1. However, phase 2 is not yet funded, so in the meantime somewhat restricted conditions will prevail. The Q3D, which is already sited at T10, will be inaccessible, the isotope separator will only be on-line via a He-jet coupling and the remaining experimental equipment will be strung end-to-end at the interim target line to offer as broad a scope as possible for utilizing the new beams as they appear.

Although TASCC is not the first in its energy range, it joins a very select group of facilities worldwide with versatile capabilities at transitional energies. As such, TASCC opens up many exciting experimental possibilities. For the purposes of this symposium, I shall focus on those in the field of nuclei far from stability.

Accelerator Facility

The acceleration process begins with negative ion generation in the 300 kV injector followed by acceleration in the tandem at voltages up to 13 MV. The beam is bunched into timed pulses by passing it through a gridded-gap prebuncher before the tandem and a two-gap drift-tube buncher between the tandem and the cyclotron. The latter produces a time focus when the beam bunches reach a stripper foil located in a magnetic valley near the center of the K=520 cyclotron (see figure 2). The foil, whose position is radially adjustable, intercepts the beam just where it becomes tangent to the radius appropriate for its energy and resulting charge state. Once the beam has been accelerated to a radius of 650 mm, turn separation is increased and it enters an electrostatic channel where it is deflected into the magnetic extraction channel.

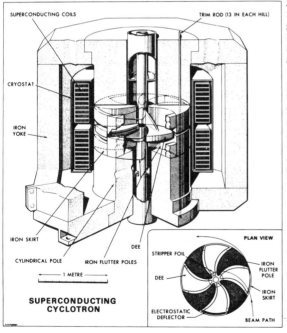

SUPERCONDUCTING COILS

TRIM ROD (13 IN EACH HILL)

CRYOSTAT

IRON YOKE

IRON SKIRT

DEE

CYLINDRICAL POLE   IRON FLUTTER POLES

1 METRE

SUPERCONDUCTING CYCLOTRON

PLAN VIEW

STRIPPER FOIL

IRON FLUTTER POLE

DEE

IRON SKIRT

ELECTROSTATIC DEFLECTOR

BEAM PATH

Fig. 2. Cutaway view of the Chalk River superconducting cyclotron.

The cyclotron itself [HUL85] forms a single 4-dee resonant structure. Because the central axis is not required for ion sources, it accommodates two coaxial-line tuners, each supporting a pair of dees located in the magnetic valleys. The magnet is excited by two superconducting niobium-titanium coils, which are copper stabilized. Fine tuning of the field is accomplished by adjusting the positions of 104 saturated-steel trim rods, which penetrate through the upper and lower magnetic hills. Small superconducting coils are used in the magnetic extraction channel.

So far, beams have not yet been accelerated in the cyclotron, although that should happen within days. A beam of $^{127}I^{7+}$ has been injected to the stripper foil and the resulting charge states were distinguished on a radial probe by their deflection in the cyclotron magnetic field. At 71 MeV, the most probable charge state was observed to be 23, in reasonable agreement with expectation [HEI85].

The On-line Isotope Separator

The Chalk River isotope separator has been operating on-line with the tandem accelerator since April 1979. When the tandem shut down in 1982, the separator was completely dismantled and reassembled with some improved components at its present location, where it will be on-line with TASCC (see figure 1). It is now working again and, although recommissioning is not yet complete, it has already achieved some of the excellent performance characteristics met before the move. A picture of the relocated instrument appears as figure 3.

Like most other on-line isotope separators (ISOLs), the Chalk River device was previously operated with its ion source essentially in line with the primary accelerator beam so that nuclei produced by reactions in a target would recoil directly into the ISOL ion source. We used a FEBIAD ion source [KIR76], which was characterized by long cathode lifetime and good efficiency for a wide variety of elements.

Fig. 3. View of a portion of the relocated ISOL. The image plane of the 135° inhomogeneous magnet is in the large central chamber; the selected beam is deflected 20° and conveyed through the beam line at the top left to a counting area.

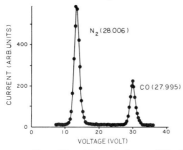

Fig. 4. The CO-N₂ doublet at mass 28 measured with the Chalk River ISOL and a FEBIAD ion source; the resolving power is 20,000. With a slit-geometry source 16,000 has been obtained. (The separation between the two peaks is 10.5 MeV.)

For the time being with TASCC, such a direct coupling is impossible since phase 1 does not provide magnetic beam transport elements to the ISOL ion source. Instead, a He-jet system, in which recoil nuclei are thermalized in helium gas, has been coupled by capillary to an ion source through a skimmer [HAR84]. A special source has been built for the purpose with an extraction slit of 1 mm x 30 mm (cf. a 1 mm dia. hole for the FEBIAD), which allows for the necessarily high throughput of helium. The ISOL itself was originally designed to accommodate slit geometry, high beam currents and a large gas load. Preliminary tests of this system before the tandem shut down indicated that efficiency of a few percent is achievable.

The quality of separated beams from the Chalk River ISOL has always been very high. Both the 40 kV accelerating voltage and the field in the 135° inhomogeneous magnet are highly stabilized ($<10^{-5}$). The most remarkable item employed in the separator, however, is the pair of correction coils with which the magnetic sector is equipped. The radial field distribution in any magnet is given by

$$B = B_0 \ [1-\alpha(r-r_0)/r_0 + \beta \ (r-r_0)^2/r_0 + ...]$$

where B_0 is the field at the mean radius r_0. For the Chalk River magnet, α is nominally 0.5 and $\beta \approx \alpha^2$, but the correction coils, called α- and β-coils, produce field contributions that independently adjust the respective indices. The α-coil is largely responsible for the position of the image along the beam axis and β can be used to minimize second order aberrations. With two simple adjustments the optimum beam profile can readily be obtained. As illustrated in figure 4, a resolving power ($M/\Delta M$) of 20,000 has been obtained, the highest ever reported for an ISOL.

Of equal importance to high resolution is high enhancement of a selected beam. Contamination, even at the percent level, from adjacent isotopes could be catastrophic to a precise measurement. For the Chalk River ISOL, cross contamination has been measured to be $< 10^{-6}$ at A=20, and, by scaling, it must be $<10^{-5}$ at A=200.

Prospects

The combination of a high quality ISOL and a versatile heavy-ion accelerator capable of energies up to 50 MeV/u is a very powerful one. When compared with intermediate-energy proton beams, heavy ions have both advantages and disadvantages in the production of exotic nuclei. At bombarding energies E < 10 MeV/u they afford a highly specific production technique which can, with the right combination of target and projectile, be focussed very far from stability. Yet they are handicapped by requiring targets thinner than a few mg/cm². It is only for the most remote products that the production rate can surpass that of a proton-based separator like ISOLDE using targets four orders of magnitude thicker.

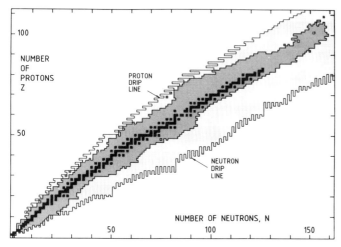

Fig. 5. Chart of the nuclides. The dark stippled area represents nuclides identified by 1981; the lighter areas, nuclides predicted to exist for an observable length of time [HAR85].

It is interesting to note, however, that at higher energies heavy ions can produce comparable yields to protons and at much lower specific energy. The main reason is that thick-target techniques now become feasible for heavy ions as well. The consequence is that, microampere-for-microampere at 50 MeV/u, ^{12}C projectiles should be the equivalent of 600 MeV protons for isotope production, and are possibly better for the most exotic products [HAR84].

In principal at least, a heavy-ion accelerator with the capabilities of TASCC should be ideally suited both to the specific synthesis of a selected group of isotopes and to high current production runs of a wide range of isotopes.

Undoubtedly the Chalk River ISOL will be used, as others are, in the identification and spectroscopy of exotic nuclei. The nuclear chart in figure 5 illustrates the scope for such studies. However, the extreme purity of isotopes separated by our ISOL has been essential in the past to precision studies of the weak interaction, in one case the lifetimes of superallowed $0^{+} \to 0^{+}$ transitions [KOS83], in another β-ν-α triple correlation coefficients in the decay of ^{20}Na [CLI83]; both yielded measurements of the weak vector coupling constant. These types of measurements will be extended to other nuclei, since they exploit the best qualities of the accelerator and separator.

Of particular interest to nuclear and astrophysicists alike are the masses of nuclei far from stability. Although the masses of some nuclei in the shaded region of figure 5 have been deduced from measured decay energies, the only direct mass determinations are from a mass spectrometer connected on-line to ISOLDE [THI81] and from the Chalk River ISOL, which because of its high resolving power can be used directly as a stand-alone mass-measurement device [SHA84]. With the former, quite extensive isotopic chains of masses were determined for alkali elements, but the techniques involved cannot easily be extended to elements with different chemistry. The latter was developed more recently and has been used to determine only a few masses so far, but it is potentially useful for measuring the mass of any isotope

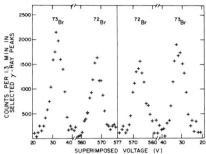

Fig. 6. Mass spectral peaks of ^{72}Br and ^{73}Br measured as a function of the voltage superimposed on the 40 kV ISOL acceleration voltage; each plotted point represents the total area of selected γ-ray peaks in measured decay spectra.

produced in the ISOL. Furthermore, since it detects isotopes by their decay radiation (rather than by direct ion counting, as with conventional mass spectrometers) there is usually no difficulty in distinguishing ground from isomeric states. An example of the power of the technique is shown in figure 6: ^{72}Br, whose mass was determined for the first time from these data [SHA84], has an isomer at 100 keV; its effects were easily removed from the data through known differences in the β-delayed γ-ray spectra.

We expect direct mass measurements to form an important component of the exotic-nuclei program at TASCC. Indeed, we already hope to upgrade our facilities through a proposal made in collaboration with the University of Manitoba to build a high-transmission mass spectrometer, which could accept beam from the ISOL directly without reionization, again making use of the ISOL's excellent beam quality. This could improve the precision of measurements to a level rivalling that now possible with stable isotopes.

Conclusions

Chalk River is not the only laboratory with an isotope separator on-line to a heavy-ion accelerator. Berkeley, Oak Ridge and GSI Darmstadt have already proven the value of such a combination in studying exotic nuclei. We believe that the versatility and energy range of TASCC, together with the proven quality of separated beams from the ISOL, will make the Chalk River facility among the best in the field.

References

[AND84] H.R. Andrews et al, AECL report AECL-8329 (1984).
[CLI83] E.T.H. Clifford et al, Phys. Rev. Lett. 50. 23 (1983).
[HAR84] J.C. Hardy, H. Schmeing and E. Hagberg, Proceedings of the Conference on Instrumentation for Heavy-Ion Nuclear Research, Oak Ridge, October 1984, to be published.
[HAR85] J.C. Hardy, Science 227, 993 (1985).
[HEI85] E.A. Heighway, J.A. Hulbert and J.H. Ormrod, private communication.
[HUL85] J.A. Hulbert et al, Proceedings of 1985 Particle Accelerator Conference, to be published in IEEE Trans. on Nucl. Sci. NS32. Oct. 1985.
[KIR76] R. Kirchner and E. Roeckl, Nucl. Instr. & Meth. 133, 187 (1976).
[KOS83] V.T. Koslowsky et al, Nucl. Phys. A405, 29 (1983).
[SCH81] H. Schmeing et al, Nucl. Instr. & Meth. 186, 47 (1981).
[SHA84] K.S. Sharma et al in 7th Int. Conf. on Atomic Masses and Fundamental Constants, edited by O. Klepper, GSI Darmstadt (1984) p68.

RECEIVED July 25, 1986

63

Recent Developments at TRISTAN

Nuclear Structure Studies of Neutron-Rich Nuclei

R. L. Gill

Brookhaven National Laboratory, Upton, NY 11973

The nuclear physics program at the fission product mass separator, TRISTAN, has greatly expanded, both in the types of experiments possible and in the range of nuclei available. Surface ionization, FEBIAD, high-temperature thermal, high-temperature plasma, and negative surface ionization ion sources are routinely available. Experimental facilities developed to further expand the capabilities of TRISTAN include a superconducting magnet for g-factor and Q_β measurements, a windowless Si(Li) detector for conversion electron measurements, and a colinear fast-beam dye laser system for hyperfine interaction studies. This combination of ion sources, experimental apparatus, and the long running time available at a reactor makes TRISTAN a powerful tool for nuclear structure studies of neutron-rich nuclei. The effect of these developments on the nuclear physics program at TRISTAN will be discussed and recent results from some of these facilities will be presented.

The TRISTAN Facility

A schematic layout of the TRISTAN ISOL facility is shown in Fig. 1. A neutron beam of about 3×10^{10} $n_{th}/cm^2/sec$ is provided by the 60 MW High Flux Beam Reactor at Brookhaven National Laboratory. Targets which typically contain 5g ^{235}U inside an ion source are positioned in the neutron beam to provide intense beams of short-lived neutron-rich nuclei produced by fission. To provide for maximum versatility, a variety of ion sources are used, each of which is best suited to producing certain elements [SHM83, GIL85, PIO84].

0097-6156/86/0324-0420$06.00/0

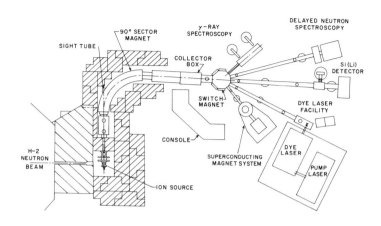

Fig. 1. Layout of the TRISTAN facility.

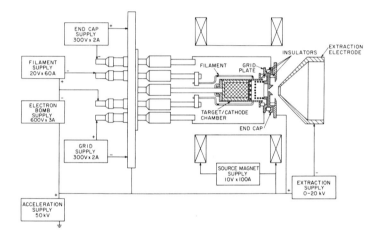

Fig. 2. Schematic diagram of the high temperature plasma ion source.

High Temperature Plasma Ion Source

The most recently developed ion source is the high temperature plasma ion source. The source is shown in Fig. 2. It is constructed in a fashion similar to the thermal and negative ion sources in that these three sources share a common design for the target, heating filamental heat shields. Thus, mass production and assembly of the sources has been simplified owing to standardization of the designs. Off-line studies have shown that this ion source exhibits ionization efficiencies of up to 60% for Kr and Xe in the low pressure limit. This source produces the same variety of elements as the FEBIAD source, but due to the larger target, higher temperature and high ionization efficiency, the intensities of the radioactive beams are significantly higher than with the FEBIAD source.

Experimental Facilities

In Fig. 1, the facilities available at each of the five beam ports at TRISTAN are labeled. The facilities are described in [GIL85] and references therein. The Si(Li) detector and superconducting magnet systems have become fully operational and widely used in the past year. These two facilities will now be described in some detail.

Si(Li) Detector. A windowless Si(Li) detector for conversion electron spectroscopy is housed in a cryostat with a compressable bellows connected to a vacuum chamber and moving tape collector. The detector can be moved to within 5 mm of the surface of an aluminized mylar tape on which radioactive ion beams are implanted. The tape system can be configured so that the ion beam deposition point is directly viewed by the detector or is located in an area shielded from the detector. In the latter configuration, the tape must be moved before the Si(Li) detector can view the source. One problem encountered when doing conversion electron spectroscopy with neutron-rich nuclei is the tendency of the β continuum to obscure the desired peaks. This interference can be suppressed by employing various gating techniques. A thin plastic scintillator (with very low γ-ray response) is used to suppress the β signal in the Si(Li) detector. The system is designed such that a standard Ge detector cryostat also views the same source as the Si(Li) detector. Thus, γ-ray electron and x-ray electron coincidences can also be employed to enhance the electron signal, as well as provide for element identification. In addition to these coincidence capabilities, electron and γ-ray singles can be simultaneously acquired, making it possible to measure internal conversion coefficients. The system has been used to identify weakly populated 0^+ states in ^{96,98}Zr.

Superconducting Magnet. A split-pole, Nb-Ti superconducting magnet system, manufactured by Oxford Instruments, is available at TRISTAN for g-factor and Q_β measurements. The magnet has a field strength of 6.23 T at 108 A at 4°K with inhomogeneity of $<1\%$ over 1 cm^2. A rectangular cross section room temperature tube penetrates the magnet bore to provide access for the tape from a moving tape collector. The magnet was designed to

provide for maximum flexibility of detector positioning. An angle of 125°
on each side of the base tube is free of significant γ-ray absorbing
material. Detectors on opposite sides of the bore tube can be positioned as
close as 90° to one another and as close as 9 cm to the source. An alignment jig was constructed to provide accurate angular positioning of each
detector and support for shielding.

The magnet can also be used for Q_β measurements. By placing a
hyperpure Ge detector in a magnetic field, electrons from the source can be
focussed onto the surface of the detector, thus providing a large effective
solid angle for betas. At the same time, the detector can be far from the
source, giving a low solid angle for γ rays. Such an arrangement is
possible with the TRISTAN magnet. The lower pole piece can be removed and a
LN_2 cooled heat shield is replaced by a cold finger containing the
detector. Thus the magnet configuration can be changed without major
disassembly of the system. Using this system, it is possible to obtain
enhancements of electrons over γ rays by a factor of 15.

The tape collector for the superconducting magnet facility was
designed and constructed at Brookhaven. It consists of a box with a large
number of low friction rollers which guide the tape through the deposit
port, into the magnet, through a series of loops, and to a stepping motor
drive sprocket. The system contains about 15 m of tape. The tape is
standard 16-mm movie film leader which has a layer of aluminum evaporated
onto one side. Repairs to the tape are made with a standard film splicer.
The drive sprocket was custom made since a much larger than standard
diameter was desired to obtain maximum speed of the tape. The tape moves
362 mm to move a source from the deposit box to the center of the magnet.
This can be accomplished in about 0.3 s with an error of less than 0.5 mm.
A small surface barrier detector is provided to monitor the ion beam
intensity.

The superconducting magnet system is equipped with several inter-
locks. One senses the external magnetic field and will give a signal if the
field diminishes below a preset value, another senses a loss of vacuum in
the tape collector and a third senses a broken tape condition. Each of the
sensors will sound an alarm, stop the tape collector cycling, isolate the
vacuum system, and stop data acquisition on the appropriate channels upon
receiving a fault signal.

G-factor Measurements at TRISTAN

Using the superconducting magnet described above, g-factors of the 2^+_1 states in ^{146,148}Ce have been measured. In previous measurements [WOL83] at TRISTAN, using a conventional magnet, 2^+_1 states in Ba isotopes were measured and interpreted in terms of changes in the number of valence protons due to a subshell closure at Z=64. It was realized that the Ce isotopes would provide a more severe test of these ideas, since they would be the lowest Z isotopes for which an abrupt change in valence proton number could be observed in this region. However, due to the short lifetimes for the levels (0.25 ns for ^{146}Ce and 1.07 ns for ^{148}Ce) and the lower yields, a superconducting magnet system was crucial to the measurement.

The g-factors in ^{146,148}Ce were measured using the integral perturbed angular correlation technique where the intensity of the cascading γ rays (in these cases for a 0^+-2^+-0^+ cascade) are measured at an appropriate angle for field up and field down. The standard double ratio technique is used [WOL83] since this technique eliminates most systematic errors. The results obtained are: $g(2^+_1) = (0.24 \pm 0.05)$ for ^{146}Ce and $g(2^+_1) = (0.37 \pm 0.06)$ for ^{148}Ce. In IBA-2 formalism, the g-factor of the first 2^+ state in an even-even nucleus can be described by [SAM84] $g(2^+_1) = g_\pi N_\pi/N_t + g_\nu N_\nu/N_t$, where $N_t = N_\pi + N_\nu$. If the experimental values of $g(2^+_1)$ for isotopes of Ba-Dy are used to plot $g(2^+_1) N_t/N_\nu$ versus N_π/N_ν, a straight line is obtained with significant deviations being observed only for those isotopes expected to exhibit a truncated proton valence space due to the subshell closure at Z=64. Additionally, all of the deviations are in a direction that indicates a smaller proton valence space. If only those isotopes with N>90 are included in the fit $g_\pi = 0.63 \pm 0.04$ and $g_\nu = 0.05 \pm 0.05$ is obtained. If these empirically determined values of g_π and g_ν are accepted as being constant over a wide range of nuclei in this region, then the experimentally determined g-factors can be used as a measure of the effective number of valence protons (N_π in IBA-2 formalism). The results of this analysis are shown in Fig. 3 for Ba, Ce, Nd, and Sm isotopes. The dashed line shows the value of N_π expected for major shell closures from Z=50-82 (Z=82) and from Z=50-64 (Z=64) and the solid line indicates the region of transition between shells [GIL82]. The figure shows not only the distinct change in N_π, but also a tendency of the N_π change to shift to lower neutron number as Z increases. This

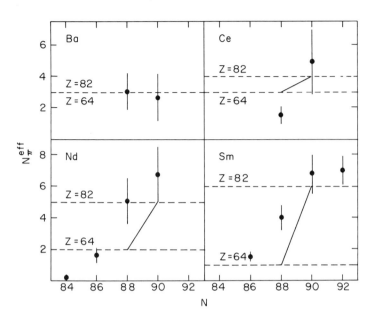

Fig. 3. N_π^{eff} deduced for isotones of Ba, Ce, Nd, and Sm. The Ba, Nd, and Sm data are from [WOL83] and references therein. The Ce data is from this work.

effect is primarily due to earlier occupation of the $\pi h_{11/2}$ orbital becoming possible as higher orbitals are filled with increasing Z. The neutron-proton interaction can then operate to the fullest extent (neutrons are occupying the $h_{9/2}$ orbital) resulting in a deformed nucleus having the lowest energy.

Acknowledgment

Research supported under contract DE-AC02-76CH00016 with the U.S. DOE.

References

[SHM83] M. Shmid et al., NIM 211 287 (1983).

[GIL85] R. L. Gill and A. Piotrowski, NIM 234 213 (1985).

[PIO84] A. Piotrowski et al., NIM 224 1 (1984).

[WOL83] A. Wolf et al., Phys. Lett. 123B 165 (1983).

[SAM84] M. Sambataro et al., Nucl. Phys. A423 333 (1984).

[GIL82] R. L. Gill et al., Phys. Lett. 118B 251 (1982).

RECEIVED July 14, 1986

64

Potential Capabilities at Los Alamos Meson Physics Facility To Study Nuclei Far from Stability

W. L. Talbert, Jr., and M. E. Bunker

Los Alamos National Laboratory, Los Alamos, NM 87545

Feasibility studies have shown that a He-jet activity transport line, with a target chamber placed in the LAMPF main beam line, will provide access to short-lived isotopes of a number of elements that cannot be extracted efficiently for study at any other type of on-line facility. The He-jet technique requires targets thin enough to allow a large fraction of the reaction products to recoil out of the target foils; hence, a very intense incident beam current, such as that uniquely available at LAMPF, is needed to produce yields of individual radioisotopes sufficient for detailed nuclear studies. We present the results of feasibility experiments on He-jet transport efficiency and timing. We also present estimates on availability of nuclei far from stability from both fission and spallation processes. Areas of interest for study of nuclear properties far from stability will be outlined.

1. Introduction

A He-jet coupled on-line mass separator, used in conjunction with a target chamber placed in the LAMPF main beam, offers an especially attractive approach for the study of nuclei far from stability. Such a facility would provide access to isotopes of a number of elements that cannot be efficiently extracted for study at any other type of on-line separator system, and the use of a long capillary transport line would allow the separator ion source to be located outside the accelerator beam-line shielding, greatly reducing the installation cost. The He-jet technique requires thin targets in order for the reaction products to escape, and to produce a sufficient yield of radioisotopes far from stability for detailed nuclear studies, a beam intensity comparable to that available at LAMPF is needed.

The mass-separated ion beams extracted from the proposed system would be directed to various experimental devices capable of determining basic nuclear properties such as half-life, spin, nuclear moments, mass, and nuclear structure. The data acquired would have broad application to theories of nuclear matter and to such related topics as nucleosynthesis of the elements. We

0097–6156/86/0324–0426$06.00/0

estimate that several hundred previously unobserved nuclei, both neutron-deficient and neutron-rich, would become available for study.

This report presents a brief summary of the results of our He-jet feasibility studies, along with estimates of which nuclei would become available with our proposed facility. Specific areas for initial studies are also suggested.

2. He-jet activity transport studies

The concerns addressed in our feasibility studies of He-jet activity transport were: 1) Will the He-jet technique work at the beam intensities that exist at LAMPF?; 2) What transport efficiencies can be expected for both fission and spallation products?; 3) What is the time dependence of the activity transported?; and 4) What aerosols and/or aerosol conditions are optimum?

Using both spallation- and fission-product targets in the LAMPF H⁻ beam, we determined that the He-jet technique should work well at LAMPF beam intensities (~800μA). Absolute efficiencies for transport of refractory-element activities through a 22-m long capillary were found to average about 60%. Transit time measurements appear convincing that activities as short as 300 ms could be made accessible for study. We found that $PbCl_2$ aerosols provided more efficient transport than KCl or NaCl aerosols.

Optimization of the target chamber configuration resulted in a design employing two radially-directed inlets at ± 135° to a single radially-directed capillary outlet. We have chosen an inside capillary diameter of 2.4 mm as a reasonable compromise between target chamber purge rate and helium flow rate.

Two target chambers have been designed, one for fission targets and one for spallation targets. These chambers, which are designed for remote servicing, would be located at the end of a vertical shield plug near the LAMPF beam-stop. The target chambers incorporate the inlet and outlet geometries determined from our feasibility experiments and feature double containment to allow use of actinide targets.

One concern remains in our considerations. Despite the considerable experience in coupling a He-jet to a mass separator ion source [MOL81,OKA81],

the reported total efficiencies are characteristically less that those
achieved with normal on-line separator systems. Current efforts to improve
the He-jet coupling and ionization efficiencies are, however, showing progress. An ion optical design has been made that incorporates a non-dispersive
intermediate image to allow for correction of ion source fluctuations. A con-
ceptual layout for the on-line system with this design is shown in Fig. 2.1.

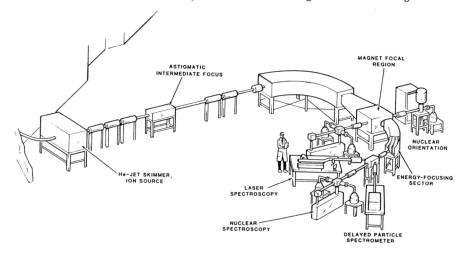

Fig. 2.1 Layout of the proposed on-line mass separator system at
          LAMPF, showing conceptual ion optical design for a
          single magnetic stage.

3. Production estimates for nuclei far from stability

To define more accurately the boundaries of the mass regions that could
be accessed with the proposed He-jet coupled mass separator system, production
cross sections for both neutron-deficient and neutron-rich nuclei far from
stability have been estimated for 800-MeV proton reactions. The spallation-
product cross sections were estimated through use of the Rudstam systematics
[RUD66]. For estimation of the fission-product cross sections, however, there
is no established, similar approach. Thus, an empirical approach was taken in

which two overlapping Gaussian distributions were fitted to existing rubidium and cesium isotopic distributions obtained at 156 MeV, 170 MeV, and 1 GeV [TRA72,BEL80] -- one Gaussian for the neutron-rich portion of the distribution and one for the neutron-deficient portion. The parameters of the Gaussians were then varied with A and Z to account for the mass-yield variations and other differences between the rubidium and cesium data. Adjustment of the cross-section distributions to 800 MeV was accomplished by interpolation.

If we assume that 1000 atoms/s of a mass-separated radionuclide are needed for spectroscopic measurements and that the mass separator system has only a one-percent overall efficiency, a partial production cross section of about 0.7 μb is required for the nuclide in question, assuming a LAMPF beam intensity of 800 μA. According to our cross-section estimates, essentially all neutron-deficient nuclei with calculated [TAK73] half-lives >300 ms will be produced in 800-MeV spallation with a cross section of >1 μb (assuming the use of at least 10 different target materials). Interesting nuclear regions that could be reached include those near ^{76}Sr and ^{100}Sn. On the neutron-rich side of stability, most nuclei with half-lives >300 ms and mass 60-140 can be produced in high-energy fission of ^{238}U with a cross section of >1 μb. Table 3.1 summarizes the enormity of our current ignorance of nuclei far from stability accessible by high-energy spallation and fission reactions, despite over two decades of prolific on-line studies.

Table 3.1. Numbers of nuclei with unknown properties between known limits and 300-ms (according to the gross theory of beta decay[9]) or 1-μb cross-section limits; Z=10-90.

Property	Neutron-rich [^{238}U(p,f)]	Neutron-deficient [Spallation]
Mass	281	436
Half-life	170	198
Decay scheme	243	630
Spin-parity	300	484

Of those nuclei included in Table 3.1, we expect that unique access to about half would come from a He-jet coupled mass separator at LAMPF. We plan to use this unique access to advantage in proposed studies, to complement (rather than compete with) similar studies at other on-line facilities.

4. Possible Specific Studies

Given the large variety of new nuclei that would become available with the proposed on-line separator system, a focus is needed for possible initial studies. On the other hand, the importance of making systematic studies of nuclear properties over sizeable regions must be recognized. For example, the novel feature of shape coexistence was established only through detailed, systematic studies of nuclear decays. A notable feature of the proposed on-line mass separator system is the capability to make unique nuclear structure studies in several interesting regions.

One such region is that around N=Z=38, postulated to be a region of strong deformation [MÖL81]. Although this prediction has some experimental support [LIS82,PRI83,HEY84], other theories predict nuclei in this region to be spherical, with some softness toward deformation [ÅBE82,BUC79]. Recent work in this region suggests an apparent quenching of pairing correlations in ^{84}Zr, resulting in moments of inertia at about rigid-body values [PRI83].

The neutron-rich nuclei near ^{100}Zr comprise another recently established deformed region [KHA77,WOL77,SCH80,AZU79]. Here, the onset of deformation is especially abrupt, and strong quenching of pairing correlations seems indicated [PEK85]. The He-jet technique offers the capability of mapping the extent of this unusual deformed region in the refractory-element area above A=100 -- a region presently inaccessible at other on-line facilities.

The region around ^{100}Sn also offers exciting possibilities. Studies of nearby nuclei have been unable to determine the applicable coupling scheme or the interplay of the nearly symmetric neutron-proton configurations. Furthermore, heavy-ion reaction cross sections pose a severe limit in extending the previous studies in this region.

Last, but not least, it is important to search for other regions besides the Pt region in which the nuclear structure can be described in terms of supersymmetric boson-fermion theory. The proven existence of several such regions would bring us one step closer to a comprehensive theory of all nuclei.

Acknowledgments

This work was supported by the U.S. Department of Energy.

References

[ÅBE82] S.Åberg, Phys. Scr. <u>25</u>, 23 (1982).
[AZU79] R. E. Azuma, G. L. Borchert, L. C. Carraz, P. G. Hansen, B. Jonson, S. Mattsson, O. B. Nielsen, G. Nyman, I. Ragnarsson, and H. L. Ravn, Phys. Lett. <u>86B</u>, 5 (1979).
[BEL80] B. N. Belyaev, V. D. Domkin, Yu. G. Korbulin, L. N. Androneko, and G. E. Solyakin, Nucl. Phys. <u>A348</u>, 479 (1980).
[BUC79] D. Bucurescu, G. Constantinescu, and M. Ivascu, Rev. Roum. Phys. <u>24</u>, 971 (1979).
[HEY84] K. Heyde, J. Moreau, and M. Waroquier, Phys. Rev. C <u>29</u>, 1859 (1984).
[KHA77] T. A. Khan, W. D. Lauppe, K. Sistemich, H. Lawin, G. Sadler, and H. A. Selic, Z. Phys. <u>A283</u>, 105 (1977).
[LIS82] C. J. Lister, B. J. Varley, H. G. Price, and J. W. Olness, Phys. Rev. Lett. <u>49</u>, 308 (1982).
[MÖL81] P. Möller and J. R. Nix, At. Data Nucl. Data Tables <u>26</u>, 165 (1981).
[MOL81] D. M. Moltz, Nucl. Instr. and Meth. <u>186</u>, 135 (1981) and references therein.
[OKA81] K. Okano, Y. Kawase, K. Kawade, H. Yamamoto, M. Hanada, and T. Katoh, Nucl. Instr. and Meth. <u>186</u>, 115 (1981).
[PEK85] L. K. Peker, F. K. Wohn, J. C. Hill, and R. F. Petry, to be published.
[PRI83] H. G. Price, C. J. Lister, B. J. Varley, W. Gelletly, and J. W. Olness, Phys. Rev. Lett. <u>51</u>, 1842 (1983).
[RUD66] G. Rudstam, Z. Naturforsch. <u>21a</u>, 1027 (1966).
[SCH80] F. Schussler, J. A. Pinston, E. Monnand, A. Moussa, G. Jung, E. Koglin, B. Pfeiffer, R. V. F. Janssens, and J. van Klinken, Nucl. Phys. <u>A339</u>, 415 (1980).
[TAK73] K. Takahashi, M. Yamada, and T. Kondoh, At. Data and Nucl. Data Tables <u>12</u>, 101 (1973).
[TRA72] B. L. Tracy, J. Chaumont, R. Klapisch, J. M. Nitschke, A. M. Poskanzer, E. Roeckl, and C. Thibault, Phys. Rev. C <u>5</u>, 222 (1972); J. Chaumont, Ph.D. thesis, Faculté des Sciences Orsay, 1970 (unpublished).
[WOL77] H. Wollnik, F. K. Wohn, K. D. Wunsch, and G. Jung, Nucl. Phys. <u>A291</u>, 355 (1977).

RECEIVED July 15, 1986

65

Developing an Accelerated Radioactive Beams Facility Using an On-Line Isotope Separator as an Injector

John M. D'Auria

Department of Chemistry, Simon Fraser University, Burnaby, British Columbia V5A 1S6, Canada

High yields of separated, radioactive ions (up to 10^{10} atoms/sec µA of incident protons at on-line isotope separators (ISOL), e.g. ISOLDE at CERN (SC), make it feasible to consider using such secondary ions as projectiles for nuclear reactions. A pressing need for reaction rate data involving radioactive species exists in nuclear astrophysics. This requires having available projectiles (A ≤ 60) in the energy range from about 200 keV/amu to 1.5 MeV/amu. It has been proposed to install an ISOL device at the TRIUMF facility to utilize the available intermediate energy (200-500 MeV), intense (≤100 µA) proton beam as the primary production source. The mass analyzed, radioactive beam (RB), extracted from the target/ion source at 60 keV, will be transported vertically to experimental areas and/or injected into a post-accelerator. Although other possibilities are being considered, most attention so far has been devoted to an RFQ/drift tube LINAC combination. The radioactive projectiles of initial interest are ^{13}N, ^{15}O, ^{18}F, ^{19}Ne and ^{21}Na, and it is estimated that RB intensities as high as nA are obtainable for some of these with presently available technology. These would interact with gaseous targets of H or He to perform (p,γ), (α,γ) and other simple fusion reactions of interest. A small on-line test ISOL is being designed to perform engineering research and development on these systems to optimize all production conditions to achieve highest yields.

A proposal has been made to install an intense radioactive beams (RB) facility at the TRIUMF (200-500 MeV), proton cyclotron, located in Vancouver, Canada. The proposed facility consists of an on-line isotope separator (ISOL), capable of producing intense, mass-separated, radioactive beams (RB), and a post-accelerator (PA), for further acceleration of the separated, radioactive ions. A key component of this proposed facility is to utilize these accelerated radioactive beams to perform measurements of rates of simple fusion reactions on gaseous hydrogen and helium targets. Such data is of great interest and importance in the area of nuclear synthesis in "hot" explosive astrophysical phenomena. The purpose of this report is to present a brief description of the proposed TRIUMF-ISOL facility, outline briefly the nuclear astrophysics interest in such low energy, reaction studies, and summarize briefly the approach proposed to perform such studies including design specifications. A discussion of present status and future plans with a small test system (TISOL) are also included.

NOTE: The author of this chapter worked with members of the TRIUMF-ISOL Collaboration, 4004 Wesbrook Mall, Vancouver, British Columbia V6T 2A3, Canada.

THE ISOL SECTION

General
    An ISOL is in essence a device consisting of a target, whose thickness
reflects the penetrating power of the production projectile, coupled to an
ion-source in which the species are ionized. This is then followed by a
magnetic mass analyzer to separate (A) the extracted radioisotopes. Electrostatic elements direct the beam to some experimental area. The proposed
system here is based upon the ISOLDE.[CERN] and ISOCELE [ORSAY] models. Some
specificity in Z will be obtained by appropriate selection of targets and ion-
source. In general, thick targets, and in particular, ion- sources can minimize or eliminate the presence of most elements. For a complete description
of the present status of proton based ISOL devices, consult, for example,
[RAV79], [RAV84], and [SAU84]. An excellent description of types of ion
sources used at ISOL facilities can be found in [KIR81].

Location
    The target of the ISOL will be located at the end of TRIUMF beam line 4A
as indicated in Fig. 1. In this position the intensity of the available variable energy (200-500 MeV) proton beam is now limited to 10 μA, although the
proposed new shielding would allow intensities up to 100 μA. Such an intensity coupled with the long range of the incident protons, allowing the use of
thick targets ($\lesssim$ mole/cm^2), indicate that the production rates of almost any
isotope would be unmatched by any present facility in the world.

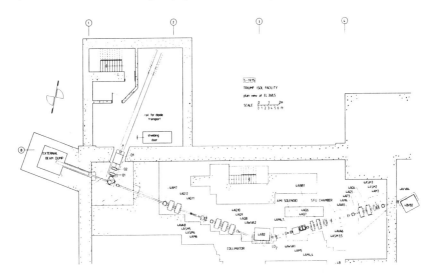

Fig. 1. Location of the ISOL target in front of the beam
dump on TRIUMF beam line 4A.

ISOL
    The isotope separator is a vertical plane system designed to transport
the ionized radioisotopes from the subterranean production site to ground
level in as short a distance as possible, and is made up of two stages (see
Fig. 2). The first stage is a medium resolution ($\sim$635 for 30 π mm/mrad emittance), mass separator (composed of a QQD) which will transport two beams.
One is referred to as the high intensity beam (HIB) and is directed into the
lower hall of a new building, for injection in the post-accelerator. The
other beam is directed into a dispersion-cancelling matched magnet set (DQQ)
which creates the high-brightness image source necessary for entrance into a

Remote Handling
     A key feature in a facility advertising to use such high proton currents
and thick targets is the question of handling of components and T/IS changes.
A novel approach is proposed here using long columns to hold separately the
T/IS system, the movable extraction electrode (EE), and the pumping components, and position them in a triangularly shaped, removable vacuum box. The
T/IS can be detached above the thick shielding block surrounding the "hot"
irradiation area using a remotely operated machine, and taken to a hot-cell
area for manual manipulations, or for storage and ultimate disposal. A novel
method of moving the EE reasonably accurately and reproducibly with 5 deg of
freedom has been devised using off-centered, circular controlling discs.
     A more complete description of the proposed facility can be found else-
where [PRO85].

THE POST-ACCELERATOR SECTION

Elemental Synthesis and Nuclear Astrophysics
     The present emphasis of astrophysics is to understand the related nucle-
ar processes leading to the synthesis of elements in star. Reaction networks
incorporating simple nuclear fusion-like reactions can be postulated which
show how the elements and isotopes are formed. Important parameters to such
cycles are the reaction rates based upon the cross sections (excitation function) and the Q-values. For such reactions the rates are strongly influenced
by the presence of narrow resonances, but the strength of these resonances can
not be accurately calculated using any theoretical models especially in the
low Z region, where the statistical model is not applicable. Thus each reaction requires input from experimental measurements. Otherwise the calculated
estimates of the various, relative isotopic abundances in the reaction
networks for stars could be wrong by many orders of magnitude [FOW84].
     It is now widely accepted that in "hot" astrophysical explosive phenome-
na such as novae, supernovae, accreting neutron stars, etc. radioactive
species can be involved as reactants in the reaction networks, i.e., the reactions are not beta-decay limited, due to the higher density of unstable
species. For example, a hot CNO cycle is proposed which uses several unstable
species. Critically important reactions for which no data exist are, e.g.
$^{13}N(p,\gamma)^{14}O$ and $^{18}F(p,\gamma)^{19}Ne$. At very hot stellar temperature Wallace and
Woosley have predicted a new process (rp-process) involving the rapid capture
of protons on unstable nuclei [WAL81]. Indeed, a number of different cycles
and instances can be described indicating the importance of obtaining reaction
rate data at low energies for radioactive species. More complete discussions
and early attempts can be found in [BUC84], [FOW84], [HAI83], and [BOY83].
Table I presents a series of nuclear reactions for which it would be of
interest and importance to measure the reaction cross section in the energy
range from 0.2 to 1.5 MeV/amu [EEC84].

A Novel Approach
     As indicated earlier the approach to be taken here to obtaining these
reaction rates will be to accelerate the heavy radioactive isotope, onto a
gaseous hydrogen (for p,γ and p,α studies) or a helium (for α,γ and α,p
studies) target, and then detect appropriately the reaction product (either
light or heavy). The generation and use of a RB for such studies has been
pioneered elsewhere by [BOY83] and Haight et al. [HAI83] but this linkage with
an intense beam, ISOL device is unique. With a 100 µA, 500 MeV proton beam,
production rates of specific interest, e.g. ^{13}N from a MgO, target (50 g/cm^2)
can be as high as 3×10^{12} atoms/sec. Cross sections in the tens of µb region
for these radiative proton capture reactions are then definitely measurable,

given a gaseous target of about 10^{19} atoms/cm^2. A key component in this
approach is the method of achieving acceleration of the radioactive ISOL
beams.

Table I

Reaction number	Reaction of interest	Reactant $T_{1/2}$	Astrophysical interest
1	$^{13}N(^{1}H,\gamma)^{14}O$	10 min	hot CNO cycle isotopic abundances
2	$^{15}O(\alpha,\gamma)^{19}Ne$	2.03 m	proposed rp-process
3	$^{18}F(p,\alpha)^{15}O$	110 min	hot CNO cycle
4	$^{18}F(p,\gamma)^{19}Ne$	110 min	hot CNO cycle
5	$^{19}Ne(p,\gamma)^{20}Na$	17.2 s	break-out rp-process leakage – hot CNO
6	$^{21}Na(p,\gamma)^{22}Mg$	22.5 s	hot NeNa cycle rp-process

The Post-Accelerator
    Various options of accelerator configurations to accelerate efficiently
the 60 keV, singly charged beam from the ISOL to a useful output energy range
have been examined. Based upon the requirements of the experimental program
for studies in nuclear astrophysics this accelerator should produce external
beams of ions up to an A value of at least 60, with energies in the range from
about 100 keV/amu to at least 1.5 MeV/amu, continuously variable with a resolution of 10^{-2} or better. Transmission efficiencies should be high so that
useful beam intensities in the nA region are available; acceptance of either
positive or negative ions from the ISOL would be a desirable feature. Based
upon these specifications, a LINAC solution was considered more suitable than
other means of acceleration including cyclotrons and Tandem accelerators, and
has received more attention [PRO85].
    A critical part of the post-accelerator is the first stage which must
capture, bunch, and accelerate the singly charged (+/-), very low velocity
($\beta \geqslant 0.15\%$), dc beam from the ISOL, with good efficiency. This is best
accomplished by some form of radiofreqency quadrupole (RFQ) such as one built
at GSI for q/A $\geqslant$ 1/130 [MUL84], or a low frequency version (~23 MHz) of the
Los Alamos four vane structure. A preliminary design of the latter version,
done at CRNL for q/A $\geqslant$ 1/40, a peak field of 1.5 Kilpatricks, and an operating
frequency of 23 MHz, has a calculated capture efficiency of about 86% for a
0.1 π cm/mrad (normalized emittance) ISOL beam [BUC85]. At some point between
60 and 100 keV/amu the beam may be stripped and injected into a post-stripper
section. This latter section can be based on existing LINAC designs such as
the UNILAC, HILAC or RILAC.
    As an illustration only, a possible post-accelerator layout, initiated
following discussions with H. Klein [KLE84], is shown in Fig. 4, along with
the associated experimental area. A foil stripper is inserted at 100 keV/amu
to increase the charge-to-mass ratio and minimize the length of the accelerator. The single gap sections shown can be switched on or off individually as
well as adjusted by their relative RF phases to give the required ion energy
variation. A debuncher is also provided to tailor the beam to meet the

required energy resolution. This system, operating at CW, would require an estimated 4 MW of RF power. Other similar solutions can be envisaged which use less power but may not meet all of the specifications. A more complete discussion of such initial concepts of the TRIUMF-ISOL post-accelerator is found elsewhere [PRO85], [DAU85]

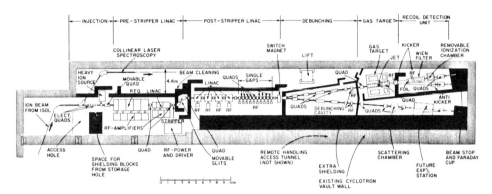

Fig. 4.   A possible RFQ LINAC Post-Accelerator layout and experimental area.

At present, a detailed examination of the LINAC configuration is being carried out. This includes a realistic estimate of the RFQ accelerator effi- ciency, the detailed arrangement of the subsequent LINAC sections including intertank matching, the positioning (in energy), the type and efficiency of the stripper, and the energy resolution that can be expected.   This will be followed by preparation of realistic cost estimates and an implementation schedule.   The results of this study, supplemented by the proceedings of the recently completed TRIUMF Radioactive Beams Workshop held in September 1985, will be available as a full proposal late this year 1985. It is expected that the entire TRIUMF-ISOL proposal will be reviewed for funding in 1986.

TISOL

The final selection of the appropriate target and ion source for the ISOL to produce such intense beams of the initial projectiles of interest, namely ^{13}N, ^{15}O, ^{18}F, ^{19}Ne, and ^{21}Na, requires considerable research and development. In general these species have not been of interest at present ISOL facilities and conditions to optimize their yields require investigation. Table II presents the highest yields observed along with target/ion source (TIS) combinations used [RAV85].   Tests are presently underway to develop new TIS systems for these species. Due to the rather non-specific nature of their decay emissions, and the absence of any long-lived isotopes of these elements to be used as tracers, a small ISOL is being installed at TRIUMF to perform such development studies, along with other proposed uses. This is scheduled for commissioning in the summer of 1986. The system will be used for limited beam current (1 μA)/target thickness ($\leqslant$1 g/cm^2) combinations to minimize radi- ation problems.

SUMMARY

In summary, a proposal for a TRIUMF-ISOL facility including a prelimi- nary version of a post-accelerator to boost the radioactive beam energies up

high resolution separator (30,000). The resultant high resolution beam (HRB) is directed to a second floor experimental hall for more traditional ISOL types of experiments. Figure 3 is an artist's conception of what the completed facility might look like. Included in the proposed new building are space for an off-line system, a target-ion source (TIS) testing and conditioning station, and experimental areas for set-ups using either the HRB or the HIB.

QUADRUPOLES (Q1, Q2, Q3, Q4):
Effective Length = 20 cm
Aperture = 10 cm

DIPOLES (D1, D2):
Bend Angle = 58°
Radius of Curvature = 1 m
Field Gradient = 0

DIPOLE (D3):
Bend Angle = 116°
Radius of Curvature = 1 m
Field Gradient = 1/2

INTERMEDIATE FOCUS:
Bend Plane Magnification = -2.5
Non-Bend Plane Magnification = -2.48
Dispersion = 3.86 cm/%
Resolution = 635
    For an ion source with a bend
    plane emittance of
    30 π mm/rad

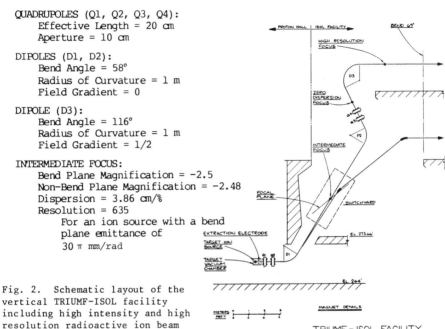

Fig. 2.   Schematic layout of the vertical TRIUMF-ISOL facility including high intensity and high resolution radioactive ion beam lines.

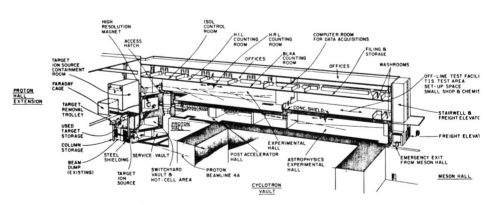

Fig. 3.   Artist's conception of proposed TRIUMF-ISOL Facility including new building, north of TRIUMF hall.

to 1.5 MeV/amu has been formally submitted to the Director of TRIUMF. The installation of such a facility would be unique in the world. The total capital cost of such a system is expected not to be greater than about 13.5 M. Early international review will be sought. Scientists in disciplines in which such a facility would be useful are requested to submit their experimental proposals to the author so that the future facility can accommodate their requirements.

Table II

Ion beam	Intensity atoms/sec (per μA of proton)	Target material	Target thickness (g/cm^2)	Ion source yield (%)
^{13}N	4.1 E + 05	CaO	2.5	0.01
^{15}O	—	—	—	—
^{18}F	—	—	—	—
^{19}Ne a)	3.0 E + 07	MgO	3	0.7
^{21}Na	1.0 E + 05	Ta	122	50

a) [RAV85]

REFERENCES

[RAV79] Ravn, H., 1979. Physics Reports 54, 1.
[RAV84] Ravn, H., 1984. In Proceedings of the TRIUMF-ISOL Workshop, eds. J. Crawford and J.M. D'Auria, Mt. Gabriel, Québec, 1984; TRI-84-1.
[SAU84] Sauvage, J., et al., 1984; ibid.
[KIR81] Kirchner, R., 1983. Nucl. Instr. Meth. 186, 275.
[PRO85] The TRIUMF-ISOL FACILITY, A Proposal for an Intense Radioactive Beams Facility, June 1985.
[FOW84] Fowler, W.A., 1984. Rev. Mod. Phys. 56, 149.
[WAL81] Wallace, R.K., and Woosley, S.E., 1981. Astrophys. J. Suppl. 45, 389.
[HAI83] Haight, R.C. et al., 1983. Nucl. Instr. Meth. 212, 245.
[BOY83] Boyd, R. et al., 1981. IEEE Trans. Nucl. Sci. NS30, No. 2, 1387.
[BUC84] Buchmann, L., 1984. In Proceedings of the TRIUMF-ISOL Workshop, eds. J. Crawford and J.M. D'Auria, Mt. Gabriel, Québec, 1984; TRI-84-1.
[EEC84] Buchmann, L., and D'Auria, J.M., 1984. Experiment 311 presented to TRIUMF Experiment Evaluation Committee, December Meeting.
[BUC85] Buchmann, L., D'Auria, J.M., Mackenzie, G., Schneider, H., and Moore, R.B., 1985. IEEE Trans. Nucl. Sci., in press.
[KLE84] Klein, H., 1984. Private communication, to be published in Proceedings of the Radioactive Beams Workshop, Parksville, B.C., Canada.
[DAU85] D'Auria, J.M. et al., 1985; ibid.
[RAV85] Ravn, H., D'Auria, J.M., Garcia-Borge, M., Huang, Y., Jonson, O.C., Nikkinen, L., and Sundell, S., 1985; ibid.

RECEIVED July 15, 1986

Proton-Rich and Neutron-Rich Nuclei Produced by Heavy Ions

Ernst Roeckl

Gesellschast für Schwerionenforschung Darmstadt, Postfach 110541, D-6100 Darmstadt, Federal Republic of Germany

Recent nuclear-structure studies of proton-rich and neutron-rich nuclei, performed at the GSI on-line mass separator, are surveyed and an outlook is given towards a continuation of such investigations with heavy-ion beams of higher energies.

In this report, recent nuclear-structure studies will be described as performed at the GSI mass-separator on-line to the heavy-ion accelerator UNILAC. We shall first discuss properties, such as masses, moments, charge radii and Gamow-Teller beta transitions, of very proton-rich nuclei obtained in fusion-evaporation reactions with ^{16}O, ^{40}Ca and ^{58}Ni beams. Secondly, measured beta decay half-lives of neutron-rich isotopes, produced in multinucleon transfer reactions with (9-15) MeV/u beams of ^{76}Ge, ^{82}Se, ^{136}Xe, ^{186}W and ^{238}U, will be presented in comparison with theoretical predictions. Finally, an outlook will be given to the extension of such experiments at the heavy-ion synchrocyclotron SIS and the experimental storage ring ESR being under construction at GSI. In order to be brief, details on the various topics are not given here, but should rather be taken from the references given at the end of this paper.

PROTON-RICH NUCLEI

The $f_{7/2}$-Shell Region

Very recent experiments attempted to produce and investigate proton-rich nuclei in the $f_{7/2}$ region using ^{40}Ca + ^{12}C and ^{32}S + ^{28}Si fusion-evaporation reactions. The new $T_z=-1$ nucleus ^{48}Mn was identified, using the p3n channel, with a half-life of 150±10 ms and strong gamma-lines at 752.1 keV ($2^+\to0^+$) and 1106.1 keV (4^+-2^+) [KOS85]. From gamma-gamma coincidence data, measured in particular for higher transition energies, higher-lying levels were identified including the 4^+, T=1 isobaric analogue state at 5792.4 keV in ^{48}Cr. Preliminarily, beta-delayed proton emission was

0097–6156/86/0324–0439$06.00/0

observed for ^{48}Mn, which would extend the A=4n, T_z=-1 family of precursors beyond the so far heaviest member ^{44}V [CER71]. The known T_z=-3/2 nucleus ^{45}Cr($T_{1/2}$=40 ms) was produced through the 2p5n channel with an intensity of approximately 1 atom/s, and there is preliminary evidence for beta-delayed proton decay of T_z=-3/2 ^{47}Mn, produced through the p4n channel. (^{47}Mn would represent the heaviest T_z=-3/2, A=4n+3 nucleus observed experimentally.) Our conclusions from these pilot experiments are, that inverse reactions such as ^{40}Ca + ^{12}C are indeed suitable for producing new proton-rich $f_{7/2}$ nuclei, and that sufficient source strength for spectroscopy of mass-separated samples can be reached even for half-lives as short as 40 ms.

"Southeast" of ^{100}Sn

Nuclear-structure studies around the expected double shell-closure at ^{100}Sn have concentrated recently to study "quenching" of Gamow-Teller (GT) beta transitions, and to investigate moments and charge radii by means of collinear laser spectroscopy. After giving some recent examples from these investigations, a novel ion source with bunched beam-release is briefly mentioned at the end of this section.

Within the series of proton-rich N=50 and Z=50 nuclei, β^+/EC delayed γ-rays were studied recently for the decay of ^{96}Pd [RYK85] and ^{104}Sn [RAT85]; from measured $\beta^+/(\beta^+$+EC) ratios, Q_{EC} values were deduced to be 3450(150) keV and 4000$\pm^{650}_{300}$ keV, respectively. These Q values together with the ones of neighbouring decays can be used to determine ground-state masses of very proton-rich isotopes, including the nuclei ^{109}I and ^{113}Cs beyond the proton drip line [RAT85]. Alpha decay-energy differences for Z=50 show that this measure of the proton shell strength tends to be enforced by neutron shell closures, with a distinct effect occurring also at N=64 and N=66 due to $d_{5/2}$ and $s_{1/2}$ neutron subshell closures. We shall return to this point when discussing results from collinear laser spectros-copy. The measured GT transition strengths of even N=50 isotones and tin (Z=50) isotopes are shown in Fig. 1 in comparison with predictions from shell-model calculations [TOW85]: The experimental results for the N=50 nuclei ^{94}Ru and ^{96}Pd agree well with the calculations of Towner, whereas the measurements for even tin isotopes appear to contradict the slope expected from a standard pairing approach.

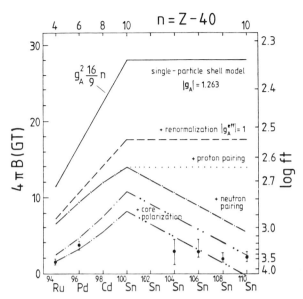

Fig. 1. Comparison of the experimentally observed GT transition strength B(GT) of even N=50 isotones and tin isotopes with predictions. The corresponding log ft value are given at the right hand scale. The full line represents predictions from the extreme single-particle shell model as function of the number n of protons in the $g_{9/2}$ shell. Lower transition probabilities are obtained by taking into account (applying in successive order): renormalization of the free neutron value of the coupling constant g_A - to account for "higher order effects" (dashed line), pairing correlations of protons (dotted line), pairing correlations of neutrons (dot-dash line) and core polarization from particle-hole interactions with a finite and a zero range force (top and bottom double-dot dash line, respectively) [6,7] See refs. [RYK85] and [KLE85] for details.

In collaboration with G. Huber et al. (University of Mainz), collinear laser spectroscopy of short-lived indium and tin isotopes [KIR85, ULM85, LOC85] was extended in recent experiments, both at GSI and at the ISOLDE facility at CERN [EBE85a]. Systematical data on magnetic dipole moments, electric quadrupole moments and isotope shifts were obtained in the mass range A=105 to A=127. The mean square charge-radii, deduced from measured isotope shifts, show clearly onset and disappearance of deformation, if compared to a simple two-parameter formula (see Fig.2). The latter ansatz approximates $\delta<r^2>$ as the addition of a first term, which describes the smoothly varying change of nuclear charge radii for spherical shape as given by the droplet model [MYE83], and a second term representing the effect of change in quadrupole deformation. A differential plot of mean

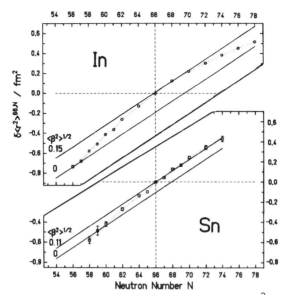

Fig. 2. Change of the nuclear mean square charge radii $\delta\langle r^2 \rangle$ for indium and tin isotopes relative to N=66. The solid lines correspond to the pure volume effect at constant shape according to the droplet model [MYE83].

square charge-radii of indium and tin isotopes (Brix-Kopfermann diagram) reveals a subshell structure around N=64 [LOC85], which relates to similar observations made on the basis of mass differences as mentioned above. The disappearance of quadrupole deformation for light indium isotopes, however, is in disagreement with the result of constant deformation as deduced from experimental quadrupole moments, and the relation to measured B(E2) values of light cadmium isotopes is also not understood at present.

In order to improve the signal-to-noise ratio in collinear laser spectroscopy, an ion source with bunched beam release was tested successfully. For this purpose, the temperature of a "cold trap" inside the ion source is reduced for storage of reaction products, which are released from the trap during a subsequent period of increased temperature. The release of indium was found to occur with a FWHM of approximately 0.5s, corresponding to a substantial improvement in signal-to-noise ratio for the fluorescence spectrum of ^{108}In [EBE85b] obtained during this release period. Moreover, such a storage-release offers the possibility of doing "ion-source chemistry": Setting time windows due to the element-specific release profiles, it was possible to enrich secondary beams of e.g. 21-s ^{104}Sn with respect to the isobaric contaminants 1.5-min ^{104}In and 58-min ^{104}Cd.

"Northwest" of ^{146}Gd

A programme of detailed decay-spectroscopic investigations in the region "northwest of ^{146}Gd" is pursued in collaboration with P. Kleinheinz et al. (KFA Jülich). These experiments, which also use the on-line separator ISOCELE at Orsay and ISOLDE at CERN, cannot be covered in this brief report. It may suffice to mention here the studies of protons and gamma rays emitted after beta decay of N=81,82 nuclei, these decays being dominated by $\pi h_{11/2} \rightarrow \nu h_{9/2}$ GT transitions. For the decay of ^{147}Dy, deexcitation of high-lying levels in the daughter nucleus ^{147}Tb$_{82}$ was investigated by observing beta-delayed protons and gamma-rays. In a preliminary evaluation of corresponding proton and gamma energies, the proton separation energy of $1/2^+$ ^{147}Tb was determined to be 1975 keV [SCH85]. This is of interest in view of the discrepancies between recent mass measurements of this nucleus.

NEUTRON-RICH NUCLEI

In collaboration with W.-D. Schmidt-Ott et al. (University of Göttingen), decay properties of neutron-rich isotopes were investigated in the manganese-to-zinc region [RUN83,RUN85], for rare-earth elements [KIR82,RYK83,ESC84], and for radium and actinium [GIP85]. The beta decay half-lives, measured in these experiments and in particular in a recent one [BOS85], are compared to theoretical predictions in Fig. 3. In the chromium-to-nickel region, the measured half-lives are systematically shorter than the predicted ones. This observation is calling for improved half-life calculations, and it is also of interest to study the influence of this effect on the astrophysical rapid neutron-capture (r) process. Preliminary conclusions from such considerations [BOS85] are, that even with shorter half-lives the A~80 r-process abundance peak cannot be built-up from iron-group nuclei during explosive helium burning; therefore explosive helium burning cannot be the only site of the r-process, and a significantly higher neutron density is required in order to reproduce the observed A~80 r-process abundance peak.

STUDIES OF FAR-UNSTABLE NUCLEI AT SIS-ESR

In May 1985, the GSI proposal [KIE85] of building an 18 Tm synchrotron (SIS) and a 10 Tm experimental storage ring (ESR) was formally accepted. The plans of using this accelerator complex for studying nuclei far from stability [ROE84] are thus becoming more realistic. Heavy ion beams from

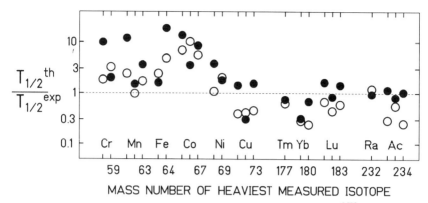

MASS NUMBER OF HEAVIEST MEASURED ISOTOPE

Fig. 3. Ratio between measured beta decay half-lives ($T_{\frac{1}{2}}^{exp}$) and predictions ($T_{\frac{1}{2}}^{th}$) from a microscopic model [KLA84] (open circles) and from the
gross theory of beta decay [TAK73] (full circles). For the latter calculations, beta decay energies were used from the mass formula of [MÖL81].

SIS and in ESR between neon and uranium with energies up to 2 GeV/u, $2 \cdot 10^{11}$
s^{-1} and 1 GeV/u, $4 \cdot 10^{10} \ s^{-1}$, respectively, and considerably higher intensities at lower energies, offer interesting possibilities to extend the
above-mentioned nuclear-structure studies to regions even further away
from the beta stability line. This statement is made on somewhat shaky
ground, since the (tails of the) production cross section distributions,
and their dependence on the N/Z ratio of projectile and target, have been
measured in a quite incomplete way so far. On the other hand, recent
results obtained with 30 to 200 MeV/u beams at GANIL, LBL and MSU show great
promise in this respect. In particular the most recent experiments at GANIL
have proven successfully that heavy-ion induced fragmentation reactions
combined with recoil separation can be used for producing new isotopes, in
impressive numbers and yields, both of neutron-rich (carbon-to-cobalt) and
of proton-rich ($T_z = -5/2$) species [LAN85, GUI85, BER85]. Investigations
with secondary beams of unstable nuclei are being planned downstream a
recoil separator [GEI85], using this fascinating possibilities with
stored, cooled and slowed-down beams [KIE85], but also applying fixed (sin-
gle-turn) target-detector arrays. This may lead to novel expeditions
towards and beyond the nucleon drip lines, which allow refined studies of
interesting phenomena in known decay modes (e.g. GT quenching, studied then
for larger beta-decay windows) as well as search for new modes of radioactivity.

REFERENCES

[BER85]   M. Bernas, private communication (1985).
[BOS85]   U. Bosch et al., "Beta Half-Lives of Neutron-Rich
          Chromium-to-Nickel Isotopes Used for New Calculation of the
          Astrophysical r-Process", to be published.
[CER71]   J. Cerny et al., Phys. Lett. 37B 380 (1971).
[EBE85a] J. Eberz, private communication (1985).
[EBE85b] J. Eberz et al., "Collinear Laser Spectroscopy on ^{108}g,^{108m}In Using
          an Ion Source with Bunched Beam Release", to be published.
[ESC84]   W. Eschner et al., Z. Phys. A317 281 (1984)
[GEI85]   H. Geißel, "Fragment Separation at SIS", contribution to the
          Accelerated Radioactive Beam Workshop, Parksville, September 5-7,
          1985.
[GIP85]   K.-L. Gippert, "Kernspektroskopie projektilähnlicher Produkte aus
          ^{238}U-induzierten Kernreaktionen", Ph. D. Thesis, University of
          Göttingen, GSI-85-20 (1985).
[GRA84]   I. S. Grant et al., in Proc. 7th Intern. Conf. on At. Masses and
          Fundamental Constants (AMCO-7), Darmstadt-Seeheim (1984), ed.
          O.Klepper, p. 170.
[GUI85]   D. Guillemaud-Mueller et al., "Production and Identification of
          New Neutron-Rich Fragments from 33MeV/u ^{86}Kr Beam in the 18≤Z≤27
          Region", GANIL P.85.08 (1985).
[KIE85]   P. Kienle,"The SIS-ESR Project of GSI", GSI-85-16 (1985).
[KIR82]   R. Kirchner et al., Nucl. Phys. A378 549 (1982).
[KIR85]   R. Kirchner et al., Nucl. Instr. and Meth. A234 224 (1985).
[KLA84]   H. V. Klapdor et al., At. Nucl. Data Tables 31 81 (1984).
[KLE85]   O. Klepper, in Proc. XXIII Intern. Winter Meeting on Nucl. Phys.,
          Bormio (1985), p. 747.
[KOS85]   V. T. Koslowsky et al., in GSI Scientific Rep. 1984, GSI 1-85
          (1985),p. 90.
[LAN85]   M. Langevin et al., Phys. Lett. 150B 71 (1985).
[LOC85]   H. Lochmann, "Kernmomente und Änderungen der mittleren quadratis-
          chen Kernladungsradien bei Zinn und Indium", Ph. D. Thesis, Uni-
          versity of Mainz, GSI-85-8 (1985).
[MÖL81]   P. Möller and J. R. Nix, Nucl. Phys. A361 117 (1981).
[MYE83]   W. D. Myers and K.-H. Schmidt, Nucl. Phys. A410 61 (1983).
[RAT85]   G.-E. Rathke et al., Z. Phys. A321 599 (1985).
[ROE84]   E. Roeckl, in Proc. of the TRIUMF-ISOL Workshop, Mont Gabriel,
          TRI-84-1 (1985), p. 6.
[RUN83]   E. Runte et al., Nucl. Phys. A399 163 (1983).
[RUN85]   E. Runte et al., Nucl. Phys. A441 237 (1985).
[RYK83]   K. Rykaczewski et al., Z. Phys. A309 273 (1983).
[RYK85]   K. Rykaczewski et al., Z. Phys. A321 (1985),in print.
[SCH85]   D. Schardt et al., in Proc. XX Winter School on Physics, Zakopane
          (1985).
[TAK73]   K. Takahashi et al., At. Nucl. Data Tables 12 101 (1973).
[TOW85]   I. S. Towner, "A Mass Dependence in Hindrance Factors for Favoured
          Gamow-Teller Transitions", to be published.
[ULM85]   G. Ulm et al., Z. Phys. A321 395 (1985).

RECEIVED July 29, 1986

Section III: New Facilities and Techniques

Part 2: Techniques

Unit A: In-Beam
Chapters 67–71
Unit B: Radioactive Targets and Radiochemistry
Chapters 72–73

67

Trends in the Study of Light Proton-Rich Nuclei

D. M. Moltz[1], J. Aysto[2], M. A. C. Hotchkis[1], and Joseph Cerny[1]

[1]Department of Chemistry and Lawrence Berkeley Laboratory, University of California, Berkeley, CA 94720
[2]Department of Physics, University of Jyväskylä, SF-40100 Jyväskylä, Finland

Recent work on light proton-rich nuclei is reviewed. Evidence for the first $T_z = -5/2$ nuclide, ^{35}Ca, is presented. Future directions in this field are discussed.

Introduction

Advances in the study of the limits of known nuclei have made possible investigations of nuclear structure in nuclei with unusual neutron to proton ratios. Questions, such as the limits of nuclear stability, the role of charge-dependent effects in nuclear systems and the existence of new radioactive decay modes, have guided the recent research on the most proton-rich light nuclides. Results from these studies have also made possible tests of advanced shell model analyses [WIL 83] far from the valley of beta-stability. Figure 1 shows a portion of the chart of nuclides through the titanium isotopes. It summarizes our present experimental knowledge of light nuclei. The most important advances in the studies of proton-rich nuclei since the last reviews [CER 77, AYS 80] have been the discoveries of the β-delayed proton decays of ^{27}P [AYS 85b], ^{31}Cl [AYS 82], and ^{36}Ca [AYS 81], the mass measurements of the $T_z = -2$ nuclides ^{24}Si [TRI 80], ^{28}S, ^{32}Ar and ^{40}Ti [BUR 80] and the discoveries of the new $T_z = -2$ nuclides ^{22}Al [CAB 82] and ^{26}P [CAB 83b]. The latter two were found to decay via β-delayed two-proton emission [CAB 83a, HON 83], a new mode of radioactivity predicted by Goldanskii [GOL 80]. Also, the first $T_z = -5/2$ nuclide, ^{35}Ca, has been discovered via its β-delayed 2p-decay [AYS 85a]. Most recently a complete β-decay study of the $T_z = -2$ nucleus ^{32}Ar by the ISOLDE-group [BJO 85]has been reported.

0097-6156/86/0324-0448$06.00/0

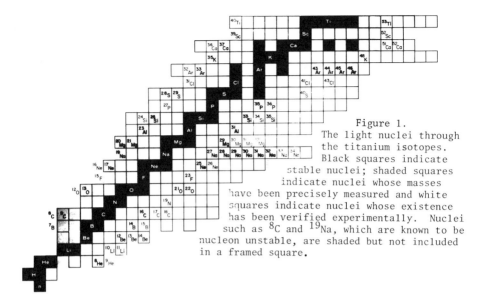

Figure 1.
The light nuclei through the titanium isotopes. Black squares indicate stable nuclei; shaded squares indicate nuclei whose masses have been precisely measured and white squares indicate nuclei whose existence has been verified experimentally. Nuclei such as ^8C and ^{19}Na, which are known to be nucleon unstable, are shaded but not included in a framed square.

Mechanism of two-proton emission following beta decay

Two-proton emission could in principle proceed via sequential emission, ^2He emission or simultaneous uncoupled emission. A decomposition of the observed two-proton spectra should indicate which mechanism dominates. Monte Carlo simulations of the first two mechanisms have been performed for ^{22}Al [CAB 84b] and have shown that ^2He emission would yield a proton energy continuum, with an angular correlation peaked at 40°, whereas sequential emission would be essentially isotropic and would exhibit two distinct peaks – one constant in energy corresponding to the first proton and a second one with appropriate kinematical shift.

The distinct peak structure of the individual proton spectra in the decay of ^{22}Al suggested that the major part of the decay occurred sequentially [CAB 84b]. Since this experiment could not completely exclude the possibility of ^2He emission, an experiment was performed utilizing position sensitive detectors to measure the relative angular correlation of the two emitted protons [JAH 85]. The essentially isotropic angular correlation shown in Figure 2 confirms that the two-proton decay of the 4^+, $T = 2$ isobaric analog state in ^{22}Mg to the first excited state in ^{20}Ne is predominantly a sequential process. The observed minor

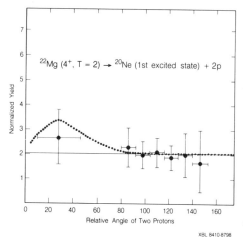

Figure 2. Normalized angular
correlation of the two protons
following ^{22}Al beta deacy. The
dotted line corresponds to a 15%
admixture of ^{2}He emission to an
otherwise isotropic distribution.
See text.

enhancement at small relative
angles cannot be interpreted as
positive evidence for correlated
diproton (^{2}He) emission due to
the poor statistics in this low
yield reaction. However, a 15%
admixture of this process cannot be
excluded. The dotted line in Figure 2 has been calculated assuming the
break-up properties of ^{2}He observed in reaction studies.

Observation of the $T_Z = -5/2$ nucleus, ^{35}Ca

By exploiting the relatively unusual decay mode of beta-delayed
two-proton emission,we have discovered the first $T_Z = -5/2$ nucleus, ^{35}Ca
[AYS 85a]. Figure 3 shows the two-proton sum spectrum of ^{35}Ca collected
after bombarding a natural Ca target with a 135 MeV ^{3}He beam from the
Lawrence Berkeley Laboratory 88-Inch Cyclotron for an integrated beam of 2.1
C and using helium-jet techniques. The distribution of individual proton
energies suggests a sequential decay process via intermediate states in
^{34}Ar. Since both of the individual proton spectra comprising G and X have
a peak at the same energy, 2.21 MeV, this indicates that the decays proceed
via the same state in ^{34}Ar. The ^{35}Ca decay scheme is shown in Figure 4.

The assignment of the observed 2p-activity to ^{35}Ca is primarily based
(i) on the agreement with the known energy difference for decays to the
ground and the first excited state in ^{33}Cl, (ii) on the agreement with
predicted 2p-energies, (iii) on various reaction energetics and (iv) on the
expected absence or non-existence of nearby $T_Z = -2$ and $-5/2$ nuclides with
similar predicted decay modes. Also see [AYS 85a].

Since three members of the A = 35, T = 5/2 isospin sextet (^{35}K*,
^{35}S*, ^{35}P) are now known, the ground state mass of ^{35}Ca can be
predicted with the Isobaric Multiplet Mass Equation, IMME. The resulting

value 4453 ± 60 keV is more than 300 keV better bound than predicted by the Kelson–Garvey relations with the most recent input masses [KEL 66].

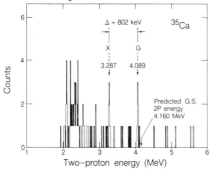

Figure 3. Beta-delayed two-proton spectrum of ^{35}Ca. G and X refer to the transitions to the ground and first excited states in the ^{33}Cl daughter, respectively.

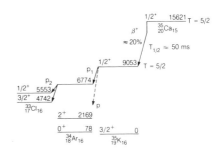

Figure 4. Proposed partial decay scheme for the beta-delayed two-proton emission of ^{35}Ca.

Future studies

Additional examples of β–delayed two-proton emitters may be sought among the A = 4n+2, T_Z = -2 nuclei such as ^{46}Mn and ^{50}Co [CAB 84a]. However, our preliminary searches for β2p emitters among the products of ^{14}N + ^{40}Ca reactions at 130 and 180 MeV have proven inconclusive.

The β–delayed two-proton decay of ^{35}Ca yields a value for the mass for the T = 5/2 state in ^{35}K; however, this only partially completes the isospin sextet, leaving three members (^{35}Ca, ^{35}Ar* and ^{35}Cl*) unmeasured. Use of the isospin–conserving (p,t) reaction to locate the T = 5/2 state in ^{35}Cl would therefore provide the first test of the IMME for an isospin sextet.

Further isospin sextets in the sd shell, with A = 23, 27 and perhaps 31, might be investigated via the β–delayed particle emission of the T_Z = -5/2 nuclei ^{23}Si, ^{27}S and ^{31}Ar. In this context the recent discovery of the lowest T = 5/2 state in ^{23}Na [EVE 85] will contribute towards the A = 23 sextet. The Kelson–Garvey mass formula [KEL 66] predicts the T_Z = -3 nuclide ^{22}Si to be effectively particle stable. Studies of this nucleus and its decay would provide data for possibly the only sd-shell example of an isospin septet, of which one member is already known (the mass of ^{22}O).

Also of great interest is the possibility of ground state two-proton radioactivity. Kelson–Garvey mass predictions suggest that two (three) T_Z = – 5/2 nuclei are potential candidates for this new mode of

radioactivity. These are ^{19}Mg, (^{31}Ar) and ^{39}Ti, with predicted 2p
separation energies of -1170, -190 and -790 keV, respectively. However, any
such observations would require specialized techniques due to their very
short predicted half-lives (< 10 ms).

Techniques for the study of exotic nuclei

Traditionally, light-ion induced reactions have been used to produce
light proton-rich nuclei, both for decay studies via (light-ion, xn)
reactions and for mass measurements from reaction Q-values such as (α,
^{8}He). Although the helium-jet technique has proven effective for the
former, fast on-line mass separation will be useful in the future. Recent
developments in the area [ARJ 85] may have an important impact on such
studies. With the observation of ^{35}Ca from the ^{40}Ca(^{3}He, α4n)^{35}Ca
reaction at extremely low levels, it is possible that this kind of reaction
has reached its limits. While the (^{3}He,^{8}He) reaction could be used to
measure the masses of the A = 4n, T_Z = -5/2 series, the cross-sections are
likely to be forbiddingly small.

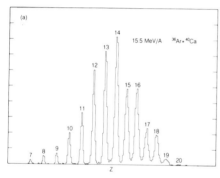

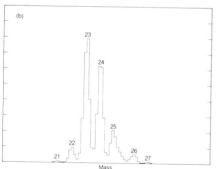

Highly neutron-rich nuclei have
been produced in heavy-ion deep
inelastic and fragmentation reactions
(for example, [GUI 85]). We are
investigating the possibility of
using this technique to produce new
proton-rich nuclei. A 15.5 MeV/A
^{36}Ar beam from the 88-Inch
Cyclotron injected by the ECR ion
source has been used to bombard a
thick calcium target. Reaction
products emitted at 5° were detected
in the focal plane of a magnetic
spectrometer. Measurements of
magnetic rigidity, time-of-flight and
differential energy loss allowed

Figure 5. a) Charge distribution for
the reaction 15.5 MeV/A ^{36}Ar + ^{40}Ca
and b) the corresponding mass distribution for Z = 12 (Mg isotopes).

identification of the wide range of nuclei so produced. Figure 5 shows a preliminary analysis of a small portion of the data. These initial results are encouraging. The collection for decay studies of new proton-rich nuclides from such reactions presents a further experimental challenge. Multinucleon transfer reactions with heavy ions might be used to measure the masses of very proton-rich nuclei. Recent results for neutron-rich nuclei [FIF 82] indicate that similar reactions for proton-rich nuclei may have substantially greater cross-sections than the light ion reactions used in the past. One might envisage, for example, using the $^{40}Ca(^{28}Si,^{28}S)^{40}Ar$ and $^{40}Ca(^{28}Si,^{27}S)^{41}Ar$ reactions to remeasure the mass of ^{28}S and then to investigate ^{27}S.

Acknowledgment

This work was supported by the D.O.E. under Contract DE-AC03-76SF00098.

References

[ARJ 85] J. Arje, et al., Phys. Rev. Lett. 54, 99 (1985).
[AYS 80] J. Aysto and J. Cerny, in Future Directions in Studies of Nuclei Far from Stability, J.H. Hamilton, et al. (eds.), North-Holland, Amsterdam, 1980, p.247.
[AYS 81] J. Aysto, et al., Phys. Rev. C 23, 879 (1981).
[AYS 82] J. Aysto, et al., Phys. Rev. Lett. 110B, 437 (1982).
[AYS 85a] J. Aysto, D.M. Moltz, X.J. Xu, J.E. Reiff and J. Cerny, Phys. Rev. Lett. in press, Report No. LBL-19732.
[AYS 85b] J. Aysto, X.J. Xu, D.M. Moltz, J.E. Reiff, J. Cerny and B.H. Wildenthal, submitted to Phys. Rev. C, Report No. LBL-19757.
[BJO 85] T. Bjornstad, et al., CERN Report EP/85-23, February 1985, to be published in Nucl. Phys. A.
[BUR 80] G.R. Burleson, et al., Phys. Rev. C 22, 1180 (1980).
[CAB 82] M.D. Cable, et al., Phys. Rev. C 26, 1978 (1982).
[CAB 83a] M.D. Cable, et al., Phys. Rev. Lett. 50, 404 (1983).
[CAB 83b] M.D. Cable, et al., Phys. Lett. 123B, 25 (1983).
[CAB 84a] M.D. Cable, et al., in Proc. of the 5th Nordic Meeting on Nuclear Physics, Jyvaskyla, Finland, March 1984, p. 119.
[CAB 84b] M.D. Cable, et al., Phys. Rev. C 30, 1276 (1984).
[CER 77] J. Cerny and J.C. Hardy, Ann. Rev. Nucl. Sci. 27, 333 (1977).
[EVE 85] D.D. Eversheim, et al., Phys. Lett. 153B, 25 (1985).
[FIF 82] L.K. Fifield, et al., Nucl. Phys. A 385, 505 (1982).
[GOL 80] V.I. Goldanskii, Pisma Zh. Eksp. Teor. Fiz. 32, 572 (1980); JETP Lett. 32, 554 (1980).
[GUI 85] D. Guillemaud-Mueller, et al., submitted to Z. Phys.
[HON 83] J. Honkanen, et al., Phys. Lett. 133B, 146 (1983).
[JAH 85] R. Jahn, et al., Phys. Rev. C 31, 1576 (1985).
[KEL 66] I. Kelson and G.T. Garvey, Phys. Lett. 23, 689 (1966).
[TRI 80] R.E. Tribble, et al., Phys. Rev C 22, 17 (1980).
[WIL 83] B.H. Wildenthal, et al., Phys. Rev. C 28, 1343 (1983).

RECEIVED July 21, 1986

68

Measuring β Lifetimes with the Michigan State University Recoil Product Mass Separator

L. H. Harwood, D. Mikolas, J. A. Nolen, Jr., B. Sherrill, J. Stevenson, and Z. Q. Xie

National Superconducting Cyclotron Laboratory and Department of Physics and Astronomy, Michigan State University, East Lansing, MI 48824

Beta-decay half-lives were measured for eight neutron rich isotopes produced by fragmentation of E/A=30MeV ^{18}O ions. The first measurements of the half-lives of ^{14}Be(4.2±0.7 ms.)and ^{17}C(202±17 ms.) have been made along with the half-lives of ^{9}Li,^{11}Li,^{12}Be,^{14}B,^{15}B. The lifetime of ^{14}Be is the shortest known beta lifetime.This is the first experiment to use the MSU Reaction Product Mass Separator

We report here measurements of the beta-decay half-lives of eight neutron rich isotopes.This experiment was the first to be performed using the Reaction Product Mass Separator(RPMS)at Michigan State University.The MSU RPMS is designed to separate exotic nuclei produced in intermediate energy heavy-ion collisions.The RPMS separates fragments of different m/q at the focal plane making it possible to study neutron rich isotopes with m/q=3 without contamination from stability line m/q=2 nuclei which are 10^6 times more abundent.The RPMS,shown in figure 1,is a triple focusing device(in x,y,and velocity) with dispersion in m/q.

Fig.1. Pictorial drawing of the RPMS. The beam inflector magnet is the first component at the left,followed by the bellows system and target chamber,aperature holder,quadrupole doublet,Wein filter(5 m. long),magnetic dipole and pivoting "tail" with quadrupole doublet and focal plane detectors at the right end.

The RPMS is designed[Nol84] to separate heavy ion reaction products with energies in the range E/A=10-100 MeV. By eliminating interfering m/q species with defining slits and concentrating the isotopes under investigation into small detectors the RPMS provides a clean environment for decay studies.
    Using a 30 MeV/A ^{18}O beam we made the first measurements of the lifetimes of ^{14}Be and ^{17}C along with the half-lives of ^{9}Li,^{11}Li,^{12}Be,^{12}B,and ^{15}B during a single 48 hour run.Beta lifetimes were measured by measuring the time interval between detection of a heavy ion and its subsequent beta decay.

0097-6156/86/0324-0454$06.00/0
© 1986 American Chemical Society

This technique has been previously demonstrated by a group at
LBL[Mur82].Beryllium and tantalum targets were used with thicknesses chosen
to stop the 30 MeV/A ^{18}O beam but allow lower charge neutron rich fragments
to exit the target into the RPMS.The tantalum target was found to give better
yields of neutron rich isotopes and was used for the measurements.

The focal plane detector consisted of a two dimensional position
sensitive proportional counter followed by a silicon ΔE-E telescope using a
300 mm²x100μm.(Si)ΔE and 500mm²x5mm.(Si-Li)E detectors. Lifetime measurements
were performed by turning the beam off whenever an ion with z≳3 was detected
in the telescope.The beam was left off for a preset time equal to several
half-lives of the longest lived isotope being collected,and all beta decays
in the interval were recorded.The beam is turned off in about 40μs. by
changing the phase of the rf on one of the dees of the K500 cyclotron.Isotope
identification was acomplished by plotting(Fig. 2.) the deflection in the
RPMS(proportional to m/q)vs the silicon telescope particle identification
function.

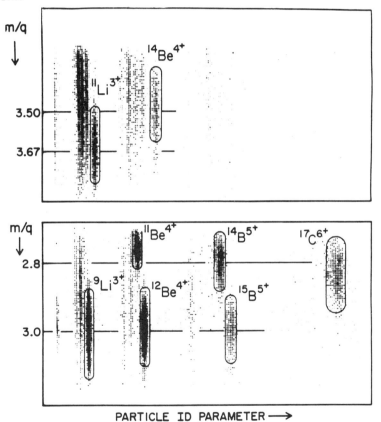

PARTICLE ID PARAMETER ⟶

Fig.2. Plot of m/q determined from deflection in the RPMS vs silicon
detector particle identification parameter.

For m/q=3 data ⁶He is a longlived beta background which is eliminated by setting the beta threshold at 2.5 MeV. ⁶He is eliminated from the m/q=3.5 data by stopping them in front of the focal plane detectors with brass defining slits.

The decay curves are shown in figure 3.The curves are best fits to an exponential plus a constant background.

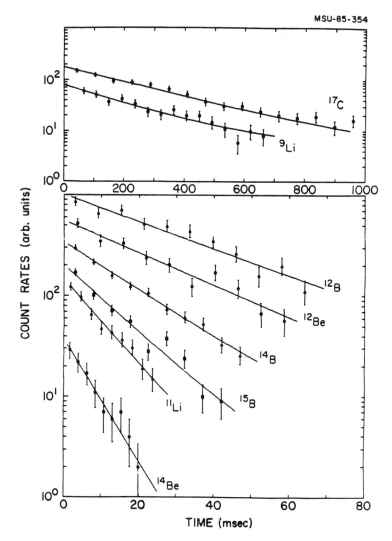

Figure 3. Beta lifetime plots,the curves are best fits to an exponential plus constant background form.

The analysis of ^{14}Be and ^{12}Be is complicated by the daughter nucleus being rather shortlived.For these two isotopes events were required to have a second beta within two daughter half-lives in order to be included in the lifetime spectrum.

The results of this experiment are compared with previous measurements in table I.

Table I Lifetime measurements and predictions

Isotope	This Measurement	Previous Measurement	Prediction[f]
^{9}Li	173±14 ms	175±1 ms[a]	·76 ms
^{11}Li	7.7±0.6 ms	8.5±0.2 ms[a]	2.3 ms
^{12}Be	21.3±2.2 ms	24.4±3.0 ms[b]	8.8 ms
^{14}Be	4.2±0.7 ms	---	3.7 ms
^{12}B	20.0±1.5 ms	20.41±.06 ms[c]	13.6 ms
^{14}B	12.8±0.8 ms	16.1±1.2 ms[d]	11.1 ms
^{15}B	8.8±0.6 ms	11.0±1.0 ms[e]	6.2 ms
^{17}C	202±17 ms	---	(414,292,238)ms

Beta-decay half-lives for the neutron-rich isotopes measured in this experiment are compared with previous measurements when they exist. Theoretical predictions for Gamow-Teller beta decay are also provided for comparison. The three theoretical half-lives for ^{17}C correspond to the assumption of $(1/2^{+}, 3/2^{+}, 5/2^{+})$ ground-state J^{π} values for ^{17}C in the calculation.

a) [Roe74]   d) [Alb74]
b) [Alb78]   e) [Duf]
c) [Ajz75]   f) [Bro85]

There are two cases of discrepancies outside of one standard deviation uncertainties ^{14}B and ^{15}Be. In both cases there was only one previous measurement and there is no obvious reason for the discrepancies.It is possible that the background under the ^{14}B decay curve of ref.[Alb74] was underestimated leading to the larger value of the extracted lifetime.In the case of the ^{15}B lifetime of ref.[Duf] no decay curve was shown so it is hard to compare quality of fits.

For the nuclei studied in this experiment the dominant decay mode is Gamow-Teller beta decay.Partial halflives corresponding to Gamow-Teller beta decay calculated in a spherical shell model formalism [Bro85] are shown in Table I.The calculated lifetimes are with one exception(^{17}C) all shorter than the measured lifetimes.

In conclusion thre first half-life measurements of light neutron rich nuclei using the MSU Reaction Product Mass Separator has resulted in the measurement of eight half-lives,two of which represent first time measurements and three of which are second measurements.The RPMS coupled with fast beam switching has proven to provide a very clean environment in which to study the decays of neutron rich nuclei.

References

[Ajz75] F.Ajzenberg-Selove,Nucl. Phys. A248,1(1975).
[Alb74] D.E. Alburger and D.R. Goosman,Phys. Rev. C10,912(1974).
[Alb78]D.E. Alburger,et. al.,Phys. Rev. C17,1525(1978).
[Bro85]B.A. Brown private communication
[Duf]J.P.Dufour,et.al.,Preprint CENBG 8430.
[Mur82]M.J. Murphy,T.J.M. Symons,G.D. Westfall and H.J. Crawford,Phys.Rev.Lett.49,455(1982).
[Nol84]J.A.Nolen,Jr.,L.H.Harwood,M.S. Curtin,E. Ormand and S. Bricker,Proceedings of the Conference on Instrumentation for Heavy-Ion Nuclear Research,Oak Ridge National Laboratory,Oct. 1984,to be published.
[Roc74]E.Roeckl,P.F. Dittner,C.Detraz,R.Klapisch,C.Thibault,and C.Rigaud,Phys. Rev. C10,1181(1974).

RECEIVED July 15, 1986

Time-of-Flight Isochronous Spectrometer for Direct Mass Measurements of Exotic Light Nuclei

G. W. Butler, D. J. Vieira, J. M. Wouters, H. Wollnik[1], K. Vaziri[2], F. K. Wohn[3], and Daeg S. Brenner[4]

Isotope and Nuclear Chemistry Division, Los Alamos National Laboratory, Los Alamos, NM 87545

In recent years there has been a rapid increase in our knowledge of the atomic-mass surface far from the valley of β stability [BEN80,CER81]. However, even with these advances there remain more than 60 neutron-rich nuclei with A < 70 for which the only known information is that they are stable with respect to neutron emission. It is clear that more detailed information about these exotic nuclei is essential. Especially important are the measurements of ground state atomic masses, since the ground state mass is one of the most fundamental properties of a nucleus.

The binding energy, and hence the mass, of a nucleus is dependent on the exact details of the nuclear force, and its prediction requires an understanding of these details. In a strong sense, an atomic-mass model includes everything that we know about the nuclear force and nuclear interactions. A systematic study of accurately determined masses encompassing a wide variety of nuclei far from β stability would provide a most challenging test of current atomic-mass theories and should yield new insight into the nuclear structure of such exotic nuclei.

Interest in the neutron-rich light mass nuclei increased with the discovery by Thibault [THI75] that, beginning with ^{31}Na, the sodium isotopes become strongly prolate deformed. This deformation, which is due to the partial filling of the $f_{7/2}$ shell [CAM75], was unexpected since for ^{31}Na the $d_{3/2}$ neutron shell closure was predicted to occur at N=20. Mapping this new region of deformation more fully is of great interest since the deformation could possibly be driven by the odd $d_{5/2}$ proton present in the sodium isotopes. This disappearance of a neutron magic number at N=20 has led to speculation that new magic numbers at N=10,14 and/or 16 might appear in such deformed nuclei [SHE81]. An experimental search for these new magic numbers and their discovery will allow more detailed investigations of the strong correlation of proton and neutron shell strengths that occurs near double shell closures and which might occur in ^{24}O.

2. Time-of-Flight Isochronous (TOFI) Spectrometer

TOFI is a recoil time-of-flight spectrometer that is being constructed at the Los Alamos Meson Physics Facility (LAMPF) for the purpose of making direct mass measurements of neutron-rich light nuclei that are produced via proton-induced fragmentation of U or Th targets. The design of TOFI will allow systematic mass measurements for a large number of neutron-rich nuclei below A=70 with accuracies of 30 keV to 1 MeV, depending on production rates (see Fig. 1). In this paper we outline the basic features of the TOFI spectrometer and its associated transport line, discuss the mass measurement

[1]Permanent address: University of Giessen, D-63 Giessen, Federal Republic of Germany
[2]Permanent address: Department of Physics, Utah State University, Logan, UT 84322
[3]Permanent address: Ames Laboratory, Iowa State University, Ames, IA 50011
[4]Permanent address: Office of Academic Affairs, Clark University, Worcester, MA 01610

0097–6156/86/0324–0459$06.00/0
© 1986 American Chemical Society

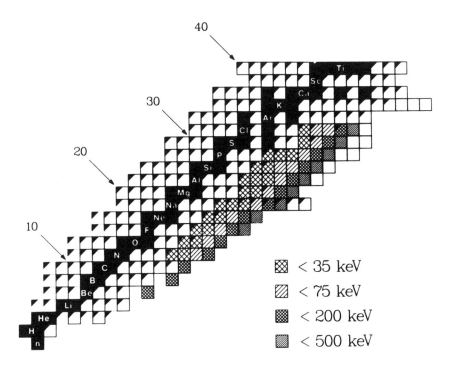

Fig. 1.   Partial chart of the nuclides showing:
(1) those nuclei whose mass has been measured [WAP85]
(black triangles) and (2) the nuclei we propose to
measure (shaded boxes; shading indicates expected
accuracy of mass measurements).

capabilities of the system, and highlight progress made on the installation
and initial testing to date.

The basic principle of the TOFI spectrometer is that it is designed to
be isochronous, which means that the transit time of a particle passing
through the spectrometer is independent of its velocity.   Thus the measured
transit time of an ion passing through the system provides a precise
measurement of the mass-to-charge ratio.   Since charge is a quantized
entity, only moderately accurate measurements (at the 1-2% level) of the
energy and velocity, together with the known momentum acceptance of the
spectrometer, are necessary to define uniquely the charge state of an ion.
An accurate mass can then be determined from the mass-to-charge ratio as
obtained from the measured transit time of the ion through the spectrometer.

Another important feature of the spectrometer is that it is focusing in
both energy and angle, thus allowing a long flight path (14 m) with a
relatively large solid angle (2.5 msr) and momentum acceptance (±2%).   Since
the spectrometer is nondispersive overall, there is no physical separation
between different ions, which allows the simultaneous measurement of many
nuclei (especially important are isobaric members with well-known masses
that will be used as internal calibration points).

The spectrometer consists of four identical, mirror symmetric unit cells that contain a 81° sector magnet located between identical drift lengths [WOU85]. The symmetry of the system results in a unity, first order transfer matrix in which only the time term dependent on the mass-to-charge ratio remains. Furthermore, all higher order aberrations are small, resulting in a total time deviation of ~80 ps for different trajectories through the spectrometer. Assuming timing uncertainties of ~200 ps from detector time jitter, ~200 ps from magnetic field inhomogeneities and ~80 ps from optical aberrations for a total of ~300 ps, the resolving power of the spectrometer is conservatively estimated to be $m/\Delta m = 2000$.

Extensive magnetic field mapping and optimization of the four spectrometer dipole magnets is currently nearing completion. Installation of the dipoles in their final configuration will begin soon, and initial testing with alpha sources is scheduled to begin in October 1985. The first experiment, in which we will measure the masses of heavy Ne, Na and Mg isotopes, will begin in November 1985.

3. TOFI Transport Line

The spectrometer is connected via a secondary beam transport line to the production target (see Fig. 2). This transport line consists of 4 quadrupole triplets that are arranged as two pairs with an intermediate

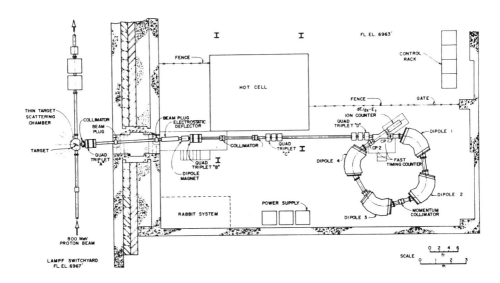

Fig. 2. Schematic of the TOFI spectrometer and its associated secondary beam transport line.

focus between the pairs. The transport line is magnifying overall with a
vertical and horizontal magnification of ~2 so that a larger solid angle
could be matched to the acceptance of the spectrometer. In addition, the
line includes a crude mass-to-charge filter (an electrostatic deflector
followed by a dipole magnet) that eliminates the high-yield light mass
reaction products such as protons, deuterons, alphas and all neutral
recoils. The transport line provides a very important advantage in that the
spectrometer can be operated independently of the high-intensity LAMPF beam
operation.

    All of the components of the transport line have been installed and
tests with alpha particles and light reaction products are nearing completion. In order to characterize its operation, the transport line has been
tuned with various alpha sources and low Z reaction products from 800-MeV
protons on thorium. A multiwire proportional counter was used to obtain a
focus, and a good image (1.0 x 1.4 cm^2) was produced at the first focal
point of the transport line from a ^{241}Am source (0.6 cm diam.) located 30 cm
behind the normal target position in the scattering chamber. To determine
the energy-transmission characteristics of the transport line, an Si detector was used in conjunction with a multiple energy alpha source (a combination of ^{148}Gd and ^{229}Th). Through a comparison of alpha intensities
measured at different energy or momentum settings of the transport line, the
momentum transmission of the first half of the transport line was measured
to be $\delta p/p$ = 26% (FWHM), which is much larger than required for the spectrometer ($\delta p/p$ = 4%). The mass filter was characterized by using Z=1 and 2
nuclides produced in 800-MeV protons on a thorium target. For this test,
the transport line was set for a particular momentum-to-charge value, and
then the mass filter was tuned for various mass-to-charge ratios. By comparing the intensity of the alphas and tritons at various mass-to-charge
settings, we obtained the mass-to-charge transmission for the first half of
the transport line (see Fig. 3). The results showed that at a setting of

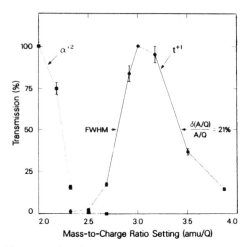

Fig. 3. The mass-to-charge transmission of the first half
of the transport line for alphas (dashed line) and tritons
(solid line), as emitted from a thorium target. The transport line was set for a central momentum-to-charge value of
190 MeV/c/Q.

A/Q = 3 (tritons), those species with mass-to-charge ratios of 2 or less
were reduced by more than 2 orders of magnitude. (Scattered background
events limited the measurement of this reduction factor.) This mass-to-
charge filtering is particularly important in our future experiments because
the protons, deuterons, and alpha particles are produced with yields that
are, respectively, 10^5, 10^4, and 10^3 times larger than those of the ions for
which measurements are planned.

4. Summary

The overall goal of the TOFI spectrometer project is to perform, in a
systematic fashion, direct mass measurements of light nuclei that are far
from stability and thus identify general, as well as isolated, trends in
nuclear structure. This isochronous, nondispersive spectrometer will be the
first high resolution recoil spectrometer to use the time-of-flight technique for direct mass measurements. The TOFI spectrometer will be able to
measure the masses of ~60 nuclei, two-thirds of which will be measured for
the first time. This experimental program will begin during the fall of 1985
with the completion of the installation of the spectrometer.

Acknowledgments

   We would like to thank both the Isotope and Nuclear Chemistry and
Meson Physics divisions of the Los Alamos National Laboratory for the
support they have provided towards the development and construction of the
TOFI spectrometer. This work was conducted under the auspices of the U.S.
Department of Energy.

References

[BEN80] W. Benenson and J. Nolen, Eds., "Proceedings of the Sixth Inter-
         national Atomic Mass Conference", East Lansing, Michigan, September
         1979 (Plenum Press, New York, 1980).
[CAM75] X. Campi, H. Flocard, A. Kerman, and S. Koonin, Nucl. Phys. A251 193
         (1975).
[CER81] "Proceedings of the Fourth International Conference on Nuclei Far
         From Stability", Helsingor, Denmark, June 1981, CERN Report 81-09
         (1981).
[SHE76] R. K. Sheline, Proceedings of the Third International Conference on
         Nuclei Far From Stability, Cargese, Corsica, May 1976, CERN Report
         76-13, 351 (1976).
[THI75] C. Thibault, R. Klapisch, C. Rigaud, A. M. Poskanzer, R. Prieels, L.
         Lessard, and W. Reisdorf, Phys. Rev. C12 644 (1975).
[WAP85] A. H. Wapstra and G. Audi, Nucl. Phys. A432 1 (1985).
[WOU85] J. Wouters, D. Vieira, H. Wollnik, H. Enge, S. Kowalski and K.
         Brown, accepted for publication in Nucl. Instr. and Meth. in Phys.,
         (1985).

RECEIVED July 15, 1986

70

Spectrometers for Light-Ion, xnγ Spectroscopy

J. Kern, J.-Cl. Dousse, and V. Ionescu

Physics Department, University of Fribourg, CH-1700 Fribourg, Switzerland

There is a large variety of processes which can be studied by the nuclear spectroscopist. Many instrumental methods can be used in each case, like singles and coincidence measurements, angular distributions and correlations. The complementarity of various approaches has been discussed by several authors, e.g., direct and (n,γ) reactions by Cizewski [CIZ82] or neutron capture and heavy-ion reactions by Schult [SCH79]. In this paper we will discuss some of the merits of (light ion, xnγ) spectroscopy, desired experimental improvements and report on special techniques and on recent results.

In comparing (n,γ) and (LI,xnγ) spectroscopy, we note for each method very important differences. Charged particle reactions are superior, in particular, with respect to the number of nuclei which can be studied and in the range of spin for which levels can be observed. To be more specific on this point we compare in Fig. 1 and 2 E_{exc} vs I diagrams showing the region of levels observed in the $^{113}Cd(n,\gamma)^{114}Cd$ reaction by A. Mheemeed et al. [MHE84] and in the $^{108}Pd(\alpha,2n)^{110}Cd$ reaction by the Fribourg group [KUS85]. In both cases we have only reported levels for which spin assignments are presently known. Several more levels have but been observed. Because ^{113}Cd has a ground state spin 1/2, only levels up to spin 4 have been observed in the (n,γ) study. The region covered by the (α,2nγ) reaction is broader and overlaps only partly with the (n,γ) region. Very important is that in many cases various projectiles with different energies can be used in the study of a particular nucleus, populating levels in different regions: (p,n) or (α,n) reaction close to threshold will populate, for instance, low spin levels in a rather nonselective way similar to (n,γ) [BRE84]. Levels in a very large region can thus be observed.

Neutron capture as well as charged particle reactions produce in general very dense γ-ray spectra. The high resolution electron [MAM78] and curved crystal spectrometers [KOC80] at ILL in Grenoble present excellent solutions for the observation of (n,e⁻) and (n,γ) singles spectra. The

0097-6156/86/0324-0464$06.00/0

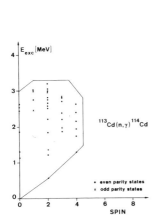

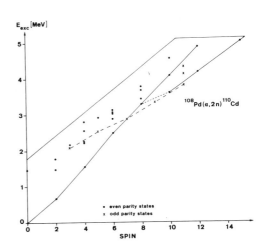

Fig. 1. Levels populated in the $^{113}Cd(n,\gamma)^{114}Cd$ reaction [MHE84] presented in a E_{exc} vs I diagram.

Fig. 2. Levels populated in the $^{108}Pd(\alpha,2n)^{110}Cd$ reaction [KUS85] presented in a E_{exc} vs I diagram.

situation is far less satisfactory for charged particle reactions where better energy resolutions would be needed in many cases (see e.g., [SCH79]). We will present the merits of two spectrometers designed to obtain better quality singles spectra.

A COMPTON SUPPRESSION SPECTROMETER FOR SINGLES MEASUREMENTS

Heavy-ion nuclear spectroscopists are installing at several laboratories arrays of BGO anti-Compton spectrometers to be used for multi-coincidence experiments [DIA85]. A very important motivation for such facilities is that coincidence experiments become less performing for small target-to-detector distances when the γ-ray multiplicity increases [SUJ80]. The performances of the new systems are very exciting, the costs, demands on electronics and computing are, however, quite high. Besides this, the extraction of precise singles intensities from coincidence experiments is not always trivial.

In (p,xnγ) like in (n,γ) spectroscopy, the neutron reactions occuring in the scintillator used as anti-Compton detector induce parasitic γ-ray lines [BEE77] in the central detector. The high counting rate in the scintillator impairs, in addition, the performances of the instrument. We have constructed an anti-Compton spectrometer where the complete detector system is enclosed in a cabinet lined with 5 to 10 cm thick lead bricks and surrounded by 30 to 50 cm thick borated paraffine blocks [ION79]. The whole is placed on a cart rolling on a rail, so that it can be turned around the target. The spectrometer is used for single measurements (excitation functions), prompt-delayed multispectrum experiments (determination of the half-lives of delayed transitions) and angular distributions. The first two kinds of experiments are performed at θ = 90°, where the spectrometer performances are best, since the beam stop can then be placed at a relatively large distance.

For precise angular distributions, accurate intensity determinations are needed. It has to be noted that the low intensity transitions are as important as the intense ones. The appreciable reduction of the background, especially at low energy (see Fig. 3) is an important factor for obtaining good quality data. At higher energy the suppression of the annihilation line and of transitions induced by neutrons in germanium is very useful [ION79].

We are using a plastic scintillator in our spectrometer. This has the advantage not to get neutron activated, to be a good neutron thermalizer and to yield fast timing signals. It is also relatively cheap. An inorganic scintillator would certainly also perform satisfactorily.

AN ON-LINE CURVED-CRYSTAL SPECTROMETER

Since the early days of diffraction γ-ray spectroscopy [DUM47], on-line applications have been considered. The success in neutron capture work is obvious [KOC80]. Application to the study of (LI,xnγ) reactions has met with considerable difficulties [JET74]. The main problems to be solved are to obtain:

- good background conditions. We use a Phoswich detector (produced by Harschaw) surrounded by a NaI(Tℓ) annulus as anti-Compton and active shield. Since most transitions are prompt, we register separately the detector pulses emitted in- and out-of-burst.
- a good energy resolution. We obtain peaks about 5-6 arc sec wide with

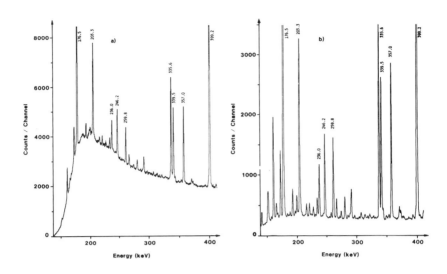

Fig. 3. Portion of the ^{108}Pd(α,2nγ)^{110}Cd gamma-ray spectrum observed a) with
    a single diode and b) with the Compton suppression spectrometer.
    Some transitions are identified by their energy in keV.

quartz crystal plates reflecting on the (110) planes, corresponding to an
energy resolution FWHM $\Delta E \cong 0.01\ E^2/n$, where n is the diffraction order,
ΔE being expressed in eV and E in keV. This requires high quality material, exact polishing and very precise bending.
- the highest possible luminosity. This is the reason for chosing a relatively small bending radius of 3 m. The target is gas cooled. The maximum
  current depends on the target characteristics (in particular the melting
  temperature) and can be also limited by the cyclotron capabilities.
- precise energy determinations. One condition is to have any mechanical
  shift of the target under control at a level of μm's. We are using the
  method proposed by Koch et al. [KOC69]: an X-ray line is continuously observed by a simplified crystal spectrometer placed above the main apparatus. Its intensity is enhanced by high frequency piezoelectric vibrations
  of the crystal which increases the reflectivity [DOU76].
In a first application of the system we have studied the ^{176}Yb(p,3n)^{174}Lu

reaction. Several parts of the emitted γ-ray spectrum have been observed. Combined with other experiments, these results have provided an important contribution for establishing a level scheme [BRU86]. The major result of this study was the determination of anomalous g-factors in the rotational band based on the 142d isomeric $K^{\pi} = 6^-$ level : this has required precise intensity measurements of the cascade and cross-over transitions emitted by the $I^{\pi} = 8^-6$ level. In both cases the γ-ray lines are components of close multiplets which have been resolved with the crystal spectrometer [KER84]. More recently we have observed portions of the ^{165}Ho(α,3nγ)^{166}Tm spectrum. Fig. 4 shows one such section and in the inset the same part observed with a good resolution Ge detector. Note that about twice as many transitions are seen with the crystal spectrometer.

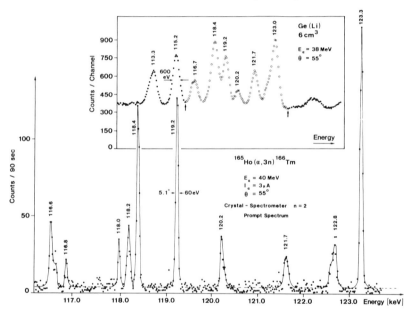

Fig. 4. Portion of the ^{165}Ho(α,3nγ)^{166}Tm spectrum observed with the on-line bent crystal spectrometer and with a good resolution germanium detector in the inset (the corresponding region is represented with open circles). Transitions are identified by their approximate energy in keV.

CONCLUSION

Detail tests on nuclear models require not only a knowledge of energy, spin and parity of many levels, but also the determination of transition multipolarities and branching ratios. Precise intensities are thus needed. The well shielded anti-Compton spectrometer offers a rather simple solution especially for accurate angular distribution measurements. When the spectra are very complex, like in the case of final doubly odd nuclei, intensities cannot be determined without use of high resolution instruments. The curved crystal spectrometer provides a powerful solution at, unfortunately, non negligible cost.

REFERENCES

[BEE77] R. Beetz et al., Nucl. Instr. Meth. 145 353 (1977)

[BRE84] P.v.Brentano et al., Proc. Int. Symp. on In-Beam Nuclear Spectroscopy Debrecen 1984, ZS. Dombrádi and T. Fényes ed., p. 189

[BRU86] A. Bruder et al., to be published

[CIZ82] J.A.Cizewski, Inst. Phys. Conf. Ser. 62 133 (1982)

[DIA85] R.M.Diamond, this Conference

[DOU76] J.-Cl.Dousse and J.Kern, Phys.Lett. 59A 159 (1976); Acta Cryst. A36 966 (1980)

[DUM47] J.W.M.DuMond, Rev. Sci. Instr. 18 626 (1947)

[ION79] V.Ionescu et al., Nucl. Instr. Meth. 163 395 (1979)

[JET74] J.H.Jett et al., Nucl. Instr. Meth. 114 301 (1974)

[KER84] J.Kern et al., Phys. Lett. 146B 183 (1984)

[KOC69] H.R.Koch et al., Proceedings of the Int. Symp. on Neutron Capt. γ-ray Spectroscopy (Studsvik 1969) p. 65

[KOC80] H.R.Koch et al., Nucl. Instr. Meth. 175 401 (1980)

[KUS85] D.Kusnezov et al., submitted to Phys. Rev. C and unpublished results

[MAM78] W.Mampe et al., Nucl. Instr. Meth. 154 127 (1978)

[MHE84] A.Mheemeed et al., Nucl. Phys. A412 113 (1984)

[SCH79] O.W.B.Schult, Proc. 3rd Int. Symp. on Neutron Capt. γ-Ray Spectroscopy, Brookhaven 1978, R.E. Chrien and W. Kane ed., (Plenum, New York 1979) p. 85

[SUJ80] Z.Sujkowski and S.Y. van der Werf, Nucl. Instr. Meth. 171 445 (1980)

RECEIVED July 15, 1986

71

Nuclear Shape Transitions and Dynamical Supersymmetries in the Pt Region

Studies by $(n,n'\gamma)$ Reaction Spectroscopy

Steven W. Yates

Department of Chemistry, University of Kentucky, Lexington, KY 40506-0055

We have studied many nuclei in the platinum region (both odd- and even-A) by the $(n,n'\gamma)$ reaction on large enriched isotopic samples in an effort to characterize the structural properties of nuclei undergoing a shape transition. Levels are placed from γ-ray excitation functions, while the shapes of these excitation functions, combined with γ-ray angular distribution measurements, can be used for making spin assignments. Complementary time-of-flight neutron detection measurements have been performed in favorable cases. Even- and odd-A nuclei can be grouped into supersymmetric multiplets, and critical comparisons between the properties of even-even nuclei and the various nuclear models are possible.

Neutron-induced reactions have played a major role in the study of nuclear structure; however, only a small number of researchers have employed such reactions on-line at accelerators for these purposes. At the University of Kentucky, we have used pulsed beam time-of-flight (TOF) methods in γ-ray and neutron spectroscopy to study the structure of heavy transitional nuclei by the neutron inelastic scattering reaction. The power of this reaction arises from the fact that Coulomb effects are absent, so that levels are strongly excited with low incident energies. Low-lying levels can thus be studied without the attendant complications of the presence of radiation from higher levels. For the purpose of obtaining structural information, low-energy neutrons have an advantage over both β-decay and charged-particle excitation in that they excite the low-spin levels of a nucleus without great selectivity. The population of any level is predominantly determined by the penetrability of neutrons going into and then out of the nucleus. The angular distributions for γ-rays contain information about the multipolarity of the decays, and thus indirect information about the spins and nucleon configurations of decaying levels.

The most serious problem associated with the use of neutron scattering for nuclear spectroscopy comes from the fact that the resolution for neutron detection is typically rather poor, and the sensitivity to small transition probabilities is also poor when neutron detection is being employed. These difficulties can be alleviated by observing the γ rays which de-excite the excited levels rather than the inelastically scattered neutrons.

EXPERIMENTAL TECHNIQUES

In the work to be described, the University of Kentucky 6.5 MV Van de Graaff accelerator was used with a proton beam incident on a ^{3}H gas target.

0097–6156/86/0324–0470$06.00/0

The beam is pulsed in the terminal at 2 MHz and accelerated in 7 ns bursts. After acceleration, it may be compressed to bursts of <1 ns. Details are shown in Fig. 1. By adjusting the gas target pressure, typically 1 atm, the neutrons resulting from the $^3H(p,n)^3He$ reaction have a spread in energy which is compatible with good spectroscopy, and yet broad enough to excite a large number of compound states. The neutrons impinge on cylindrical samples whose typical size is 1 cm in diameter and 2 cm in length. The sample is placed at 2 to 6 cm from the end of the ^3H target such that the axis of the cylinder is perpendicular to the beam axis.

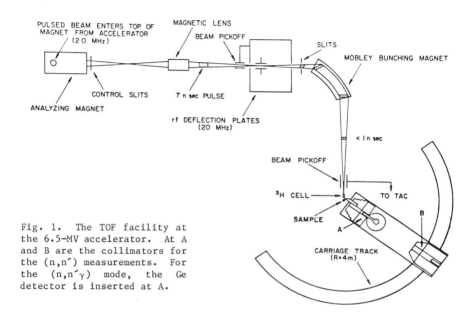

Fig. 1. The TOF facility at the 6.5-MV accelerator. At A and B are the collimators for the (n,n′) measurements. For the (n,n′γ) mode, the Ge detector is inserted at A.

The best overall combination of sensitivity and accuracy is achieved with samples of 0.2 to 0.3 moles. Recent experience [YAT81] shows us that we can work effectively with samples as small as 0.025 mole, and work in (n,n′γ) detection is now in progress with samples of <0.01 mole. Sample size corrections must be made for incident flux attenuation, multiple neutron scattering, and outgoing neutron or γ-ray attenuation. For the (n,n′γ) measurements these three effects are approximated by three factors. That this procedure works well has been confirmed through Monte-Carlo calculations [VEL75] developed for making these corrections as well as those for angular spread subtended by the sample at the source, anisotropy of the flux from the source, and energy spread of neutrons from the source.

As noted earlier, spectroscopic information can be obtained by observing either the inelastically scattered neutrons or the de-excitation γ rays. In the γ-ray detection mode, i.e. (n,n′γ), an intrinsic HpGe or Ge(Li) detector is placed at approximately 100 cm from the cylindrical sample (see Fig. 1). Since the beam is pulsed, the neutron-induced events in the detector and time-independent background can be rejected through the use of the TOF difference between neutrons and γ rays. In the neutron detection mode, elastically- and inelastically-scattered neutrons are detected in a liquid scintillator at 2 to 4 m from the cylindrical sample; the neutron TOF is measured. A typical TOF spectrum is shown in Fig. 2.

In both neutron and γ-ray detection, the shielding of the detector is extremely important. Especially in the neutron detection measurements, the long target-to-detector distance (2-4 m) which is required to obtain velocity resolution via the neutron TOF technique means many more neutrons are produced than are actually scattered from the sample and then detected. These extraneous neutrons create a disastrous background unless the detector is adequately shielded. We have accomplished this with a large cylindrical shield which contains a lead cavity surrounded by Li_2CO_3-loaded paraffin. The entrance collimator has steel and lead liners; the main detector shield weighs about 2000 kg (see B in Fig. 1).

For the case of γ-ray detection, attention must be paid to the γ rays which are produced by the capture of neutrons after they have been slowed to thermal energies. Because of the damage produced by neutrons on Ge detectors, adequate shielding is imperative to prolong detector life. From experience, the conclusion has been reached that the best procedure is to interpose as much matter as possible between the Ge detector and the neutron source; this led to the selection of metal as the material. In the quantity which was required, copper was the practical best choice. Lead was rejected due to associated radioactivity and sparse level density for degrading neutrons through inelastic scattering; tungsten was probably the first choice but was too expensive. As shown in Fig. 1, the detector is shielded by 50 cm of copper and 30 cm of boron-loaded polyethylene. Typical γ-ray spectra are acquired with pulsed beam and TOF, which enables rejection of the events induced by neutrons in the detector. The success of these shielding and TOF

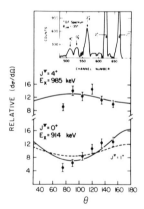

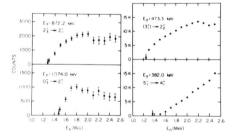

Fig. 2. A typical neutron TOF spectrum, above. Below, (n,n´) angular distributions which show the sensitivity to the spin emitting spin-4 and spin-0 final states.

Fig. 3. Gamma-ray excitation curves for [198]Pt displaying the spin dependence of the shape.

background rejection schemes is attested to by our sensitivity range, which is from about 1 mb to the largest inelastic scattering cross sections of 500-700 mb. This high sensitivity permits us to study weak transitions in great detail.

GAMMA-RAY EXCITATION FUNCTIONS

An important measurement is an accurate determination of the γ-ray yield threshold for exciting a particular level. Thresholds can be determined with uncertainties of only a few keV, when necessary to guarantee certain placement of a γ-ray in a level scheme. Usually, thresholds determined to within 30 to 50 keV are sufficient for this purpose, even in fairly heavy nuclei. The reason for this is that, typically, excited levels will have one or more decays to low-lying levels, which are spaced 50 keV or more apart, even in heavy deformed nuclei.

Excitation functions of γ-ray yields are useful also for inferring limits on the angular momentum transfer to the excited level, and therefore the spin of that level. Within 100 or 200 keV of threshold, cascades to that level from higher levels are insignificant, so that the shape of the excitation function limits spins sharply in heavy nuclei, where one finds sufficient neutron angular momentum present even at low energies. The effects of the angular momentum transfer are illustrated in Fig. 3 for scattering to several levels of ^{198}Pt. For low-spin levels, with 0 or 1 unit of angular momentum, the excitation function rises sharply above threshold, with negative curvature. It reaches a peak close to threshold and then declines sharply. For intermediate transfer of 2 or 3 units, the excitation function rises linearly above threshold for several hundred keV. For large transfer, 5 to 7 units, the curve has positive curvature and continues to rise far above threshold. We see that the shape of the excitation functions near threshold is useful information when assessing spin assignments to excited levels. This behavior is precisely that expected and can be reproduced by statistical model calculations.

ANGULAR DISTRIBUTIONS

The angular distributions of both neutrons and γ-rays from inelastic scattering have valuable spectroscopic information. In the case of neutron distributions, the anisotropy of scattering reflects the spin of the excited level, but usually weakly. Thus to use that information confidently for spin limitations would require very precise measurements, which are often not readily obtainable. Two exceptions occur to this rule of small anisotropies, for spin-0 and spin-1 levels of even-A nuclei. This is illustrated in Fig. 2 where angular distributions are presented for (n,n´) scattering from ^{198}Pt. The spin-0 excited level shows very strong anisotropy, a simple consequence of the fact that at 0° and 180° both spin-flip and non-spin-flip components of the scattering process can contribute fully. As one moves away from those angles the non-spin flip process contributes less strongly. Thus neutron distributions are a powerful tool for picking out spin-0 excited levels and have been so used by several authors [YAT81,MCE74].

The (n,n´) reaction aligns excited levels, so that the γ-ray angular distributions from their decays show anisotropies reflecting the alignment, the spins of the levels, and the multipolarities of the transitions. Especially in even-A nuclei one often finds that a level decays through more than one transition, one of known pure multipolarity (usually E1 or E2) and others with unknown mixtures of multipolarities. In these cases the pure multipole transition can be used to deduce the excited state alignment, and then the other distributions can be analyzed to determine multipole mixing ratios. The advantage of this method is that it is completely independent of the neutron scattering process -- it is a model independent way of determining mixing ratios and spins of excited levels.

In practice, the excited state alignments produced in neutron scattering are essentially independent of the reaction process. Calculations showed [SHE66] that direct interaction inelastic scattering would produce virtually the same alignments as compound nucleus formation. Thus the alignments calculated with statistical models are sufficient, and many tests have shown that mixing ratios determined using statistical model alignments agree well with those determined from the model-independent methods cited.

REPRESENTATIVE EXPERIMENTAL RESULTS

In Fig. 4 we show the platinum region of the chart of the nuclides. Those nuclei which have been studied in our laboratories in the past few years are indicated by the triangle in the upper right corner of the appropriate box. This diversity of nuclei has allowed us to study a variety

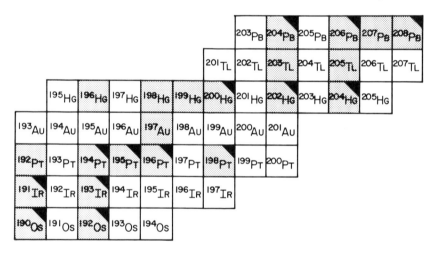

Fig. 4. Platinum region of the nuclidic chart. Nuclei which we have studied are indicated by a triangle in the upper right corner.

of phenomena and to obtain unique insights into the level structure of this heavy transitional region. For example, the nuclei ^{190}Os-^{191}Ir-^{192}Pt and ^{192}Os-^{193}Ir-^{194}Pt represent two supersymmetric multiplets of the U(6/4) supersymmetry model [IAC80]. We were able to examine experimentally five of these nuclei and to determine that this description offers a good first approximation to the structure of transitional nuclei [YAT84].

The combined use of the $(n,n'\gamma)$ and (n,n') measurements has proven particularly valuable in establishing the low-lying level structure of ^{198}Pt. Prior to our studies [YAT81,YAT83] of ^{198}Pt, the only tentative evidence for the 914.5-keV 0_2^+ level came from the β-decay of ^{198}Ir where this level was assigned [SCH72] as (0^+) on the basis of decay systematics. It should also be noted that this state is not populated in either the (p,p') or (t,p) reactions [DEA81,CIZ79] and apparently not in Coulomb excitation, as suggested previously [BOL81]. The 914.5-keV level was, however, clearly observed in the spectrum of neutrons scattered from ^{198}Pt and, as shown in Fig. 2 the neutron angular distribution carries the unique signature of a 0^+ level. The location and properties of the low-lying 0^+ states are extremely important in our attempts to assess the applicability of the various nuclear models. Additional details of these studies may be found in [KER84].

Acknowledgments

This work was supported in part by the National Science Foundation.

References

[BOL81]   H.H. Bolotin, A.E. Stuckbery, I. Morrison, D.L. Kennedy, C.G. Ryan, and S.H. Sie, Nucl. Phys. A370 146 (1981).

[CIZ79]   J.A. Cizewski, E.R. Flynn, R.E. Brown, and J.W. Sunier, Phys. Lett. 88B 207 (1979); J.A. Cizewski, E.R. Flynn, R.E. Brown, D.L. Hanson, S.D. Orbesen, and J.W. Sunier, Phys. Rev. C23 1453 (1981).

[DEA81]   P.T. Deason, C.H. King, R.M. Ronnigen, T.L. Khoo, F.M. Bernthal, and J.A. Nolen, Phys. Rev. C23 1414 (1981).

[IAC80]   F. Iachello, Phys. Rev. Lett. 44 772 (1980).

[MCE74]   M.T. McEllistrem, J.D. Brandenberger, K. Sinram, G.P. Glasgow, and K.C. Chung, Phys. Rev. C9 670 (1974).

[SCH72]   F.W. Schweden and N. Kafrell, BMBW-FB Bonn Report No. K72-15 88 (1972).

[SHE66]   E. Sheldon and D.M. van Patter, Rev. Mod. Phys. 38 143 (1966) and references cited therein.

[VEL75]   D.E. Velkley, J.D. Brandenberger, D.W. Glasgow, and M.T. McEllistrem, Nucl. Instr. and Methods 129 231 (1975).

[YAT81]   S.W. Yates, A. Khan, M.C. Mirzaa, and M.T. McEllistrem, Phys. Rev. C23 1993 (1981).

[YAT83]   S.W. Yates, A. Khan, A.J. Filo, M.C. Mirzaa, J.L. Weil, and M.T. McEllistrem, Nucl. Phys. A406 519 (1983).

[YAT84]   S.W. Yates and E.W. Kleppinger, Proc. Symp. on In-Beam Nuclear Spectroscopy, Debrecen, Hungary, p. 381 (1984).

[KER84]   B.D. Kern, M.T. McEllistrem, J.L. Weil and S.W. Yates, Proc. Symp. on In-Beam Nuclear Spectroscopy, Debrecen, Hungary, p. 163 (1984).

RECEIVED June 5, 1986

Special Targets for Nuclear Reaction and Spectroscopic Studies

Robert G. Lanier

Nuclear Chemistry Division, Lawrence Livermore National Laboratory, Livermore, CA 94550

Strongly focused and monoenergetic charged-particle beams from modern accelerators and targets fabricated from quantities of isotopically enriched and stable materials are the essential components from many current nuclear physics experiments. Although a large body of this kind of experimental work requires substantial amounts of target material, an important subset of such experiments can be done with as little as a few μg of material. Experiments where charged particles or electrons can be focused on or transported to a detector are examples of accelerator-based studies which can be made with targets that contain relatively small amounts of material. For these kinds of studies, it then becomes possible to extend the domain of potential target materials to species which are very rare or which are unstable and undergo radioactive decay. At our laboratory during the last ten years, we have made targets for nuclear spectroscopy studies of ^{152}Eu (13.4y), ^{154}Eu (8.5y), ^{249}Bk (320d), ^{151}Sm (90y), and ^{148}Gd (75y). We will report our experience with fabricating these and other kinds of stable targets and discuss our plans for preparing additional targets which offer interesting and exciting prospects for future nuclear research studies.

Although there are always a variety of technical difficulties associated with doing any kind of nuclear science research, the problem of obtaining a reasonable target for various accelerator or reactor based experiments is the most pervasive. In principle, the idea of bringing together focused particle beams and stationary target nuclei is easily conceptualized. In practice, however, once an interesting physics study is identified, the preparation of a suitable target usually involves significant technical challenge and ingenuity. In this report, we describe some of our experience in our Nuclear Properties Group at the Lawrence Livermore National Laboratory (LLNL) with preparing special targets for nuclear research studies.

Our studies have involved the use of both neutron and charged-particle beams and Table I presents a partial list of the targets prepared. The Nobel gas targets were prepared for in-beam γ-ray studies which used an external thermal neutron beam. Large amounts of target material were required for these experiments and a special device - a cryostat - was constructed to isolate and contain a large quantity of gas (in the solid state) so that the measurements could be done. The details of the construction of this device, as well as the various studies performed with it have been described previously.[1,2,3]

0097–6156/86/0324–0476$06.00/0
© 1986 American Chemical Society

The ^{176}Lu target is unique because this material has an extremely low natural isotopic abundance (2.61%). A sample of several tens of milligrams of Lu_2O_3 was obtained from the Oak Ridge National Laboratory and had an isotopic enrichment of ∼ 70% in ^{176}Lu. An additional isotope separation step[4] was done on this sample at LLNL. By this procedure we were able to make a target of isotopically pure ^{176}Lu (∼ 99.9%) that had a thickness of ∼ 22 mg/cm^2 and which was supported on a Th substrate. This target was subsequently used for ^{16}O Coulomb-excitation experiments.[5]

The remainder of the list in Table I involves radioactive targets which were prepared for charged-particle spectroscopy experiments. These targets were generally thin (10-40 µg/cm^2) and were used for studies where light-ion reaction products were measured with a magnetic spectrograph. In general, the procedure for preparing such targets involves three steps: (1) production of the target material, (2) material purification and (3) depositing the material on a suitable substrate. We describe here in detail our preparation of a ^{148}Gd target. A similar procedure was used for the remaining radioactive targets. The results from the various studies performed with these targets have been published elsewhere.[6,7,8]

Table I. Targets for nuclear studies prepared
by the LLNL Nuclear Properties Group.

Isotopic Material	t1/2	Material Production Technique	Approx. Thickness	Nuclear Studies (reference)
^{86}Kr	--	--	2 g/cm^2	2
^{136}Xe	--	--	2 g/cm^2	3
^{148}Gd	75y	Ta + P spallation	25 µg/cm^2	10
^{151}Sm	90y	^{150}Sm + n	35 µg/cm^2	7
^{152}Eu	13y	^{151}Eu + n	30 µg/cm^2	6
^{154}Eu	8.6y	^{153}Eu + n	15 µg/cm^2	6
^{176}Lu	--	--	22 mg/cm^2	5
^{249}Bk	311d	^{247}Bk + 2n	30 µg/cm^2	8

Raw material for the ^{148}Gd target was produced by spallation reactions in a tantalum metal target by ∼ 750-MeV protons at the Isotopes Production Facility at LAMPF. The details of the irradiation and the chemistries associated with separating the hafnium and lanthanide fractions have been reported previously.[9]

The source material we obtained had been separated from the lanthanide fraction and contained gadolinium dissolved in an acidic solution. An assay of this material showed that it contained ~ 170 μg of ^{148}Gd and high levels of europium, samarium, and terbium contamination. In addition, a high level of lead contamination was present in ~ 20% of the sample because of leakage into a shielded container during shipping. We used standard ion-exchange chemistries to remove the lead and rare-earth contaminants and obtained a clean gadolinium fraction that contained ~ 132 μg of ^{148}Gd.

We prepared samples for the isotope separator from the clean gadolinium fraction by electrodeposition. The pure gadolinium fraction was evaporated to dryness, converted to the nitrate, and dissolved in ~ 200 mL of 0.1N HNO$_3$. A 20-μL aliquot was taken from this solution and added to about 1 mL of isopropyl alcohol in a small-volume (2 mL) quartz electrodeposition cell. A small platinum disk was sealed to the base of the cell and served as the anode. The cathode, which acted as the collector electrode, was a small strip of tungsten (10 mm x 2 mm x 0.03 mm); it was immersed to a depth of ~ 3 mm in the nonaqueous plating solution. We began the deposition by using an applied voltage of ~ 200 V and typicaly observed initial cell currents in the range 0.2 to 0.4 mA. We concluded the plating procedure when the current dropped below ~ 0.1 mA with an applied voltage of ~ 400 V. The time to complete the deposition was ~ 1 h. During plating, we adjusted the voltage to maintain cell current in the range 0.2 to 0.5 mA, but did not allow it to exceed 400 V. The deposit on the cathode was not chemically identified; it had a white powdery appearance and may have been a hydrated oxide of gadolinium. Eight tungsten strips coated with gadolinium were prepared in this way, and the total amount of ^{148}Gd plated was ~ 99 μg.

We prepared thin targets by collecting a ^{148}Gd beam from the Nuclear Chemistry Isotope Separator[4] on thin (~ 50-μg/cm^2) carbon foils. The gadolinium-coated tungsten foils each contained 10 to 15 μg of ^{148}Gd. We placed the individual tungsten foils in the isotope separator ion source and collected ^{148}Gd on the carbon foil until the beam became depleted. Approximately two coated tungsten foils were required to deposit ~ 1 μg of ^{148}Gd on a single carbon foil. We prepared four targets in this manner. We collected the ^{148}Gd over an area of about 3 mm x 1.5 mm to get target thicknesses ranging between about 20 and 30 μg/cm^2. We estimate that the total amount of ^{148}Gd finally recovered as usable targets was about 3 to 4 μg or 2.1% of the original material.

We used the ^{148}Gd targets for a number of charged-particle reaction experiments for nuclear structure studies.[10] For one of the targets, a sample triton spectrum from the ^{148}Gd (p,t) ^{146}Gd reaction is shown in Fig. 1 which is a computer reconstruction of the original data on a linear energy scale and combines the results of five separate experiments. Each experiment revealed only a portion (~ 20%) of the total spectrum. The spectrum is of excellent quality and, in particular, is free from contaminating peaks caused by other rare-earth materials.

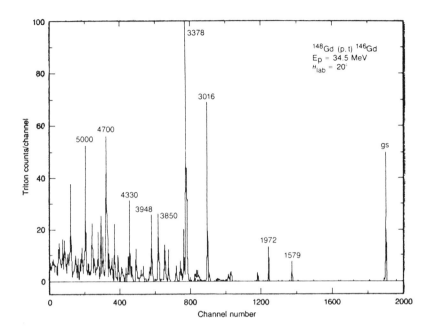

Fig. 1

Table II summarizes the losses we experienced at each step of the target preparation procedure. As the table shows, the most inefficient aspect of our technique is the isotope separation step. Normally, we expect ~ 30% recovery with isotope separation of gadolinium, but this level was obviously not obtained in this particular case. With other rare-earth materials, such as europium and samarium, the expected isotope separation recovery is > 50% and can reach levels as high as 70%. Therefore, allowing for similar chemistry losses, we expect that in those selected cases where isotope separation recoveries are ~ 50%, the amount of initial material required to make a reasonable number of usuable targets can be as low as ~ 100 µg. The amount of intial material required can also be reduced somewhat by paying more attention to increasing the efficiencies of the wet-chemistry yields.

Proposed future work in our group will continue to involve the need for special targets. In particular, we are investigating techniques to prepare radioactive targets of ^{150}Eu (30y) and ^{227}Ac (22y) for charged-particle spectroscopy work. Each of these targets present special problems which are outside our range of experience and offer considerable technical challenge. Beyond that, we have begun a program to measure light-ion charged-particle cross sections by activation and in-beam techniques. Activation studies require the use of many separate target foils. Moreover, each foil must be homogeneous and each must have a thickness known accurately to a few percent.

Table II. Yield summary for ^{148}Gd target preparation.

	Original material[a]	Initial cleanup[b]	Electro-deposition[b]	Isotope separation[c]
Mass (μg)	170	132	99	4
Relative yield (%)	--	78	75	4
Cumulative yield (%)	--	78	58	2

[a]The initial amount of ^{148}Gd was determined by alpha-counting a small aliquot from the new sample.

[b]The chemical yields were determined by gamma-counting techniques. The sample contained a number of radioactive isotopes of gadolinium prior to isotope separation.

[c]The assay of the final targets was obtained by back alpha-counting and x-ray fluorescence techniques.

We are experimenting with two schemes to prepare targets for cross-section measurements. The first is forming a self-supporting "ceramic" target which can be prepared from a metal oxide. This procedure has been used by Quinby,[11] who prepared self-supporting oxide targets with thicknesses in the range 150-2500 μg/cm^2 from several materials. An alternate technique involves producing thin Kapton[12] plastic sheets (1-2 mil) which contain accurately known amounts of any material. The only requirement for the target material is that it is in the form of a powder and that the natural particle size is small enough so that it can be homogeneously mixed (coated) with the liquid plastic base before spreading and curing. Both techniques show considerable promise for making cheap, high quality targets for light-ion studies when low-Z element contamination can be tolerated.

Acknowledgments

This work was performed under the auspices of the U.S. Department of Energy by Lawrence Livermore National Laboratory under contract No. W-7405-Eng-48.

I am grateful to Mr. David Sisson for his work with the electro-deposition procedures and to Mr. Thomas Massey for conducting the final purification chemistries.

References

1. C. M. Jensen, W. M. Buckley, R. G. Lanier, G. L. Struble, S. G. Prussin, and D. H. White, Nuclear Instrum. and Methods 135, 21 (1976).

2. C. M. Jensen, R. G. Lanier, G. L. Struble, L. G. Mann, and S. G. Prussin, Phys. Rev. C 15, 1972 (1977).

3. S. G. Prussin, R. G. Lanier, G. L. Struble, L. G. Mann, and S. M. Schoenung, Phys. Rev. C 16, 1001 (1977).

4. R. J. Dupzyk, C. M. Henderson, W. M. Buckley, G. L. Struble, R. G. Lanier, and L. G. Mann, Nuclear Instrum. and Methods 153, 53 (1978).

5. Jean Kern, University of Fribourg, Fribourg, Switzerland, private communication (1985).

6. See H. E. Martz, R. G. Lanier, G. L. Struble, L. G. Mann, R. K. Sheline, and W. Stöffl, Nucl. Phys. A439, 299 (1985) and references therein.

7. D. Mueller, R. T. Kouzes, R. A. Naumann, G. L. Struble, R. G. Lanier, and L. G. Mann, Bull. Am. Phys. Soc. 23, 91 (1978).

8. R. A. Dewberry, R. T. Kouzes, R. A. Naumann, R. G. Lanier, H. Börner and R. W. Hoff, Nucl. Phys. A399, 1 (1983).

9. K. E. Thomas, Radiochimica Acta 34, 135 (1983).

10. A preliminary target of ^{148}Gd was prepared earlier with a somewhat less rigorous technique. Experimental studies with this target have been published in E. R. Flynn, J. van der Plicht, J. B. Wilhelmy, L. G. Mann, G. L. Struble, and R. G. Lanier, Phys. Rev. C 28, 97 (1983).

11. "Fabrication of Self-Supporting Oxide Targets by Cationic Adsorption in Cellulosic Membranes and Thermal Decomposition," Thomas C. Quinby in Proceeding of Workshop 1983 of the International Nuclear Target Development Society, Argonne National Laboratory, Argonne, IL, Sept. 7-9, 1983. ANY/PHY-84-2.

12. Trademark E. I. Du Pont de Nemours & Co. (Inc.) Wilmington, DL 19898.

RECEIVED July 15, 1986

73

Rapid, Continuous Chemical Separations

J. V. Kratz, N. Trautmann, and G. Herrmann

Institut für Kernchemie, Universität Mainz, D-6500 Mainz, Federal Republic of Germany

We report on a number of on-line chemical procedures which were developed for the study of short-lived fission products and products from heavy-ion interactions. These techniques combine gas-jet recoil-transport systems with I) multistage solvent extraction methods using high-speed centrifuges for rapid phase separation and II) thermochromatographic columns. The formation of volatile species between recoil atoms and reactive gases is another alternative. We have also coupled a gas-jet transport system to a mass separator equipped with a hollow cathode- or a high temperature ion source. Typical applications of these methods for studies of short-lived nuclides are described.

1. Introduction

Nuclear reactions producing exotic nuclei at the limits of stability are usually very non-specific. For the fast and efficient removal of typically several tens of interferring elements with several hundreds of isotopes from the nuclides selected for study mainly mass separation [Han 79, Rav 79] and rapid chemical procedures [Her 82] are applied. The use of conventional mass separators is limited to elements for which suitable ion sources are available. There exists a number of elements, such as niobium, the noble metals etc., which create problems in mass separation due to restrictions in the diffusion-, evaporation- or ionization process. Such limitations do not exist for chemical methods. Although rapid off-line chemical methods are still valuable for some applications, continuously operated chemical procedures have been advanced recently since they deliver a steady source of activity needed for measurements with low counting efficiencies and for studies of rare decay modes. The present paper presents several examples for such techniques and reports briefly actual applications of these methods for the study of exotic nuclei.

0097–6156/86/0324–0482$06.00/0

2. Combination of a gas-jet with the solvent extraction system SISAK

Nuclear reaction products recoiling out of a thin solid target are stopped in nitrogen gas containing aerosols and the thermalised products attached to these aerosols (usually inorganic clusters like potassium chloride [Ste 80] are used) are transported along with the carrier gas out of the stopping chamber through a capillary to the SISAK system. The latter provides dissolution of the clusters in an aqueous solution and subsequent multistage solvent extraction in static mixers with separation of the two phases in H-centrifuges operated at 20000 rpm. Fig. 1 shows schematically the set-up for the isolation of technetium with the centrifuge system SISAK II [Ska 80] .

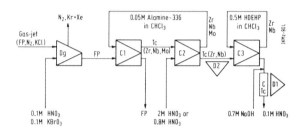

<u>Fig. 1</u> Flow-sheet for the continuous separation of technetium [Bro 81] from fission products with the centrifuge system SISAK II. DG = degassing unit, D_1, D_2 = detector positions, C1, C2, C3 = mixer-centrifuge units, C = column, FP = fission products

The reaction products from the thermal neutron induced fission of ^{239}Pu (700 μg ^{239}Pu; 6×10^{11} n/cm^2·s) carried by a KCl/N$_2$-jet are dissolved in a static mixer with a solution consisting of 0.1 M nitric acid and 0.1 M potassium bromate. The solution is heated to 80° C to accelerate the dissolution of the clusters together with the reaction products and the oxidation of technetium to the pertechnetate. The gas-liquid mixture is fed into a degassing unit where the carrier gas and the fission noble gases are separated and removed by suction. In the first mixer-centrifuge unit technetium is extracted into 0.05 M Alamine-336 in chloroform. Traces of molybdenum, zirconium and niobium are co-extracted. In the next step (C2) technetium is back extracted with 2 M or 0.8 M nitric acid. In a third stage (C3) zirconium and niobium contaminations are removed by extraction into di(2-ethylhexyl) orthophosphoric acid (HDEHP). The outgoing aqueous phase contains only technetium nuclides, their daughter products and a small contamination of molybdenum.

The decay of 36-s ^{106}Tc and of 5-s ^{108}Tc into excited levels of the even-even isotopes ^{106}Ru and ^{108}Ru was studied in detail. Extended level schemes, spin assignments, multipole mixing ratios and B(E2) values for ^{106}Ru and ^{108}Ru were obtained. With this rather complete set of data the collective structure of these transitional nuclei could be confronted in detail with several models Sta 84 .

Recently, the SISAK technique has been applied for the separation of new, neutron-rich neptunium isotopes formed in direct transfer reactions between ^{136}Xe projectiles and targets of ^{244}Pu at the UNILAC heavy-ion accelerator. The SISAK system consisted of three mixer-centrifuge units and a degasser.

In the first step of the separation procedure the KCl-clusters together with the attached reaction products are dissolved in a static mixer in 0.6 M HCl containing $TiCl_3$. With $TiCl_3$ neptunium is reduced to the 3^+ state. The mixture is pumped through the degassing unit and is fed into a mixer-centrifuge and contacted with 5% HDEHP in CCl_4. Thorium and uranium together with several fission products are extracted into the organic phase whereas neptunium remains in the aqueous phase. The aqueous phase is mixed with 6 M HNO_3. The nitric acid causes the oxidation of neptunium to the 4^+ state. In this form it is extracted into 7% HDEHP in CCl_4 in the second mixer-centrifuge. Complexing agents keep the fission products in the aqueous phase. In the last step the organic phase is pumped into the third mixer centrifuge, where neptunium is back-extracted with H_3PO_4 Tet 85 .

In the γ-ray spectra four new γ-lines that fit into the known level schemes of ^{243}Pu and ^{244}Pu have been observed. The half-life of the new isotope ^{243}Np was determined as 61 ± 7 s. The decay curves of ^{244}Np are complex and can be explained by two isomers of ^{244}Np with half-lives of 7 ± 4 s and 140 ± 20 s. These should be compared with theoretical (microscopic) predictions [Kla 84] of 195 s for ^{243}Np and 8 s for ^{244}Np. The gross theory [Tak 73] predicts 25 min and 3 min.

So far on-line chemical separations with the SISAK system have been developed and applied for short-lived isotopes of As, Br, Zr, Nb, Te, Ru, Pd, I, La, Ce, Pr, and Np [Ska 80, Bro 81, Ska 83, Rog 84, Aro 74, Tet 85].

3. Combination of a gas-jet with thermochromatographic methods

In thermochromatography volatile species are formed in reactions with reactive gases and are deposited in a column with a temperature gradient - the less volatile species in the high temperature region and the more volatile species in the low temperature region. The combination of a gas-jet with a thermochromatographic column allows the continuous separation of the reaction products. This approach was applied for the on-line separation of fission products forming volatile halides [Hic 80, Hic 84].

It was also applied for the on-line isolation of volatile metals such as Hg, Tl, Pb, Bi, Po, At and of noble gases in a recent, rigorous attempt to synthesize superheavy elements in the ^{48}Ca + ^{248}Cm reaction [Süm 84, Arm 85]. In the latter case the volatile species were allowed to leave the column, see Fig. 2, and were condensed on ultra-thin nickel foils. These were rotated stepwise between pairs of surface barrier detectors to detect α-particles and spontaneous-fission fragments.

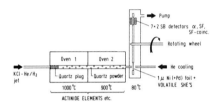

Fig. 2: On-line gas phase chemistry
procedure for the separation
of volatile superheavy
elements [Süm 84].

4. Continuous separations by chemical reactions in the carrier gas

Instead of using a cluster-jet for the transportation of the nuclear reaction products to the chemistry apparatus, it is possible to use chemical reactions of the recoils in the carrier gas: Either the recoils are stopped in a reactive gas where highly volatile products are formed which are continuously swept out of the stopping chamber, or volatile compounds are formed by thermal synthesis. The volatile species are then separated by selective absorbents. The thermal synthesis of volatile compounds in a gas-jet was used for the separation and investigation of short-lived tellurium and selenium isotopes [Zen 78]. Through recoil into a mixture of nitrogen and hydrochloric acid volatile chlorides are formed inside the reactor and are separated by selective absorption. This technique has been applied in on-line investigations of short-lived germanium isotopes [Zen 81]. Nitrogen and hydrofluoric acid were used for the on-line separation of arsenic [Hic 84].

5. On-line mass separation

The HELIOS mass separator at the Mainz TRIGA reactor is one of the few facilities where a gas-jet transportation system has been coupled successfully to a mass separator. Details of the gas-jet system, the integrated skimmer-ion sources, the separator and its collector facilities have been published elsewhere [Maz 80, Maz 81]. Recently, HELIOS was applied for spectroscopic studies of neutron-rich isotopes of praseodymium [Bru 85] and neodymium [Kar 85]. Work is in progress in order to replace the existing ion sources by an inductively coupled plasma ion source (ICP). In case of a successful coupling of the gas-jet with the ICP source it should be possible to ionize any element in the periodic table because temperatures of 6000 - 10000 K are reached in the plasma.

6. Outlook

Continuous chemical separations have been proven to be powerful techniques allowing detailed measurements of the decay properties of exotic nuclei with half-lives down to a second over long periods of time. For chemical procedures this time scale is close to the limit set by diffusion of chemical species through boundary layers between two phases and by the velocity of phase separations. Further explorations of yet unknown short-lived nuclides will require the development of procedures on the time scale of milliseconds. Promising ways seem to be combinations of specific nuclear effects with chemically selective reactions. On-line mass separations in the millisecond range have been performed [Ärj 84] by charge exchange and thermalization processes in helium. It is conceivable that the choice of other gases may introduce chemical selectivity already into the thermalization process.

References

[Arm 85]   P. Armbruster et al., Phys. Rev. Lett. 54, 406 (1985)

[Aro 74]   P.O. Aronsson et al., Inorg. Nucl. Chem. Lett. 10, 499 (1974)

[Ärj 84]   J. Ärje et al., Department of Physics, University of Jyväskylä, Research Report No. 13 (1984) and J. Äystö et al., Phys. Lett. 138B, 369 (1984)

[Bro 81]   K. Brodén et al., J. Inorg. Nucl. Chem. 43, 765 (1981)

[Brü 85]   M. Brügger et al., Nucl. Instr. Meth. A234, 218 (1985)

[Han 79]   P.G. Hansen, Ann. Rev. Nucl. Part. Sci. 29, 69 (1979)

[Her 82]   G. Herrmann, N. Trautmann, Ann. Rev. Nucl. Part. Sci 32, 117 (1982)

[Hic 80]   U. Hickmann et al., Nucl. Instr. Meth. 174, 507 (1980)

[Hic 84]   U. Hickmann et al., GSI Scientific Report 1983, GSI 84-1, 230 (1984) and U. Hickmann, Doctoral thesis, Universität Mainz (1984)

[Kar 85]   T. Karlewski et al., Z. Physik, in press

[Kla 84]   H.V. Klapdor et al., Atomic Data Nucl. Data Tables 31, 81 (1984)

[Maz 80]   A.K. Mazumdar et al., Nucl. Instr. Meth. 174, 183 (1980)

[Maz 81]   A.K. Mazumdar et al., Nucl. Instr. Meth. 186, 131 (1981)

[Moo 85]   K.J. Moody et al., GSI Scientific Report 1984, GSI 85-1, 93 (1985)

[Rav 79]   H. Ravn, Phys. Rep. 54, 203 (1979)

[Rog 84]   J. Rogowski et al., to be published

[Ska 80]   G. Skarnemark et al., Nucl. Instr. Meth. 171, 323 (1980)

[Ska 83]   G. Skarnemark et al., Radiochim. Acta 33, 97 (1983)

[Sta 84]   J. Stachel et al., Z. Physik A316, 105 (1984)

[Ste 80]   E. Stender et al., Radioanalyt. Lett. 42, 291 (1980)

[Süm 84]   K. Sümmerer et al., Preprint GSI-84-17 (1984)

[Tak 73]   K. Takahashi et al., Atomic Data Nucl. Data Tables 12, 101 (1973)

[Tet 85]   H. Tetzlaff et al., GSI Scientific Report 1984, GSI 85-1, 273 (1985)

[Zen 78]   M. Zendel et al., Nucl. Instr. Meth. 153, 149 (1978)

[Zen 81]   M. Zendel et al., Radiochim. Acta 29, 17 (1981)

RECEIVED August 28, 1986

Section III: New Facilities and Techniques
Part 3: Technique Comparison (Investigations of ^{138}Sm)

First Results at Système Accélérateur Rhône-Alpes (SARA) with a He-Jet Coupled to a High-Current Mass Separator Source

R. Beraud [1], A. Charvet [1], R. Duffait [1], A. Emsallem [1], M. Meyer [1], T. Ollivier [1], N. Redon [1], J. Genevey [2], A. Gizon [2], N. Idrissi [2], and J. Treherne [2]

[1]Institut de Physique Nucléaire (et IN2P3), Université Lyon 1, 43 boulevard du 11 Novembre 1918, F. 69622 Villeurbanne Cédex, France
[2]Institut des Sciences Nucléaires (IN2P3 et USMG), 53 avenue des Martyrs, F. 38026 Grenoble Cédex, France

We report on the performances of a He-jet connected to a Bernas-Nier ion source. After a brief description of the system we give the first spectroscopic results on very neutron deficient rare-earth isotopes.

Among the techniques used to study the nuclear structure all over the mass surface, in-beam γ-spectroscopy and isotope separator on-line (ISOL) are the two most important ones. The first one gives access to high spin physics whereas the second one is mainly related with low-spin states at lower energy depending on available decay energy. The ISOL systems allow to investigate nuclear species produced with very small cross-sections since the products can be studied in a very low background environment. The use of intense beams and the development of fast and efficient ISOL systems are both necessary to study nuclei far off the valley of stability.
In this paper I will explain our motivations concerning the technical choice of a He-jet coupled to a mass separator and then give our first results.

2. Experimental facility

2.1 SARA (Système Accélérateur Rhône-Alpes)

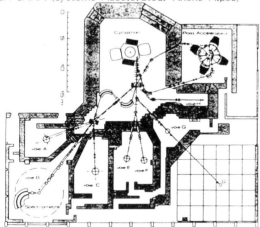

This facility is composed of two cyclotrons, a K=90 compact one and a K=160 separated sector one which is operating since 1982. An ECR ion source is now commonly used for the production of C,N,O and Ne beams up to 40 MeV/u energy. The layout of the beam lines, as shown in fig.1, allows to use also the first cyclotron alone when high intensity, moderate energy beams are needed. This is a common mode of operation to carry out spectroscopic studies by means of fusion-evaporation reactions.

Fig. 1 : General layout of the SARA machine and experimental areas.

0097-6156/86/0324-0490$06.00/0

2.2 Layout of the on-line isotope separator

A schematic plan view of the new on-line separator located on beam line C is given in figure 2. The beam from the accelerator enters the He-jet recoil chamber through a thin Ni window (2mg/cm²), bombards the target and is then stopped in a beam-stopper composed mainly of iron, lead and polystyrene in order to prevent from both neutron and γ background.
The ion-source of the separator is fed by a capillary transporting the recoils thermalized in the He-jet chamber. The separator beam is extracted at right angle to the axis of the beam line and then travels in a 120° magnet of index n=1/2. From the focal plane the mass-separated beam is then transported by means of a 6m long Einzel lens to a well-shielded collection chamber. A programmable tape-transport device carries the activity to the counting station where X-rays, γ-rays and particles are detected.
The overall performance of the separator is mainly determined by the quality of the magnet. The resolution is nowadays 1000 FWHM at mass 100 and the separation 17 mm between A=100 and A=101. It is planned to improve the resolution by changing mechanically the entrance and exit magnet boundaries.

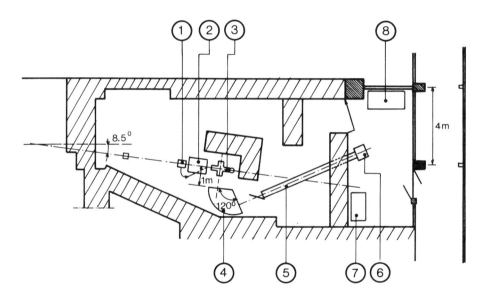

Fig. 2 : Layout of isotope separator in beam line C. The numbers indicate : (1) He-jet recoil chamber, (2) beam-stopper, (3) ion-source, (4) analysing magnet, (5) Einzel lens for transport of mass-separated beam, (6) tape-transport and counting station, (7) power/control rack of tape-transport, (8) power/control of mass-separator.

2.3. He-jet coupled to ion-source

A schematic view of the He-jet recoil chamber and integrated skimmer ion-sour
ce is shown in figure 3. The recoil chamber is either of multicapillary or sheet
type as described in [PLA 81].

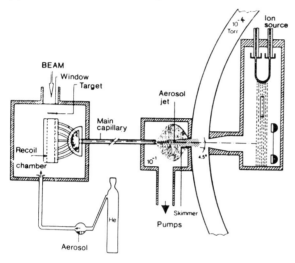

The collection volume ($\emptyset$ = 12mm, length=100 mm) is made of an Aluminium cylinder and connected to the main stainless capillary. It is placed in the middle of the He pressurized reaction chamber and He gas is fed through a temperature controled bubbler. Diffusion pump aerosols were mainly used in the experiments reported here. For monitoring of the beam, a Faraday cup fixed behind a 4 mg/cm² Havar window was used.

The ion-source is of Bernas-Nier type and is described in more details in [CHA 67]. The extracted

Fig.3 : Schematic view of He-jet recoil chamber and
integrated skimmer ion-source

current may be $\sim$ 5 mA, allowing thus outgasing or target evaporation. An important application is the use of carrier gas and/or chemical products which may
give volatile compounds of refractory elements. A high flow pumping system composed of two roots and a primary pump (3000 m³/h- 350 m³/h - 120 m³/h) is used
to skim off the He from the skimmer box where the pressure is maintained around
10^{-1} torr. The skimmer is a flat stainless steel plate with a circular hole $\emptyset$ =
1,2 mm in the middle.

3. Results

3.1 The decay of 12s ^{143}Tb [OLL 85]

The activity was produced by bombarding a self-supporting 2mg/cm² ^{112}Sn
target (enriched to 88 %) with a 191 MeV ^{35}Cl beam. The recoils, stopped in
1 bar He pressure, were transported via a 6 m long capillary to the tape-transport
system. At the counting position, measurements including γ-ray multianalysis
decays (typically 8x2 s and 8x10 s), γ-X and γ-γ coincidences were carried out
with intrinsic Ge dectectors. Particle spectra were simultaneously measured with
a 500 μm Si detector ; no significant particle activity was observed.
The γ-rays were ascribed to ^{143}Tb decay on the basis of following experimental
evidence : i) all these γ-rays are in coincidence with K_α X-rays of Gd (see figure
4). ii) the average measured half-life of the most intense γ-lines is (12 ± 1) s.

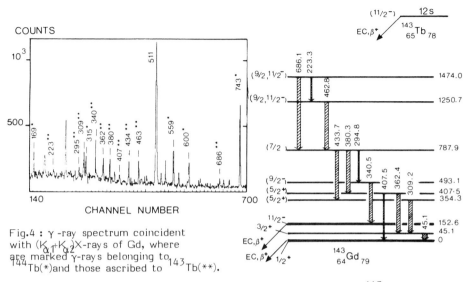

Fig.4 : γ-ray spectrum coincident with $(K_{\alpha_1}+K_{\alpha_2})$X-rays of Gd, where are marked γ-rays belonging to ^{144}Tb(*)and those ascribed to ^{143}Tb(**).

Fig.5 : Decay scheme of ^{143}Tb.

Using γ-γ coincidence relations, the first decay scheme of ^{143}Tb → ^{143}Gd is shown in figure 5. No γ transition decaying the 152.6 keV level was observed and it is therefore assumed to be the 11/2⁻ isomeric level (1.83 min) of ^{143}Gd [WIS 76]. The analogeous state was observed at higher energy (748.7 keV) in ^{145}Gd [NOL 82]

However it is worth noting the similarity between both level schemes. The log ft value to the 1474 keV state is 4.8 which is typical of an allowed β transition. Therefore we propose a tentative spin-parity assignment 11/2⁻ to ^{143}Tb according to an $h_{11/2}$ proton state.

3.2 Identification and decay of 12 s ^{138}Eu [CHA 85]

In this experiment the same 191 MeV ^{35}Cl beam has been used to bombard a 2 mg/cm² enriched ^{106}Cd target. Using the same He-jet system as in section 3.1 a set of measurements including γ-ray multianalysis (8x3 s) γ-X ray and γ-γ coincidences were performed. A group of strong γ-rays, with an average half-life of (12.1 ± 0.6) s, was found to be coincident with K_α X-rays of Sm. In order to determine which Eu → Sm mass chain was concerned, the coupling of the He-jet to the ion-source of the separator was realized. As shown in figure 6, the main γ-rays with $T_{1/2}$ = 12 s are present in the mass-separated spectrum at A=138.

From these spectra, the (skimmer + ion-source + magnetic analyzer + lens transport) efficiency was roughly estimated to be better than 1.0 %.

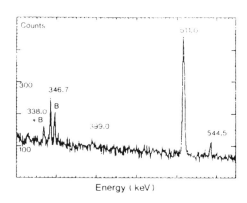

Fig.6 : Part of the γ-ray spectrum recorded
with the He-jet coupled to mass separator
at A=138 (B=background)

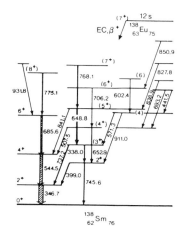

Fig.7 : Decay scheme of
12s ^{138}Eu.

The first detailed level scheme of ^{138}Sm shown in figure 7 was constructed from
this work. The most strongly populated sequence involving Δ I=2 transitions is
identified as the g.s. band excited up to 8$^+$ in good agreement with a recent in-beam
experiment [LUN 85].
A tentative spin parity assignment of 7$^+$ has been made for the 12 s ^{138}Eu state
in view of direct feeding, via β$^+$+EC decay, of levels with spins 6,7 or 8. As
shown in figure 8 the rapid lowering of the first 2$^+$ excited state from N=82 shell
closure to N=76 corresponds to the increase of nuclear deformation predicted
by Möller and Nix [MOL 81].
Another level sequence with ΔI=1, exhibiting strong ΔI=2 transitions and based
on a 2$^+$ state excited at 745.6 keV can be interpreted as a quasi-gamma band.
It is worth noting the position of the 2$'^+$ band head located more than 200 keV
under the first 4$^+$ level. In the framework of a rigid rotator model, this leads
to an asymetry parameter γ≈25°. Additional data, especially on ^{140}Sm, are needed
to give a coherent interpretation of the structure of these nuclei.

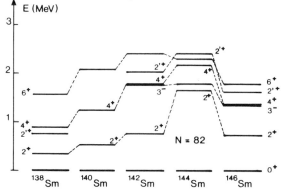

Fig.8 : Partial level
schemes for ^{138,140,142,}
^{144,146}Sm showing
the first excited
states.

4. Conclusion

From these first results obtained by means of our new facility we have the feeling that a lot of physics may be gained via similar experiments. The new H.I. metallic beams (Al, Si and Ca) available in a very near future will increase the number of investigations to be made. Regarding the detectors used, an electron detection station is under construction and will be of particular interest for the study of low-energy transitions.

Acknowledgments

The authors wish to thank V. Boninchi, G. Margotton, S. Morier, J.P. Richaud, S. Vanzetto, L. Vidal, J.L. Vieux-Rochaz and especially A. Plantier for valuable technical assistance in carrying out the experiments. We are also indebted to the SARA staff.   J. Äystö, J. Honkanen (University of Jyväskylä) and K. Deneffe (I.K.S. Leuven) are greatly acknowleged.

References

[CHA 67] I. Chavet, R. Bernas, Nucl. Instr. and Meth. 51 77 (1967)
[CHA 85] A. Charvet, T. Ollivier, R. Béraud, R. Duffait, A. Emsallem, N. Idrissi, J. Genevey, A. Gizon, Z. Phys. A 321 697 (1985)
[LUN 85] S. Lunardi, F. Scarlassara, F. Soramel, S. Beghini, M. Morando, C. Signorini, W. Meczynski, W. Starzecki, G. Fortuna, A.M. Stefanini, Z. Phys. A 321 177 (1985)
[MOL 81] P. Möller, J.R. Nix, At. Data and Nucl. Data Tables 26 165 (1981)
[OLL 85] T. Ollivier, R. Béraud, A. Charvet, R. Duffait, A. Emsallem, M. Meyer, N. Idrissi, A. Gizon, J. Tréherne, Z. Phys. A 320 695 (1985)
[PLA 81] A. Plantier, University Thesis, Lyon-1 (1981)
[WIS 76] K. Wisshak , A. Hanser, H. Klewe-Nebenius, J. Buschmann, H. Rebel, H. Faust, H. Toki, A. Fässler, Z. Phys. A 277 129 (1976)

RECEIVED July 15, 1986

Silicon Box: New Tool for In-Beam γ-Ray Spectroscopy Through Heavy-Ion Fusion Reactions

M. Ishii[1], A. Makishima[2], M. Hoshi[2], and T. Ishii[1]

[1]Japan Atomic Energy Research Institute, Tokai, Ibaraki-ken, Japan
[2]Tokyo Institute of Technology, Meguro-ku, Tokyo, Japan

A new instrument "silicon box" has been developed which works as a charged-particle multiplicity filter. It permits us to sort out in-beam gamma-rays by the proton number of the fusion residues. It has been applied to studies on Sm-138, Sm-136 and Nd-132. Early results of Sm-134, Ce-126 and Ce-124 are briefly reported.

1. Introduction

Heavy-ion fusion reactions provide us with a means for studying very proton-rich nuclei. Energetic heavy projectiles induce nuclear fusion with the target nuclei and transfer their energies and angular momenta to the compound nuclei. The nuclear fusion is followed by the emission of several nucleons and light nuclei with a part of the excitation energy and angular momentum. The residual nuclei, still with high excitation energies and spins, de-excite electromagnetically. Generally speaking, the charged-particle emission is disadvantageous for the production of proton-rich nuclei. However there are many proton-rich combinations of the target and the projectile nuclides which more than compensate for the disadvantage. Those heavy-ion fusion reactions lead us further from the line of stability than light-ion fusion reactions.

However this does not mean that gamma-ray transitions in such nuclei are easy to detect. Usually we confront the competition of many nuclear reaction channels, which makes in-beam gamma-ray spectra too composite to analyze. In-beam gamma-ray spectroscopy through heavy-ion fusion is longing for a new method of selectively observing gamma-rays emitted via a nuclear reaction channel of particular interest.

"A silicon box", titled in the present paper, is a challenge to this problem. It works as a charged-particle multiplicity filter, that is, an

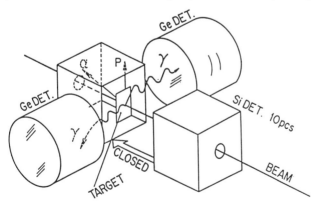

Fig.1. A silicon box. It consists of ten silicon detectors which take count of protons and alpha-particles emitted in each fusion events.

0097–6156/86/0324–0496$06.00/0

instrument of taking count of charged-particles emitted in individual fusion events. It permits us to sort out in-beam gamma-rays by the proton number of the fusion residues. In the present paper we first describe the silicon box. Then we talk about some experiments, laying stress on technical aspects. Finally a brief discussion is given on physical aspects of the experiments, the nuclear deformation in the ground band states of nuclei with the proton and neutron numbers ranging from 50 to 82.

2. Description of silicon box

The silicon box comprises ten silicon detectors surrounding the target with almost 4π steradian, as shown in fig.1; two of them are of an annular, square type and the others are plain, rectangular; either of them has no marginal insensitive area. A half of the box subtends the front 2π with respect to the target and the other half the rear 2π. Each silicon detector works as a partially depleted ΔE counter and discriminates between protons and alpha-particles. Its depletion depth is chosen such that the energies deposited by protons do not exceed the minimum energy deposited by alpha-particles. A typical depletion depth employed was 0.4 mm.

The silicon detectors were made of n-typed single crystal of 1 mm thick. They have a MOS structure of gold, tungsten oxide, n-typed silicon and aluminum back contact. Since these layers can be deposited on the silicon wafer by evaporation techniques, the fabrication process is so simple as to be applicable to fabrication of the detectors for a special use. No surface treatment for passivation is given to them so that their performance is affected by ambient gases. For example, some good detectors show a leakage current of half micro-ampere at room temperature in the atmosphere, but a few micro-ampere in vacuum. So, in order to stabilize their performance, the silicon detectors were operated at the dry ice temterature.

A practical unit of detector is composed of a silicon wafer, an indium foil of 0.5 mm thick and a substrate of machinable glass. The indium foil makes electric contact with the aluminum layer on the wafer. The dimensions of the wafers and the substrates are as follows:

Annular type:
Si wafer   35 mm x 35 mm with a hole of 12 mm in diameter.
Substrate   37.5 mm x 37.5 mm x 5 mm with a hole of the same size.
Plain type:
Si wafer   30 mm x 35 mm.
Substrate   37.5 mm x 37. 5 mm x 5 mm.

The foil was fixed on the substrate with cyanoacrylate adhesive, and then the wafer on the foil. Finally the gold layer of the detector was covered with a 10 μm thick copper, both ends of which were fixed on the sides of the substrate. The copper foil conducts the current from the detector to the ground and, on the other hand, protects the detector from the scattered beam and low energy delta-rays from the target. Five detectors are assembled on an aluninum holder: The annular detector is fastened with two screws by the back of the substrate and the four plain others are put upright arround the annular detectors.

The reliability of the multiplicity measurement depends on the number of the detectors employed, a total of solid angles subtended by them and the efficiency and accuracy of the particle identification. The probability of two charged-particles entering a detector decreases with the number of the detectors, while the solid angle uncovered and the complexity of electronic processing of the signals from the detectors increase. As a result, ten detectors to cover the target is a reasonable compromise. The efficiency of

of charged-particle detection exceeds 95 % in the silicon box. However the
ΔE pulse heights of protons and alpha-particles do not completely separate
from each other. Additionally the ΔE spectra contain continuums of positrons
and high energy delta-rays. These degrade the quality of the particle identification. The silicon box has shown a reliability of 70 to 80 % in multiplicity measurements.

3. Experiments
        In-beam experiments have been made on light isotopes of Sm, Nd and Ce
by the use of the silicon box at the JAERI tandem accelerator laboratory.
The silicon box was mounted in a small vacuum chamber. In the gamma-gamma
coincidences a pair of germanium detectors of 60 c.c. were placed 4 cm from
the target at 90°with respect to the direction of incident beams. This configuration is a practical choice for the small germanium detectors. Several
detectors, if available, will allow another choice which avoids the Doppler
broardening of gamma-lines and favors the anisotropy of the angular distributions of E2 transitions. The target used in most experiments consisted of
two layers; one was about 3 mg/cm² of enriched isotope, and the other 4 mg
/cm² of the backing material which stops recoiling fusion residues but does
not fuse with the projectiles. If the target foils of about 0.5 mg/cm² were
available that had a high melting temperature enough to withstand the beam
irradiation, the Doppler-shifted gamma-rays were observed for the lifetime
measurements. Experiments on Sm-138, Sm-136 and Nd-132 have been finished,
but those on Sm-134, Ce-126 and Ce-124 are continued. In the following the
experiments carried out are summarized:
        Sm-138:   Ag-107(Cl-35, 2p xn), E(Cl-35)=165 MeV, 2p-γ-γ coincidences,
                  2p-γ(θ), 2p-Doppler shifted γ.
        Sm-136 and Nd-132: Ag-107(S-32, p xα yn), E(S-31)=160 MeV, p-γ-γ,
                  p-γ(θ), p-Doppler shifted γ, p-γ-n.
        Sm-134:   Cd-106(S-32, 2p xn), E(S-32)=170 MeV, 2p-γ-γ, 2p-γ-n.
        Ce-126:   Mo-92(Cl-37, p xn), E(Cl-37)=165 MeV, p-γ-γ, p-γ-n.
        Ce-124:   Mo-92(Cl-35, p xn), E(Cl-35)=165 MeV, p-γ-γ, p-γ-n.
        Now we illustrare the performance of the silicon box with some exper-
imental results. Fig.2 shows a gamma-ray spectrum coincident with two protons emitted in the reaction Cd-106(S-32, xp yn). The ALICE [PLA75] predicts

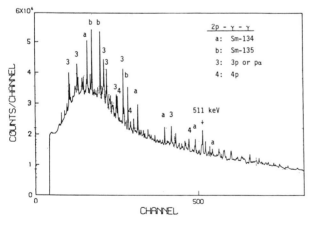

Fig.2. Gamma-rays
coincident with two
protons in reaction
Cd-106(S-32, xp yn).

that in this reaction fusion residues Nd-134, Nd-132, Pm-135, Pm-134, Sm-135 and Sm-134 have the yields over 30 mb at projectile energies of 140 to 170 MeV; Pm-135 and Nd-134 are strongly populated and the other yields are of comparable magnitude. In fact fig.2 includes gamma-ray transitions emitted from Pm and Nd isotopes which should be found in the gamma-ray spectra with multiplicities 3 and 4, respectively. From the contamination by the gamma-rays with higher multiplicities the reliability of the multiplicity is estimated to be 70 to 80 %. Furthermore fig.2 exemplifies the isotopes identification. The silicon box makes it possible to identify induced nuclei within a few isotopes. Usually, as is the case for this, one is an even-even nucleus and another an odd one. So we can distinguish between them by a peculiar pattern of gamma-ray spectra from even-even nuclei. Such one is illustrated in fig.3 by a spectrum of the ground band transitions in Sm-134 which is reduced from event-by-event record 2p-γ-γ.

The silicon box has been employed for the purpose of the lifetime measurements as well. In the case where the recoil distance is up to 6 mm, the target and the plunger foils, suported on thin, flat frames, are placed between the two halves of the box. Beyond a recoil distance of 6 mm, one of the side detectors of the rear half is removed and the target foil is set there. This is a minimum expense of the multiplicity measurement. The flatness of the target and the plunger limitted the accuracy of measured lifetimes to ±3 ps.

4. Results and discussion

Many well-deformed nuclei have been found in rare earth and actinide regions. In those nuclei the ratio of the excitation energies of the 4^+ to the 2^+ states E_4/E_2 are almost equal to the rotational limit 10/3. The lowest values of the E_2 are about 72 keV and 43 keV, respectively. Can we find such well-deformed nuclei in the region of the proton and neutron numbers ranging from 50 to 82 ? What the minimum E_2 there ? these questions have not been answered yet because well-deformed nuclei, if any, are too proton-rich and have too poor yields to observe by conventional in-beam spectroscopic methods. The present studies on light isotopes of Sm, Nd and Ce aim at finding a clue to the questions.

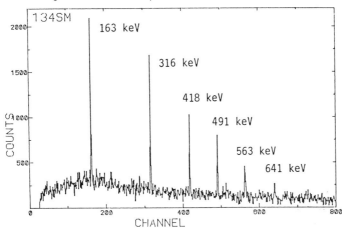

Fig.3: Gamma-rays in Sm-134. The spectrum was projected from event-by-event record 2p-γ-γ in the same reaction as in Fig.2.

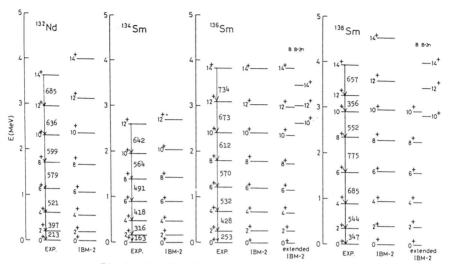

Fig.4. Level schemes of Sm-138, Sm-136,
Sm-134 and Nd-132.

Level schemes of Sm-138, Sm-136, Sm-134 and Nd- 132 are given in fig.
4 and the lifetimes of some excited states are summarized in table 1. The
excitation energies in these nuclei have been computed on the basis of the
interacting boson model, IBM-2 ⌊OTS78⌋. The computations are in good agreement with the experimental results of Sm-136, Sm-134 and Nd-132 but not of
Sm-138. The extended IBM-2 which includes the interaction between the bosons
and a two-quasiparticle 10^+ state can reproduce the experimental situation:
In Sm-138 an isomeric 10^+ state and the successive high spin states appear at
lower excitation energies than the ground band states with the corresponding
spins which the IBM-2 predicts. This indicates the fact that the collectivity in the ground band states rapidly evolves with the decrease of the neutron
number. The B(E2)'s deduced from the lifetimes of the 2^+ states in Sm-138,
Sm-136 and Nd-132 give another support to the above viewpoint; the enhancements of the respective B(E2)'s are 80, 120 and 180 in Weisskopf units.
Details are given in reference [MAK78].

Table 1. Half-lifetimes of excited states in Sm-138, Sm-136
and Nd-132 in units of ps.

	2^+	4^+	10^+	12^+
Sm-138	45(6)		$5.5(0.2) \times 10^2$	33(2)
Sm-136	$1.3(0.1) \times 10^2$	8 - 13	< a few ps	
Nd-132	$2.2(0.2) \times 10^2$	14 - 19	< a few ps	

Other accessible candidates of well-deformed nuclei are Ce-126 and
Ce-124. Early results revealed their level schemes as depicted in fig.5
which have been built on the p-γ-γ coincidences. These nuclei, together with
other light Ce isotopes, form smooth systematic curves of the E_2 and the
E_4/E_2. Especially the E_2 of 142 keV and the E_4/E_2 of 3.15 in Ce-124 are the

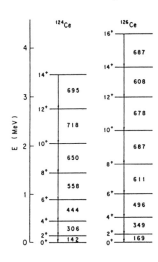

Fig.5. Level schemes of
Ce-126 and Ce-124.
These are built on the
p-γ-γ coincidences in
reaction Mo-92(Cl-35/37,
xp yn). The spins are
tentatively assigned.

extremes that have ever been measured. It is worthy of note that the former
gives an estimate of the nuclear deformation β=0.31 [DAV66] and the latter
is very close to the rotational limit 10/3. However a detailed discussion on
nuclear deformation needs the lifetimes of the first excited state and others.
Another interesting fact is that backbending takes place in the level scheme
of Ce-124 but not in those of Sm-134 and Nd-132.

Finally we ask further question: Can we find any more well-deformed
nuclei in this region ? We may think of more proton-rich nuclei such as Sm-
132, Nd=126 and Ce-122. However these nuclei are situated on the verge of
the proton dripping so that their yields are a few mb or less. The observation of in-beam gamma-rays from them requires a higher quality of charged-
particle multiplicity filter.

Acknowledgments

We wish to thank the staff at the JAERI tandem accelerator facility.
We thank also Drs. K.Harada, N,Shikazono and M.Maruyama of JAERI and
Professors H.Taketani and M.Ogawa of Tokyo Institute of Technology for their
kind advice and fruitful discussions.

References

[DAV66] J.P.Davidson, Nucl. Phys. 86, 561(1966).
[MAK85] A.Makishima, Ph.D thesis, Tokyo Institute of Technology (1985).
[OTS78] T.Otsuka et al., Phys. Lett. 76B, 139(1978).
        Refer also to JAERI-M REPORT 85-094(1985) by T.Otsuka and N.Yoshida.
[PLA75] F.Plasil and M.Blann, Phys. Rev. C11, 508(1975). The CODE "ALICE"
        was developed by M.Blann and F.Plasil.

RECEIVED July 14, 1986

76

Decay of ^{138}Eu and ^{136}Eu and Deformation in the Light Sm Region

R. L. Mlekodaj[1], R. A. Braga[2], B. D. Kern[3], G. A. Leander[1], K. S. Toth[4], and B. Gnade[5]

[1]UNISOR, Oak Ridge Associated Universities, Box 118, Oak Ridge, TN 37831
[2]Georgia Institute of Technology, Atlanta, GA 30332
[3]University of Kentucky, Lexington, KY 40506
[4]Oak Ridge National Laboratory, Oak Ridge, TN 37831
[5]Texas Instruments Corporation, Dallas, TX 75265

The remote region of deformation in the very light rare earth region has been investigated by observation of the decay of ^{138}Eu and ^{136}Eu. These nuclei were produced with ^{48}Ti beams incident on ^{92}Mo targets and mass separated on-line at the UNISOR facility. Fifteen gamma rays associated with a 12±1 s activity attributed to ^{138}Eu decay have been identified and ten of these placed in a decay scheme. The 2+→0+ transition in ^{136}Sm has also been identified. These results are compared to IBM2 and cranking model predictions.

It has long been recognized [SHE61] that a region of permanently deformed nuclei exists in the region where Z and A are both between 50 and 82. This region, however, is located well away from the line of β-stability with the centroid lying beyond the proton drip line. The remote location has made experimental investigations difficult. The characterization of these nuclei has recently been facilitated, however, by the development of heavy-ion accelerators capable of producing these nuclei coupled with the continuing development of on-line isotope separator systems and multi-detector arrays. In addition, theoretical impetus for investigation of this region has been provided by a recent deformed shell model calculation which shows that a promontory of strong deformation directed back toward the line of β-stability may exist around the samarium and promethium nuclei [LEA82]. The recent development of an on-line thermal-ionization ion source [MLE86] at UNISOR [SPE81] capable of efficiently ionizing the elements in this region coupled with the availability of favorable targets with Z from 40 to 47 (Zr to Ag) as well as good beams of ^{32}S, ^{35}S, ^{35}Cl, ^{48}Ti and ^{46}Ti

0097–6156/86/0324–0502$06.00/0

available from the Holifield Heavy-Ion Facility have led to the initiation of a program to study the unknown or little known nuclei in this region. The specific experiment described here is the production of ^{136}Eu and ^{138}Eu and the first observation of gamma rays from their decay.

The ion source used in these investigations is shown in Fig. 1. It is a very small volume ion source heated with a tungsten filament to internal temperatures of up to 3000°K. The target/window, open to the accelerator beam line, reaches a maximum temperature of 2200°K with no incident beam. Heating by the incident beam increases the target/window temperature during operation. For the present experiment, a target/window of ^{92}Mo of 1.85 mg/cm^2 was used with a beam of ^{48}Ti at an energy of 225 MeV. The recoiling product nuclei were stopped in a layer of graphite felt from where they were evaporated, ionized and extracted to form a beam. The reactions ^{92}Mo(^{48}Ti,pn)^{138}Eu and ^{92}Mo(^{48}Ti,p3n)^{136}Eu were chosen because very low cross sections for xn reactions relative to pxn using targets and projectiles of about these masses, were predicted by the evaporation code OVERLAID ALICE [BLA76] and also confirmed in test experiments.

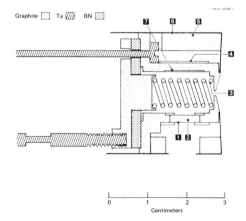

Graphite ☐ Ta ▨ BN ▦

Fig. 1. UNISOR thermal-ionization source. 1) target/window, 2) graphite felt catcher, 3) tungsten filament, 4) Ta target retainer, 5) graphite felt heat shielding, 6) Ta heat shield, 7) Ta support for graphite felt.

Centimeters

The mass-separated beam of appropriate mass was directed down one of the UNISOR beam lines to a fast tape collection system. After a collection time of 24 s, the tape moved in 0.5 sec to a counting station where multi-spectrum scaling and gamma-gamma coincidence data were simultaneously acquired. The ^{138}Eu decay was observed for about 20 hr and the ^{136}Eu for only about 2 hr at a typical ^{48}Ti beam level of 30 particle-nA.

Two gamma rays attributed to ^{138}Eu,   347 keV and 544 keV, were intense enough to perform a half-life analysis. The results are indicated in Fig. 2; the best value of the ^{138}Eu half-life is 12±1 s. No indication of the previously reported half-lives of 1.5 s and 35 s [BOG77] could be found even though data were also acquired at shorter and longer collection times. The 12 s half-life agrees with a value obtained in an earlier reported analysis of X-rays and attributed to ^{138}Eu [NOW82].

The total coincidence spectrum for ^{138}Eu decay is shown in Fig. 3. The level scheme derived for ^{138}Sm based on a complete analysis of the coincidence relationships is given in Fig. 4. The identification of the ground state band in ^{138}Sm up to a spin of 8+ is in agreement with recent in-beam experiments [LIS85]. The population of the 8+ state in ^{138}Sm indicates a high spin for the decaying state in ^{138}Eu and may be related to the 7- level identified in the decay of ^{142}Eu [KEN75]. No indication of the low spin decay (i.e. a 1+ level as in ^{140}Eu and ^{142}Eu) could be identified. This low spin isomer is expected to have a much shorter half-life and may not escape the ion source in time to be detected. On the other hand, a low spin isomer may not be populated to a great extent in this heavy-ion reaction. The placement of a second 2+ level from in-beam work [LUN85] at 745 keV is also substantiated by our data although the spin cannot be uniquely determined.

In addition, the ^{136}Eu decay was investigated for a brief time. Although no gamma lines could be discerned in the singles data, the 256 keV (2+ → 0+ transition in ^{136}Sm [LIS85]) line was observed in the total coincidence spectrum, Fig. 5.

In Fig. 6, the 2+ energies from this work along with other known 2+ energies for even-even Sm isotopes are plotted along with the results of some theoretical predictions. The IBM2 calculations were carried out as described by Scholten [SCH79] using parameters for the light Sm isotopes derived from Puddu et al. [PUD80]. The cranking calculations were made with

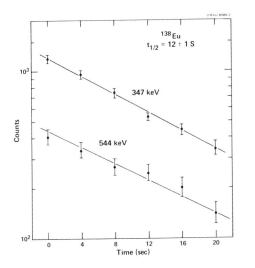

Fig. 2. Half-life determination for ¹³⁸Eu.

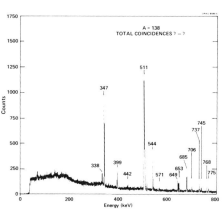

Fig. 3. Total coincidence spectrum for mass 138.

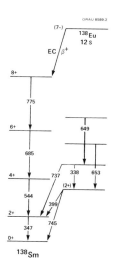

Fig. 4. Derived level scheme for ¹³⁸Sm.

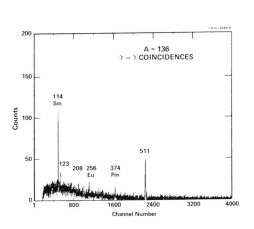

Fig. 5. Total coincidence spectrum for mass 136.

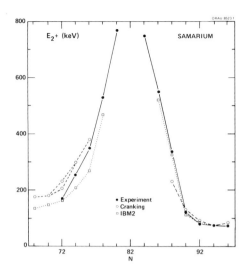

Fig. 6. Comparison of
experimental data with
cranking and IBM2 models.

the modified oscillator (Nilsson) single-particle model at the Strutinsky
equilibrium deformations. BCS pairing was treated by the average gap method.
The average gap was obtained by scaling the global formula of Jensen et al.
[JEN84]. The present cranking calculations are slight extensions and
modifications of calculations by Ragnarsson et al. [RAG74].

The experimental points for ^{138}Sm (N=76) and ^{136}Sm (N=74) fall between
the predictions of the two calculations. The trend of the 2+ energies shows a
smooth rapid decrease out to N=74. The 2+ energy at N=72 as determined by
Lister et al. [LIS85] is 163 keV, again within the range of the calculations,
and continues the rapid decrease. The trend from N=76 to 72 appears more
compatible with the recently predicted [LEA82] large increase in deformation
between these neutron numbers (lower branch of the dashed curve in Fig. 6)
than with the smooth evolution of deformation obtained earlier [RAG74]
(upper branch) and also suggested by the IBM2 results. Finally, we note that
the theoretical 2+ energies are higher for N<82 than for N>82 and this prediction is borne out by the data.

Acknowledgment

        Support for this work was provided by the U.S. Department of Energy
under contract numbers DE-AC05-76OR00033, DE-AS05-76ERO-3346 and
DE-AC05-84OR21400.

References

[BLA76] M. Blann, OVERLAID ALICE, U.S. Energy Research and Development Administration Report No. COO-3493-29, 1976 (unpublished).

[BOG77] D.D. Bogdanov, A.V. Demyanov, V.A. Karnaukhov, L.A. Petrov, A. Plohocki, V.G. Subbotin and J. Voboril, Nucl. Phys. A275 229 (1977).

[JEN84] A.S. Jensen, P.G. Hansen and B. Jonson, Nucl. Phys. A431 393 (1984).

[KEN75] G.G. Kennedy, S.C. Gujrathi and S.K. Mark, Phys. Rev. C12 553 (1975).

[LEA82] G.A. Leander and P. Möller, Phys. Lett. 110B (1982) 17.

[LIS85] C.J. Lister, B.J. Varley, R. Moscrop, W. Gelletly, P.J. Nolan, D.J.G. Lover, P.J. Bishop, A. Kirwan, D.J. Thornley, L. Ying, R. Wadsworth, M. O'Donnell, H.G. Price and A.H. Nelson, Phys. Rev. Lett. 55 810 (1985).

[LUN85] S. Lunardi, F. Scarlassara, F. Soramel, S. Beghini, M. Morando, C. Signorini, W. Meczynski, W. Starzecki, G. Fortuna and A.M. Stefanini, Z. Phys. A321 177 (1985).

[MLE86] R.L. Mlekodaj, to be submitted, Nucl. Instr. & Meth.

[NOW82] M. Nowicki, D.D. Bogdanov, A.A. Demyanov and Z. Stachura, Act. Phys. Pol. B12 879 (1982).

[PUD80] G. Puddu, O. Scholten and T. Otsuka, Nucl. Phys. A348 109 (1980).

[RAG74] I. Ragnarsson, A. Sobiczewski, R.K. Sheline, S.E. Larsson and B. Nerlo-Pomorska, Nucl. Phys. A233 329 (1974).

[SCH79] O. Scholten, in "Interacting Bosons in Nuclear Physics", ed. F. Iachello, Plenum Press, NY, 1979, p. 17.

[SHE61] R.K. Sheline, T. Sikkeland and R.N. Chanda, Phys. Rev. Lett. 7 446 (1961).

[SPE81] E.H. Spejewski, R.L. Mlekodaj and H.K. Carter, Nucl. Instr. & Meth. 186 71 (1981).

RECEIVED September 10, 1986

Section IV: Survey of Current Research

Abstracts of papers that could not be included in this book but were pertinent to the symposium are included here. Copies of these presentations can be obtained by writing the editors, if a complete set is desired, or by contacting the author for individual papers.

Abstracts

The International Activities of the National Science Foundation

Gerson S. Sher

Division of International Programs, National Science Foundation, Washington, DC 20550

The National Science Foundation supports cooperation
between U.S. scientists and engineers and their foreign
colleagues to achieve three related objectives:
facilitating access to unique research facilities and
specialists; facilitating study of large-scale natural
phenomena that transcend national boundaries; and helping
advance specific U.S. foreign-policy objectives through
activities in science and engineering. The NSF is also
responsible for the collection, analysis, and
dissemination of a wide variety of international R&D
data. To help promote these objectives, the NSF's
Division of International Programs manages more than 30
formal bilateral agreements for S&T cooperation with
other countries.

New Directions at UNISOR and the Importance of Reinforcing Spherical and Deformed Shell Gaps

J. H. Hamilton

Physics Department, Vanderbilt University, Nashville, TN 37235

An on-line nuclear orientation facility under construction for
UNISOR is described. The strong competition between shell gaps at
spherical, prolate and oblate deformation is shown to give rise to
various structures from spherical double closed shell, to coexisting
near-spherical and deformed shapes to deformed double closed shell
nuclei in the region of A = 70-104. The importance of the reinforcing
of the shape driving forces when the nucleus has shell gaps for
the protons and neutrons at the same deformation on nuclear shapes
and the switching of magic numbers is described.

0097-6156/86/0324-0510$06.00/0

Local and Global Nuclear Structure Correlations

I. Kelson

School of Physics and Astronomy, Tel-Aviv University, Tel-Aviv 69978, Israel

Many nuclear structure effects can be identified and described in terms of correlations of functionals of nuclear entities. Within such a framework, correlations between relatively shifted nuclear mass relationships are evaluated and are shown to display an anomalous behavior. The statistical significance of this puzzling effect is established, and possible physical explanations for its occurrence are mentioned. The implications of this observation for extrapolations based on parametrizations related to the solution of difference equations are discussed.

Nuclear Structure Information Obtained from Charge Radii

C. E. Alonso and J. M. Arias

Depto. Física Nuclear, Apdo. 1065, 41080 Sevilla, Spain

Charge radii provide stringent tests of nuclear models. We have analyzed isotope and isomer shifts in the even-even and odd-even $_{54}$Xe and odd-even $_{55}$Cs isotopes. These shifts show a dramatic deviation from the simple law, $R=R_o A^{1/3}$. Calculations based on the interacting boson model-2 appear to describe the experimental results reasonably well. Of particular importance is the occurrence of the intruder configuration in $_{55}$Cs originating from the $1g_{9/2}$ single particle orbital. When this becomes the ground state configuration, one observes a sudden jump in the isotope shift.

Nuclear Data and Related Services

Jagdish K. Tuli

National Nuclear Data Center, Brookhaven National Laboratory, Upton, NY 11973

National Nuclear Data Center (NNDC) maintains a number of data
bases containing bibliographic information and evaluated as well
as experimental nuclear properties. An evaluated computer file
maintained by the NNDC, called the Evaluated Nuclear Structure
Data File (ENSDF), contains nuclear structure information for all
known nuclides. The ENSDF is the source for the journal *Nuclear
Data Sheets* which is produced and edited by NNDC. The Evaluated
Nuclear Data File (ENDF), on the other hand is designed for
storage and retrieval of such evaluated nuclear data as are used
in neutronic, photonic, and decay heat calculations in a large
variety of applications. Some of the publications from these
data bases are the Nuclear Wallet Cards, Radioactivity Handbook,
and books on neutron cross sections and resonance parameters. In
addition, the NNDC maintains three bibliographic files: NSR - for
nuclear structure and decay data related references, CINDA - a
bibliographic file for neutron induced reactions, and CPBIB - for
charged particle reactions. Selected retrievals from evaluated
data and bibliographic files are possible on-line or on request
from NNDC.

$B(E2)\uparrow$ for the 2_1^+ States

S. Raman and C. W. Nestor, Jr.

Oak Ridge National Laboratory, Oak Ridge, TN 37831

We have nearly completed a compilation of experimental results for the reduced
electric quadrupole transition probability $[B(E2)\uparrow]$ between the 0^+ ground state
and the first 2^+ state in even-even nuclei. This compilation together with certain
simple relationships noted by other authors can be employed to make reasonable
predictions of unmeasured $B(E2)\uparrow$ values.

Nuclear Level Densities and Level Spacing Distributions

T. von Egidy, A. N. Behkami[1], and H. H. Schmidt

Technische Universität München, D-8046 Garching, Federal Republic of Germany

Extensive and complete level schemes and neutron resonance
densities of 72 nuclides have been used to determine the spin

[1]Permanent address: University of Shiraz, Shiraz, Iran

cut-off parameter, parameters for level density formulae and level spacing distributions. The level spacing distributions indicate the existence of a further good quantum number at low excitation energies in addition to spin and parity.

The (t, ^{3}He) Reaction

F. Ajzenberg-Selove

Department of Physics, University of Pennsylvania, Philadelphia, PA 19104

A review will be presented of the nuclear spectroscopic evidence for nuclei off the line of stability reached in (t, ^{3}He) reactions. The importance of some of the results to astrophysics will be discussed as will the possibilities for future work.

Fragmentation Theory of Mass Distributions in Superheavy Elements and Nuclear Decay by Heavy-Ion Emission

K. Depta[1], R. Herrmann[1], V. Schneider[1], J. A. Maruhn[1,2], W. Greiner[1,2], D. Poenaru[3], M. Ivascu[3], and A. Sandulescu[4]

[1]Institut für Theoretische Physik der Universität Frankfurt, Postfach 117 932, D-6000 Frankfurt 11, Federal Republic of Germany
[2]Joint Institute for Heavy Ion Research, Holifield Heavy Ion Research Facility, Oak Ridge, TN 37831
[3]Central Institute of Physics, 76900 Bucharest, Romania
[4]Joint Institute for Nuclear Research, 141 980 Dubna, USSR

The fragmentation theory, which treats the fission process in terms of collective coordinates and the two-center shell model, was applied to predicting mass distributions for the fission of superheavy elements. The model also predicted the decay of heavy nuclei through heavy ion emission, a process which has recently been observed experimentally. We give an overview of predicted properties of this decay mode.

Particle Decay Studies at or near Closed Shells

K. S. Toth

Oak Ridge National Laboratory, Oak Ridge, TN 37831

We summarize briefly two sets of investigations. In the first one α-decay rates of neutron-deficient even-even Pb isotopes were studied by using the UNISOR separator on-line at the Holifield Heavy Ion Research Facility. These data indicate that midway between $N = 82$ and $N = 126$, the $Z = 82$ shell gap may not exist. The second investigation, carried out at the Lawrence Berkeley Laboratory 88-Inch Cyclotron and SuperHILAC, has dealt with the delayed-proton decays of the $N = 81$ precursors ^{147}Dy, ^{149}Er, and ^{151}Yb. All three proton spectra have sharp peaks and thus provide evidence for greatly reduced level densities in the $N = 82$ proton emitting nuclei ^{147}Tb, ^{149}Ho, and ^{151}Tm.

The Approach to Sphericity:
An (n,n′γ) Study of ^{202,204}Hg

E. W. Kleppinger, R. A. Gatenby, and Steven W. Yates

Department of Chemistry, University of Kentucky, Lexington, KY 40506-0055

We have studied ^{202,204}Hg, the heaviest of the stable mercury isotopes, by the (n,n′γ) reaction on isotopically enriched samples. Gamma-ray excitation functions and angular distribution measurements have enabled us to confirm previously placed energy levels as well as to clarify several spin assignments. The results of this work, along with those from complementary studies, will provide a detailed picture of the changes taking place with the approach of double shell closure.

Are There Isotopes of Element Zero?

R. L. Kozub

Department of Physics, Tennessee Technological University, Cookeville, TN 38505

An experimental search for barely-bound multi-neutron systems using kinematically-reversed heavy-ion reactions is described.

Composite Detectors

H. C. Griffin

Department of Chemistry, University of Michigan, Ann Arbor, MI 48109

The highly complex spectra of radiations emitted by nuclides off the line of stability require detectors with good discrimination. These nuclides are generally produced in low yield, which must be offset by high efficiency of a detector system. These criteria lead to composite detectors in the sense of both physical parts and discrimination criteria.

Detection Characteristics of a NaI Array Used with a Ge Detector

Robert N. Ceo and H. C. Griffin

Department of Chemistry, University of Michigan, Ann Arbor, MI 48109

We have investigated the use of an array of up to 12 rectangular (5 cm x 5 cm x 15 cm) NaI(Tl) detectors in 4 configurations: a total energy spectrometer, a multiplicity filter, and (with a Ge detector) an anti-Compton shield and a pair spectrometer. The array is particularly suitable for multiplicity and pair effects. It is useful, but less effective than the same volume of scintillator in a custom configuration for the other two applications.

Simulation of Beam-Target Interactions for Isotope Production

E. A. Hugel[1], H. C. Griffin[1], and R. D. Hichwa[2]

[1]Department of Chemistry, University of Michigan, Ann Arbor, MI 48109
[2]Cyclotron/PET Facility, University of Michigan, Ann Arbor, MI 48109

The efficient production of short-lived radionuclides used in nuclear medicine is strongly dependent on the factors of beam current, incident particle energy, and target size. The interdependencies among the parameters, as well as the nonlinearities associated with changes in the target environment due to beam heating, impede optimization of production systems. We are developing an interactive production simulation program to predict practical target yields. The computer program, written in Fortran-77, considers factors such as beam intensity and distribution, beam divergence and scattering, range straggling, and heating effects along the beam path. The simulation is being used in the design of targetry (degraders, windows, gas pressures, and dimensions) for the production of isotopes (F-18, C-11, O-15) used in positron emission tomography studies.

Table of Radioactive Isotopes

R. B. Firestone and E. Browne

Lawrence Berkeley Laboratory, University of California, Berkeley, CA 94720

The *Table of Radioactive Isotopes* will be published in 1986. It will contain recommended radiation data for all radioactive isotopes and is produced from the LBL Isotopes Project's extended Evaluated Nuclear Structure Data File [ENSDF] database. The book will include mass-chain decay schemes updated from the 7th Edition of the *Table of Isotopes* [LED78] and tabular data with adopted γ-ray, alpha particle, atomic-electron, beta, and internal bremsstrahlung energies and intensities. Average radiation energies per disintegration will be provided when possible. The data presented are generally derived from ENSDF with some updating. In addition to the tabular data, there will be text and appendices. The database from which the *Table of Radioactive Isotopes* is being produced is available to users through the LBL computers.

Nuclear Moments of ^{85m}Rb Measured by Using Laser-Induced Nuclear Orientation

M. S. Otteson, G. Shimkaveg, D. Smith, W. W. Quivers, Jr., J. E. Thomas, R. R. Dasari, and M. S. Feld

George R. Harrison Spectroscopy Laboratory, Massachusetts Institute of Technology, Cambridge, MA 02139

Laser induced gamma ray anisotropy is used to obtain the D_1 hyperfine (hf) structure of the 1 usec isomer ^{85m}Rb. An isomer shift of -52 ± 9 MHz relative to ^{85}Rb, and a nuclear magnetic dipole moment of $(6.046 \pm 0.010)u_n$ has been obtained. Continuing work may lead to sub-Doppler resolution.

The Structure of Odd-Odd Lathanum Fission Product Nuclides

W. B. Walters, C. Chung, S. Faller, C. A. Stone, and J. D. Robertson

Department of Chemistry, University of Maryland, College Park, MD 20742

The conclusions drawn from a series of investigations of the structure of odd-odd La nuclides are discussed.

Fast Radiochemical Separations

Krishnaswamy Rengan

Department of Chemistry, Eastern Michigan University, Ypsilanti, MI 48197

In the last decade nuclear chemists have been exploring highly unstable nuclides which have half lives in the order of a few seconds to few minutes. Study of neutron-rich nuclides formed in fission often involved fast radiochemical separation of element of interest. Production of transactinides also necessitated fast separation procedures. A number of automated, batch separation procedures have been developed. In addition, gasjet systems have been developed for the continuous delivery and separation of nuclear reaction products. This paper reviews the developments in the area of fast chemistry.

The Leap To Produce Heavy Nuclei at the Limits of Nuclear Stability

Darleane C. Hoffman

Nuclear Science Division, Lawrence Berkeley Laboratory, University of California, Berkeley, CA 94720

The Large Einsteinium Activation Program (LEAP) has been proposed by a consortium of four national laboratories. Central to the proposal is the preparation, for the first time, of a large target of Es-254 in order to accomplish a unique scientific program. Progress and activities to date and plans for the future will be discussed.

Actinide Production in ^{136}Xe + ^{249}Cf Bombardments

Kenneth E. Gregorich, Kenton J. Moody, D. Lee, Robert B. Welch, Wing Kot, Phillip A. Wilmarth, and Glenn T. Seaborg

Lawrence Berkeley Laboratory, Berkeley, CA 94720

Excitation functions for the production of isotopes of elements 93 through 101 were measured in ^{136}Xe bombardments of ^{249}Cf at energies 1.02, 1.09, and 1.16 times the Coulomb barrier. The experimental procedures are presented and the actinide production cross sections are compared to those from similar reactions. A procedure for modeling the actinide cross section distributions for the products of damped transfer reactions is also presented. The results are interpreted with emphasis on the usefulness of heavy ion transfer reactions in producing new actinide (and trans-actinide) isotopes.

INDEXES

Author Index

520

Subject Index

A

Production by Cara Aldridge Young
Jacket design by Pamela Lewis

Elements typeset by Hot Type Ltd., Washington, DC
Printed and bound by Maple Press Co., York, PA

Recent ACS Books

Writing the Laboratory Notebook
By Howard M. Kanare
145 pages; clothbound ISBN 0-8412-0906-5

Polymeric Materials for Corrosion Control
Edited by Ray A. Dickie and F. Louis Floyd
ACS Symposium Series 322; 384 pp; ISBN 0-8412-0998-7

Porphyrins: Excited States and Dynamics
Edited by Martin Gouterman, Peter M. Rentzepis, and Karl D. Straub
ACS Symposium Series 321; 384 pp; ISBN 0-8412-0997-9

Agricultural Uses of Antibiotics
Edited by William A. Moats
ACS Symposium Series 320; 189 pp; ISBN 0-8412-0996-0

Fossil Fuels Utilization
Edited by Richard Markuszewski and Bernard D. Blaustein
ACS Symposium Series 319; 381 pp; ISBN 0-8412-0990-1

Materials Degradation Caused by Acid Rain
Edited by Robert Baboian
ACS Symposium Series 318; 449 pp; ISBN 0-8412-0988-X

Biogeneration of Aromas
Edited by Thomas H. Parliment and Rodney Croteau
ACS Symposium Series 317; 397 pp; ISBN 0-8412-0987-1

Formaldehyde Release from Wood Products
Edited by B. Meyer, B. A. Kottes Andrews, and R. M. Reinhardt
ACS Symposium Series 316; 240 pp; ISBN 0-8412-0982-0

Evaluation of Pesticides in Ground Water
Edited by Willa Y. Garner, Richard C. Honeycutt, and Herbert N. Nigg
ACS Symposium Series 315; 573 pp; ISBN 0-8412-0979-0

Water-Soluble Polymers: Beauty with Performance
Edited by J. E. Glass
Advances in Chemistry Series 213; 449 pp; ISBN 0-8412-0931-6

Historic Textile and Paper Materials: Conservation and Characterization
Edited by Howard L. Needles and S. Haig Zeronian
Advances in Chemistry Series 212; 464 pp; ISBN 0-8412-0900-6

For further information and a free catalog of ACS books, contact:
American Chemical Society, Sales Office
1155 16th Street, NW, Washington, DC 20036
Telephone 800-424-6747